Plant biochemistry and molecular biology

Plant biochemistry and molecular biology

Hans-Walter Heldt

Institute of Plant Biochemistry, Göttingen

with the collaboration of Fiona Heldt

Oxford   New York   Tokyo
OXFORD UNIVERSITY PRESS
1997

Oxford University Press, Great Clarendon Street, Oxford OX2 6DP

Oxford   New York
Athens   Auckland   Bangkok   Bogota   Bombay   Buenos Aires
Calcutta   Cape Town   Dar es Salaam   Delhi   Florence   Hong Kong
Istanbul   Karachi   Kuala Lumpur   Madras   Madrid   Melbourne
Mexico City   Nairobi   Paris   Singapore   Taipei   Tokyo   Toronto
Warsaw

and associated companies in
Berlin   Ibadan

Oxford is a trade mark of Oxford University Press

Published in the United States
by Oxford University Press Inc., New York

Original title: Pflanzenbiochemie by Hans W. Heldt
Published by Spektrum Akademischer Verlag
Copyright © 1996 Spektrum Akademischer Verlag
English language edition © Hans-Walter Heldt, 1997

A catalogue record for this book is available from the British Library

Library of Congress Cataloging in Publication Data
Data available

 ISBN 0 19 850180 3 (Hbk)
ISBN 0 19 850179 X (Pbk)

Typeset by Footnote Graphics, Warminster, Wilts
Printed in Great Britain by The Bath Press, Bath, Avon

preface

Plant sciences, like other fields of science, suffer from the problem that both the students and researchers have become more and more specialized and have only detailed knowledge of their own sector. In plant biochemistry there is the additional complication that most of the present general textbooks on biochemistry concentrate on the metabolism of animals and micro-organisms and even the most fundamental aspects of plant biochemistry find no mention. This may be the reason why knowledge of the biochemical processes in the plant—ultimately the basis of life on our planet—is not very widespread. My goal in writing this book was to try to awaken the interest of students in this subject.

I have taught and carried out research in both animal and plant biochemistry. Thereby I have come to the conviction that to deal with plant biochemistry in a productive way, it is very important to 'look over the fence' in order to be aware that certain metabolic processes in animals often take place in a very similar, or even in the same way, in plants and micro-organisms. Since research into the biochemical processes in animals and micro-organisms is often more intensive than that in plants, familiarity with general biochemistry is indispensable for a student of plant biochemistry.

Since there are many excellent textbooks on general biochemistry, I have deliberately omitted dealing with elements such as the structure and function of amino acids, carbohydrates, and nucleotides, the function of nucleic acids as carriers of genetic information, and the structure and function of proteins and basis of enzyme catalysis. I have dealt with topics of general biochemistry only when it seemed necessary for enhancing understanding of the problem in hand. Thus, in the end this book is a compromise between a general and a specialized textbook. When choosing the topics my aim was to enable the reader to understand thoroughly certain basic reactions in plant biochemistry, such as photosynthesis. Therefore I have attached importance to an easily understood description of the principles of metabolism.

I intended to give the reader a general view of the whole subject of plant biochemistry. I have tried to show how it can be applied and also to incorporate molecular biology. It is my aim that every student interested in biochemistry also knows which steps are necessary to produce a transgenic plant and is aware of the goals of plant gene technology.

I am very grateful to the many colleagues who, through discussions and the reprints they have sent me on their special fields, have given me the information which made it possible for me to write this book. It was especially helpful for me that the colleagues in the following list took the great trouble to read critically one or more chapters, to draw my attention to mistakes, and to suggest improvements, for which I am particularly grateful. Without their help I would not have dared to write this book alone. I would specially like to thank my Australian colleagues for their advice in the preparation of the English version. I thank the students and co-workers in our institute for pointing out mistakes and also what was not clear or difficult to understand in the text. Many thanks also to my colleagues, named in the legends, who have made figures available to me. I would especially like to mention Professor David Robinson from whom I have received most of the electron micrographs shown in the book.

The idea of writing this book was suggested by Mrs Karin von der Saal, project editor at Spektrum-Verlag. I would like to thank her here for her intensive technical help, constructive criticism, and advice during the development of this book. My special thanks goes also the graphic artist Mrs Christiane Solodkoff, who drew my drafts of metabolic schemes better than I could ever have wished for. I thank Mr Wolfgang Zettlmeier for the excellent illustration of the chemical formulae. Thanks also to Spektrum-Verlag and its staff for the successful design and production of the book.

Finally my special thanks go to my co-worker and wife Fiona Heldt. Without her intensive support it would not have been possible for me to write this book and to prepare the English version. In the course of translation the content of this textbook has been updated; the English edition therefore resembles the forthcoming second edition in German.

I have tried to eradicate as many mistakes as possible but probably not with complete success. I am therefore grateful for any suggestions and comments.

Hans-Walter Heldt
Göttingen, January 1997

Thanks to the following colleagues for
looking through one or more chapters
of this book:

Prof. Jan Anderson, Canberra, Australia
Prof. John Andrews, Canberra, Australia
Prof. Tom ap Rees, Cambridge, UK
Prof. Kozi Asada, Uji, Kyoto, Japan
Dr Tony Ashton, Canberra, Australia
Dr Murray Badger, Canberra, Australia
Prof. Peter Böger, Konstanz, Germany
Dr Sieglinde Borchert, Göttingen, Germany
Prof. Axel Brennicke, Ulm, Germany
Prof. David Day, Canberra, Australia
Prof. Gerry Edwards, Pullman, Washington, USA
Prof. Walter Eschrich, Göttingen, Germany
Prof. Ulf-Ingo Flügge, Cologne, Germany
Prof. Margrit Frentzen, Hamburg, Germany
Prof. Wolf Frommer, Tübingen, Germany
Dr R. T. Furbank, Canberra, Australia
Prof. Jan Eiler Graebe, Göttingen, Germany
Prof. Peter Gräber, Stuttgart, Germany
Prof. Erwin Grill, Munich, Germany
Dr Bernhard Grimm, Gatersleben, Germany
Prof. Wolfgang Haehnel, Freiburg, Germany
Dr Iris Hanning, Göttingen, Germany
Dr Marshall D. Hatch, Canberra, Australia
Prof. Ulrich Heber, Würzburg, Germany
Dr Dieter Heineke, Göttingen, Germany
Prof. E. Heinz, Hamburg, Germany
Dr Klaus-Peter Heise, Göttingen, Germany
Dr Frank Hellwig, Göttingen, Germany
Dr Gieselbert Hinz, Göttingen, Germany
Prof. Steven C. Huber, Raleigh, USA
Dr Graham Hudson, Canberra, Australia
Dr Colin Jenkins, Canberra, Australia
Prof. Wolfgang Junge, Osnabrück, Germany
Prof. Werner Kaiser, Würzburg, Germany
Dr Steven King, Canberra, Australia
Prof. Martin Klingenberg, Munich, Germany
Prof. Gotthard H. Krause, Düsseldorf, Germany

Dr Silke Krömer, Osnabrück, Germany
Dr Anne Kruse, Göttingen, Germany
Prof. Werner Kühlbrandt, Heidelberg, Germany
Dr Toni Kutchan, Munich, Germany
Dr Gertrud Lohaus, Göttingen, Germany
Dr John Lunn, Canberra, Australia
Prof. Enrico Martinoia, Poitiers, France
Prof. Hartmut Michel, Frankfurt, Germany
Prof. K. Müntz, Gatersleben, Germany
Prof. Walter Neupert, Munich, Germany
Prof. Lutz Nover, Frankfurt, Germany
Prof. C. Barry Osmond, Canberra, Australia
Prof. Birgit Piechulla, Rostock, Germany
Prof. Klaus Raschke, Göttingen, Germany
Dr Günter Retzlaff, Limburger Hof, Germany
Dr Sigrun Reumann, Göttingen, Germany
Prof. Gerhard Röbbelen, Göttingen, Germany
Prof. David Robinson, Göttingen, Germany
Dr Hans Rurainski, Göttingen, Germany
Prof. Eberhard Schäfer, Freiburg, Germany
Prof. Dierk Scheel, Halle, Germany
Prof. Hugo Scheer, Munich, Germany
Prof. Renate Scheibe, Osnabrück, Germany
Dr Karin Schott, Göttingen, Germany
Dr Ulrich Schreiber, Würzburg, Germany
Prof. Gernot Schultz, Hannover, Germany
Dr Danja Schünemann, Göttingen, Germany
Prof. Jens D. Schwenn, Bochum, Germany
Prof. Jürgen Soll, Kiel, Germany
Prof. Martin Steup, Potsdam, Germany
Prof. Heinrich Strotmann, Düsseldorf, Germany
Dr Gerhard Thiel, Göttingen, Germany
Prof. Rudolf Tischner, Göttingen, Germany
Prof. Achim Trebst, Bochum, Germany
Prof. Hanns Weiss, Düsseldorf, Germany
Prof. Dietrich Werner, Marburg, Germany
Prof. Peter Westhoff, Düsseldorf, Germany

Dedicated to my teacher, Martin Klingenberg

brief contents

contents

introduction

Plant biochemistry examines the molecular mechanisms of plant life. One of the main topics is photosynthesis, which in higher plants takes place mainly in the leaves. Photosynthesis utilizes the energy of the sun to synthesize carbohydrates and amino acids from water, carbon dioxide, nitrate, and sulfate. A major part of these products is transported via the vascular system from the leaves through the stem into other regions of the plant, where they are required, for example, to build up the roots and supply them with energy. Hence the leaves have been given the name source, and the roots the name sink. The reservoirs in seeds are also an important group of sink tissues, and, depending on the species, act as a store for many agricultural products, such as carbohydrates, proteins, and fat.

In contrast to animals, plants have a very large surface, often with very thin leaves in order to keep the diffusion pathway for carbon dioxide as short as possible and to catch as much light as possible. In the finely branched root hairs the plant has an efficient system for extracting water and inorganic nutrients from the soil. However, this large surface exposes plants to all the changes in their environment. They must be able to withstand extreme conditions such as drought, heat, cold or even frost, as well as an excess of radiated light energy. The leaves have to contend with the change between photosynthetic metabolism during the day and oxidative metabolism during the night. Plants encounter these extreme changes in external conditions with an astonishingly flexible metabolism, in which a variety of regulatory processes take part. Since plants cannot run away from their enemies, they have developed a whole arsenal of defence substances to protect themselves from being eaten.

Plant agricultural production is the basis for human nutrition. Plant gene technology, which can be regarded as a section of plant biochemistry, makes a contribution to combating the impending global food shortage due to the enormous growth of the world population. The use of environmentally compatible herbicides and protection against viral or fungal infestation by means of gene technology is of great economical importance. Plant biochemistry is also instrumental in breeding productive varieties of crop plants.

Plants are the source of important industrial raw material such as fat and starch, but they are also the basis for the production of pharmaceuticals. It is to be expected that in future gene technology will lead to the extensive use of plants as a means of producing sustainable raw material for industrial purposes.

The aim of this short list is to show that plant biochemistry is not only an important field of basic science, explaining the molecular function of a plant, but is also an applied science which, now in a revolutionary phase of its development, is in a position to contribute to the solution of important economic problems.

To reach this goal it is necessary that sectors of plant biochemistry such as bioenergetics, the biochemistry of intermediary metabolism and the secondary plant compounds, as well as molecular biology and other sections of plant sciences such as plant physiology and the cell biology of plants co-operate closely with one another. Only the integration of the results and methods of working of the different sectors of plant sciences can help us to understand how a plant functions and to put this knowledge to economical use. This book will try to describe how this could be achieved.

Since there are already very many good general textbooks on biochemistry, the elements of general biochemistry will not be dealt with here.

Hans-Walter Heldt
in cooperation with Fiona Heldt

chapter 1

A leaf cell consists of several metabolic compartments

In higher plants photosynthesis occurs mainly in the *mesophyll*, the chloroplast-rich tissue of leaves. Figure 1.1 shows an electron micrograph of a mesophyll cell and Fig. 1.2 a diagram of the cell structure. The cell contents are surrounded by a *plasma membrane* called the plasmalemma, and enclosed by a *cell wall*. The cell contains organelles, each with its own characteristic shape, which divide the cell

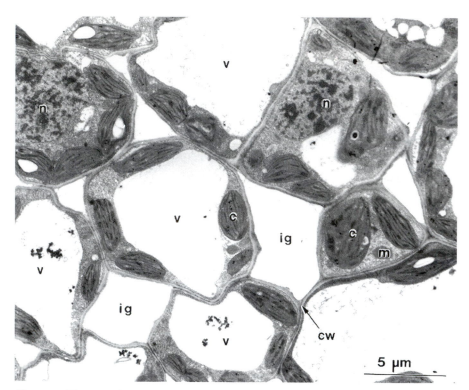

Figure 1.1 Electron micrograph of mesophyll tissue from tobacco. In most cells the large central vacuole is to be seen (v). Between the cells are the intercellular gas spaces (ig), which are somewhat enlarged by the fixation process. c, Chloroplast; cw, cell wall; n, nucleus; m, mitochondrion. (By D. G. Robinson, Göttingen.)

into various compartments (subcellular compartments). Each compartment has specialized metabolic functions, which will be discussed in detail in the following chapters (see also Table 1.1). The largest organelle, the vacuole, usually fills about 80 per cent of the total cell volume. Chloroplasts represent the next largest compartment and the rest of the cell volume is filled with mitochondria, peroxisomes, the nucleus, the endoplasmic reticulum, the Golgi bodies and, outside these organelles, the cell plasma, called cytosol.

The *nucleus* is surrounded by the *nuclear envelope* which consists of the two membranes of the endoplasmic reticulum. The space between the two membranes is known as the *perinuclear space*. The nuclear envelope is interrupted by *nuclear pores* with a diameter of about 50 nm. The nucleus contains chromatin, consisting of DNA double strands which are stabilized by being bound to basic proteins (histones). The genes of the nucleus are collectively referred to as the *nuclear genome*. Within the nucleus, usually off-centre, lies the *nucleolus*, where ribosomal subunits are formed. These ribosomal subunits, and also the messenger RNA formed

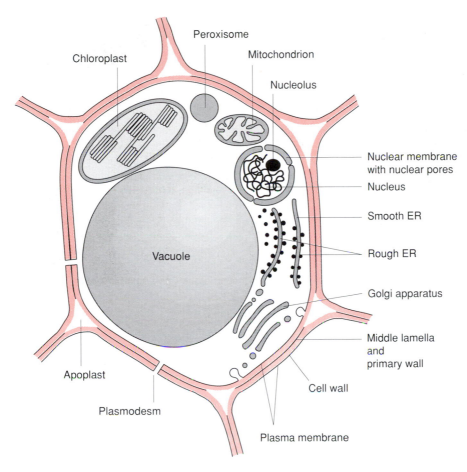

Chloroplast

Peroxisome

Mitochondrion

Nucleolus

Nuclear membrane with nuclear pores

Nucleus

Smooth ER

Rough ER

Golgi apparatus

Vacuole

Middle lamella and primary wall

Apoplast

Cell wall

Plasmodesm

Plasma membrane

Figure 1.2 Diagram of a mesophyll cell.

by transcription of the DNA in the nucleus, migrate through the nuclear pores to the ribosomes in the cytosol, the site of protein biosynthesis. The synthesized proteins are distributed between the different cell compartments according to their final destination.

1.1 The cell wall gives the plant cell mechanical stability

A major difference between plant and animal cells is that plant cells have a cell wall. This wall limits the volume of the plant cell. Water taken up into the cell by osmosis presses the plasmalemma against the inside of the cell wall, thus giving the cell mechanical stability.

Table 1.1 Subcellular compartments in a mesophyll cell[a] and some of their functions

Compartment	Per cent of the total cell volume	Functions (incomplete)
Vacuole	79	Maintenance of cell turgor, store and waste depository
Chloroplasts	16	Photosynthesis, synthesis of starch and lipids
Cytosol	3	General metabolic compartment, synthesis of sucrose
Mitochondria	0.5	Cell respiration
Nucleus	0.3	Contains the genome of the cell. Reaction site of replication and transcription
Peroxisomes		Reaction site for processes in which toxic intermediates are formed
Endoplasmic reticulum		Storage of Ca^{2+} ions, participation in the export of proteins from the cell and in the transport of proteins into the vacuole
Golgi body		Processing and sorting of proteins destined for export from the cells or transport into the vacuole

[a] Mesophyll cells of spinach, data from Winter *et al.* (1994).

The cell wall mainly consists of carbohydrates and proteins

The cell wall of a higher plant is made up of about 90 per cent carbohydrates and 10 per cent proteins. The main carbohydrate constituent is *cellulose*. Cellulose is an unbranched polymer consisting of D-glucose molecules which are connected to each other by glycosidic ($\beta1\rightarrow4$) linkages (Fig. 1.3a). Each glucose unit is rotated by 180° from its neighbour, so that very long, straight chains can be formed with a chain length of 2000–8000 glucose residues. About 150 cellulose chains are associated by inter-chain hydrogen bonds to a crystalline lattice structure known as a *microfibril*. These crystalline regions are impermeable to water. The microfibrils have an unusually high tensile strength, are very resistant to chemical and biological degradations and are in fact so stable that they are very difficult to hydrolyse. However, many bacteria and fungi have cellulose-hydrolysing enzymes (cellulases). These bacteria can be found in the digestive tract of some animals (e.g. ruminants) enabling them to digest grass and straw. It is interesting to note that cellulose is the most abundant organic substance on earth, representing about half of the total organically bound carbon.

Hemicelluloses are also important constituents of the cell wall. They are defined as those polysaccharides which can be extracted by alkaline solutions. The name is derived from an initial belief, which later turned out to be incorrect, that hemicelluloses were precursors of cellulose. Hemicelluloses consist of a variety of unbranched polysaccharides which contain, beside D-glucose, other carbohy-

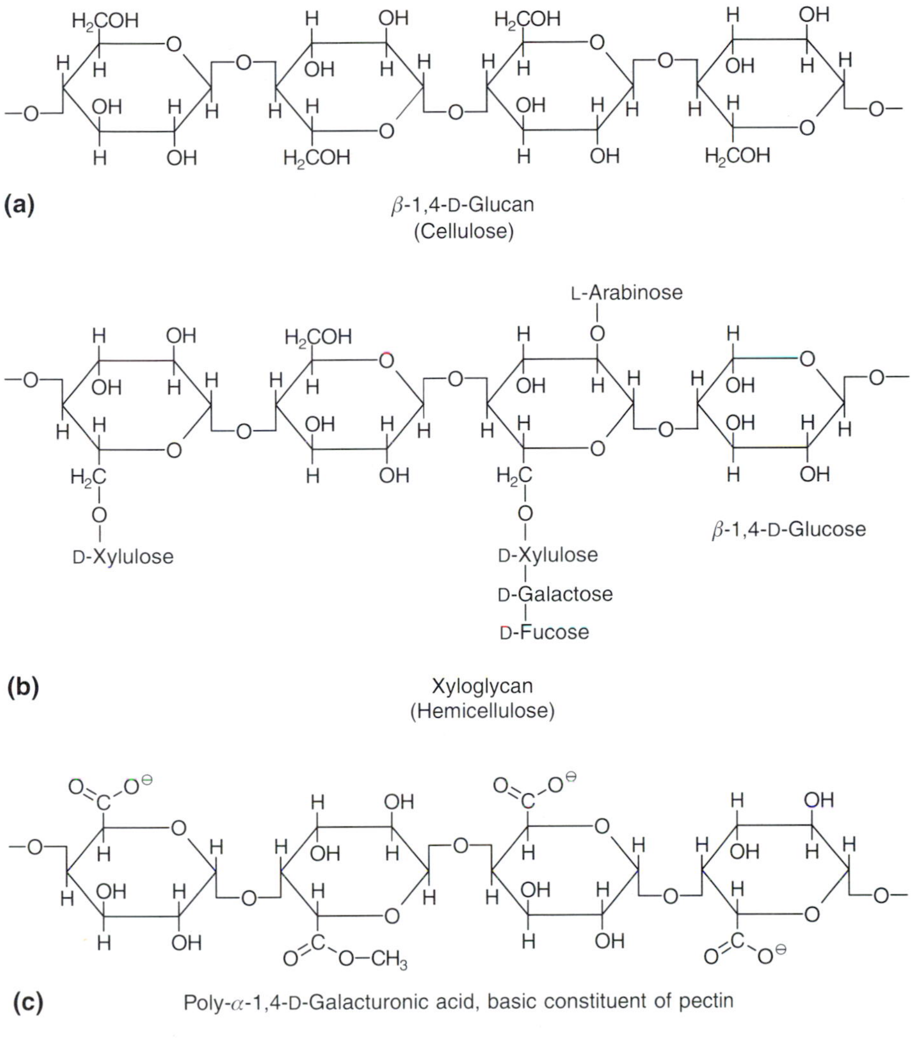

(a) β-1,4-D-Glucan (Cellulose)

(b) Xyloglycan (Hemicellulose)

(c) Poly-α-1,4-D-Galacturonic acid, basic constituent of pectin

Figure 1.3 Main constituents of the cell wall. (a) Cellulose; (b) a hemicellulose; (c) constituent of pectin.

drates such as the hexoses D-mannose, D-galactose, D-fucose, and the pentoses D-xylose and L-arabinose. Figure 1.3b shows xyloglycan as an example of a hemicellulose. The basic structure is a β-1,4-glucan chain to which xylose residues are bound via (α1→6) glycosidic linkages, which in turn are linked to D-galactose and D-fucose. In addition to this, L-arabinose residues are linked to the 2-OH group of the glucan chain.

Another major constituent of the cell wall is *pectin*, a mixture of polymers from sugar acids, such as D-galacturonic acid, which are connected by (α1→4) glycosidic links (Fig. 1.3c). Some of the carboxyl groups are esterified by methyl groups. The free carboxyl groups of adjacent chains are linked by Ca^{2+} and Mg^{2+} ions

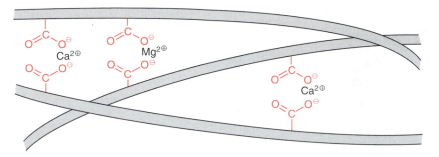

Figure 1.4 Ca^{2+} and Mg^{2+} ions mediate electrostatic interactions between pectin strands.

(Fig. 1.4). When Mg^{2+} and Ca^{2+} ions are not present, pectin is a soluble compound. The Ca^{2+}/Mg^{2+} salt of pectin forms an amorphous, deformable gel which is able to swell. The food industry and cooks make use of this property of pectin when preparing jellies and jams.

The structural proteins of the cell wall are connected by glycosidic linkages to the branched polysaccharide chains, and belong to the class of proteins known as *glycoproteins*. The carbohydrate portion of these glycoproteins varies from 50 per cent to over 90 per cent. Cell walls also contain waxes (Chapter 15), cutin and suberin (Chapter 18).

In a monocot plant the *primary wall*, i.e. the wall initially formed after the growth of the cell, consists of 20–30 per cent cellulose, 25 per cent hemicellulose, 30 per cent pectin, and 5–10 per cent glycoprotein. It is permeable to water. Pectin makes the wall elastic and, together with the glycoproteins and the hemicellulose, forms the matrix in which the cellulose microfibrils are embedded. When the cell has reached its final size and shape, another layer, the *secondary wall*, which consists mainly of cellulose, is added to the primary wall. The microfibrils in the secondary wall are arranged in a layered structure like plywood (Fig. 1.5).

The incorporation of lignin into the secondary wall gives the cell a woody structure. The cells of some plant parts eventually die, leaving the dead cell wall with only a supporting function, for example for forming the branches and twigs of trees or the stems of herbaceous plants. Section 18.3 describes in detail how lignin is formed by the polymerization of the phenylpropane derivatives cumaryl alcohol, coniferyl alcohol and sinapyl alcohol, resulting in a very solid structure. Dry wood consists of about 30 per cent lignin, 40 per cent cellulose, and 30 per cent hemicellulose. After cellulose, lignin is the most abundant natural substance on earth.

Plasmodesmata form a connection between neighbouring cells

Neighbouring cells are often connected by *plasmodesmata* thrusting through the cell walls. The plasmodesmata allow the passage of molecules up to a molecular

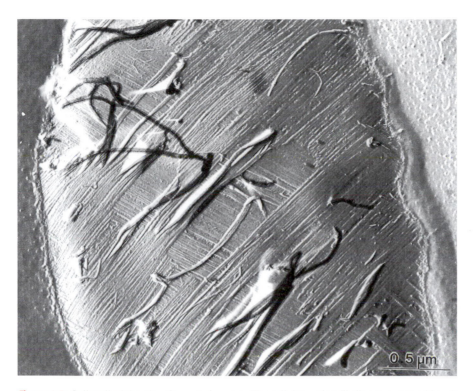

Figure 1.5 Cell wall of the alga *Oocystis lacustris*. The cellulose microfibrils are arranged in a layered pattern, in which parallel layers are arranged one above the other. (Freeze etching; by D. G. Robinson, Göttingen.)

weight of about 800–900 Da. They are permeable to the various intermediates of metabolism such as soluble sugars, amino acids, and free nucleotides, but they are generally impermeable to macromolecules such as proteins and nucleic acids. An exception to this rule will be dealt with at the end of this section. A single plant cell may contain from 1000 to more than 10 000 plasmodesmata. These plasmodesmata connect many plant cells to form a single, large metabolic compartment where the metabolites in the cytosol can move between the various cells by diffusion. This continuous compartment formed by different plant cells (Fig. 1.6) is called the *symplast*. There are also spaces between cells which are often continuous and are known as the extracellular space or the *apoplast* (Fig. 1.2).

Figure 1.7 shows a diagram of a plasmodesma. The tube-like opening through the cell wall is lined by the plasmalemma, which is continuous between the neighbouring cells. In the interior of this tube there is another tube-like membrane structure, the *desmotubule*. The latter forms the connection between the endoplasmic reticulum in neighbouring cells. The space between plasmalemma and desmotubule represents the diffusion pathway between the cytosol of neighbouring cells (Fig. 1.6). Protein particles, which are connected to each other, are

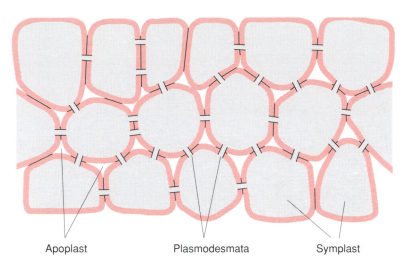

Apoplast   Plasmodesmata   Symplast

Figure 1.6 Plasmodesmata connect neighbouring cells, forming a symplast. The extracellular spaces between the cell walls form the apoplast.

attached to the outer tube formed by the plasma membrane and the desmotubule. It is assumed that the free space between these protein particles determines the aperture of the plasmodesma. A number of plant viruses, including the tobacco mosaic virus, cause the synthesis of *virus movement proteins* which can widen the plasmodesmata to such an extent that a virus particle can slip through. Thus after infecting a single cell a virus can spread through the entire symplast. The possibility is being discussed that the plasmodesmata can also be widened by putative movement proteins of the plant cell and that in this way macromolecules, such as transcription factors, are distributed as signals through the entire symplast.

The plant cell wall can be hydrolysed by cellulose and pectin hydrolysing enzymes obtained from micro-organisms. After leaf pieces have been incubated with these enzymes, plant cells can be obtained without their surrounding cell wall. These naked cells are called *protoplasts*. However, protoplasts are only stable in an *isotonic medium* in which the osmotic pressure corresponds to the osmotic pressure of the cell fluid. In pure water the protoplasts, devoid of the cell wall, swell so much that they burst. In appropriate media the protoplasts are viable for cell culture: in some cases it has been possible to regenerate a whole plant from such protoplasts.

1.2 Vacuoles have multiple functions

The vacuole is surrounded by a membrane, called the *tonoplast*. The number and the size of the vacuoles in different plant cells varies greatly. Young cells contain a larger number of smaller vacuoles but all, taken together, occupy only a small part

A

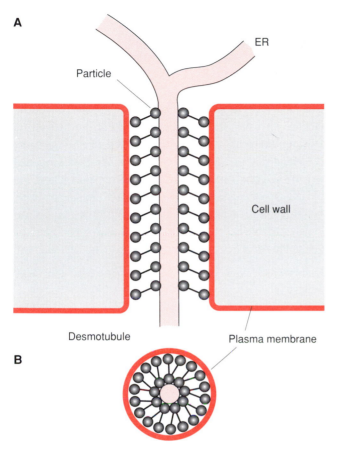

Figure 1.7 Diagram of a plasmodesm. The plasmalemma of the neighbouring cells is connected by a tube-like membrane invagination. Inside this tube is the desmotubule, a continuation of the endoplasmic reticulum. A, cross-sectional view of membrane; B, vertical view.

of the cell volume. When cells mature, the individual vacuoles join together to form a *central vacuole* (Figs 1.1, 1.2). The increasing volume of the cell is, in the main, due to the enlargement of the vacuole. In cells of storage or epidermal tissues, the vacuole often takes up almost the whole cellular space.

An important function of the vacuole is to maintain *cell turgor*. For this purpose salts, mainly of inorganic and organic acids, are accumulated in the vacuole. The accumulation of these osmotically active substances draws water into the vacuole which, in turn, causes the tonoplast to press the protoplasm of the cell against the surrounding plasma membrane. Plant turgor is responsible for the rigidity of non-woody plant parts. The plant wilts when the turgor decreases due to lack of water.

Vacuoles also have a *storage function*. Many plants use the vacuole to store reserves of nitrate and phosphate. Some plants store malate temporarily in the vacuoles in a diurnal cycle (section 8.5). Vacuoles of storage tissues contain carbohydrates (section 13.3) and storage proteins (Chapter 14).

In addition to this, vacuoles have an important function in *recycling* those cellular constituents which are defective or no longer required. Vacuoles contain hydrolytic enzymes for degrading various macromolecules, such as proteins, nucleic acids, and many polysaccharides. Structures such as mitochondria can be transferred by endocytosis to the vacuole and are digested there. The resulting degradation products, such as amino acids and carbohydrates, are then made available to the cell. This is especially important during *senescence* (section 19.5) when, prior to abscission, part of the constituents of the leaves are mobilized, for instance to form seeds.

Last, but not least, vacuoles also function as *waste deposits*. With the exception of gaseous substances, leaves are unable to rid themselves of waste products or xenobiotics such as herbicides. These are ultimately deposited in the vacuole (Chapter 12).

1.3 Plastids have evolved from cyanobacteria

Plastids are cell organelles which only occur in plant cells. They multiply by division and are in most cases inherited *maternally*. This means that all the plastids in a plant have usually descended from the *proplastids* in the egg cell. During cell differentiation the proplastids can differentiate into green *chloroplasts*, coloured *chromoplasts*, or colourless *leucoplasts*. Plastids possess their own circular chromosome and also the enzymes for gene duplication, gene expression, and protein synthesis. The plastid genome (*plastome*) has similar properties to that of the prokaryotic genome, for instance in the cyanobacteria, but encodes only a minor part of the plastid proteins; the majority of these proteins are encoded in the nucleus. Some of the proteins of photosynthetic electron transport and of ATP synthesis are encoded by the plastome.

As early as 1883 the botanist Andreas Schimper postulated that plastids are evolutionary descendants of intracellular symbionts, thus founding the basis for the *endosymbiont hypothesis*. According to this hypothesis the plastids are descended from certain cyanobacteria called *protochlorophytes*, which were taken up by phagocytosis into a host cell (Fig. 1.8) and lived there in a symbiotic relationship. Through time these endosymbionts lost the ability to live independently because a large portion of the genetic information of the plastid genome was transferred to the nucleus. A comparison of the sequences of proteins from chloroplasts and from cyanobacteria has provided convincing evidence supporting the endosymbiont hypothesis. The use of the term *endosymbiont theory* is therefore now justified.

Proplastids (Fig. 1.9A) are very small organelles (diameter 1–1.5 μm). They are undifferentiated plastids found in the meristematic cells of the shoot and the root.

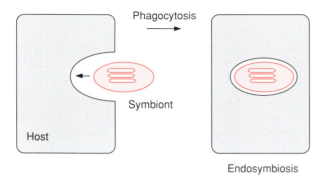

Figure 1.8 A cyanobacterium forms a symbiosis with a host cell.

They, like all other plastids, are surrounded by two membranes forming an envelope. According to the endosymbiont theory, the inner envelope membrane derives from the plasma membrane of the protochlorophyte and the outer envelope membrane from plasmalemma of the host cell.

Chloroplasts (Fig. 1.9B) are formed by differentiation of the proplastids (Fig. 1.10). A mature mesophyll cell contains about 50 chloroplasts. By definition, chloroplasts contain chlorophyll. However, they are not always green. In red and brown algae other pigments mask the green colour of the chlorophyll. Chloroplasts are lens-shaped and can adjust their position within the cell to receive an optimal amount of light. In higher plants their length is 3–10 μm. The two *envelope membranes* surround the *stroma*. The stroma contains a system of membranes arranged as flattened sacs (Fig. 1.11) which were given the name *thylakoids* (in Greek, sac-like) by Wilhelm Menke in 1960. During differentiation of the chloroplasts the inner envelope membrane invaginates to form thylakoids which are subsequently sealed off. In this way a large membrane area is provided as the site for the photosynthesis apparatus. The thylakoids are connected with each other by tube-like structures, forming a continuous compartment inside the thylakoid space. Many of the thylakoid membranes are squeezed very closely together; they are said to be stacked. These stacks can be seen by light microscopy as small particles within the chloroplasts and have been named *grana*.

There are three different compartments in chloroplasts; the *intermembrane space* between the outer and inner envelope membranes, the *stroma space* between the inner envelope membrane and the thylakoid membrane, and the *thylakoid lumen*, which is the space within the thylakoid membranes. The *inner envelope membrane* is a permeability barrier for metabolites and nucleotides, which can pass through only with the aid of specific translocators (section 1.9). In contrast, the *outer envelope membrane* is permeable to metabolites and nucleotides (but not to macromolecules such as proteins or nucleic acids). This permeability is due to the presence of specific membrane proteins called *porins*,

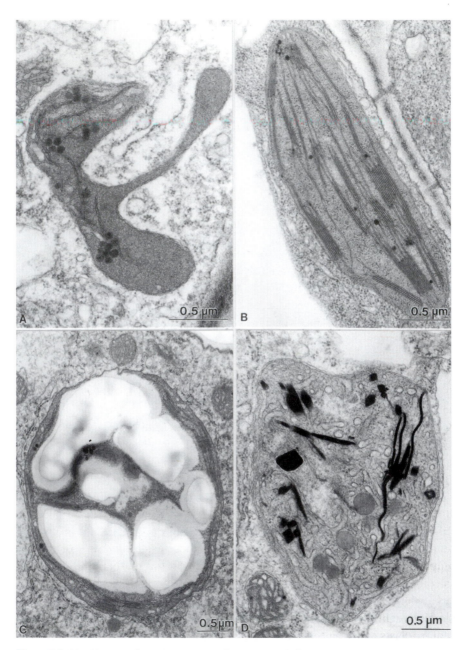

Figure 1.9 Plastids occur in various differentiated forms. (A) Proplastid from young primary leaf of *Curcubita pepo* (zucchini); (B) chloroplast from mesophyll cell of tobacco leaf fixed at the end of the dark period; (C) leucoplast: amyloplast from the root of *Cestrum auranticum*; (D) chromoplast from petal of *C. auranticum*. (By D. G. Robinson, Göttingen.)

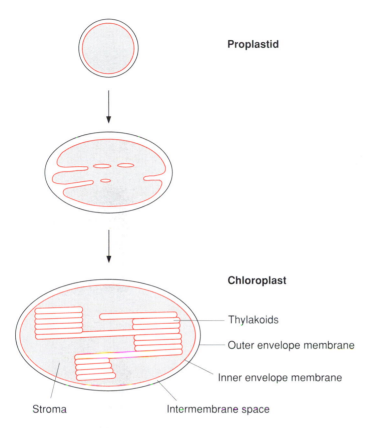

Proplastid

Chloroplast

Thylakoids

Outer envelope membrane

Inner envelope membrane

Stroma   Intermembrane space

Figure 1.10 Scheme showing the differentiation of a proplastid to a chloroplast.

which form pores permeable to substances with a molecular weight below 10 000 Da (section 1.11). Thus, the inner envelope membrane is the actual boundary membrane of the metabolic compartment of the chloroplasts, so that the chloroplast stroma can be regarded as the protoplasm of the plastids. In comparison with this, the thylakoid lumen represents an external space which functions primarily as a compartment for partitioning protons to form a proton gradient (Chapter 3).

The stroma of chloroplasts contains *starch grains*. This starch serves mainly as a diurnal carbohydrate store, the starch formed during the day being a reserve for the following night (section 9.1). Therefore at the end of the day the starch grains in the chloroplasts are usually very large and, during the following night, become very small again. The formation of starch in plants always takes place in plastids.

Often structures are found inside the stroma, which are not surrounded by a membrane. They are known as *plastoglobuli* and contain, amongst other substances, lipids and plastoquinone. A particularly high amount of plastoglobuli is found in the plastids of senescent leaves, containing degraded products of the thylakoid membrane. About 20–50 identical plastid genomes are localized in a

Figure 1.11 The grana stacks of the thylakoid membranes are connected by tubes, forming a continuous thylakoid space (thylakoid lumen). (After Weier *et al.* 1963.)

special region of the stroma known as the *nucleoid*. The ribosomes present in the chloroplasts are either free in the stroma or bound to the surface of the thylakoid membranes.

In leaves grown in the dark (etiolated plants) the plastids have a yellowish colour and are termed *etioplasts*. These etioplasts contain some, but not all, of the chloroplast proteins. They are devoid of chlorophyll but contain instead membrane precursors, termed *prolaminar bodies*, which probably consist of lipids. The etioplasts are regarded as an intermediate stage of chloroplast development.

Leucoplasts (Fig. 1.9C) are a group of plastids which include many differentiated, colourless organelles with very different functions, e.g. the *amyloplasts*, which act as a store for starch in non-green tissues such as root, tubers, or seeds. Leucoplasts are also the site of lipid biosynthesis in non-green tissues. Lipid synthesis in plants is generally located in plastids. The reduction of nitrite to ammonia, a partial step of nitrate assimilation (Chapter 10), is also always located in plastids. In those cases where nitrate assimilation takes place in the roots, leucoplasts are the site of nitrite reduction.

Chromoplasts (Fig. 1.9D) are plastids which, due to their high carotenoid content (Fig. 2.9), are coloured red, orange, or yellow. They are the same size as chloroplasts but have no known metabolic function. Their main function may be to house the pigments of some flowers and fruit, for instance the red colour of tomatoes.

1.4 Mitochondria also result from endosymbionts

Mitochondria are the site of cellular respiration, where substrates are oxidized for generating ATP (Chapter 5). Mitochondria, like plastids, multiply by division and are maternally inherited. They also have their own genome (consisting in plants of a large circular DNA strand and often several small circular DNA strands), and their own machinery for gene duplication, gene expression, and protein synthesis. The mitochondrial genome encodes only a small number of the mitochondrial proteins (Table 20.6); most of them are encoded in the nucleus. Mitochondria are of endosymbiotic origin and probably derive from anaerobic bacteria.

The endosymbiotic origin (Fig. 1.8) explains why the mitochondria are surrounded by two membranes (Fig. 1.12). Similar to chloroplasts, the *mitochondrial outer membrane* contains *porins* (section 1.11) which make this membrane permeable to molecules below a mass of 4000–6000 Da, including also metabolites and free nucleotides. The permeability barrier for these substances and the site of specific translocators (section 5.8) is the *mitochondrial inner membrane*. Therefore the *intermembrane space* between the inner and the outer membrane has to be considered as an external compartment. The 'protoplasm' of the mitochondria, surrounded by the inner membrane, is called the *mitochondrial matrix*. The mitochondrial inner membrane contains the proteins of the respiratory chain (section 5.5). In order to enlarge the surface area of the inner membrane, it is invaginated in folds (cristae mitochondriales) or tubuli (Fig. 1.13) into the matrix. The structure of these membrane invaginations corresponds to the structure of the thylakoids, the only difference being that in the mitochondria these invaginations are not separated as a distinct compartment from the inner membrane. Also, the mitochondrial inner membrane is the site for formation of a proton gradient. Therefore the mitochondrial intermembrane space and the chloroplastic thylakoid lumen correspond to each other in functional terms.

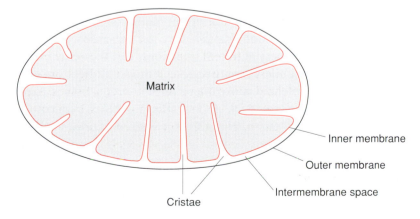

Figure 1.12 Diagram of the structure of a mitochondrion.

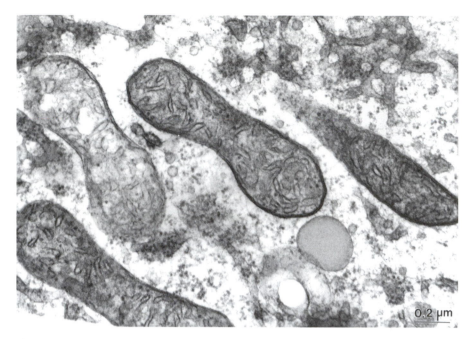

Figure 1.13 In mitochondria, invaginations of the inner membrane result in an enlargement of the membrane surface. The figure shows mitochondria in a barley aleurone cell. (By D. G. Robinson, Göttingen.)

1.5 Peroxisomes are the site of reactions in which toxic intermediates are formed

Peroxisomes, also termed microbodies, are small, spherical organelles with a diameter of 0.5–1.5 μm (Fig. 1.14) which, in contrast to plastids and mitochondria, are surrounded by only a single membrane. This membrane also contains porins. The peroxisomal matrix represents a specialized compartment for reactions in which toxic intermediates are formed. Thus peroxisomes contain enzymes catalysing the oxidation of substances accompanied by the formation of H_2O_2, and also contain catalase, which immediately degrades this H_2O_2 (section 7.4). Peroxisomes are a common constituent of eukaryotic cells. In plants there are two important differentiated forms, the *leaf peroxisomes* (Fig. 1.14A), which participate in photorespiration (Chapter 7), and the *glyoxysomes* (Fig. 1.14B), which are present in seeds containing oil and play a role in the conversion of oils to carbohydrates (section 15.6). The origin of peroxisomes is unresolved. In the past it was assumed that peroxisomes are synthesized *de novo* from invaginations of the membranes of the endoplasmic reticulum. Now it is widely accepted that peroxisomes are formed by division of pre-existing peroxisomes, like plastids and mitochondria, with the one difference that they have no genetic apparatus. A comparison of protein sequences has shown that peroxisomes from plants, fungi,

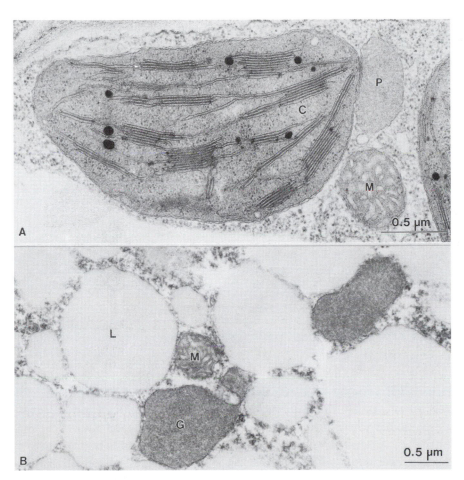

Figure 1.14 Peroxisomes. (A) Peroxisomes from the mesophyll cells of tobacco. The proximity of peroxisome (P), mitochondrion (M), and chloroplast (C) reflects the rapid metabolite exchange between these organelles in the course of photorespiration (dealt with in Chapter 7). (B) Glyoxysomes from germinating cotyledons of *Curcubita pepo* (zucchini). The lipid degradation described in section 15.6 and the accompanying gluconeogenesis require a close contact between lipid droplets (L), glyoxysome (G), and mitochondrion (M). (By D. G. Robinson, Göttingen.)

and animals have a common ancestor. Whether this was also an endosymbiont, as in the case of mitochondria and plastids, is still not clear.

1.6 The endoplasmic reticulum and Golgi apparatus form a network for the distribution of biosynthesis products

In an electron micrograph the *endoplasmic reticulum* (ER) appears as a labyrinth traversing the cell (Fig. 1.15). Two structural types of ER can be differentiated: the

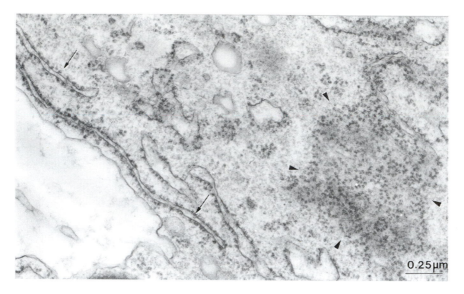

Figure 1.15 Rough endoplasmic reticulum, cross-section (arrows) and tangential sections (arrowheads). The ribosomes temporarily attached to the membrane occur as polysome complexes (ribosome + mRNA). Section from the cell of a maturing pea cotyledon. (By D. G. Robinson, Göttingen.)

rough and the smooth forms. The *rough ER* consists of flattened sacs, which are sometimes arranged in loose stacks of which the outer side of the membranes is occupied by *ribosomes*. The *smooth ER* consists primarily of branched tubes without ribosomes. Despite these morphological differences, the rough and the smooth ER are constituents of a continuous membrane system. The presence of ribosomes on the outer surface of the ER is temporary. Ribosomes are only attached to the ER membrane when the protein that they form is destined for the vacuole or for export from the cell. These proteins contain an amino acid sequence (signal sequence) which causes the peptide chain during its synthesis to enter the lumen of the ER (section 14.5). A snapshot of the ribosome complement of the ER would show only those ribosomes which at the moment of fixation of the tissue are involved in the synthesis of proteins destined for import into the ER lumen. Membranes of the ER are also the site of membrane lipid synthesis, where the necessary fatty acids are provided by the plastids.

Proteins channelled into the ER lumen are transferred to the *cis* side of the *Golgi apparatus* by vesicles budding off from the ER (Fig. 1.16). The Golgi apparatus, discovered in 1898 by the Italian Camillo Golgi, consists of up to 20 curved discs arranged in parallel, the so-called Golgi cisternae or dictyosomes, which are surrounded by smooth membranes (not occupied by ribosomes) (Fig. 1.17). On both sides of the discs, vesicles of various sizes bud off. The Golgi apparatus consists of the *cis* compartment, the middle compartment, and the *trans* compartment. The proteins pass through the different cisternae by budding off and vesicle transfer

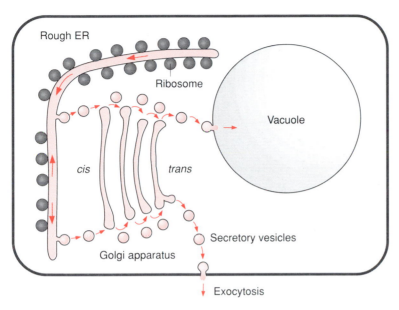

Figure 1.16 Scheme of the interplay between the endoplasmic reticulum and the Golgi apparatus in the transfer of proteins from the ER to the vacuoles and in the secretion of proteins from the cell.

and are often processed when passing through the Golgi apparatus. On the *trans* side special transfer vesicles are formed by budding off; these then diffuse to other locations. Thus vesicles diffusing to the plasma membrane can fuse with it and their contents are excreted by exocytosis into the extracellular space (Fig. 1.16). Alternatively, vesicles can move from the *trans* side of the Golgi apparatus to the vacuoles.

The amino acid signal sequences of the corresponding proteins determine whether proteins are directed by vesicle transfer into the vacuole or secreted via the plasma membrane. The collective term for the ER membrane, the membranes of the Golgi apparatus (derived from the ER), the transfer vesicles, and also the nuclear envelope is the *endomembrane system*.

1.7 Functionally intact cell organelles can be isolated from plant cells

In order to isolate cell organelles, the cell has to be disrupted to such an extent that its organelles are released into the isolation medium. This forms what is known as a *cell homogenate*. In order to prevent the free organelles from swelling and finally rupturing, the isolation medium must be isotonic, i.e. by the presence of an *osmoticum* (e.g. sucrose) an osmotic pressure is generated in the medium, which corresponds to the osmotic pressure of the aqueous phase within the

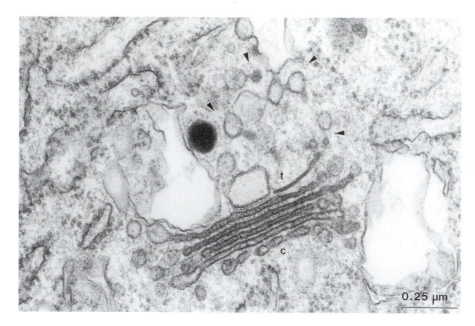

Figure 1.17 Golgi apparatus in a cell of a maturing pea cotyledon; c, *cis* side; t, *trans* side. Arrowheads point to the *trans* Golgi network. (By D. G. Robinson, Göttingen.)

organelle. Usually media containing 0.3 mol l^{-1} sucrose or sorbitol are used for cell homogenization.

Figure 1.18 shows the protocol for the isolation of chloroplasts as an example. Small leaf pieces are homogenized by cutting them up within seconds using blades rotating at high speed, such as in a food mixer. It is important that the homogenization time is short, otherwise the cell organelles released into the isolation medium would also be destroyed. However, such homogenization only works with leaves with soft cell walls, such as spinach. In the case of leaves with more rigid cell walls (e.g. cereal plants), protoplasts are first prepared from leaf pieces as described in section 1.1. These protoplasts are then ruptured by forcing the protoplast suspension through a net with a mesh smaller than the size of the protoplasts.

The desired organelles can be separated and purified from the rest of the cell homogenate by differential or density gradient centrifugation. In the case of *differential centrifugation*, the homogenate is suspended in a medium with a density much lower than that of the cell organelles. In the gravitational field of the centrifuge the sedimentation velocity of the particles depends primarily on the particle size (large particles sediment faster than small ones). As shown in Fig. 1.18, taking the isolation of chloroplasts as an example, relatively pure organelle preparations can be obtained within a short time by a sequence of centrifugation steps at increasing speeds.

Isolation of chloroplasts from
spinach leaves (all steps at 0°C)

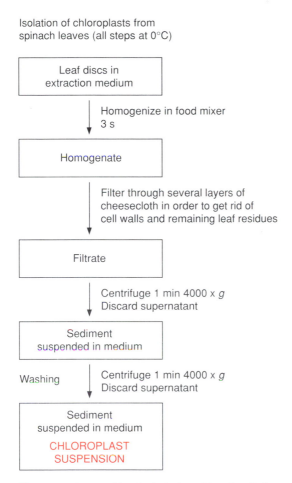

Figure 1.18 Protocol for the isolation of functionally intact chloroplasts.

In the case of *density gradient centrifugation* (Fig. 1.19), the organelles are separated according to their density. Media of differing densities are assembled in a centrifuge tube so that the density increases from top to bottom. To prevent altering the osmolarity of the medium, heavy macromolecules, e.g. Percoll (silica gel), are used to achieve a high density. The cell homogenate is layered on the *density gradient* prepared in the centrifuge tube and centrifuged until all the particles of the homogenate have reached their zone of equal density in the gradient. As this density gradient centrifugation requires high centrifugation speed and long running times, it is often used as the final purification step after preliminary separation by differential centrifugation.

By using these techniques it is possible to obtain functionally intact chloroplasts, mitochondria, peroxisomes, and vacuoles of high purity, in order to study their metabolic properties in the test tube.

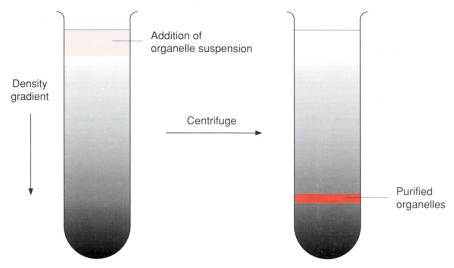

Density
gradient

Addition of
organelle suspension

Centrifuge

Purified
organelles

Figure 1.19 Particles are separated by density gradient centrifugation according to their different density.

1.8 Various transport processes facilitate the exchange of metabolites between different compartments

Each of the cell organelles mentioned in the preceding sections has a specific function in cell metabolism. The interplay of the metabolic processes in the various compartments requires a transfer of substances across the membranes of these cell organelles as well as between the different cells. This transfer of material takes place in various ways: by specific translocators, channels, pores, via vesicle transport, and in a few cases (e.g. CO_2 or O_2) by non-specific diffusion through membranes. Vesicle transport and the function of the plasmodesmata have already been described.

Figure 1.20 describes various types of transport processes according to formal criteria. When a molecule moves across a membrane independent of the transport of other molecules, the process is called *uniport*, and when counter-exchange of molecules is involved, it is called *antiport*. The mandatory simultaneous transport of two substances in the same direction is called *symport*. A transport via uniport, antiport, or symport, in which a charge is also moved simultaneously, is termed *electrogenic transport*. A vectorial transport, which is coupled to a chemical or photochemical reaction, is called *active* or *primary active transport*. Examples of active transport are the transport of protons driven by the electron transfer of the photosynthetic transport chain (Chapter 3) or the respiratory chain (Chapter 5) or by the consumption of ATP (Fig. 1.20c). Such proton transport is electrogenic: the transfer of a positive charge results in the formation of a membrane potential. Another example of primary active transport is the ATP-dependent transport of glutathione conjugates into vacuoles (section 12.2).

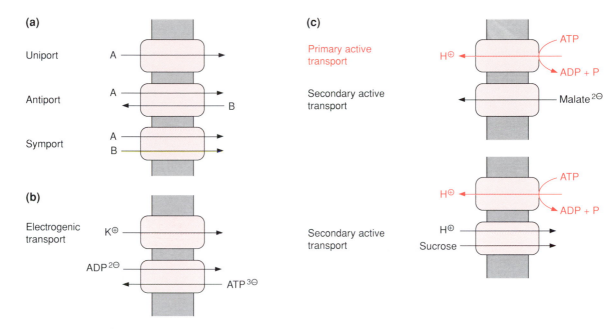

Figure 1.20 Classification of membrane transport processes.

In *secondary active transport* the only driving force is an electrochemical potential across the membrane. In the case of electrogenic uniport the membrane potential can be the driving force by which a substrate is transported across the membrane against the concentration gradient. An example of this is the accumulation of malate in the vacuole (Fig. 1.20c, see also Chapter 8). Another example of secondary active transport is the transport of sucrose via an H⁺–sucrose symport in which a proton gradient, formed by primary active transport, drives the accumulation of sucrose (Fig. 1.20c). This transport plays an important role in loading sieve tubes with sucrose (Chapter 13).

1.9 Translocators catalyse the specific transport of substrates and products of metabolism

Specialized membrane proteins catalyse specific transport across membranes. In the past these proteins were called carriers, as it was assumed that after binding the substrate at one side of the membrane, they would diffuse through the membrane to release the substrate on the other side. We now know that this simple picture does not apply. Instead, transport can be visualized as a process by which a molecule moves through a specific pore. The proteins catalysing such a transport are termed *translocators*. The triose phosphate–phosphate translocator of chloroplasts will be used as an example to describe the structure and function of such a translocator. This translocator enables the export of photoassimilates

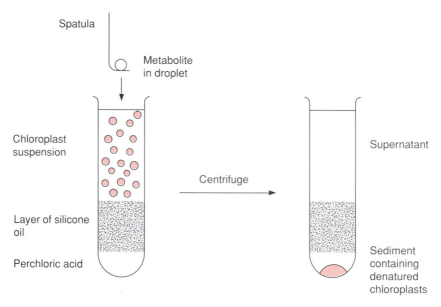

Figure 1.21 Silicone oil filtering centrifugation: measurement of the uptake of substances into isolated chloroplasts. For the measurement, the bottom of a centrifuge tube contains perchloric acid on which silicone oil is layered. The substance to be transported is added to the chloroplast suspension above the silicone layer, using a small spatula. To simplify detection, metabolites labelled with radioactive isotopes (e.g. ^{32}P or ^{14}C) are usually used. The uptake of metabolites into the chloroplasts is terminated by centrifugation in a rapidly accelerating centrifuge. Upon centrifugation the chloroplasts migrate within a few seconds through the silicone layer into the perchloric phase where they are denatured. That portion of the metabolite which has not been taken up remains in the supernatant. The amount of metabolite which has been taken up into the chloroplasts is determined by measurement of the radioactivity in the sedimented fraction. The amount of metabolite carried non-specifically through the silicone layer, either by adhering to the outer surface of the plastid or present in the space between the inner and the outer envelope membranes, can be evaluated in a control experiment in which a substance is added (e.g. sucrose) which is known not to permeate the inner envelope membrane.

from the chloroplasts by catalysing a counter-exchange of phosphate with triose phosphate (dihydroxyacetone phosphate or glyceraldehyde-3-phosphate) (Fig. 9.12). It is the most abundant transport protein in plants.

Silicone layer filtering centrifugation is a very useful tool (Fig. 1.21) for measuring the uptake of substrates into chloroplasts or other cell organelles. To start measurement of transport, the corresponding substrate is added to a suspension of isolated chloroplasts, and it is terminated by separating the chloroplasts from the surrounding medium by centrifugation through a silicone layer. The amount of substrate taken up into the separated chloroplasts is then quantitatively analysed.

A hyperbolic curve is observed (Fig. 1.22) when this method is used to measure the uptake of phosphate into chloroplasts at various external concentrations of

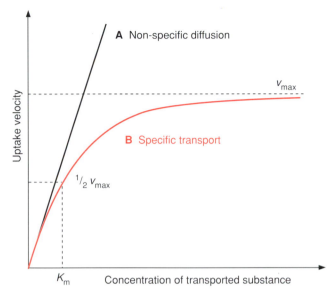

Figure 1.22 By measuring the concentration dependence of the rate of uptake for a substance, it can be decided whether the uptake occurs by non-specific diffusion through the membrane (A) or by specific transport (B).

phosphate. At very low phosphate concentrations the rate of uptake rises proportionally to the external concentration, whereas at a higher phosphate concentration the curve levels off until a *maximal velocity* is reached (V_{max}). These are the same characteristics as seen in enzyme catalysis. During enzyme catalysis the substrate (S) is first bound to the enzyme (E). The product (P) formed on the enzyme surface is then released:

$$\text{Catalysis}$$
$$E + S \rightarrow ES \quad \rightarrow \quad EP \rightarrow E + P$$

The transport by a specific translocator can be depicted in a similar way:

$$\text{Transport}$$
$$S + T \rightarrow ST \quad \rightarrow \quad TS \rightarrow T + S$$

The substrate is bound to a specific binding site of the translocator protein (T), transported through the membrane, and then released from the translocator. The maximal velocity, V_{max}, corresponds to a state in which all the binding sites of the translocators are saturated with substrate. As is the case for enzymes, the K_m for a translocator corresponds to the substrate concentration at which transport occurs at half-maximal velocity. Also in analogy to enzyme catalysis, the translocators usually show high *specificity* for the transported substrates. For instance, the chloroplast triose phosphate–phosphate translocator of C_3 plants (see Fig. 9.12) transports orthophosphate, dihydroxyacetone phosphate, glyceraldehyde-3-

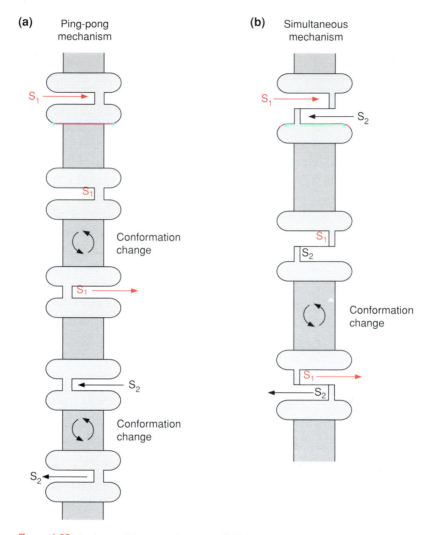

Figure 1.23 Antiport. Diagram of two possibilities for the counter-exchange of two substrate molecules (S_1, S_2). (a) A translocator molecule catalyses the transport of S_1 and S_2 sequentially; (b) S_1 and S_2 are transported simultaneously by two translocator molecules tightly coupled to each other. See text.

phosphate, and 3-phosphoglycerate, but not 2-phosphoglycerate. The various substrates compete for the binding site. Therefore, one substrate, such as phosphate, will be a *competitive inhibitor* for the transport of another substrate, such as 3-phosphoglycerate. The triose phosphate–phosphate translocator of chloroplasts is an antiport, so that for each molecule transported inward (e.g. phosphate), another molecule (e.g. dihydroxyacetone phosphate) must be transported out of the chloroplasts.

Figure 1.23 shows a scheme of the transport process. The triose phosphate–

phosphate translocator from chloroplasts, like translocators from mitochondria and other compartments, consists of two subunits which form a *gated pore*. Both subunits form a common substrate binding site, which is accessible either from the inside or from the outside, depending on the conformation of the translocator protein. The first step of the transport process is to bind a substrate (S_1) to the substrate binding site accessible from the outside. A *conformational change* then occurs and the substrate is finally released to the inner side. Another substrate (S_2) can now bind to the free binding site and thus be transported to the outside. An obligatory counter-exchange may be due to the fact that the shift of the binding site from one side of the membrane to the other can only occur when the binding site is occupied by a substrate. This mode of antiport is called a *ping-pong mechanism* (Fig. 1.23a). However, in many cases counter-exchange proceeds by a *simultaneous mechanism* (Fig. 1.23b), in which two translocators are linked to each other in such a way that a change of conformation can occur only if both binding sites are either occupied or unoccupied. In the case of the chloroplast triose phosphate–phosphate translocator the counter-exchange occurs according to the ping-pong mechanism. In contrast, the known mitochondrial translocators operate by a simultaneous mechanism.

Translocators have a common basic structure

Translocators are *integral membrane proteins*. They span the lipid bilayer of the membrane in the form of α-helices. In such *transmembrane α-helices* the side chains have to be hydrophobic, therefore they contain hydrophobic amino acids such as alanine, valine, leucine, isoleucine, or phenylalanine.

Membrane proteins are not soluble in water due to their high hydrophobicity. Mild non-ionic detergents such as octylglucoside (Fig. 1.24) can be used to dissolve these proteins from the membrane. The hydrophobic carbon chain of the detergent binds to the hydrophobic protein and, due to the glucose residues, the micelle thus formed is water soluble. If the detergent is then removed, the membrane proteins aggregate, forming a sticky mass which cannot be solubilized again.

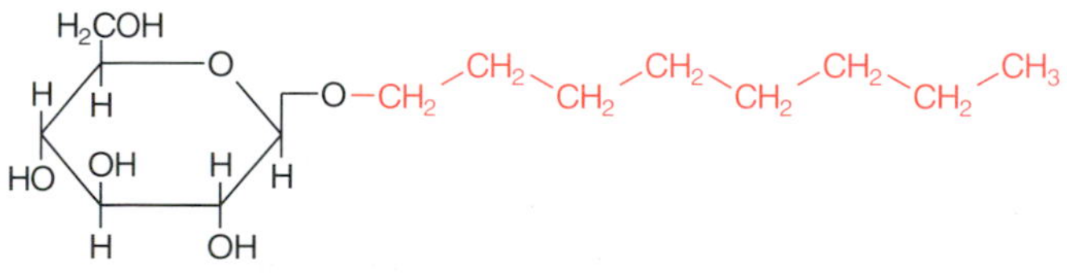

Octylglucoside

Figure 1.24 Octylglucoside, a glycoside composed from α-D-glucose and octyl alcohol, is a mild non-ionic detergent which allows membrane proteins to be solubilized from the membranes without being denatured.

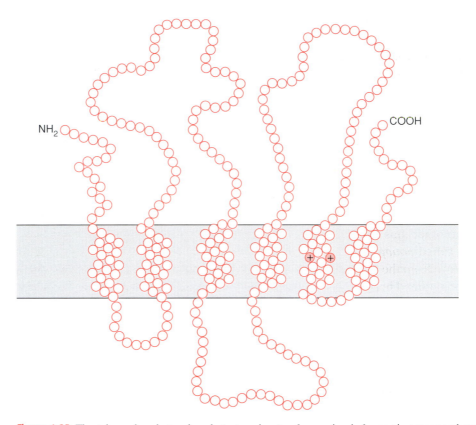

Figure 1.25 The triose phosphate–phosphate translocator from spinach forms six transmembrane helices. Each circle represents one amino acid. The likely positions of the transmembrane helices were evaluated from the hydrophobicity of the single amino acid residues. The amino acids containing a positive charge in helix 5 (marked with red) are an arginine and a lysine. These amino acids probably provide the binding sites for the anionic substrates of the triose phosphate–phosphate translocator. (Data from Flügge *et al.* 1989.)

Using this method, a number of translocator proteins have been isolated and purified. This has allowed analysis of their amino acid sequences but as yet, analysis of their three-dimensional structure by X-ray crystallography (section 3.3) has not been achieved. However, from the hydrophobicity of the amino acids in a peptide sequence, it can be predicted which sections of the chain are likely to form transmembrane helices (hydropathy analysis). This procedure has led to the prediction that the 324 amino acids of the subunit of the chloroplast triose phosphate–phosphate translocator (Fig. 1.25) fold to form six transmembrane helices. Each contains about 20 amino acids and thus is large enough to span the envelope membrane with its cross-sectional distance of about 6 nm. In its functional form the chloroplast triose phosphate–phosphate translocator consists of two identical subunits. It has been suggested that the translocation pore, by which the substrates pass through the membrane, is made up of all 12 transmembrane helices of the dimer. A comparison of sequences revealed that almost all the translocators from bacteria, plants, and animals, known so far, consist either of a dimer where each monomer has six transmembrane helices or of a single monomer with 12 transmembrane helices. Apparently 12 transmembrane helices form the basic structure for a large variety of translocators.

Water channel proteins make cell membranes permeable to water

The osmotic behaviour of cells and cell organelles indicates that their membranes are permeable to water. For a long time there has been a controversy about whether water moves through membranes via special channels. Recently, proteins have been discovered which form channels in membranes through which water, but no other ions or metabolites, can diffuse. These water channel proteins have been called *aquaporins*. Since these aquaporins seem to differ entirely in their structure from the porins discussed in section 1.11, the term *water channel proteins* will be retained here. Results so far indicate that water channel proteins include a very large family of similar proteins which are widespread in the membranes of animals and plants. According to predictions from amino acid sequences, these proteins also contain six transmembrane helices. In membranes they form tetramers, of which each monomer apparently forms a channel, transporting 10^9–10^{11} water molecules per second. This very effective water transport involves cysteine residues of the protein. Mercury compounds inhibit the water transport by binding to these cysteine residues.

1.10 Ion channels have a very high transport capacity

The chloroplast phosphate translocator mentioned above has a turnover number of $80 \, s^{-1}$ at $20°C$, which means that it transports 80 substrate molecules per second. The turnover numbers of other translocators are in the range of 10–$1000 \, s^{-1}$. Membranes also contain proteins which form *ion channels* that transport various ions at least three orders of magnitude faster than translocators (10^6–10^8 ions s^{-1}). They differ from the translocators in having a pore open to both sides at the same time. The flux of ions through the ion channel is so large that it is possible to determine the transport capacity of a single channel from the measurement of electrical conductivity.

  The procedure for such single channel measurements, called the *patch clamp technique*, was developed by two German scientists, Erwin Neher and Bert Sakmann, who were awarded the Nobel prize for Medicine and Physiology in 1991 for this research. The set-up for this measurement (Fig. 1.26) consists of a glass pipette which contains an electrode filled with an electrolyte fluid. The very thin tip of this pipette (diameter about 1 μm) is sealed tightly by a membrane patch. The number of ions transported through this patch per unit time can be determined by

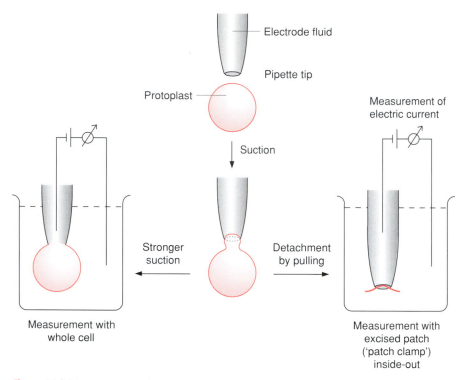

Figure 1.26 Measurement of ion channel currents by the 'patch clamp' technique. A glass pipette with a diameter of about 1 μm at the tip, containing an electrode and electrode fluid, is brought into contact with the membrane of a protoplast or a cell organelle (e.g. vacuole). By applying slight suction, the opening of the pipette tip is sealed by the membrane. By applying stronger suction the membrane surface over the pipette opening breaks, and the electrode within the pipette is now in direct electrical connection with the space inside the cell. In this way the channel currents can be measured for all the channels present in the membrane (measurement with whole cell). Alternatively, by slight pulling, the pipette tip can be removed from the protoplast or the vacuole with the membrane patch, which is sealed to the tip, being torn off from the rest of the cell. In this way, the currents are measured only for those channels which are present in the membrane patch. A voltage is applied for measurement of the channel current and, after amplification, the current is measured.

measuring the electrical current (usually expressed as conductivity in Siemens (S)). Figure 1.27 shows an example of the measurement of single channel current with membranes of parenchymal cells of barley roots. The recording of the change in current shows that the channel opens for various lengths of time and then closes again. This principle of stochastic switching between a non-conductive state and a defined conductive state is a typical property of ion channels. In the open state, various channels have different conductivities, which can range between a few picosiemens and several hundred picosiemens. Moreover, various channels have characteristic mean open and closed times which can, depending on the channel, last from a few milliseconds to seconds. The transport capacity of the channel per

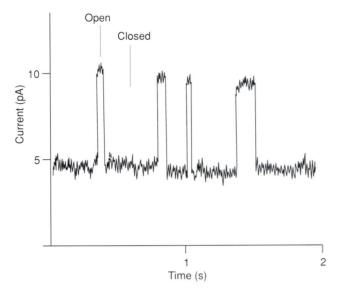

Figure 1.27 Measurement of single channel currents with protoplasts from xylem parenchyma cells from barley roots in the presence of Ca^{2+} ions (voltage, $V = -60$ mV). (By K. Raschke, Göttingen.)

unit time depends therefore on the conductivity of the opened channel as well as on the mean duration of the open state.

Many ion channels have been characterized which are more or less specific for certain ions. Plants contain highly selective ion channels for H^+, K^+, and Ca^{2+} and also selective anion channels for Cl^- and dicarboxylates such as malate. Plant membranes, in contrast to those of animals, seem to possess no specific channels for Na^+ ions. The opening of ion channels is often regulated by the electric membrane potential. This means that membranes have a very important function in the electrical regulation of ion fluxes. Thus, in guard cells (section 8.1) the hyperpolarization of the plasma membrane (< -100 mV) opens a channel which allows potassium ions to flow into the cell (K^+ inward channel), whereas depolarization opens another channel by which K^+ ions can leave the cell (K^+ outward channel). In addition, the opening of many ion channels is controlled by ligands such as Ca^{2+} ions or by phosphorylation of the channel protein. This enables regulation of the channel activity by metabolic processes and also by messenger substances (Chapter 19).

Analyses of the amino acid sequences of ion channels from plant and animal tissues, have shown that in most cases these also consist of two times six transmembrane helices, or multiples of these. This indicates that the basic structure of many channels and translocators is similar. Moreover, it has been demonstrated that translocators, such as the chloroplast triose phosphate–phosphate translocator, can be converted by reaction with chemical agents into a channel, open to both sides at the same time, with an ion conductivity similar to that of the ion channels dealt with above. The functional differences between translocators and ion channels—the translocation pore of translocators is accessible only from one side at a time and transport involves a change of conformation, whereas in ion channels the aqueous pore is open to both sides—may be due to differences in the filling of the pore by peptide chains, providing the pore with a gate.

1.11 Porins consist of β-sheet structures

The outer membrane of chloroplasts and mitochondria, as well as the membrane of peroxisomes, contain pore-forming proteins, called *porins*. These porins represent a family of pore-forming proteins which are entirely different from the channel and translocator proteins dealt with above. Porins form a large, open channel filled with water which, in many cases, allows *non-specific diffusion* of substrate molecules. Porins are also found in the outer membrane of Gram-negative bacteria.

 The size of the aperture of the pore formed by a porin can be determined by incorporating porins into an artificial lipid membrane which separates two chambers filled with an electrolyte (Fig. 1.28). Membrane proteins which have

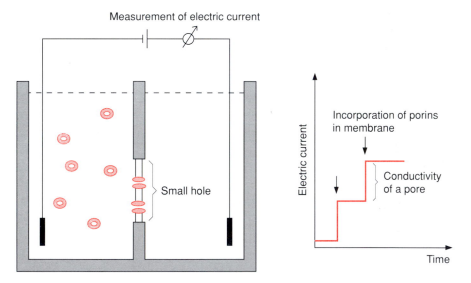

Figure 1.28 Measurement of the size of a porin aperture. Two chambers, each provided with an electrode and filled with electrolyte fluid, are separated from each other by a divider containing a small hole. A small drop of solvent containing a membrane lipid is brushed across this hole. The solvent is taken up into the aqueous phase and the remaining lipid forms a double layer, an artificial membrane. Upon the addition of a porin, which has been solubilized from a membrane, spontaneous incorporation of the single porin molecule into the artificial membrane occurs. The aqueous channel through the lipid membrane formed with each incorporation of a porin protein results in a stepwise increase of current activity.

been solubilized in a detergent, are added to one of the two chambers. Because of their hydrophobicity, the porin molecules are incorporated one after another into the artificial lipid membrane, each time forming a new channel, which can be seen in the stepwise increase in conductivity. As each stepwise increase in conductivity corresponds to the conductivity of a single pore, it is possible to evaluate the size of the aperture of the pore from the conductivity of the electrolyte fluid. Thus the size of the aperture for the porin pore of mitochondria has been evaluated as 1.7 nm and that of chloroplasts as about 3 nm.

Porins from different origins consist of subunits with a molecular mass of about 30 kDa. Porins in the membrane often occur as trimers, in which each of the three subunits forms a pore. Porins differ distinctly from the translocator proteins dealt with above in that they have no exclusively hydrophobic regions in their amino acid sequence, a requirement for forming transmembrane helices. Analysis of the three-dimensional structure of a bacterial porin by X-ray structure analysis (section 3.3) revealed that the walls of the pore are formed by *β-sheet* structures (Fig. 1.29). Altogether, 16 β-sheets, each consisting of about 13 amino acids and connected to each other by hydrogen bonds, form a pore (Fig. 1.30a). This structure resembles a barrel in which the β-sheets represent the barrel staves. Hydrophilic and hydrophobic amino acids alternate in the amino acid sequences of the β-sheets. One side of the β-sheet, occupied by hydrophobic residues, is directed towards the lipid membrane phase. The other side, with the hydrophilic residues, is directed towards the aqueous phase inside the pore (Fig. 1.30b). Compared with the ion channel proteins, the porins have an economical structure

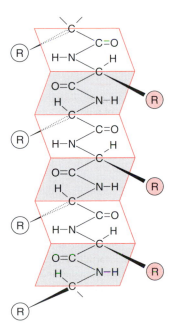

Figure 1.29 With β-sheet conformation the amino acid residues of a peptide chain are arranged alternately in front of and behind the surface of the sheet.

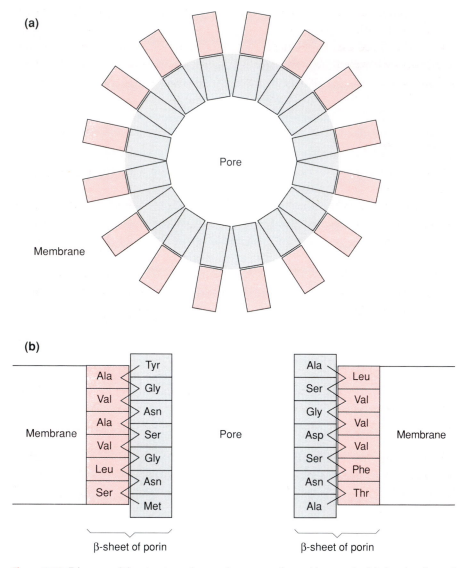

Figure 1.30 Diagram of the structure of a membrane pore formed by a porin: (a) the view from above; (b) a cross-section through the membrane. Sixteen β-sheet sequences of the porin molecules, each 13 amino acids long, form the pore. The amino acid residues directed towards the membrane side of the pore have hydrophobic character; those directed to the aqueous pore are hydrophilic. The amino acid sequence of the porin shown in the cross-section is from a porin of maize amyloplasts. (Data from Fischer *et al.* 1994.)

in the sense that a much larger channel is formed by one porin molecule than by a channel protein with a two times higher molecular mass. Up till now, X-ray analysis of the three-dimensional structure has only been determined for the bacterial porin. However, the similarity of amino acid sequences would suggest that the three-dimensional structure of porins from mitochondria in animals and plants and also from plastids will turn out to be very similar.

Further reading

Benz, R. (1994). Permeation of hydrophilic solutes through mitochondrial porins. *Biochimica et Biophysica Acta*, **1197**, 167–96.

Blatt, R. and Thiel, G. (1993). Hormonal control of ion channel gating. *Annual Review of Plant Physiology and Plant Molecular Biology*, **44**, 543–67.

Chrispeels, M. J. and Maurel, C. (1994). Aquaporins: the molecular basis of facilitated water movement through living plant cells. *Plant Physiology*, **105**, 9–13.

Fischer, K., Weber, A., Brink, S., Arbinger, B., Schünemann, D., Borchert, S., *et al.* (1994). Porins from plants: molecular cloning and functional characterization of two new members of the porin family. *Journal of Biological Chemistry*, **269**, 25754–60.

Flügge, U. I., Fischer, K., Gross, A., Sebald, W., Lottspeich, and F., Eckerskorn, C. (1989). The triose phosphate-3-phosphoglycerate-phosphate translocator from spinach chloroplasts: nucleotide sequence of a full-length cDNA clone and import of the *in vitro* synthesized precursor protein into chloroplasts. *The EMBO Journal*, **8**, 39–46.

Flügge, U. I. and Heldt, H. W. (1991). Metabolite translocators in the chloroplast envelope. *Annual Review of Plant Physiology and Plant Molecular Biology*, **42**, 129–44.

Gunning, D. E. S. and Steer, M. W. (1975). *Ultrastructure and the biology of plant cells*. Edward Arnold, London.

Hawes, C. and Satiat-Jeunemaitre, B. (1996). Stacks of questions: how does the plant Golgi work? *Trends in Plant Science*, **1**, 395–401.

Henderson, P. J. F. (1993). The transmembrane helix transporters. *Current Opinion in Cell Biology*, **5**, 708–21.

Keller, B. U., Hedrich, R., and Raschke, K. (1989). Voltage-dependent anion channels in the plasma membrane of guard cells. *Nature*, **341**, 450–3.

Krämer, R. (1994). Functional principles of solute transport systems: concepts and perspectives. *Biochimica et Biophysica Acta*, **1185**, 1–34.

Lucas, W. J. and Gilbertson, R. L. (1994). Plasmodesmata in relation to viral movement within leaf tissues. *Annual Review of Phytopathology*, **32**, 387–411.

Oparka, K. J., Boevink, P., and Santa Cruz, S. (1996). Studying the movement of plant viruses using green fluorescent protein. *Trends in Plant Science*, **1**, 412–418.

Overall, R. L. and Blackman, L. M. (1996). A model of the macromolecular structure of plasmodesmata. *Trends in Plant Science*, **1**, 307–11.

Robards, A. W. and Lucas, W. J. (1990). Plasmodesmata. *Annual Review of Plant Physiology and Plant Molecular Biology*, **41**, 369–419.

Robinson, D. (1985). *Plant membranes*. John Wiley and Sons, Chichester.

Schroeder, J. I., Ward, J. M., and Gassmann, W. (1994). Perspectives on the physiology and structure of inward rectifying K$^+$ channels in higher plants. Biophysical implications for K$^+$ uptake. *Annual Review of Biophysics and Biomolecular Structure*, **23**, 441–71.

Staehelin, L. A. and Moore, I. (1995). The plant Golgi apparatus: structure, functional organization and trafficking mechanisms. *Annual Review of Plant Physiology and Molecular Biology*, **46**, 261–88.

Tanner, W. and Carpari, T. (1996). Membrane transport carriers. *Annual Review of Plant Physiology and Plant Molecular Biology*, **47**, 595–626.

Wallmeier, H., Weber, A., Gross, A., and Flügge, U. I. (1992). Insights into the structure of the chloroplast phosphate translocator protein. In *Transport and receptor proteins of plant membranes*, (ed. D. T. Cooke and D. T. Clarkson), pp. 77–89. Plenum Press, New York.

Weier, T. E., Stocking, C. R., Thomson, W. W., and Drever, H. (1963). The grana as structural units in chloroplasts of mesophyll of *Nicotiana rustica* and *Phaseolus vulgaris*. *J. Ultrastructure Research*, **8**, 122–43.

Winter, H., Robinson, D. G., and Heldt, H. W. (1994). Subcellular volumes and metabolite concentrations in spinach leaves. *Planta*, **193**, 530–35.

chapter 2

The use of energy from sunlight by photosynthesis

Plants and cyanobacteria capture the light of the sun and utilize its energy to synthesize organic compounds from inorganic substances, such as CO_2, nitrate and sulfate, to make their cellular material: they are *photoautotrophic*. In photosynthesis photon energy splits water into oxygen and hydrogen, the latter bound as NADPH. This process, termed the *light reaction*, takes place in the photosynthetic reaction centres embedded in membranes. It involves the transport of electrons which is coupled to the synthesis of ATP. NADPH and ATP are consumed in a so-called *dark reaction* to synthesize carbohydrates from CO_2 (Fig. 2.1). The photosynthesis of plants and cyanobacteria created the biomass on earth, including the deposits of fossil fuels and atmospheric oxygen. Animals are dependent on the supply of carbohydrates and other organic substances as food; they are *heterotrophic*. They generate the energy required for their life processes by oxidizing the biomass, which has first been produced by plants. When oxygen is consumed, CO_2 is formed. It follows that the light energy captured by plants is also the source of energy for the life processes of animals.

2.1 How did photosynthesis start?

Measurements of the distribution of radioisotopes led to the conclusion that the earth was formed about 4.6 billion years ago. The earliest indicators of life on

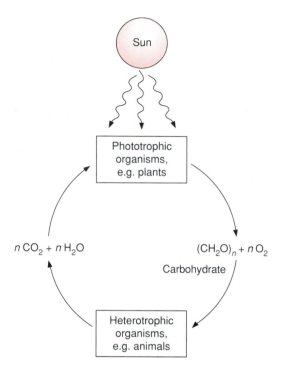

Figure 2.1 Life on Earth involves a CO_2 cycle.

earth are fossils of bacteria-like structures, estimated to be 3.5 billion years old. There was no oxygen in the atmosphere when life on earth commenced. This is concluded from the fact that in very early sedimentary rocks iron is present as Fe^{2+}. Mineral iron is oxidized to Fe^{3+} in the presence of oxygen. According to our present knowledge the earth's atmosphere initially contained components such as carbon dioxide, molecular hydrogen, methane, ammonia, prussic acid, and also water.

   In 1922 the Russian scientist Alexander Oparin presented the interesting hypothesis that organic compounds were formed spontaneously in the early atmosphere by the input of energy, e.g. in the form of ultraviolet radiation (there was no protective ozone layer), electrical discharges (lightning), or volcanic heat. These organic compounds accumulated in ancient seas and became the constituents of early forms of life. In 1953 the American scientists Stanley Miller and Harold Urey substantiated this hypothesis by simulating the postulated *pre-biotic synthesis* of organic substances. They exposed a gaseous mixture of components present in the early atmosphere, consisting of H_2O, CH_4, NH_3, and H_2, to electrical discharges for about a week at 80 °C. Amino acids (such as glycine and alanine) and other carboxylic acids (such as formic, acidic, lactic, and succinic acids) were found in the condensate of this experiment. Other investigators added substances such as CO_2, HCN, and formaldehyde to the gaseous mixture and these experiments showed that many components of living cells (e.g. carbohydrates, fatty acids, tetrapyrroles, and the nucleobases adenine, guanine, cytosine, and uracil) were formed spontaneously by exposing a postulated early atmosphere to electrical or thermal energy.

There is a hypothesis that the organic substances formed by the abiotic processes mentioned above accumulated in the ancient seas, lakes, and pools over a long period of time prior to the emergence of life on earth. There was no oxygen to oxidize the substances that had accumulated and no bacteria or other organisms to degrade them. Oparin had already speculated that a *'primordial' soup* was formed in this way, providing the building material for the origin of life. Since oxygen was not yet present, the first organisms must have been *anaerobes*, generating the energy for their metabolism by fermentative degradation of organic compounds accumulated in the 'primordial' soup.

Bacteria propagate quickly. It is feasible that after the first formation of life on earth, a situation might have arisen where substances in the 'primordial' soup were degraded by fermentation more rapidly than they could be replenished by abiotic processes. Depletion of the resources would have led to the loss of the basis of existence for the newly formed life. Thus, if evolution had not produced organisms able to utilize the sun's energy as a source for biomolecule synthesis, life on earth might just have been an episode. Photosynthesis probably occurred at a very early stage in evolution. The now widely distributed *purple bacteria* and *green sulfur bacteria* may be regarded as relics from an early period in the evolution of photosynthesis.

Prior to the description of photosynthesis in Chapter 3, the present chapter will discuss how plants capture sunlight and how the light energy is conducted into the photosynthesis apparatus.

2.2 Pigments capture energy from sunlight

The energy content of light depends on its wavelength

In Berlin at the beginning of this century Max Planck and Albert Einstein, two Nobel prize winners, carried out the epoch-making studies proving that light has a dual nature. It can be regarded as an electromagnetic wave as well as an emission of particles, which are termed *light quanta* or *photons*.

The energy of the photon is proportional to its frequency, v:

$$E = hv = h\frac{c}{\lambda} \tag{2.1}$$

where h is the Planck constant (6.6×10^{-34} J s) and c the velocity of the light (3×10^8 m s^{-1}). λ is the wavelength of light.

The mole (abbreviated to mol) is used as a chemical measure of the amount of molecules and also the amount of photons corresponding to 6×10^{23} molecules or photons (Avogadro constant, N_A). The energy of 1 mol photons amounts to:

$$E = h \frac{c}{\lambda} N_A \qquad (2.2)$$

In order to utilize the energy of a photon in a thermodynamic sense, this energy must be at least as high as the Gibbs free energy of the photochemical reaction involved. (In fact much energy is lost during energy conversion (see section 3.4), with the consequence that the energy of the photon must be higher than the Gibbs free energy of the corresponding reaction). We can equate the Gibbs free energy, ΔG, with the energy of the absorbed light:

$$\Delta G = E = h \frac{c}{\lambda} N_A \qquad (2.3)$$

The introduction of numerical values of the constants h, c and N_A yields:

$$\Delta G = 6.6 \times 10^{-34} (\text{J s}) \times \frac{3 \times 10^8 \ (\text{m})}{(\text{s})} \times \frac{1}{\lambda(\text{m})} \times \frac{6 \times 10^{23}}{(\text{mol})}$$

$$\Delta G = \frac{119\,000}{\lambda(\text{nm})} \ \ [\text{kJ (mol photons)}^{-1}]. \qquad (2.5)$$

It is often useful to state the electrical potential (ΔE) of the irradiation instead of energy when comparing photosynthetic reactions with redox reactions, which will be dealt with in Chapter 3:

$$\Delta E = -\frac{\Delta G}{F} \qquad (2.6)$$

where F = number of charges per mol = $96\,480$ A s mol^{-1}. The introduction of this value yields:

$$\Delta E = -\frac{N h c}{F \lambda(\text{nm})} = \frac{1231}{\lambda(\text{nm})} \quad [\text{Volt}]. \qquad (2.7)$$

The human eye perceives only the small range between about 400 and 700 nm of the broad spectrum of electromagnetic waves (Fig. 2.2). The light in this range, where the intensity of solar radiation is especially high, is utilized in plant photosynthesis. Bacterial photosynthesis, however, is able to utilize light in the infrared range.

According to equation 2.3 the energy of irradiated light is inversely proportional to the wavelength. Table 2.1 shows the light energy per mol photons for light of different colours. Consequently violet light has an energy of about 300 kJ (mol photons)$^{-1}$. Dark-red light, with the highest wavelength (700 nm) that can still be utilized by plant photosynthesis, contains 170 kJ (mol photons)$^{-1}$. This is only about half the energy content of violet light.

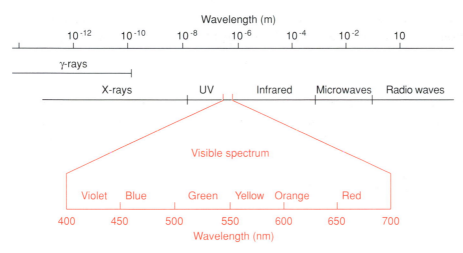

Figure 2.2 Spectrum of the electromagnetic radiation. The section shows the visible spectrum.

Table 2.1 The energy content and the electrochemical potential difference of photons of different wavelengths

Wavelength (nm)	Light colour	Energy content kJ (mol photons)$^{-1}$	ΔE (eV)
700	Red	170	1.76
650	Bright red	183	1.90
600	Yellow	199	2.06
500	Blue-green	238	2.47
440	Blue	271	2.80
400	Violet	298	3.09

Chlorophyll is the main photosynthetic pigment

In photosynthesis light is collected primarily by *chlorophylls*, pigments which absorb light at a wavelength below 480 nm and between 550 and 700 nm (Fig. 2.3). When white sunlight falls on a chlorophyll layer, the green light with a wavelength between 480 and 550 nm is not absorbed, but is reflected. This is why plant chlorophylls and whole leaves are green.

Experiments carried out between 1905 and 1913 in Zurich and in Berlin by Richard Willstätter and his collaborators, led to the discovery of the structural formula of the green leaf pigment chlorophyll, a milestone in the history of chemistry. This discovery made such an impact that Richard Willstätter was awarded the Nobel prize for Chemistry as early as 1915. There are different classes of

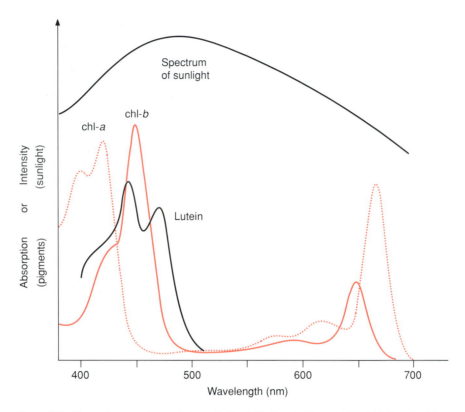

Figure 2.3 Absorption spectrum of chorophyll-*a* (chl-*a*) and chlorophyll-*b* (chl-*b*) and of the xanthophyll lutein dissolved in acetone. The intensity of the sun's radiation at different wavelengths is given as a comparison.

chlorophylls. Figure 2.4 shows the structural formula of chlorophyll-*a* and *-b* (chl-*a*, chl-*b*). The basic structure is a ring made of four pyrroles, a tetrapyrrole, which is also named *porphyrin*. Mg^{2+} is present in the centre of the ring. Mg^{2+} is covalently bound with two N- atoms and coordinately bound to the other two atoms of the tetrapyrrole ring. A cyclopentanone is attached to ring c. At ring d a propionic acid group forms an ester with the alcohol phytol. Phytol consists of a long, branched hydrocarbon chain with one C–C double bond. It is derived from an isoprenoid, formed from four isoprene units (section 17.7). This long, hydrophobic hydrocarbon tail renders the chlorophyll highly soluble in lipids and therefore promotes its presence in the membrane phase. Chlorophyll always occurs bound to proteins. In ring b, chl-*b* contains a formyl residue instead of the methyl residue in chl-*a*. This small difference has a large influence on light absorption. Figure 2.3 shows that the absorption spectra of chl-*a* and *-b* differ markedly.

In plants, the ratio of chl-*a* to chl-*b* is about three to one. Only chl-*a* is a constituent of the photosynthetic reaction centres (sections 3.6, 3.8) and therefore it

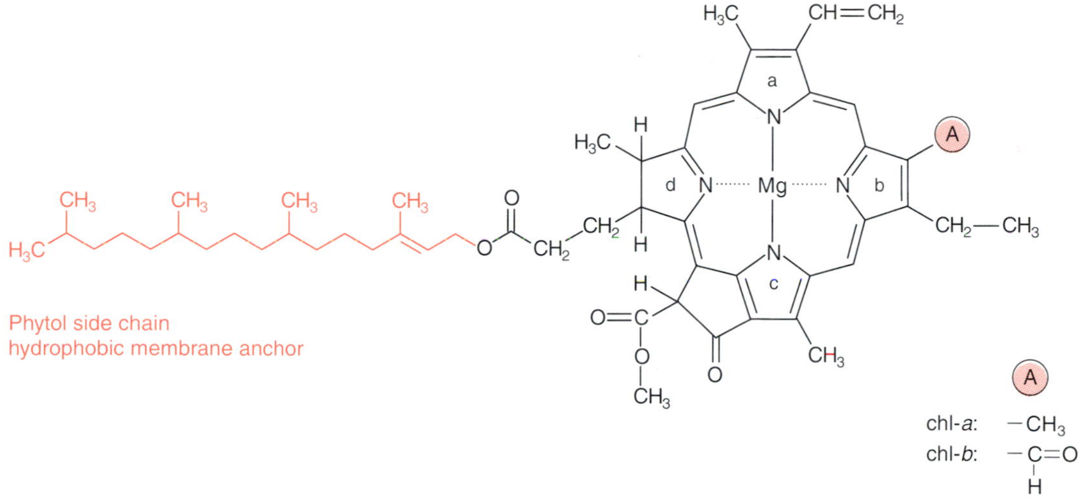

Figure 2.4 Structural formula of chlorophyll-*a*. In chlorophyll-*b* the methyl group in ring b is replaced by a formyl group (red). The phytol side chain gives chlorophyll a lipid character.

can be regarded as the central photosynthetic pigment. In a wide range of the visible spectrum, however, chl-*a* does not absorb light. This non-absorbing region is called the '*green window*'. The absorption gap is narrowed by the light absorption of chl-*b*, with its first maximum at a higher wavelength than chl-*a* and the second maximum at a lower wavelength. As shown in section 2.4, the light energy absorbed by chl-*b* can be transferred very efficiently to chl-*a*. In this way chl-*b* enhances the plant's efficiency for utilizing sunlight energy.

The structure of chlorophylls has remained remarkably constant during the course of evolution. Purple bacteria, formed probably more than 3 billion years ago, contain as photosynthetic pigment a bacteriochlorophyll-*a*, which differs from the chlorophyll-*a* shown in Fig. 2.4 only by the alteration of one side chain and by one double bond. This, however, influences light absorption; both absorption maxima are shifted outwards and the non-absorbing spectral region in the middle is broadened. This shift allows purple bacteria to utilize light in the infrared region.

The tetrapyrrole ring is not only a constituent of chlorophyll but has also attained a variety of other functions during evolution. It is involved in methane formation by bacteria with Ni as the central atom. With Co it forms *cobalamin* (vitamin B_{12}), which participates as a cofactor in reactions where hydrogen and organic groups change their position. With Fe^{2+} instead of Mg^{2+} as the central atom, the tetrapyrrole ring forms the basic structure of *haems* (Fig. 3.24), which, on the one hand, as cytochromes, function as redox carriers in electron transport processes (sections 3.7 and 5.5) and, on the other hand, as myoglobin or haemoglobin, store or transport oxygen in aerobic organisms. The tetrapyrrole ring in animal haemoglobin differs only slightly from the tetrapyrrole ring of chlorophyll-*a* (Fig. 2.4).

It seems remarkable that a substance that attained a certain function during evolution is being utilized after only minor changes for completely different functions. The reason for this functional variability of substances such as chlorophyll or haem is that their reactivity is governed to a great extent by the proteins to which they are bound.

Chlorophyll molecules are bound to chlorophyll-binding proteins. In a complex with proteins the absorption spectrum of the bound chlorophyll may differ considerably from the absorption spectrum of the free chlorophyll. The same applies to other light-absorbing substances, such as carotenoids, xanthophylls, and phycobilins, which also occur bound to proteins. These substances will be dealt with in the following sections. In this text, free absorbing substances will be called *chromophores* (Greek, carrier of colour) and the chromophore–protein complexes called *pigments*. Pigments are often named after the wavelength of their absorption maximum. Chl-a_{700} means a pigment of chl-*a* with an absorption maximum of 700 nm. Another common designation is P_{700}; this name leaves the nature of the chromophore open.

2.3 Light absorption excites the chlorophyll molecule

What happens when a pigment absorbs a photon? When a photon with a certain wavelength hits a chromophore that absorbs light of this wavelength, the energy of the photon excites electrons to a higher energy level. This occurs as an 'all or nothing' process. According to the principle of energy conservation expressed by the first law of thermodynamics, the energy of the chromophore is increased by the energy of the photon, which results in an *excited state* of the chromophore molecule. The energy required to excite a chromophore molecule depends on the chromophore structure. A general property of chromophores is that they contain many *conjugated double bonds*, 10 in the case of the tetrapyrrole ring of the chl-*a*. These double bonds are delocalized. Figure 2.5 shows two possible resonance forms.

After absorption of energy, an electron of the conjugated system is elevated to a higher orbital. Since only a single electron is present in this high orbital, this excitation state is termed *singlet*. Figure 2.6 shows a scheme of the excitation process. As a rule, the higher the number of double bonds in the conjugated state, the lower the amount of energy required to produce a first singlet state. For the excitation of chlorophyll, dark-red light is sufficient, whereas butadiene, with only two conjugated double bonds, requires energy-rich ultraviolet light for excitation. The light absorption of the conjugated system of the tetrapyrrole ring is influenced by the side chains. Thus, the differences in the absorption maxima of

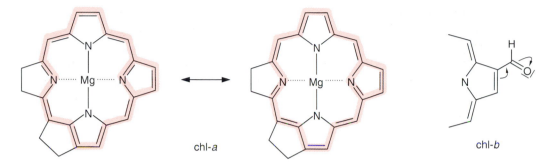

Figure 2.5 Resonance structures of chlorophyll-*a*. In the region marked in red the double bonds are not localized; the π electrons are distributed over the entire conjugated system. The formyl residue of chlorophyll-*b* attracts electrons and thus affects the π electrons of the conjugated system.

chl-*a* and chl-*b* mentioned above can be explained by an electron-attracting effect of the carbonyl side chain in ring b of the chl-*b* (Fig. 2.5).

The spectra of chl-*a* and chl-*b* (Fig. 2.3) each have two main absorption maxima, showing that each chlorophyll has two main excitation states. In addition, chlorophylls have minor absorption maxima, which for the sake of simplicity will not be dealt with here. The two main excitation states of chlorophyll are known as first and second singlet (Fig. 2.6). The absorption maxima in the spectra are relatively broad. At a higher resolution the spectra can be shown to consist of many separate absorption lines. This fine structure of the absorption spectra is due to chlorophyll molecules in the ground and in the singlet states being in various states of *rotation* and *vibration*. In the energy diagram scheme of Fig. 2.6 the various rotation and vibration energy levels are drawn as fine lines and the corresponding ground states as solid lines.

The energy levels of the various rotation and vibration states of the ground state overlap with the lowest energy levels of the *first singlet*. Analogously, the energy levels of the first and the second singlet also overlap. If a chlorophyll molecule absorbs light in the region of its absorption maximum in blue light, one of its electrons is elevated to the *second singlet* state. This second singlet state, with a half-life of only 10^{-12} s, is too unstable to use its energy for chemical work. The excited molecules lose energy in form of heat by rotations and vibrations until the first singlet state is reached. This first singlet state can also be attained by absorption of a photon of red light, which contains less energy. The first singlet state is much more stable than the second one: its half-life amounts to 5×10^{-9} s.

The return of the chlorophyll molecule from the first singlet state to the ground state can proceed in different ways

1. The most important path for conversion of the energy released when the first singlet state returns to the ground state is its utilization for *chemical work*. The

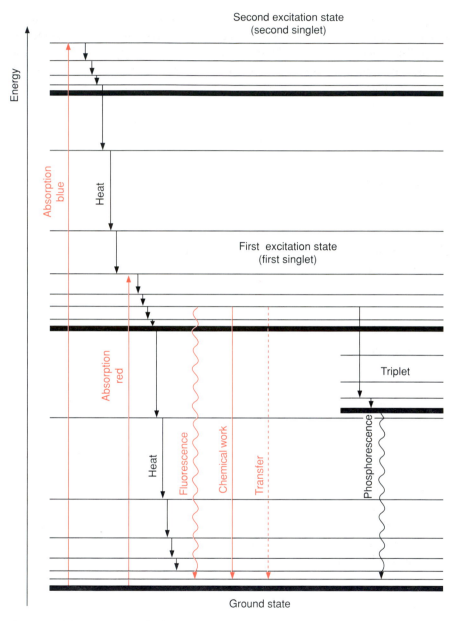

Figure 2.6 Scheme of the excitation states of chlorophyll-*a* and its return to the ground state during which the released excitation energy is converted to photochemical work, fluorescence or phosphorescence light, or dissipated into heat. This simplified scheme shows only the excitation states of the two main absorbing maxima of the chlorophylls.

chlorophyll molecule transfers the excited electron from the first singlet state to an electron acceptor and a positively charged chlorophyll radical, chl$\overset{+}{\bullet}$, remains. This is possible since the excited electron is bound less strongly to the chromato-phore molecule than in the ground state. Section 3.4 describes in detail how the electron can be transferred back from the acceptor to the chl$\overset{+}{\bullet}$ radical via an elec-tron transport chain, by which the chlorophyll molecule returns to the ground state and the free energy derived from this process is conserved for chemical work. As an alternative, the electron deficit in the chl$\overset{+}{\bullet}$, radical may be replenished by an other electron donor, e.g. water (section 3.6).

2. The excited chlorophyll can return to the ground state by releasing excitation energy in the form of light; this light emission is called *fluorescence*. Due to vibra-tions and rotations in the form of heat, part of the excitation energy is usually lost beforehand, with the result that the fluorescent light has less energy (longer in wavelength) than the energy of the excitation light, which was required for attaining the first singlet state (Fig. 2.7).

3. It is also possible that the return from the first singlet to the ground state pro-ceeds in a stepwise fashion via the various levels of vibration and rotation energy, by which the energy difference is completely converted into heat.

4. By releasing part of the excitation energy in form of heat, the chlorophyll molecule can attain an excited state of lower energy, called the first *triplet state*. This triplet state cannot be reached directly from the ground state by excitation. In the triplet state the spin of the excited electrons has been reversed. As the probability of a spin reversal is low, the triplet state does not occur frequently. By emitting so-called *phosphorescent light* the molecule can return from the triplet state to the ground state. Phosphorescent light is again lower in energy than the light required to attain the first singlet state. The return from the triplet state to the ground state requires a reversal of the *electron spin*. As this is rather improb-able, the triplet state, in comparison with the first singlet state, has a relatively long life (half-life 10^{-4}–10^{-2} s.). The triplet state of the chlorophyll has no function in photosynthesis *per se*. In its triplet state, however, the chlorophyll can excite oxygen to a singlet state, whereby the oxygen becomes very reactive with a

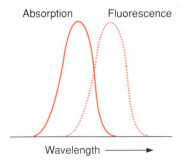

Absorption   Fluorescence

Wavelength ⟶

Figure 2.7 Fluorescent light generally has a longer wavelength than excitation light.

damaging effect on cell constituents. Section 3.10 describes how the plant manages to protect itself from the harmful singlet oxygen.

5. The return to the ground state can be coupled with the excitation of a neighbouring chromatophore molecule. This transfer is important for the function of antennae and will be described in the following section.

2.4 An antenna is required to capture light

In order to excite a photosynthetic reaction centre, a photon with a defined energy content has to react with a chlorophyll molecule in the reaction centre. The probability is very slight that a photon not only has the proper energy, but also hits the pigment exactly at the site of the chlorophyll molecule. Therefore efficient photosynthesis is only possible when the energy of photons of various wavelengths is captured over a certain surface by a so-called *antenna* (Fig. 2.8). Similarly, radio and television sets could not work without an antenna .

The antennae of plants consist of a large number of protein-bound chlorophyll molecules, which absorb photons and transfer their energy to the reaction centre. Only a few thousandths of the chlorophyll molecules in the leaf are constituents of the actual reaction centres, the remainder are contained in the antennae. Observations made as early as 1932 by Robert Emerson and William Arnold in the USA, indicated that the large majority of chlorophyll molecules are not part of the reaction centres. At that time there was a scientific controversy as to how many photons were required for the photosynthetic formation of one molecule of oxygen from water. In order to measure the quantum requirement of photosynthesis, the two researchers illuminated a suspension of the green alga *Chlorella* with light pulses of 10 µs duration, interrupted by dark intervals of 20 ms. The light

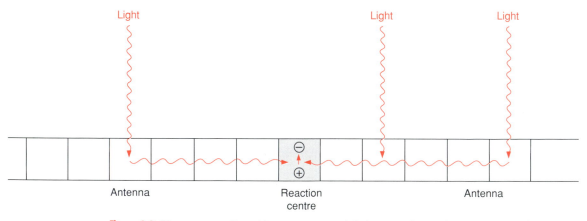

Figure 2.8 Photons are collected by an antenna and their energy is transferred to the reaction centre. In this scheme the squares represent chlorophyll molecules. The excitons conducted to the reaction centre cause a charge separation (see section 3.4).

pulses were made so short that chlorophyll could undergo only one photo-synthetic excitation cycle. When the pulses had a very low light intensity the amount of oxygen formed increased proportionally with the light intensity. From this it was calculated that the release of one molecule of oxygen required a minimum of about eight photons. At about the same time James Franck concluded from his studies that the quantum requirement for photosynthetic formation of O_2 was eight. These important observations were fundamental for the discovery of the mechanism of photosynthesis.

The experiments of Emerson and Arnold led to a further observation: as was to be expected, the amount of oxygen formed increased with increasing light intensity until the photosynthetic apparatus was saturated with photons. Analysis of the chlorophyll content of the algal suspension showed that only one molecule of O_2 was formed per 2400 chlorophyll molecules under saturating conditions. The results of Emerson and Arnold led to the conclusion that when the quantum requirement is evaluated at eight photons per molecule of O_2 formed, and (a fact which was recognized later) upon the formation of O_2 two reaction centres require four photons each, about 300 chlorophyll molecules are associated with one reaction centre. These are constituents of the antennae.

The antennae contain additional accessory pigments to utilize those photons where the wavelength corresponds to the green window between the absorption maxima of the chlorophylls. In higher plants these pigments are carotenoids, mainly *xanthophylls*, including lutein and the related violaxanthin, and also *carotene* with β-carotene as the main substance (Fig. 2.9). However, one of the main functions of these carotenoids in the antennae is to act as a protection

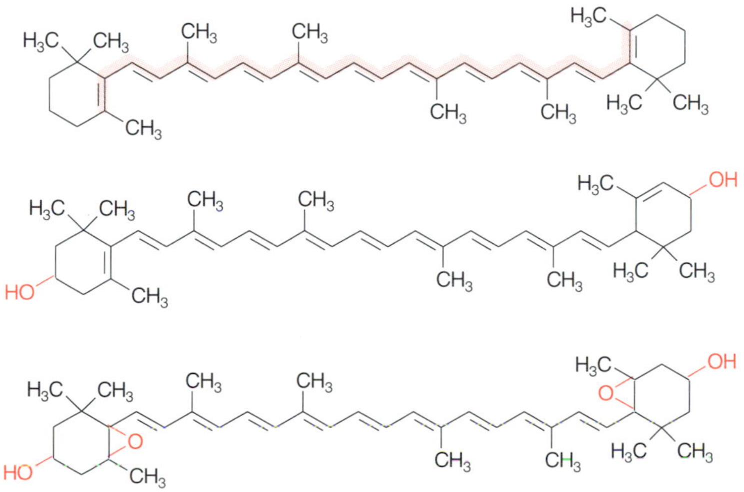

Figure 2.9 Structural formula of a carotene (β-carotene) and of two xanthophylls (lutein and violaxanthin). Due to the conjugated isoprenoid chain these molecules absorb light and also have lipid character.

against the formation of the harmful triplet state of the chlorophylls previously mentioned (see also section 3.10). Important constituents of the antennae in cyanobacteria are *phycobilins* which will be dealt with at the end of this section.

How is the excitation energy of the photons, which have been captured in the antennae, transferred to the reaction centres?

The possibility has been excluded that in the transfer of the energy in the antennae, electrons are transported from chromatophore to chromatophore in a sequence of redox processes, like in the electron transport chains of photosynthesis or of mitochondrial respiration. Such an electron transport would have considerable activation energy. This is not the case, however, since a flux of excitation energy can be measured in the antennae at temperatures as low as 1 K. At these low temperatures light absorption and fluorescence still occur, whereas chemical processes catalysed by enzymes are completely frozen. This makes it probable that the energy transfer in the antennae proceeds according to a mechanism which is related to those of light absorption and fluorescence.

When chromatophores are positioned very close to each other it is possible that the quantum energy of an irradiated photon is transferred from one chromophore to the next. Just as one quantum of light energy is called a photon, one quantum of excitation energy transferred from one molecule to the next is termed *exciton*. A prerequisite for the transfer of excitons is that the chromatophores involved are arranged in a specific way. This is achieved by proteins. Therefore the chromophores of the antennae always occur as protein complexes.

The antennae of plants consist of an inner and an outer part (Fig. 2.10). The

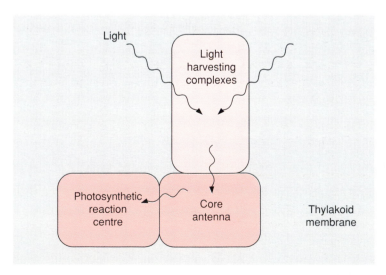

Figure 2.10 Basic scheme of an antenna.

outer part, formed by the *light harvesting complexes* (LHCs), collects the light. The inner part of the antenna, consisting of the *core complexes*, is an integral constituent of the reaction centres and conducts the excitons collected in the outer part of the antenna into the photosynthetic reaction centres.

The light harvesting complexes are formed by polypeptides, which bind chl-*a*, chl-*b*, xanthophylls, and carotenes. These proteins, termed *LHC polypeptides*, are encoded in the nucleus. A plant contains many different LHC polypeptides. In a tomato, for instance, at least 19 different genes for LHC polypeptides have been found, which are very similar to each other. They are homologous, as they have all evolved from a common ancestral form. These LHC polypeptides form a multigene family.

Plants contain two reaction centres which are arranged in sequence, photosystem II (PS II) which has an absorption maximum at 680 nm, and photosystem I (PS I) with an absorption maximum at 700 nm. The function of these reaction centres will be described in sections 3.6 and 3.8. The two photosystems have different light harvesting complexes.

The function of an antenna can be illustrated using the antenna of photosystem II as an example

The antennae of the PS II reaction centre contain primarily four light harvesting complexes, termed LHC-II*a*–*d*. The main component is LHC-II*b*; it represents 67 per cent of the total chlorophyll of the PS II antenna and is the most abundant membrane protein of the thylakoid membrane, thus lending itself to particularly thorough investigation. LHC-II*b* occurs in the membrane most probably as a trimer. The monomer (Table 2.2) consists of a polypeptide to which two molecules of lutein (Fig. 2.9) are bound. The polypeptide contains one threonine residue which can be phosphorylated by ATP via a protein kinase. Phosphorylation regulates the activity of LHC-II (section 3.10).

There has been a recent breakthrough in establishing the three-dimensional structure of LHC-II*b* by electron cyromicroscopy at 4 K of crystalline layers of

Table 2.2 Composition of the LHC-II*b* monomer (Kühlbrandt 1994)

Peptide:	232 Amino acids
Lipids:	1 Phospatidylglycerol
	1 Digalactosyldiacylglycerol
Chromophores:	7 Chl-*a*
	5 Chl-*b*
	2 Lutein

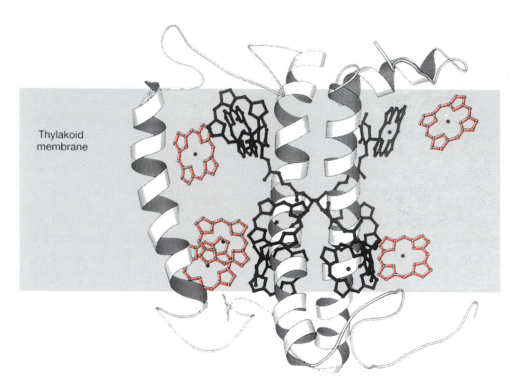

Thylakoid membrane

Figure 2.11 Sterical arrangement of the LHC-II*b* monomer in the thylakoid membrane, viewed from the side. Three α-helices of the protein span the membrane. Chlorophyll-*a* (black) and -*b* (red) are orientated almost perpendicularly to the membrane surface. Two lutein molecules (black) in the centre of the complex, act as an internal cross-brace. (By courtesy of W. Kühlbrandt, Heidelberg.)

LHC-II*b* trimers (Fig. 2.11). The LHC-II*b* protein forms three transmembrane helices. The two lutein molecules span the membrane crosswise. The chl-*b* molecules, where the absorption maximum in the red spectral region lies at a shorter wavelength than that of chl-*a*, are positioned in the outer region of the complexes. Only one of the chl-*a* molecules is positioned in the outer regions; the others are all present in the centre. Figure 2.12 shows a vertical projection of the probable arrangement of the monomers to form a trimer. The chl-*a* positioned in the outer region mediates the transfer of energy to the neighbouring trimers or to the reaction centre. The other LHC-II peptides (*a*, *c*, *d*) are very similar to the LHC-II*b* peptide: they differ in only 5 per cent of their amino acid sequence and it can be assumed that their structure is also very similar. The chl-*a*/chl-*b* ratio is much higher in LHC-II*a* and LHC-II*c* than in LHC-II*b*. Most likely LHC-II*a* and -*c* are positioned between LHC-II*b* and the reaction centre.

Figure 2.13 shows a hypothetical scheme of the probable array of the PS II antenna. The outer complexes, consisting of LHC-II*b*, are present at the periphery of the antenna. The excitons captured by chl-*b* in LHC-II*b* are transferred to chl-*a*

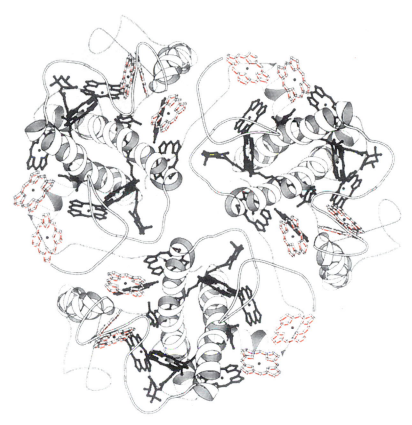

Figure 2.12 The LHC-II trimer viewed from above from the stroma side. Within each monomer the central pair of helices forms a left-handed supercoil, which is surrounded by chlorophyll molecules. The chl-*b* molecules (red) are positioned at the side of the monomers. (By courtesy of W. Kühlbrandt, Heidelberg.)

in the centre of the LHC-II*b* monomers and are then transferred further by chl-*a* contacts between the trimers to the inner antennae complexes. The inner complexes are connected by small chlorophyll-containing subunits to the core complex. This consists of the antenna proteins CP 43 and CP 47 which are closely attached to the reaction centre (Fig. 3.22), and each contain about 15 chlorophyll-*a* molecules (section 3.6). Since the absorption maximum of chl-*b* is at a lower wavelength than that of chl-*a*, the transfer of excitons from chl-*b* to chl-*a* is accompanied by loss of energy as heat. This promotes the flux of electrons from the periphery to the reaction centre. The connection between the outer light harvesting complexes (LHC-II*b*) and PS II can be interrupted by phosphorylation. In this way the actual size of the antenna can be adjusted to the intensity of illumination (section 3.10).

Photosystem I contains fewer light harvesting complexes than photosystem II since its core antenna is larger than that in PS II. The LHCs of PS I are similar to

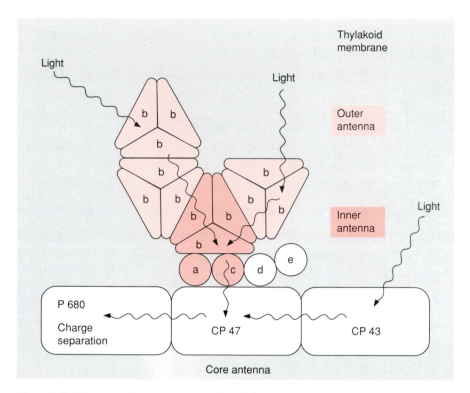

Figure 2.13 Scheme of the arrangement of the light harvesting complexes in the antenna of photo-system II in a plant, viewed from above (after Thornber). 'a' means LHC-II*a*, and so on. The inner antenna complexes are linked by LHC-II*a* and LHC-II*c* monomers to the core complex. The function of the LHC-II*d* and LHC-II*e* monomers is not entirely known.

those of PS II. It has been suggested that in the phosphorylated state LHC-II*b* can also function as an antenna of PS I (see section 3.10).

The mechanism of the movement of excitons in the antenna is not yet fully understood. The excited electron may be delocalized by being distributed over a whole group of chromophore molecules. On the other hand, the exciton may be localized initially in a certain chromophore molecule and subsequently emitted to a neighbouring chromophore. This process of exciton transfer has been termed the *Förster mechanism*. The transfer of excitons between closely neighbouring chlorophyll molecules within a light harvesting complex probably proceeds via delocalized electrons, and the transfer between these complexes and the reaction centre via the Förster mechanism. Absorption measurements using the ultra fast laser technique have shown that the exciton transfer between two chlorophyll molecules proceeds within 0.1 picoseconds (10^{-13} s). Thus the exciton transfer in the antennae is much faster than the charge separation in the reaction centre ($\approx$3.5 picoseconds) discussed in section 3.4. The reaction centre functions as an energy trap for excitons present in the antenna.

Phycobilisomes enable cyanobacteria and red algae to carry out photosynthesis even in dim light

Cyanobacteria and red algae possess antenna structures which can collect light of very low intensity. These antennae are arranged as particles on top of the membrane near the reaction centres of photosystem II (Fig. 2.14). These particles, termed *phycobilisomes*, consist of proteins (*phycobiliproteins*), which are covalently linked with phycobilins. *Phycobilins* are open-chained tetrapyrroles and therefore structurally related to the chlorophylls. Open-chained tetrapyrroles are also contained in bile, which explains the name -bilin. The phycobilins are linked to the protein by a thioether bond between an SH-group of the protein and the vinyl side chain of the phycobilin. The protein *phycoerythrin* is linked to the chromophore *phycoerythrobilin*, and the proteins *phycocyanin* and *allophycocyanin* to the chromophore *phycocyanobilin* (Fig. 2.15). The basic structure of the phycobiliproteins consists of a heterodimer (α,β). Each of these subunits contains 1–4 phycobilins as a chromophore. Three of these heterodimers aggregate to form a trimer $(\alpha,\beta)_3$ and thus form the actual building block of a phycobilisome. The so-called linker polypeptides function as 'mortar' between the building blocks.

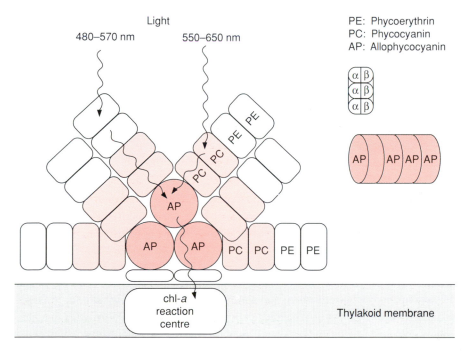

Figure 2.14 Scheme of a side-view of the structure of a phycobilisome. The units shown each consist of three α- and three β-subunits (after Glazer).

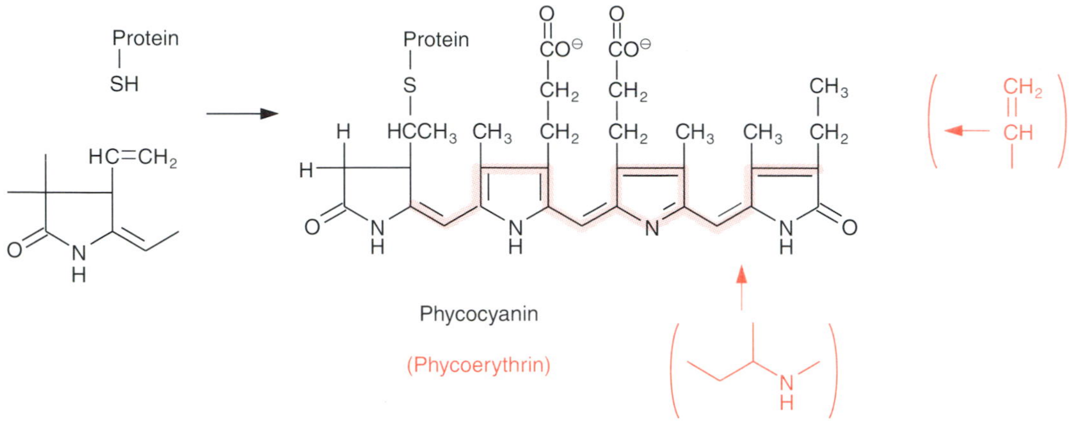

Figure 2.15 Structural formula of the biliproteins present in the phycobilisomes, phycocyanin (black) and phycoerytherin (difference from phycocyanin shown in red). The corresponding chromophores phycocyanobilin and phycoerythrobilin are covalently bound to proteins via thioether linkages formed by the addition of the SH-group of a cysteine residue of the protein to the vinyl group of the chromophore. The conjugated double bonds (marked red) give the molecule a pigment-like character.

Figure 2.14 shows the structure of a phycobilisome. The phycobilisome is attached to the membrane by anchor proteins. Three aggregates of about five $(\alpha,\beta)_3$ units form the core. This core contains the pigment allophycocyanin (AP) to which cylindrical rod-like structures are attached, each with 4–6 building blocks. The inner units mainly contain phycocyanin (PC) and the outer ones phycoerythrin (PE). The function of this structural organization is illustrated by the absorption spectra of the various biliproteins shown in Fig. 2.16. The light of shorter wavelength is absorbed in the periphery of the rods by phycoerythrin and the light of longer wavelength in the inner regions of the rods by phycocyanin. The core transfers the excitons to the reaction centre. A spatial distribution between the short wavelength absorbing pigments at the periphery and the long wavelength absorbing pigments in the centre was shown in the preceding section for the PS II antennae of higher plants.

Due to the phycobiliproteins, phycobilisomes are able to absorb green light very efficiently (Fig. 2.16), thus allowing cyanobacteria and red algae to survive in deep water. At these depths, due to the green window of photosynthesis (Fig. 2.3), only green light is available, as the light of the other wavelengths is absorbed by green algae living in the upper regions of the water. The algae in the deeper regions are obliged to invest a large portion of their cellular matter in phycobilisomes in order to carry out photosynthesis at this very low light intensity. Biliproteins can amount to 40 per cent of the total cellular protein of the algae. These organisms undertake extraordinary expenditure to collect enough light for survival.

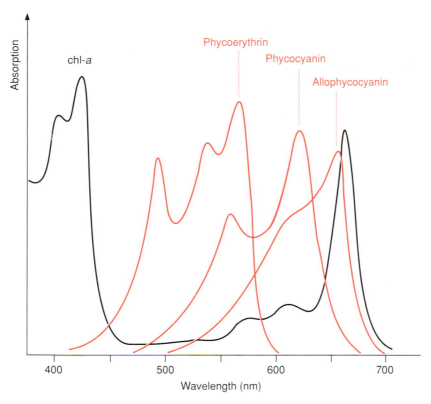

Figure 2.16 Absorption spectra of the phycobiliproteins phycoerythrin, phycocyanin, and allophyco-cyanin and, for the sake of comparison, also of chlorophyll-*a*.

Further reading

Glazer, A. N. (1981). Photosynthetic accessory proteins with bilin prosthetic groups. *The biochemistry of plants*, Vol 8 (ed. M. D. Hatch, N. K. Boardman), pp. 51–96. Academic Press, New York.

Jansson, S. (1994). Review: The light harvesting chlorophyll-*a*/*b* binding proteins. *Biochimica et Biophysica Acta*, **1184**, 1–19.

Kühlbrandt, W. (1994). Structure and function of the plant light harvesting complex, LHC-II. *Current Biology*, **4**, 519–28.

Turconi, S., Weber, N., Schweitzer, G., Strotmann, H., and Holzwarth, A. R. (1994). Energy charge separation kinetics in photosystem I. 2. Picosecond fluorescence study of various PSI particles and light-harvesting complex isolated from higher plants. *Biochimica et Biophysica Acta*, **1187**, 324–34.

Vogelmann, T. C., Nishio, J. N., and Smith, W. K. (1996). Leaves and light capture, light propagation and gradients of carbon fixation in leaves. *Trends in Plant Science*, **1**, 65–70.

chapter 3

Photosynthesis is an electron transport process

The previous chapter described how photons are captured by an antenna and conducted to the reaction centres. This chapter deals with the function of these reaction centres and describes how photon energy is converted to chemical energy to be utilized by the cell. As already mentioned in the previous chapter, plant photosynthesis probably evolved from bacterial photosynthesis, so that the basic mechanisms of the photosynthetic reactions are alike in bacteria and plants. Bacteria have proved to be very suitable objects for studying the principles of photosynthesis since their reaction centres are more simply structured than those of plants and they are more easily isolated. For this reason first bacterial and then plant photosynthesis will be described.

3.1 The photosynthetic machinery is constructed from modules

The photosynthetic machinery of bacteria is constructed from defined complexes which also appear as components of the photosynthetic machinery in plants. As will be described in Chapter 5, some of these complexes are also components of mitochondrial electron transport. These complexes can be thought of as modules which developed at an early stage of evolution and have been combined in various ways for different purposes. For easier understanding, the functions of these modules in photosynthesis will be treated first as black boxes and a detailed description of their structure and function will be given later.

Purple bacteria have only one reaction centre (Fig. 3.1). In this reaction centre the energy of the absorbed photon excites an electron, which means it is raised to a negative redox state. The excited electron is transferred back to the ground state

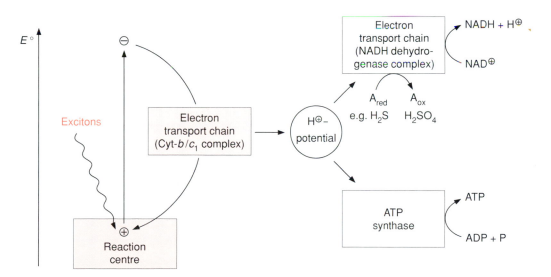

Figure 3.1 Scheme of the photosynthetic apparatus of purple bacteria. The energy of a captured exciton in the reaction centre elevates an electron to a negative redox state. The electron is transferred to the ground state via an electron transport chain including the cytochrome-b/c_1 complex. Free energy of this process is conserved by formation of a proton potential which is used partly for synthesis of ATP and partly to enable an electron flow for the formation of NADH from electron donors such as H_2S.

by an electron transport chain, called the cytochrome-b/c_1 complex, and its energy is transformed to a chemical form which is then used for the synthesis of biomass (proteins, carbohydrates). Generation of energy is based on coupling the electron transport with the transport of protons across the membrane. In this way the energy of the excited electron is conserved in the form of an electrochemical H^+-potential across the membrane. The photosynthetic reaction centres and the components of the electron transport chain are always located in a membrane.

Via ATP synthase, the energy of the H^+-potential is used to synthesize ATP from ADP and phosphate. Since the excited electrons in purple bacteria return to the ground state of the reaction centre, this electron transport is called *cyclic electron transport*. In purple bacteria the proton gradient is also used to reduce NAD via an additional electron transport chain called the NADH dehydrogenase complex (Fig. 3.1). By consuming the energy of the H^+-potential, electrons are transferred from a reduced substance (e.g. organic acids or hydrogen sulfide) to NAD. The ATP and NADH formed by bacterial photosynthesis are used for the synthesis of organic matter; especially important is the synthesis of carbohydrates from CO_2 via the Calvin cycle (Chapter 6).

The reaction centre of *green sulfur bacteria* (Fig. 3.2) is similar to that of purple bacteria, suggesting that they both evolved from a common precursor. ATP is also

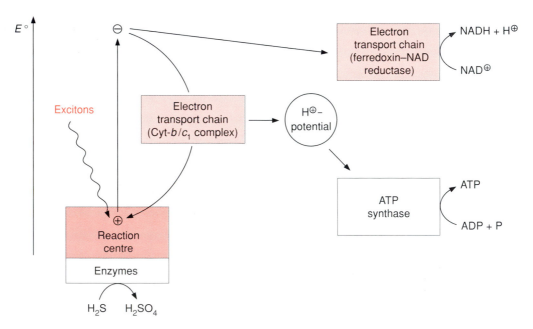

Figure 3.2 Scheme of the photosynthetic apparatus in green sulfur bacteria. In contrast to the scheme in Fig. 3.1, part of the electrons elevated to a negative redox state are transferred via an electron transport chain (ferredoxin–NAD reductase) to NAD yielding NADH. The electron deficit arising in the reaction centre is compensated for by electron donors such as H_2S.

formed in green sulfur bacteria by cyclic electron transport. The electron transport chain (cytochrome-b/c_1 complex) and the ATP synthase involved here are very similar to those in purple bacteria. However, in contrast to purple bacteria, green sulfur bacteria are able to form NADH by a *non-cyclic electron transport* process. In this case, the excited electrons are transferred to the ferredoxin–NAD reductase complex which reduces NAD to NADH. Since the excited electrons in this non-cyclic pathway do not return to the ground state, an electron deficit remains in the reaction centre and is replenished by electron donors such as H_2S, being oxidized to sulfate.

Cyanobacteria and plants use water as electron donor in photosynthesis (Fig. 3.3). As oxygen is liberated, this process is called *oxygenic photosynthesis*. Two photosystems, designated II and I, are arranged here in tandem. The machinery of oxygenic photosynthesis is made up of modules which have already been described in bacterial photosynthesis. The structure of the reaction centre of photosystem II corresponds to that of the reaction centre of purple bacteria, and that of photosystem I to the reaction centre of green sulfur bacteria. The enzymes ATP synthase and ferredoxin–NADP reductase are very similar to those of photosynthetic bacteria. The electron transport chain of the cytochrome-

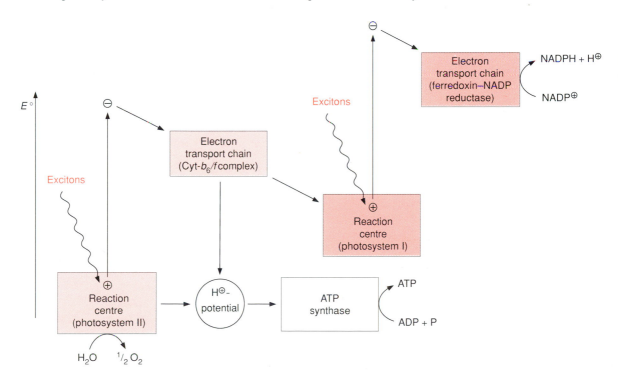

Figure 3.3 In the photosynthetic apparatus of cyanobacteria and plants two reaction centres corresponding in their function to the photosynthetic reaction centres of purple bacteria and green sulfur bacteria (shown in Figs 3.1 and 3.2) are arranged in sequence.

b_6/f complex has the same basic structure as the cytochrome-b/c_1 complex in bacteria.

Four excitons are required in oxygenic photosynthesis to split one molecule of water:

$$H_2O + NADP^+ \xrightarrow{\text{4 excitons}} {}^{1/2}O_2 + NADPH + H^+.$$

In this non-cyclic electron transport, electrons are transferred to NADP and protons are partitioned across the membrane to give the proton gradient that drives the synthesis of ATP. Thus for each mol of NADPH formed by oxygenic photosynthesis, 1.3–1.5 mol of ATP are generated simultaneously (section 4.4). Most of this ATP and NADPH is used for assimilating CO_2 and nitrate to give carbohydrates and amino acids. Oxygenic photosynthesis in plants takes place in the chloroplast, a cell organelle of the plastid family (section 1.3).

3.2 A reductant and an oxidant are formed during photosynthesis

At the beginning of this century Otto Warburg postulated that the light energy is transferred to CO_2 and that the CO_2, activated in this way, reacts with water to form a carbohydrate, accompanied by the release of oxygen. According to this hypothesis the oxygen released by photosynthesis was derived from the CO_2. In 1931 this hypothesis was opposed by Cornelis van Niel by postulating that during photosynthesis a *reductant* is formed which then reacts with CO_2. The so-called van Niel equation describes photosynthesis in the following way:

$$CO_2 + 2H_2A \xrightarrow{\text{Light}} [CH_2O] + H_2O + 2A$$

($[CH_2O]$ = carbohydrate). He proposed that a substance, H_2A, is split by light energy into a reducing compound (H) and an oxidizing compound (A). For oxygenic photosynthesis of cyanobacteria or plants, this can be rewritten as:

$$CO_2 + 2H_2O \xrightarrow{\text{Light}} [CH_2O] + H_2O + O_2.$$

In this equation the oxygen released during photosynthesis is derived from water.

In 1937 Robert Hill in Cambridge proved that a reductant is actually formed in the course of photosynthesis. He was the first to succeed in isolating chloroplasts with some photosynthetic activity, which, however, had no intact envelope membranes and consisted only of thylakoid membranes. When these chloroplasts were illuminated in the presence of Fe^{3+} compounds (initially ferrioxalate, later ferricyanide ($[Fe(CN)_6]^{3-}$)), oxygen was evolved, accompanied by reduction of the Fe^{3+} compounds to the Fe^{2+} form. Since CO_2 was not involved in this '*Hill reaction*', this experiment proved that the photochemical splitting of water can be separated from the reduction of the CO_2. The total reaction of photosynthetic CO_2 assimilation can be divided into two partial reactions:

(1) the so-called *light reaction*, in which water is split by photon energy to yield reductive power (in form of NADPH) and chemical energy (in form of ATP), and

(2) the so-called *dark reaction* (Chapter 6) in which CO_2 is assimilated at the expense of this reductive power and of ATP.

In 1952 the Dutchman Louis Duysens made a very important observation which helped to explain the mechanism of photosynthesis. When illuminating isolated membranes of the purple bacterium *Rhodospirillum rubrum* with short light pulses, he found a decrease in light absorption at 890 nm which was immediately reversed when the bacteria were placed in the dark again. The same '*bleaching*' effect was found at 870 nm in the purple bacterium *Rhodobacter sphaeroides*. Later, Bessel Kok (USA) and Horst Witt (Germany) also found similar pigment bleaching at 700 nm and 680 nm in chloroplasts. This bleaching was attributed to the *primary reaction of photosynthesis*, and the corresponding pigments of the reaction centres were named P_{870} (*Rb. sphaeroides*) and P_{680} and P_{700} (chloroplasts). When an oxidant, e.g. $[Fe(CN)_6]^{3-}$, was added, this bleaching effect could also be achieved in the dark. These results indicated that these absorption changes of the pigments were due to a *redox reaction*. This was the first indication that chlorophyll could be oxidized. Electron spin resonance measurements revealed that radicals are formed during 'bleaching'. 'Bleaching' was also observed at the very low temperature of 1 K. This showed that in the electron transfer leading to the formation of *radicals*, the reaction partners are located so close to each other that thermal oscillation of the reaction partners (normally the precondition for a chemical reaction) is not required for this redox reaction. Spectroscopic measurements indicated that the reaction partners of this primary redox reaction are two chlorophyll molecules arranged as a pair, called a '*special pair*'.

3.3 The basic structure of a photosynthetic reaction centre has been resolved by X-ray structure analysis

The reaction centres of purple bacteria proved to be especially suitable objects for explaining the structure and function of photosynthetic machinery. It was a great step forward when in 1970 Roderick Clayton (USA) developed a method for isolating reaction centres from purple bacteria. Analysis of the components of the reaction centres of the different purple bacteria (shown in Table 3.1 for the reaction centre of *Rhodobacter sphaeroides* as an example), revealed that the reaction centres had the same basic structure in all the purple bacteria investigated. The minimum structure consists of three subunits: L, M, and H (light, medium, and heavy) Subunits L and M are peptides with a similar amino acid sequence.

Table 3.1 Composition of the reaction centre of *Rhodobacter sphaeroides* (P_{870})

Component	Number present	Molecular mass (kDa)
subunit L	1	21
subunit M	1	24
subunit H	1	28
bacteriochlorophyll-*a*	4	
bacteriopheophytin-*a*	2	
ubiquinone	2	
non-haem-Fe protein	1	
carotenoid	1	

They are homologous. The reaction centre of *Rb. sphaeroides* contains four bacteriochlorophyll-*a* (Bchl-*a*, Fig. 3.4) and two bacteriopheophytin-*a* (BPhe-*a*). Pheophytin differs from chlorophyll in that it lacks magnesium as the central atom. In addition, the reaction centre contains an iron atom which is not a part of a haem ring. It is therefore called a *non-haem iron*. Furthermore, the reaction centre contains two molecules of ubiquinone (Fig. 3.5), which are designated as Q_A and Q_B. Q_A is tightly bound to the reaction centre, whereas Q_B is only loosely associated with it.

X-ray structure analysis of the photosynthetic reaction centre

If ordered crystals can be prepared from a protein, it is possible to analyse the spherical structure of the protein molecule by *X-ray structure analysis*. (In a

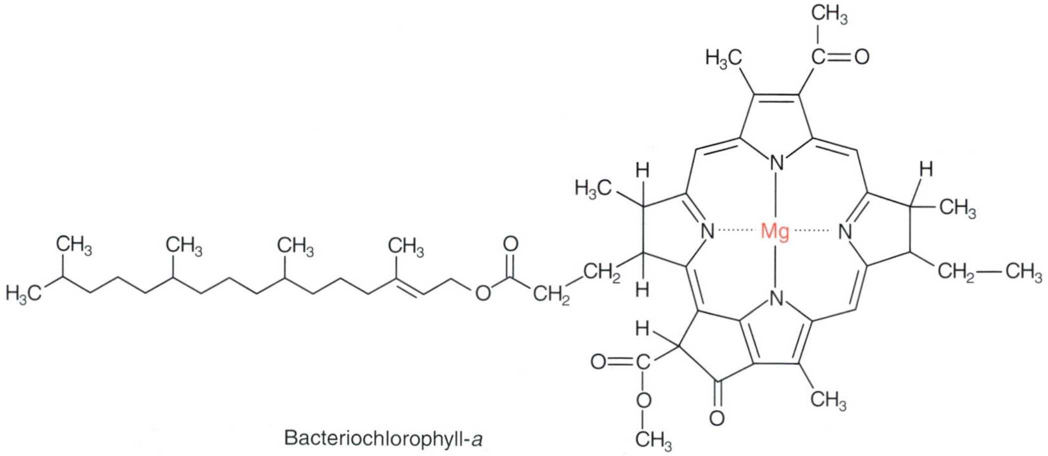

Bacteriochlorophyll-*a*

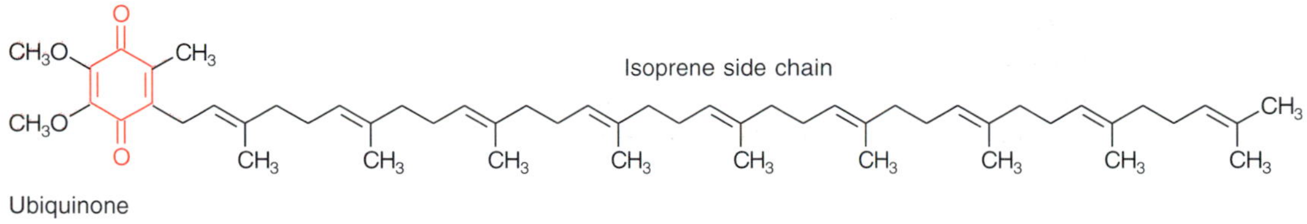

CH$_3$O
CH$_3$
O
CH$_3$O
O

Isoprene side chain

CH$_3$

CH$_3$ CH$_3$ CH$_3$ CH$_3$ CH$_3$ CH$_3$ CH$_3$ CH$_3$ CH$_3$

Ubiquinone

Figure 3.5 Ubiquinone. The long isoprenoid side chain gives the substance a lipophilic character.

similar way a structural analysis can be also obtained from crystalline molecular layers by using electron cryo-microscopy (section 2.4), but in this method the experimental expenditure is particularly high.) In X-ray structure analysis a protein crystal is irradiated by an X-ray source. The electrons of the atoms in the molecule cause a scattering of X-rays. Diffraction is observed when the irradiation passes through a regular repeating structure. The corresponding diffraction pattern, consisting of many single reflections, is measured by an X-ray film positioned behind the crystal or by an alternative detector. The principle is demonstrated in Fig. 3.6. In order to obtain as many reflections as possible, the crystal, mounted in a capillary, is rotated. From a few dozen to up to several hundred exposures are required for one set of data, depending on the form of the crystal and the size of the crystal lattice. To evaluate a new protein structure several sets of data are required in which the protein has been changed by incorporation or binding of a heavy metal ion. With the help of elaborate computer programs it is possible to reconstruct the spherical structure of the exposed protein molecules by applying the rules for scattering of X-rays by atoms of various electron density.

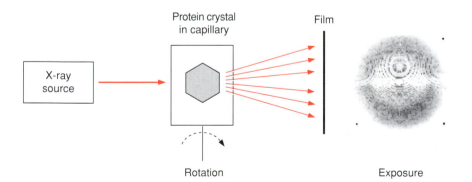

X-ray
source

Protein crystal
in capillary

Film

Rotation

Exposure

Figure 3.6 Scheme of X-ray structure analysis of a protein crystal. A capillary containing the crystal is made to rotate slowly and the diffraction pattern is monitored on an X-ray film. Nowadays much more sensitive detector systems (image platers) are used instead of films. The diffraction pattern shown was obtained by the structural analysis of the reaction centre of *Rb. sphaeroides*. (By courtesy of H. Michel, Frankfurt.)

X-ray structure analysis requires a high technical expenditure and is very time consuming, but the actual limiting factor in the elucidation of a spherical structure is usually the preparation of *suitable single crystals*. Until 1980 it was thought to be impossible to prepare crystals suitable for X-ray structure analysis from hydrophobic membrane proteins. The application of the detergent *N,N'-dimethyl-dodecylamine-N-oxide* (Fig. 3.7) was a great step forward in helping to solve this problem. This detergent forms water-soluble protein–detergent micelles with membrane proteins which can then be made to crystallize when ammonium sulfate or polyethylene glycol is added. The micelles form a regular lattice in these crystals (Fig. 3.8). The protein in the crystal remains in its native state since the hydrophobic regions of the membrane protein, which normally border on the hydrophobic membrane, are covered by the hydrophobic chains of the detergent.

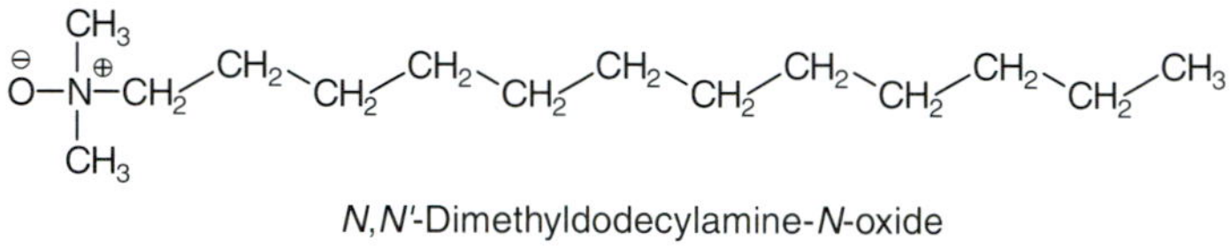

$$O^{\ominus}-\overset{\overset{\displaystyle CH_3}{|}}{\underset{\underset{\displaystyle CH_3}{|}}{N^{\oplus}}}-CH_2\diagdown^{CH_2}\diagdown_{CH_2}\diagdown^{CH_2}\diagdown_{CH_2}\diagdown^{CH_2}\diagdown_{CH_2}\diagdown^{CH_2}\diagdown_{CH_2}\diagdown^{CH_2}\diagdown_{CH_2}\diagdown^{CH_3}$$

N,N'-Dimethyldodecylamine-N-oxide

Figure 3.7 A detergent.

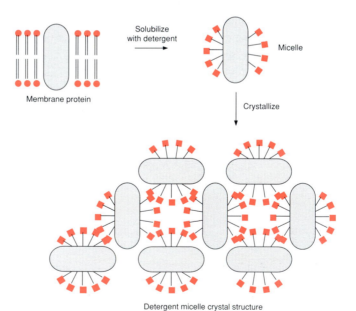

Membrane protein

Solubilize with detergent

Micelle

Crystallize

Detergent micelle crystal structure

Figure 3.8 A detergent micelle is formed after solubilization of a membrane protein by a detergent. The hydrophobic region of the membrane proteins, the membrane lipids and the detergent are marked black and the hydrophilic regions red. Crystal structures can be formed by association of the hydrophilic regions of the detergent micelle.

Using this procedure Hartmut Michel from Munich succeeded in obtaining crystals from the reaction centre of the purple bacterium *Rhodopseudomonas viridis*, and together with his colleague Johann Deisenhofer from the department of Robert Huber, performed an X-ray structural analysis of these crystals. The immense amount of time invested in these investigations is illustrated by the fact that the evaluation of the stored data sets alone took 2½ years. In 1988 the three scientists were awarded the Nobel prize for Chemistry for the elucidation of the structure of this photosynthetic reaction centre. Using the same method, the re-action centre of *Rhodobacter sphaeroides* was analysed and it turned out that the basic structure of the two reaction centres are astonishingly similar.

The reaction centre of *Rhodobacter sphaeroides* has a symmetrical structure

Figure 3.9 shows the three-dimensional structure of the reaction centre of the purple bacterium *Rhodobacter sphaeroides*. The molecule has a cylindrical shape and is about 8 nm long. The homologous subunits L (red) and M (black) are arranged symmetrically and enclose the chlorophyll and pheophytin molecules. The H subunit is attached to the lower part of the cylinder.

In the same projection as in Fig. 3.9, Fig. 3.10 shows the location of the chromo-phores in the protein molecule. All the chromophores are positioned as pairs divided by a symmetry axis. Two Bchl-*a* molecules (D_M, D_L) can be recognized in the upper part of the structure. The two tetrapyrrole rings are so close (0.3 nm) that their orbitals overlap. This proved the actual existence of the '*special pair*' of chlorophyll molecules, postulated earlier from spectroscopic investigations, as the site of the primary redox process of photosynthesis. The chromophores are arranged below this chlorophyll pair in two nearly identical branches, each con-taining a Bchl-*a* molecule (B_A, B_B) as monomer, followed in each branch by a bac-teriopheophytin (Φ_A, Φ_B). Whereas the chlorophyll pair (D_M, D_L) is bound by both subunits L and M, the chlorophyll B_A and the pheophytin Φ_A are associated with subunit L, and B_B and Φ_B with subunit M. The quinone ring of Q_A is bound via hydrogen bonds and hydrophobic interaction to subunit M, whereas the loosely associated Q_B is bound to subunit L.

3.4 How does a reaction centre function?

Structural analysis and extensive kinetic investigations allowed a detailed descrip-tion of the function of the bacterial reaction centre. The kinetic investigations included measurements by absorption and fluorescence spectroscopy after light flashes in the range of 10^{-12} s (picoseconds), as well as measurements of nuclear spin and electron spin resonance. Although the reaction centre shows a symmetry with two almost identical branches of chromophores, electron transfer proceeds only along the branch on the right in Fig. 3.10 on the *L side*. The chlorophyll

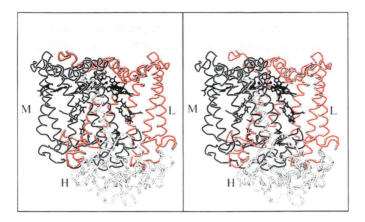

Figure 3.9 Stereo pair of the three-dimensional structure of the reaction centre of *Rb. sphaeroides*. The peptide chain of subunit L is marked red and that of subunit M black. The polypeptide chains are shown as bands and the chromophores (chlorophylls, pheophytins) and the quinones are shown as wire models. The upper part of the reaction centre borders on the periplasmatic compartment and the lower part on the cytoplasm. (By courtesy of H. Michel, U. Ermler, and R. C. R. D. Lancaster, Frankfurt.)

monomer (B_B) on the M side is in close contact with a carotenoid molecule (not shown in Fig. 3.10), which abolishes a harmful *triplet state* of chlorophylls in the reaction centre (section 2.3). The function of the pheophytin (Φ_B) on the M side and of the non-haem iron is not yet fully understood.

Figure 3.11 shows the scheme of the reaction centre in which the reaction partners are arranged according to their electrochemical potential. The primary reaction with the exciton provided by the antenna (section 2.4) excites the chlorophyll pair.

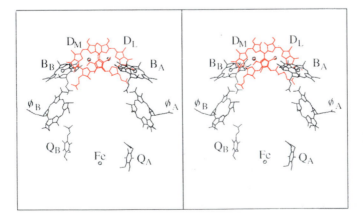

Figure 3.10 Stereo pair of the three-dimensional array of chromophores and quinones in the reaction centre of *Rb. sphaeroides*. The projection corresponds to the structure shown in Fig. 3.9. The Bchl-*a* pair, $D_M D_L$ (see text), is marked red. (By courtesy of H. Michel, U. Ermler, and R. C. R. D. Lancaster, Frankfurt. P. Kraulis, Uppsala, produced Figs 3.9 and 3.10 using the program, MOLSCRIPT.)

This primary excitation state has only a very short half-life, a charge separation occurs within femtoseconds (10^{-15} s) and, as a result of the large potential difference, within picoseconds an electron is removed to reduce bacteriopheophytin (BPhe).

$$(Bchl)_2 + 1 \; Exciton \rightarrow (Bchl)_2^*$$
$$(Bchl)_2^* + BPhe \rightarrow (Bchl)_2^+\bullet + BPhe\bar{\bullet}.$$

The electron is probably transferred first to the Bchl monomer (B_A) and then to the pheophytin molecule (Φ_A). The second electron transfer proceeds with a half-time of 0.9 picoseconds, about four times as fast as the electron transfer to B_A. The *pheophytin radical* has a tendency to return to the ground state by a return of the

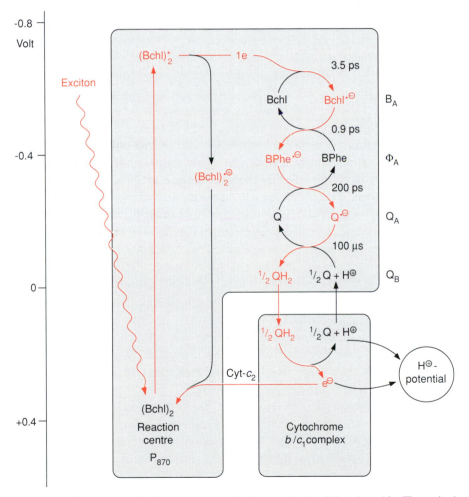

Figure 3.11 Scheme of cyclic electron transport in photosynthesis of *Rb. sphaeroides*. The excited state, symbolized by a star, results in a charge separation; an electron is transferred via pheophytin, Q_A, Q_B (both ubiquinone), and the cyt-*b/c* complex to the positively charged chlorophyll radical. Q, quinone; $Q^{\bullet-}$, semiquinone radical; QH_2, hydroquinone.

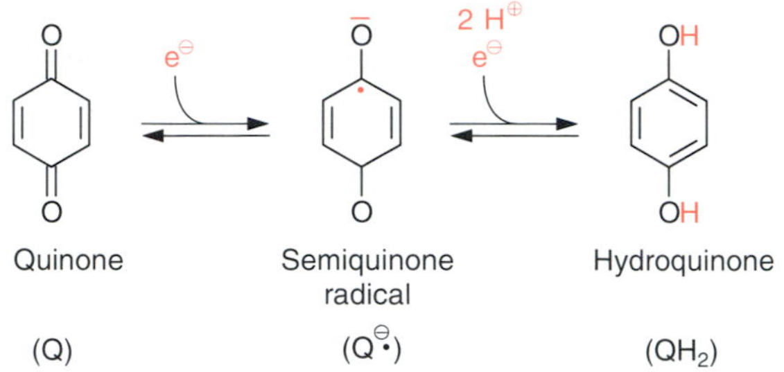

Figure 3.12 Reduction of a quinone by one electron results in a semiquinone radical and further reduction to hydroquinone.

translocated electron to the Bchl monomer (B_A). To prevent this, within 200 picoseconds a high potential difference withdraws the electron from the pheophytin radical to ubiquinone (Q_A). The *ubisemiquinone radical*, thus formed, in response to a further potential difference transfers its electron to the loosely bound ubiquinone, Q_B. In this way first *ubisemiquinone* and then *ubihydroquinone* is formed after a second electron transfer (Fig. 3.12). In contrast to the very labile radical intermediates of the pathway described so far, ubihydroquinone is a stable reductant. However, this stability has its price. The formation of ubiquinone as a first stable product from the primary excitation state of the chlorophyll costs almost two-thirds of the exciton energy, which is dissipated as *heat*.

Ubiquinone (Fig. 3.5) contains a hydrophobic isoprenoid side chain (Fig. 3.12), which makes it very soluble in the lipid phase of the photosynthetic membrane. The same function of an isoprenoid side chain has already been described for chlorophyll (section 2.2). In contrast to chlorophyll, pheophytin, and Q_A, which are all tightly bound to proteins, the ubihydroquinone Q_B is only loosely associated with the reaction centre and can be exchanged for another ubiquinone. Ubihydroquinone remains in the lipid phase, is able to diffuse rapidly along the membrane, and functions as a transport metabolite for reducing equivalents in the membrane phase. It feeds the reductive equivalents into the *cytochrome-b/c₁ complex*, also located in the membrane, and the electrons are transferred back through this complex to the reaction centre. Energy is conserved in this electron transport to form a proton potential (section 4.1), which is used for ATP synthesis. The structure and mechanism of the cytochrome-*b/c₁* complex and of ATP synthase will be described in section 3.7 and Chapter 4, respectively.

In summary, the cyclic electron transport of the purple bacteria may resemble a simple electrical circuit, as shown in Fig. 3.13. The chlorophyll pair and pheophytin, between which an electron is transferred by light energy, may be regarded as the two plates of a capacitor between which a voltage is generated, driving a flux of electrons, a current. The major part of electron energy is dissipated as heat by a voltage drop via a resistor. This resistor functions as an electron trap,

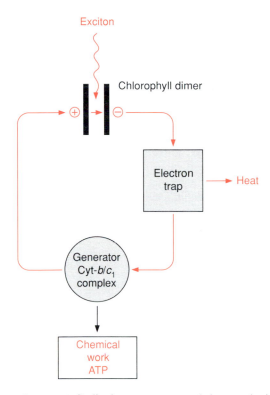

Figure 3.13 Cyclic electron transport of photosynthesis drawn as an electrical circuit.

withdrawing the electrons rapidly from the capacitor. A generator utilizes the remaining voltage to produce chemical energy.

3.5 Two photosynthetic reaction centres are arranged in tandem in photosynthesis of algae and higher plants

A *quantum requirement* (photons absorbed per molecule O_2 produced) of about eight has been determined for photosynthetic water splitting by green algae (section 2.4). Instead of the term quantum requirement, one often uses the reciprocal term, *quantum yield* (molecules O_2 produced per photon absorbed). Investigations into the dependence of quantum yield on the colour of irradiating light (action spectrum) revealed that the quantum yield dropped very sharply when the algae were illuminated with red light above a wavelength of 680 nm (Fig. 3.14). At first this effect, the '*red drop*', remained unexplained since algae contain chlorophyll which absorbs light at 700 nm. Robert Emerson and co-workers solved this problem in 1957 when they observed in an experiment that the quantum yield in the spectral range above 680 nm increased dramatically when orange-coloured light of 650 nm

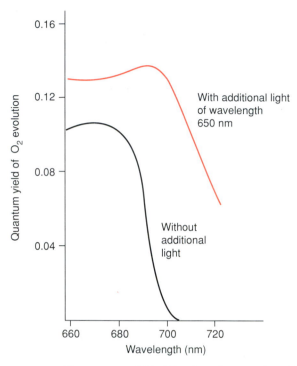

Figure 3.14 The quantum yield of O_2 release in green algae (*Chlorella*) depending on the wavelength of irradiating light. The upper curve shows the result of supplementary irradiation with 650 nm light. For this curve the O_2 evolution resulting from 650 nm light alone is subtracted. (After Emerson.)

was used together with red light. When algae were irradiated with the light of the two colours simultaneously the quantum yield was higher than the sum of the yields obtained when algae were irradiated separately with the light of each wavelength.

This *Emerson effect* led to the conclusion that two different reaction centres are involved in photosynthesis of green algae (and also of cyanobacteria and higher plants). In 1960 Robert Hill and Fay Bendall postulated a reaction scheme (Fig. 3.15) in which *two reaction centres are arranged in tandem* and connected by an electron transport chain containing cytochromes-b_6 and -f (cytochrome-f is a cytochrome of the c type; see section 3.7). Light energy of 700 nm was sufficient for the excitation of the one reaction centre, whereas excitation of the other re-action centre required light of higher energy, with a wavelength up to 680 nm. A reaction diagram according to the redox potentials shows a zigzag, leading to the name *Z scheme*. The numbering of the two photosystems corresponds to the sequence of their discovery. *Photosystem II* (PS II) can use light up to a wave-length of *680 nm*, whereas *photosystem I* (PS I) can utilize light with a wavelength up to *700 nm*. The sequence of the two photosystems makes it possible that at PS II a very *strong oxidant* is generated for oxidation of water and at PS I a very *strong reductant* is generated for reduction of NADP (see also Fig. 3.3).

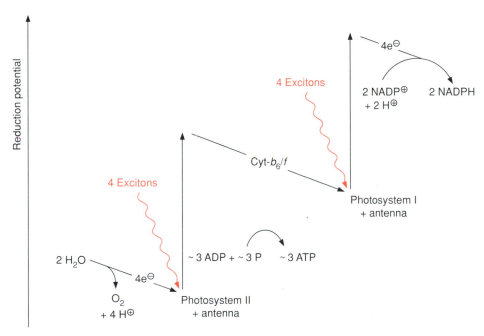

Figure 3.15 The Z scheme of photosynthesis in plants. Electrons are transferred by two photosystems arranged in tandem from water to NADP, and during this ATP is formed. The amount of ATP formed is uncertain but is probably between 2.66 and 3 per four excitons captured at each reaction centre (section 4.4).

Figure 3.16 gives an overview of electron transport through the photosynthetic complexes; the carriers of electron transport are drawn according to their electric potential as in Fig. 3.11. Figure 3.17 shows how the photosynthetic complexes are arranged in the thylakoid membrane. There is a potential difference of about 1.2 V between oxidation of water and reduction of the NADP. The two absorbed photons of 680 and 700 nm together correspond to a total potential difference of 3.45 V (see section 2.2, equation 2.7). Thus, only about one-third of the energy of the photons absorbed by the two photosystems is used to transfer electrons from water to NADP. In addition to this, about one-eighth of the light energy absorbed by the two photosystems is conserved by pumping protons into the lumen of the thylakoids via PS II and the cytochrome-b_6/f complex (Fig. 3.17). This proton transport leads to the formation of a proton gradient between the lumen and the stroma space. An H^+-ATP synthase, also located in the thylakoid membrane, uses the energy of the proton gradient for synthesis of ATP.

Thus about half the energy of the light absorbed by the two photosystems is not used for chemical work but is dissipated as heat. The significance of this loss of energy in the form of heat in photosynthetic electron transport has been discussed above.

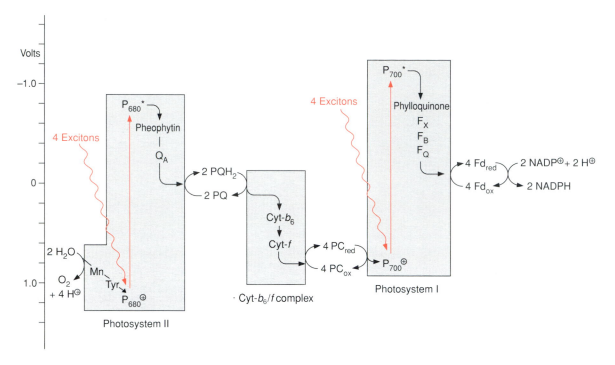

Figure 3.16 Scheme of non-cyclic electron transport in plants. The redox components are placed according to their midpoint redox potential and their association with the three complexes involved in the electron transport. A star symbolizes an excited state. The electron transport between the photosystem II complex and the cyt-b_6/f complex occurs by plastohydroquinone (PQH$_2$) which is oxidized by the cyt-b_6/f complex to plastoquinone (PQ). The electrons are transferred from the cyt-b_6/f complex to photosystem I by plastocyanin (PC). This reaction scheme is also valid for cyanobacteria, with the exception that instead of plastocyanin, cytochrome-c is involved in the second electron transfer.

3.6 Water is split by photosystem II

Photosystem II is very similar to the reaction centre of purple bacteria described in section 3.3. The photosynthesis pigment (P$_{680}$) of PS II is probably also a 'special pair' of two chlorophyll molecules, (chl-a)$_2$. Their excitation by excitons results in a charge separation (Fig. 3.18). An electron is transferred from the chloropyll pair, probably via the chl-a monomer, to pheophytin and from there further to a tightly bound plastoquinone (Q$_A$), thus forming a semiquinone radical. The electron is then transferred to a loosely bound plastoquinone (Q$_B$). This plastoquinone (PQ) (Fig. 3.19) accepts two electrons one after the other and is thus reduced to hydroquinone (PQH$_2$). The hydroquinone is released from the photosynthesis complex and may be regarded as the final product of photosystem II. This sequence, consisting of a transfer of a single electron between the 'special pair' (chl-a)$_2$ and Q$_A$ and the transfer of two electrons between Q$_A$ and Q$_B$, corresponds to the reaction sequence shown for *Rb. sphaeroides* (Fig. 3.11). The only difference is that the

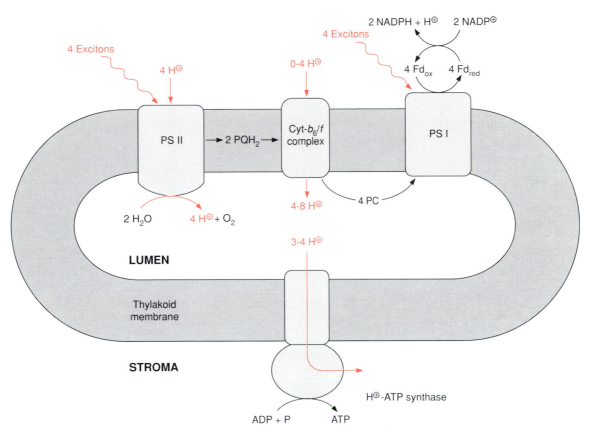

Figure 3.17 Scheme showing the positioning of the photosynthetic complexes and the H^+-ATP synthase in the thylakoid membrane. Transport of electrons between PS II and the cytochrome-b_6/f complex is mediated by plastohydroquinone (PQH_2) and that between the cytochrome-b_6/f complex and PS I by plastocyanin (PC). Splitting of the water occurs on the lumen side of the membrane and formation of NADPH and of ATP on the stroma side. The electrochemical gradient of protons pumped into the lumen drives ATP synthesis. The number of protons transported to the lumen during electron transport and the proton requirement of ATP synthesis is uncertain (section 4.4).

quinones are ubiquinone or menaquinone in bacteria and plastoquinone in photosystem II.

However, the similarity between the reaction sequence in PS II and the photosystem of the purple bacteria applies only to the electron donor region. The electron acceptor function in PS II is completely different from that in purple bacteria. The electron deficit in $(chl\text{-}a)_2^+\bullet$ caused by non-cyclic electron transport is compensated for by electrons derived from oxidation of water. Manganese cations and a *tyrosine* residue are involved in transport of electrons from water to chlorophyll. The $(chl\text{-}a)_2^+\bullet$ radical with a redox potential of about +1.1 V is such a strong oxidant that it can withdraw an electron from a tyrosine residue in the

STROMA

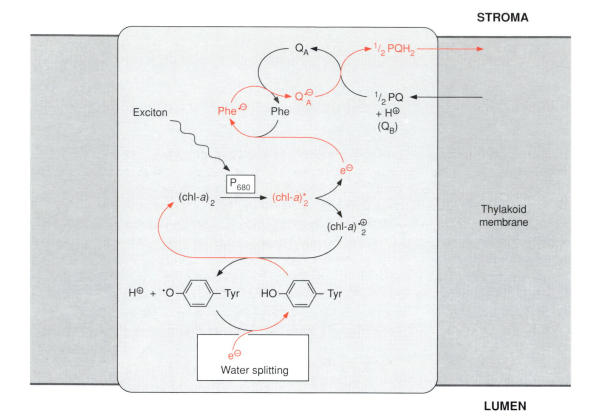

Thylakoid membrane

LUMEN

Figure 3.18 Reaction scheme of photosynthetic electron transport in the photosystem II complex. Excitation by a photon results in a charge separation between two chlorophyll molecules. The positively charged chlorophyll radical is reduced by a tyrosine residue and the latter by a cluster of probably four manganese atoms involved in the oxidation of water (Fig. 3.20). The negatively charged chlorophyll radical transfers its electron via pheophytin and a quinone, Q, of which the entire structure is not yet known, finally to plastoquinone.

protein of the reaction centre and a tyrosine radical remains. This reactive tyrosine residue is often designated as Z. The electron deficit in the tyrosine radical is restored by oxidation of a manganese ion (Fig. 3.20). The PS II complex contains several manganese ions, probably four, which are close to each other. This arrangement of Mn ions is called a *Mn cluster*. The Mn cluster depicts a redox system which can take up four electrons and release them again. During this process the Mn atoms probably change between the oxidation state Mn^{3+} and Mn^{4+}.

To liberate one molecule of O_2 from water the reaction centre must withdraw four electrons and thus capture four excitons. The time differences between the capture of the single excitons in the reaction centre depends on the intensity of illumination. If oxidation of water were to proceed stepwise, *oxygen radicals* could

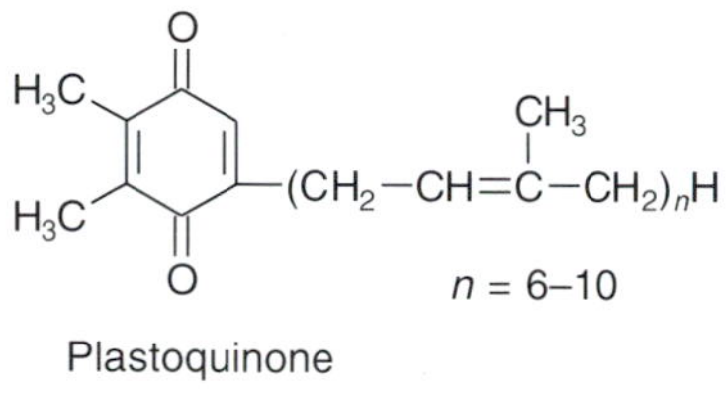

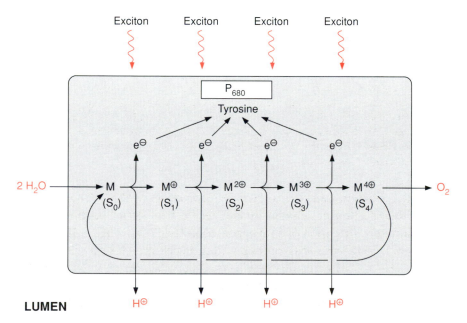

H₃C

$$H_3C$$

Plastoquinone

$n = 6\text{--}10$

Figure 3.19

be formed as intermediary products, especially at low light intensities. Oxygen radicals have a destructive effect on biomolecules such as lipids and proteins (section 3.10). The water-splitting machinery of the Mn clusters minimizes the formation of oxygen radical intermediates by supplying the reaction centre via tyrosine with four electrons one after the other (Fig. 3.20). The Mn cluster is transformed during this transfer from the ground oxidation state stepwise to four different oxidation states (these have been designated as S_0 and S_1–S_4).

Experiments by Pierre Joliot and Bessel Kok presented evidence that the water-splitting apparatus can be in five different oxidation states (Fig. 3.21). When chloroplasts kept in the dark were illuminated with a series of light pulses, an oscillation of the oxygen release was observed. Whereas after the first two light

Figure 3.20 A scheme showing the mechanism of water splitting by photosystem II. M means a cluster of four manganese atoms. The different manganese atoms are present in different oxidation states. The cluster functions as a redox unit and feeds, one after the other, a total of four electrons into the reaction centre of PS II. The deficit of these four electrons is compensated for by splitting of $2H_2O$ to O_2 and $4H^+$. M means $(4\,Mn)^{n+}$, M^+ means $(4\,Mn)^{(n+1)+}$ and so on.

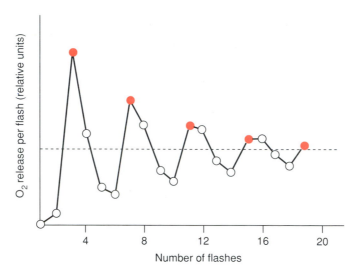

Figure 3.21 Yield of the oxygen released by chloroplasts as a function of the number of light pulses. The chloroplasts, previously kept in the dark, were illuminated by light pulses of 2 µs duration, interrupted by pauses of 0.3 s. (After Forbush *et al.* 1971.)

pulses almost no O_2 was released, the O_2 release was maximal after three pulses and then after a further four pulses, and so on. An increasing number of light pulses, however, dampened the oscillation more and more. This can be explained by some pulses not causing excitation of PS II and thus desynchronizing oscillation. In darkened chloroplasts the water-splitting apparatus is apparently in the S_1 state. After the fourth oxidation state (S_4) has been reached, O_2 is released in one reaction and the Mn cluster returns to its ground oxidation state (S_0). In this reaction protons from water are released to the lumen of the thylakoids. The formal description of this reaction is:

$$2H_2O \;\rightarrow\; 4H^+ + 2O^{2-}$$
$$2O^{2-} + M^{4+} \;\rightarrow\; O_2 + M$$

Figuratively speaking, the four electrons needed in the reaction centre are loaned in advanced by the Mn cluster and then repaid at one stroke by oxidation of water to one oxygen molecule. In this way the manganese cluster minimizes the formation of oxygen radicals by photosystem II. In spite of this safety device, probably enough oxygen radicals are formed in the PS II complex to have some damaging effect on the proteins of the complex, the consequences of which will be discussed next.

The photosystem II complex is very similar to the reaction centre in purple bacteria

Photosystem II is a complex consisting of at least 16 different subunits (Table 3.2), of which only two are involved in the actual reaction centre. The exact function of

Table 3.2 Protein components of photosystem II (list not complete) (after Vermaas 1993)

Protein	Molecular mass (kDa)	Localization	Coded in	Function
D_1	32	Intrinsic	Chloroplast	Binding of P_{680}, Phe, Q_B, Tyr, Mn cluster
D_2	34	Intrinsic	Chloroplast	Binding of P_{680}, Phe, Q_A, Mn cluster
CP 47	47	Intrinsic	Chloroplast	Core antenna, binds LHC
CP 43	43	Intrinsic	Chloroplast	Core antenna, binds LHC
Cyt-b_{559a}	9	Intrinsic	Chloroplast	Binds haem, protection of PS II against light damage
Cyt-b_{559b}	4	Intrinsic	Chloroplast	Binds haem, protection of PS II against light damage
Phosphoprotein	9	Intrinsic	Chloroplast	?
MSP	33	Peripheral: lumen	Nucleus	Stabilization of Mn cluster
P	23	Peripheral: lumen	Nucleus	?
Q	16	Peripheral: lumen	Nucleus	?
R	10	Peripheral:stroma	Nucleus	?

many subunits is not yet known. For this reason the scheme of the PS II complex shown in Fig. 3.22 contains only those subunits for which functions are known.

The centre of the PS II complex is a heterodimer made up of the subunits D_1 and D_2 with four or five chl-*a* molecules, two pheophytin and two plastoquinone molecules bound to it. The D_1 and the D_2 proteins are homologous to each other and also to the L protein and M protein from the reaction centre of purple bacteria (section 3.3). As in purple bacteria, only the pheophytin molecule bound to the D_1 protein of PS II is involved in electron transport. On the other hand, Q_A is bound to the D_2 protein whereas Q_B is bound to the D_1 protein. The Mn cluster is probably enclosed by both the D_1 and D_2 proteins. The tyrosine that is reactive in electron transfer is a constituent of D_1. The subunit MSP (manganese stabilizing protein) stabilizes the Mn cluster. The two subunits CP 43 and CP 47 (CP means *chlorophyll protein*) each bind about 15 chlorophyll molecules and form the *core complex of the antenna* shown in Fig. 2.10. Cyt-b_{559} does not seem to be involved in the electron transport of PS II; possibly its function is to protect the PS II complex from light damage.

The D_1 protein of the PS II complex has a high turnover, as it is degraded and resynthesized at a high rate. It seems that the D_1 protein wears out during its function, perhaps through damage by oxygen radicals, e.g. due to a slight imperfection of the water-splitting apparatus. It has been estimated that the D_1 protein is replaced after 10^6–10^7 catalytical cycles of the reaction centre of PS II.

Photosystem II complex

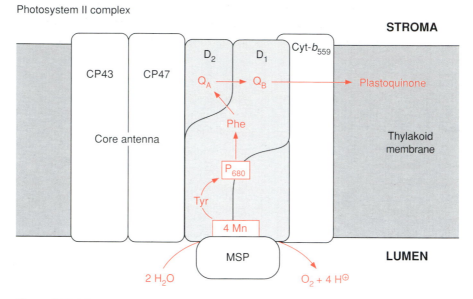

Figure 3.22 Scheme of the structure of the photosystem II complex. This scheme is based on the molecular structures predicted from the amino acid sequences of the various subunits and investigations of the binding of quinone to the subunits D_1 and D_2. Because of the homology of D_1 and D_2 to the subunits L and M in purple bacteria, the conclusion has been drawn that the structure of PS II and the structure of the reaction centres in purple bacteria resolved by X-ray analysis share the same basic features. See also Table 3.2.

A number of substances which are similar in their structure to plastoquinone can block the plastoquinone binding site at the D_1 protein, causing inhibition of photosynthesis. Such substances are used as weed killers (herbicides). Before the effect of these substances is dealt with in detail, some general aspects of the application of herbicides will be introduced.

Mechanized agriculture usually necessitates the use of herbicides

Herbicides account for about half the money spent world-wide on substances for plant protection. The high cost of labour is one of the main reasons for using herbicides in agriculture. It is cheaper and faster to keep a field free of weeds by using herbicides than by manual labour. A large number of herbicides (examples will be given at the end of this section) inhibit photosystem II by being antagonists of plastoquinone. For this the herbicide molecule has to be bound to most of the many photosynthetic reaction centres. To be effective, 125–4000 g of these herbicides have to be applied per hectare.

In an attempt to reduce the amount of herbicides needed to be effective, new herbicides have been developed which inhibit certain biosynthetic processes such as the synthesis of fatty acids, of certain amino acids, carotenoides or chloro-phyll. There are also herbicides which act as analogues of phytohormones or mitosis inhibitors. Some of these herbicides are effective in amounts as low as 5 g ha^{-1}.

Some herbicides are taken up only by the roots and others by the leaves. For example, to keep the railway tracks free of weeds, *non-selective herbicides* are employed which destroy all the vegetation. For such purposes herbicides are often used which are degraded slowly and are taken up by the roots or emerging shoots. Non-selective herbicides are also used in agriculture, e.g. to combat weeds in citrus plantations, but in this case, herbicides which are taken up only by the leaves of herbaceous plants are applied at ground level. Especially interesting are those selective herbicides which combat only weeds and affect cultivars as little as pos-sible. Selectivity can be caused by various factors, e.g. by differences between the uptake of the herbicide in different plants, between the sensitivity of metabolism in different plants towards the herbicide or between the ability of the plants to detoxify the herbicide (section 12.2). *Selective herbicides* have the advantage that weeds can be destroyed, for example, when it is opportune at a later growth stage of the cultivars and the dead weeds can form a mulch layer, conserving water and preventing erosion.

In some cases, the application of herbicides has led to the development of herbicide-resistant plant mutants (section 10.4). Conventional breeding has used such mutants to generate herbicide-resistant cultivars. In contrast to these results, which have come about by chance, genetic engineering makes it possible to generate made-to-measure herbicide-resistant cultivars. This means that selective herbicides can be used which are degraded rapidly and are effective in small amounts. Examples of this will be dealt with in sections 10.1 and 10.4.

A large number of herbicides inhibit photosynthesis: the urea derivative DCMU (Diuron, DuPont), the triazine Atrazine (Ciba Geigy), Bentazon (BASF) (Fig. 3.23) and many similar substances function as herbicides by binding to the plastoquinone binding site on the D_1 protein and thus blocking photosynthetic electron transport. Nowadays, DCMU is not often used, as the dose required is high and its degradation slow. It is, however, often used in the laboratory in order to inhibit photosynthesis, e.g. of leaves or isolated chloroplasts. Atrazine acts selectively: maize plants are relatively insensitive towards this herbicide since they have a particularly efficient mechanism for its detoxification (section 12.2). Because of its relatively slow degradation in the soil, the use of Atrazine has been restricted in some countries. In areas where certain herbicides have been used continuously over the years, some weeds have become resistant to these herbi-cides. In some cases the resistance can be traced back to a single amino acid change in the D_1 protein, caused by mutation. These changes do not markedly affect photosynthesis of these weeds but they do decrease binding of the herbi-cides to the D_1 protein.

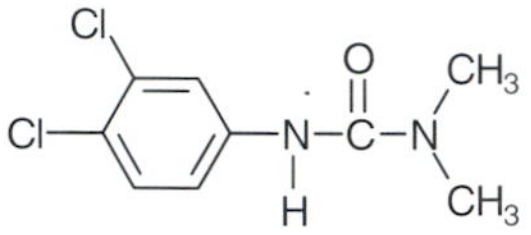

Diuron (DuPont)
3-(3,4-Dichlorophenyl)-1,1-
dimethylurea (DCMU)

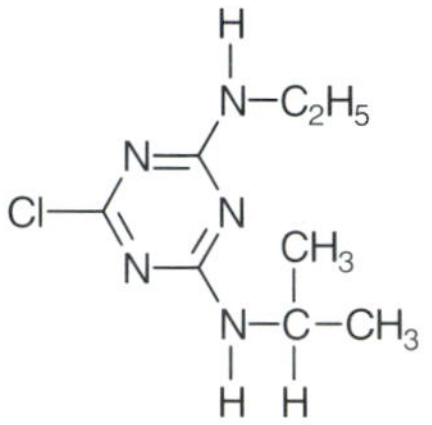

Atrazine
(Ciba Geigy)

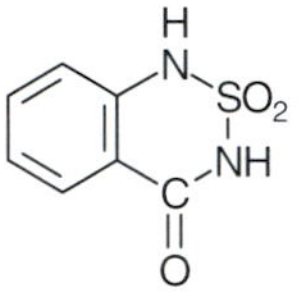

Bentazon
(BASF)

Figure 3.23 Inhibitors of photosystem II used as herbicides.

3.7 The cytochrome-b_6/f complex mediates electron transport between photosystem II and photosystem I

Iron atoms in cytochromes and in iron–sulfur centres have a central function as redox carriers

Cytochromes occur in all organisms except a few obligate anaerobes. These are proteins to which one or two *tetrapyrrole* rings are bound. These tetrapyrroles are very similar to the chromophores of chlorophylls. However, chlorophylls contain Mg^{2+} as central atom in the tetrapyrrole, whereas the cytochromes have an iron atom (Fig. 3.24). The tetrapyrrole ring of the cytochromes with iron as central atom is called *haem*. The bound iron atom can change between the oxidation states Fe^{3+} and Fe^{2+} so that cytochromes can function as electron carriers.

Cytochromes are divided into three main groups, the cytochromes-*a*, -*b*, and -*c*. These correspond to haem-*a*, -*b* and -*c*. Haem-*b* may be regarded as the basic molecule (Fig. 3.24). In haem-*c* the SH_2 group of a cysteine is added to each of the two vinyl groups of haem-*b*. In this way haem-*c* is covalently bound by a sulfur bridge to the protein of the cytochrome. Such a mode of covalent binding has already been shown for phycocyanin in Fig. 2.15 and there is a structural relationship between the corresponding apoproteins. In haem-*a* (not shown) an isoprenoid side chain consisting of three isoprene units is attached to one of the vinyl groups

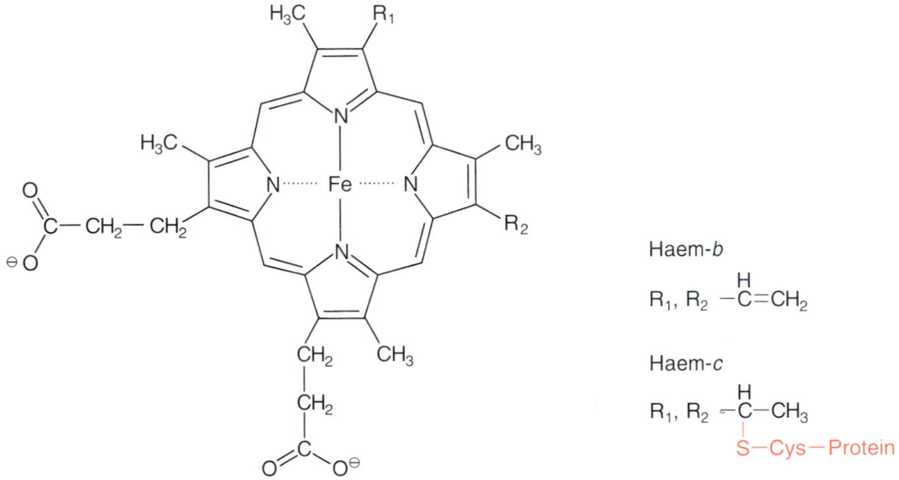

Figure 3.24 Haem-*b* and -*c* as prosthetic group of the cytochromes. Haem-*c* is covalently bound to the protein of the cytochrome by addition of two cysteine residues of the protein to the two vinyl groups of haem-*b*.

of haem-*b*. This side chain has the function of a hydrophobic membrane anchor, similar to that found in quinones (Figs. 3.5 and 3.19). Haem-*a* is mentioned here only for the sake of completeness. It plays no role in photosynthesis, but has a function in the mitochondrial electron transport chain (section 5.5).

The iron atom in the haem can form up to six coordinate bonds. Four of these bonds are formed with the nitrogen atoms of the tetrapyrrole ring. This ring has a planar structure. The two remaining coordinate bonds to the Fe atom are formed by two histidine residues, which are positioned vertically to the tetrapyrrole plane (Fig. 3.25). Cyt-*f* (*f* = foliar, in leaves) contains, like cyt-*c*, one haem-*c* and therefore belongs to the *c*-type cytochromes. In cyt-*f* one coordinative bond of the Fe atom is formed with the terminal amino group of the protein and the other with a histidine residue.

Iron–sulfur centres are of general importance as electron carriers in electron transport chains and thus also in photosynthetic electron transport. Cysteine residues of proteins within iron–sulfur centres (Fig. 3.26) are coordinatively or covalently bound to Fe atoms. These iron atoms are linked to each other by S-bridges. Upon acidification of the proteins the sulfur between the Fe atoms is released as H_2S, and for this reason has been named *labile sulfur*. Iron–sulfur centres occur mainly as 2Fe–2S or 4Fe–4S centres. The Fe atoms in these centres are present in the oxidation states Fe^{2+} and Fe^{3+}. Irrespective of the number of Fe atoms in a centre, the oxidized and reduced states of the centre differ only by a single charge. For this reason iron–sulfur centres can take up only one electron. Various iron–sulfur centres have very different redox potentials, depending on the surrounding protein.

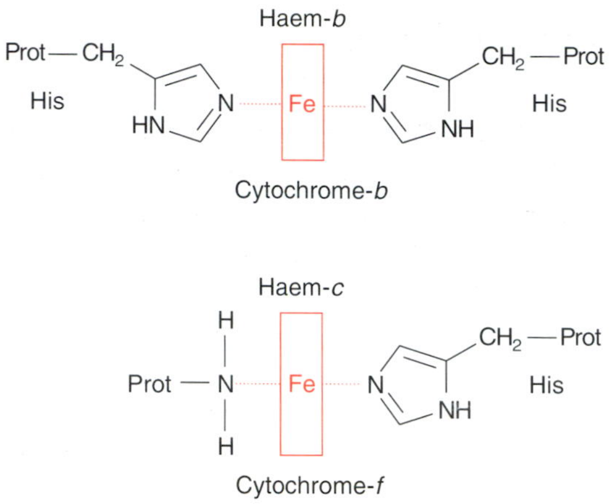

Figure 3.25 Axial ligands of the Fe atoms in the haem groups of cytochromes-*b* and -*f*. Of the six possible coordinative bonds of the Fe atom in the haem, four are saturated with the N-atoms present in the planar tetrapyrrole ring. The two remaining coordinate bonds are formed either with two histidine (His) residues of the protein, located vertically to the plane of the tetrapyrrole, or with the terminal amino group and one histidine residue of the protein.

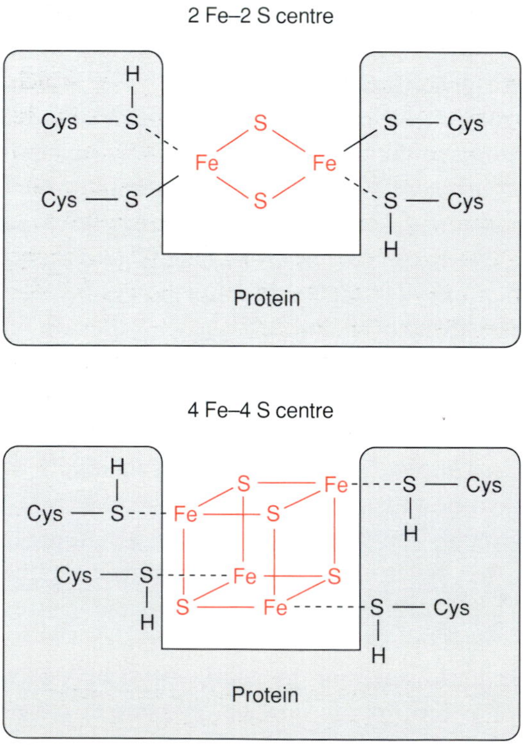

Figure 3.26 Structure of metal clusters of iron–sulfur proteins.

Electron transport by the cytochrome-b_6/f complex is coupled to proton transport

Plastohydroquinone (PQH$_2$), formed by PS II, diffuses through the lipid phase of the thylakoid membrane and transfers its electrons to the cytochrome-b_6/f complex (Fig. 3.17). This complex then transfers the electrons to *plastocyanin*, which is thus reduced. Therefore the cytochrome-b_6/f complex has also been named *plastohydroquinone–plastocyanin oxidoreductase*. Plastocyanin is a protein with a molecular mass of 10.5 kDa, containing a *copper atom* which is coordinately bound to one cysteine, one methionine, and two histidine residues of the protein (Fig. 3.27). This copper atom alternates between the oxidation states Cu$^+$ and Cu^{2+} and thus is able to take up one electron. Plastocyanin is soluble in water and is located in the thylakoid lumen.

Electron transport through the cyt-b_6/f complex proceeds along a potential difference gradient of about 0.4 V (Fig. 3.16). The energy liberated by the transfer of the electron down this redox gradient is conserved by transporting protons to the thylakoid lumen. The cyt-b_6/f complex is a membrane protein consisting of many subunits. The main components of this complex are four subunits: cyt-b_6, cyt-f, an iron–sulfur protein called *Rieske protein* after its discoverer, and a subunit IV. The Rieske protein has a 2Fe–2S centre with the very positive redox potential of +0.3 V, untypical of such iron–sulfur centres.

The cyt-b_6/f complex has an *asymmetrical structure* (Fig. 3.28). Cyt-b_6, which contains two molecules of haem-b, spans the membrane, as does subunit IV. In the cyt-b_6/f complex the two haem molecules are placed one above the other and in this way form a redox chain reaching from one side of the membrane to the other. One amino acid chain of cyt-f protrudes into the membrane, forming an anchor. The haem-c, as the functional group of cyt-f, is positioned at the periphery of the membrane on the luminal side. On the same side there is also the Rieske 2Fe–2S protein which protrudes only slightly into the membrane.

The cyt-b_6/f complex resembles in its structure the cyt-b/c_1 complex in bacteria and mitochondria (section 5.5). Table 3.3 summarizes the function of these cyt-b_6/f and cyt-b/c_1 complexes. All these complexes possess one iron–sulfur protein. The

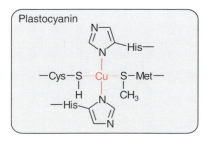

Figure 3.27 Plastocyanin: two histidine, one methionine and one cysteine residue of this protein bind one Cu-atom, which changes between the redox states Cu$^+$ and Cu^{2+} by the uptake and transmission of an electron.

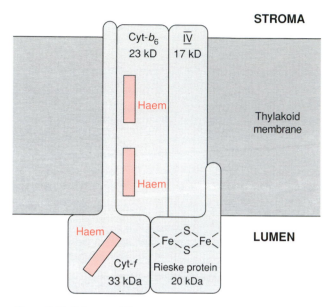

Figure 3.28 Scheme of the structure of the cytochrome-b_6/f complex. The scheme is based on the molecular structures predicted from their amino acid sequences. (After Hauska and Büttner 1995.)

amino acid sequence of cyt-b in the cyt-b/c_1 complex of bacteria and in mitochondria corresponds to the sum of the sequences of cyt-b_6 and subunit IV in the cyt-b_6/f complex. Apparently a cleavage of the cyt-b gene into the two genes for cyt-b_6 and subunit IV occurred during evolution. Whereas in plants the cyt-b_6/f complex reduces plastocyanin, the cyt-b/c_1 complex of bacteria and mitochondria reduces cyt-c. Cyt-c is a very small cytochrome molecule which is water soluble and, like plastocyanin, transfers redox equivalents from the cyt-b_6/f complex to the next complex along the aqueous phase. In cyanobacteria, which also possess a cyt-b_6/f complex, the electrons are transferred from this complex to photosystem I via cyt-c instead of plastocyanin. The great similarity between the cyt-b_6/f complex in plants and the cyt-b/c_1 complexes in bacteria and mitochondria suggests that these complexes have basically similar functions in photosynthesis and in mito-

Table 3.3 Function of cytochrome-b/c complexes

Purple bacteria	Cyt-b/c_1	Reduction of cyt-c	Proton pump
Green sulfur bacteria	Cyt-b/c_1	Reduction of cyt-c	Proton pump
Mitochondria	Cyt-b/c_1	Reduction of cyt-c	Proton pump
Cyanobacteria	Cyt-b_6/f	Reduction of cyt-c	Proton pump
Chloroplasts	Cyt-b_6/f	Reduction of plastocyanin	Proton pump

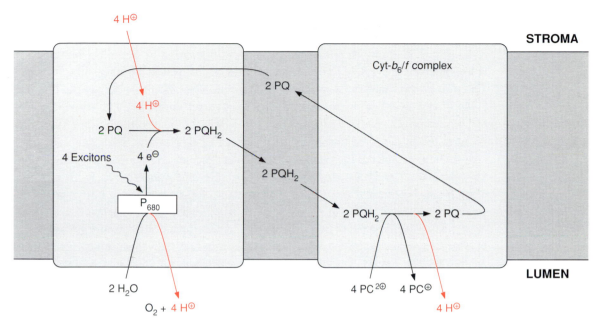

Figure 3.29 Proton transport coupled to electron transport by PS II and the cyt-b_6/f complex in the absence of a Q-cycle. The basis of this scheme is that the oxidation of water in PS II and oxidation of plastohydroquinone (PQH_2) by cyt-b_6/f occur on the luminal side of the thylakoid membrane.

chondrial oxidation: they are proton translocators which are driven by a hydro-quinone–plastocyanin (or –cyt-c) reductase.

The transfer of protons from the stroma to the thylakoid lumen is explained by reduction and oxidation of the quinones taking place on different sides of the membrane. The protons required for reduction of plastoquinone in the PS II complex come from the stroma side, whereas oxidation of plastohydroquinone by plastocyanin, with the resultant release of protons, occurs on the luminal side of the cyt-b_6/f complex (Fig. 3.29). In this way, four protons are transferred from the stromal space to the lumen after the capture of four excitons by the PS II complex. In addition to this, four protons are released into the lumen during the splitting of water by PS II.

The number of protons pumped through the cyt-b_6/f complex can be doubled by a Q-cycle

Studies with mitochondria indicated that during electron transport through the cyt-b/c_1 complex, the number of protons transferred per transported electron is larger than shown in Fig. 3.29. Peter Mitchell, who established the chemiosmotic hypothesis of energy conservation (section 4.1), also postulated a so-called *Q-cycle*, by which the number of protons transported for each electron transferred through the cyt-b/c_1 complex is doubled. It became apparent later that the Q-cycle might also have a role in photosynthetic electron transport.

Figure 3.30 shows the principle of Q-cycle operation in the photosynthesis of chloroplasts. The cyt-b_6/f complex contains two different binding sites for conversion of quinones, one located on the stromal side and the other on the luminal side of the thylakoid membrane. The plastohydroquinone (PQH_2) formed in the PS II complex is oxidized by the Rieske iron–sulfur centre at the binding site adjacent to the lumen. Due to its very positive redox potential the Rieske protein tears off one electron from the plastohydroquinone. Because its redox potential is very negative the remaining semiquinone is unstable and transfers its electron to the first haem-b of the cyt-b_6 (b_p) and from there to the second haem-b (b_n), attaining a redox potential of about -0.1 V. In this way, a total of four protons are transported to the thylakoid lumen per two molecules of plastohydroquinone oxidized. Of the two plastoquinone (PQ) molecules formed, only one molecule returns to the PS II complex. The other PQ diffuses through the lipid phase of the mem-

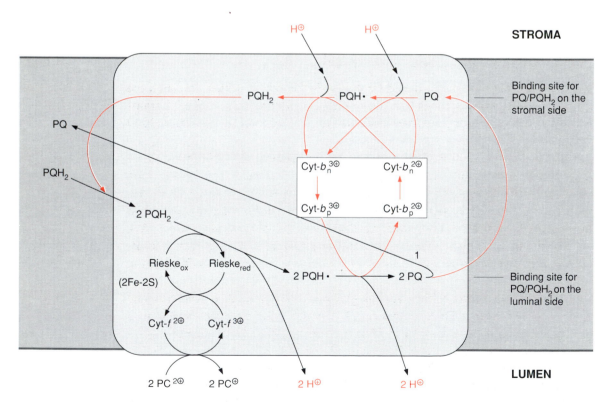

Figure 3.30 The number of protons released by the cyt-b_6/f complex into the lumen is doubled by the Q-cycle. This cycle is based on the finding that the redox reactions of the PQH_2 and PQ occur at two binding sites, one in the lumen and one in the stromal region of the thylakoid membrane. How the cycle functions is explained in the text.

brane to the other binding site on the stromal side and is reduced there by haem-b_n with its high reduction potential via semiquinone to hydroquinone. This is accompanied by the uptake of two protons from the stromal space. The hydroquinone thus regenerated is oxidized, in turn, by the Rieske protein on the luminal side, and so on. In total, the number of transported protons is doubled by the Q-cycle ($1/2 + 1/4 + 1/8 + 1/16 \ldots + 1/n = 1$). When the Q-cycle is operating fully, the transport of four electrons through the cyt-b_6/f complex leads in total to the transfer of eight protons from the stroma to the lumen. The function of this Q-cycle in mitochondrial oxidation is now undisputed. In the case of photosynthetic electron transport, however, its function is still a matter of controversy. It could be expected from the analogy of the cyt-b_6/f complex to the cyt-b/c_1 complex that the Q-cycle also plays an important role in chloroplasts. So far, the functioning of a Q-cycle in plants has been observed mainly under low light conditions. The Q-cycle is perhaps suppressed by a high proton gradient across the thylakoid membrane generated, for instance, by irradiation with high light intensity. In this way the flow of electrons through the Q-cycle could be adjusted to the energy demand of the plant cell.

3.8 Photosystem I reduces NADP

Plastocyanin that has been reduced by the cyt-b_6/f complex diffuses through the lumen of the thylakoids, binds to a positively charged binding site of PS I, transfers its electron, and then diffuses back to the cyt-b_6/f complex in the oxidized form (Fig. 3.31).

The reaction centre of PS I with an absorption maximum of 700 nm also contains a chlorophyll pair, (chl-a)$_2$ (Fig. 3.31). As in PS II excitation by a photon results in a charge separation giving (chl-a)$_2^+$, which is then reduced by plastocyanin (Fig. 3.32). It is assumed that (chl-a)$_2$ transfers its electron to a chl-a monomer (A_0) which then transfers the electron to a bound *phylloquinone* (A_1). Phylloquinone contains the same phytol side chain as chl-a. The electron is transferred from the semiquinone form of phylloquinone to an iron–sulfur centre, called F_X.

F_X is a 4Fe–4S centre with a very negative redox potential. It transfers one electron to two further Fe–S centres (F_A, F_B), which in turn reduce *ferredoxin*, a protein with a molecular mass of 11 kDa, containing a 2Fe 2S centre. Ferredoxin also takes up only one electron. The reduction occurs on the stromal side of the thylakoid membrane. For this purpose the ferredoxin binds at a positively charged binding site on subunit D of PS I (Fig. 3.33). The reduction of NADP$^+$ by ferredoxin, catalysed by *ferredoxin–NADP reductase*, yields NADPH as a primary product of the photosynthetic electron transport.

Currently, functions have been attributed to only 6 of the 12 different identified subunits of the PS I complex (Table 3.4). The centre of the PS I complex is also a

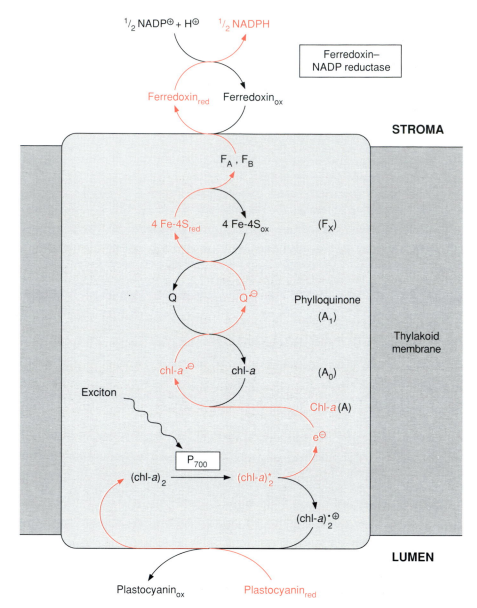

Figure 3.31 Reaction scheme of electron transport in photosystem I. The negatively charged chlorophyll radical formed after excitation of a chlorophyll pair results in reduction of NADP via chl-*a*, phylloquinone and three iron–sulfur proteins. The electron deficit in the positively charged chlorophyll radical is compensated for by an electron delivered from plastocyanin.

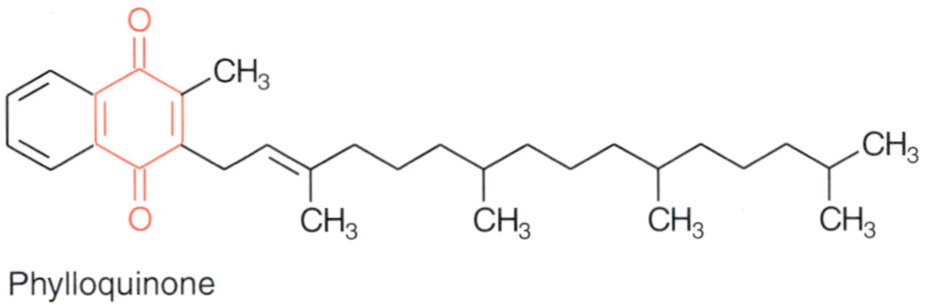

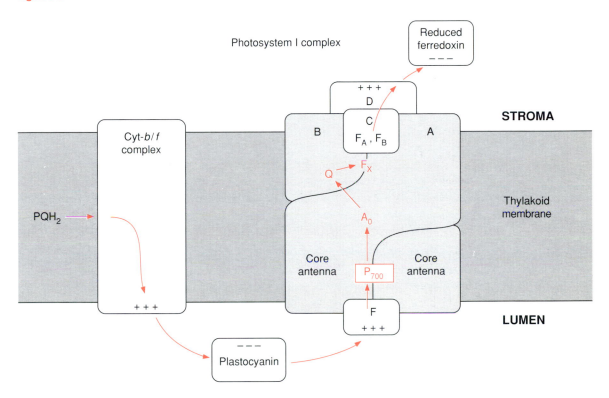

Phylloquinone

Figure 3.32

Figure 3.33 Scheme of the structure of the photosystem I complex. This scheme is based on the structures of the subunits as predicted from their amino acid sequences. According to recent results of X-ray structural analysis by H. T. Witt, the basic structure of the PS I complex is similar to the bacterial reaction centres and thus also to the PS II complex. The electron carriers are arranged accordingly.

heterodimer (as is the centre of PS II) consisting of subunits A and B (Fig. 3.33). The molecular masses of A and B (82–83 kDa) each correspond to about the sum of the molecular masses of D_1 and CP 43 and D_2 and CP 47, respectively, in PS II (Table 3.2). In fact both subunits A and B have a double function. Like D_1 and D_2 in PS II, they bind chromophores (chl-*a*) and redox carriers (phylloquinone, Fe_X) of the reaction centre and, additionally, they contain about 100 chl-*a* molecules as antenna pigments. Thus, the heterodimer of A and B carries the reaction centre

Table 3.4 Protein components of photosystem I (list not complete) (after Andersson and Franzén 1992)

Protein	Molecular mass (kDa)	localization	Coded in	Function
A	83	Intrinsic	Chloroplast	Binding of P_{700}, chl-a, A_0, A_1, F_X, antennae function
B	82	Intrinsic	Chloroplast	(As in protein A)
C	9	Peripheral: stroma	Chloroplast	Binding of F_A, F_B
D	17	Peripheral: stroma	Nucleus	Binding of ferredoxin
E	10	Peripheral: stroma	Nucleus	Binding of ferredoxin
F	18	Peripheral: lumen	Nucleus	Binding of plastocyanin
I	5	Intrinsic (?)	Chloroplast	?
J	5	Intrinsic (?)	Chloroplast	?

and also the core antenna. Recently the three-dimensional structure of photosystem I of the thermophilic cyanobacterium, *Synechococcus elongatus*, was resolved by X-ray structural analysis with a resolution of 4.5 Å. Although at this resolution some details of the structure are not noticeable, these studies showed that the basic structure of photosystem I and of the bacterial reaction centre (and thus also of photosystem II) are rather similar. The Fe–S centres F_A and F_B are ascribed to subunit C, and subunits F and D are proposed to be binding sites for plastocyanin and ferredoxin, respectively.

In cyclic electron transport by PS I light energy is used for the synthesis of ATP only

Besides the non-cyclic electron transport discussed so far, cyclic electron transfer can also take place where the electrons from the excited photosystem I are fed into the cyt-b_6/f complex and transferred back to the ground state of PS I (Fig. 3.34). The energy thus released is only used for the synthesis of ATP and NADPH is not formed. This is termed *cyclic photophosphorylation*. In intact leaves, and even in isolated intact chloroplasts, it is quite difficult to differentiate experimentally between cyclic and non-cyclic photophosphorylation. It is therefore still a matter of debate as to whether and to what extent cyclic photophosphorylation occurs in a leaf under normal physiological conditions. However, there is no doubt that cyclic photophosphorylation must operate at very high rates in the bundle sheath chloroplasts of certain C_4 plants (section 8.4). These cells have a high demand for ATP and they contain high PS I activity but very little PS II. In all likelihood, cyclic electron flow is governed by the redox state of the acceptor of photosystem I in such a way that an increased reduction of the NADP system, and

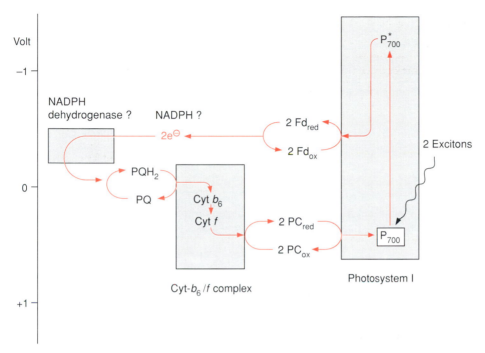

Figure 3.34 Cyclic electron transport between photosystem I and the cyt-b_6/f complex. The path of the electrons from the excited PS I to the cyt-b_6/f complex is still unclear.

consequently that of ferredoxin, enhances the diversion of the electrons in the cycle. The function of cyclic electron transport is probably to adjust the rates of ATP and NADPH formation according to demand.

Despite intensive investigations, the pathway of electron flow from PS I to the cyt-b_6/f complex in cyclic electron transport is still unresolved. Most experiments on cyclic electron transport have been carried out with thylakoid membranes which only catalyse cyclic electron transport when redox mediators, such as ferredoxin or flavin adenine mononucleotide (FMN, Fig. 5.16) have been added. Cyclic electron transport is inhibited by the antibiotic *antimycin A*. It is not clear at which site this inhibitor functions. Antimycin A does not inhibit non-cyclic electron transport.

Surprisingly, proteins of the NADP dehydrogenase complex of the mitochondrial respiratory chain (section 5.5) were identified recently in the thylakoid membrane of chloroplasts. The function of these proteius in chloroplasts is still unknown. The proteins of this complex occur very frequently in chloroplasts from bundle sheath cells of C_4 plants and, as mentioned above, these cells have little PS II but a particularly high cyclic photophosphorylation activity. These observations raise the possibility that in cyclic electron transport the flow of electrons from NADPH or ferredoxin to plastoquinone proceeds via a complex similar to the mitochondrial NADH dehydrogenase complex. As will be shown in section 5.5, the mitochondrial NADH dehydrogenase complex transfers electrons from NADH to ubiquinone. In cyanobacteria the participation of an NADPH dehydrogenase complex similar to the mitochondrial NADH dehydrogenase complex in cyclic electron transport has been proved.

3.9 In the absence of other acceptors electrons can be transferred from photosystem I to oxygen

When ferredoxin is very highly reduced it is possible that electrons are transferred from PS I to oxygen to form *superoxide radicals* ($O_2^-\bullet$) (Fig. 3.35). This process is called the *Mehler reaction*. The superoxide radical is a chemically very aggressive substance. It reduces metal ions present in the cell such as Fe^{3+} and Cu^{2+} (M^{n+}):

$$O_2^-\bullet + M^{n+} \rightarrow O_2 + M^{(n-1)+}.$$

Therefore it is important that the superoxide radical is efficiently eliminated. *Superoxide dismutase*, which is localized in the thylakoid membrane, catalyses the dismutation of superoxide into hydrogen peroxide and oxygen, accompanied by the uptake of two protons:

$$2\,O_2^-\bullet + 2H^+ \rightarrow O_2 + H_2O_2.$$

The metal ions reduced by superoxide react with hydrogen peroxide to form hydroxyl radicals:

$$H_2O_2 + M^{(n-1)+} \rightarrow OH^- + \bullet OH + M^{n+}.$$

The hydroxyl radical ($\bullet$OH) is a very aggressive substance and damages enzymes and lipids by oxidation. The plant cell has no protective enzymes against $\bullet$OH. Therefore it is essential that reduction of the metal ions is prevented by rapid elimination of $O_2^-\bullet$ by superoxide dismutase. But hydrogen peroxide also has a damaging effect on many enzymes. It is eliminated by an *ascorbate peroxidase* located in the thylakoid membrane. *Ascorbate*, an important antioxidant in plant cells (Fig. 3.36) is oxidized by this enzyme to the radical *monodehydroascorbate*, which is spontaneously reconverted by photosystem I to ascorbate via reduced ferredoxin. Monodehydroascorbate can be also reduced to ascorbate by an NAD(P)H dependent monodehydroascorbate reductase which is present in the chloroplast stroma and the cytosol.

As an alternative to the above reaction, two molecules of monodehydroascorbate can dismutate to ascorbate and dehydroascorbate. From dehydroascorbate, ascorbate is regenerated by reduction with glutathione (Fig. 3.37). *Glutathione* (GSH) occurs as an antioxidant in all plant cells. It is a tripeptide composed of the amino acids glutamate, cysteine, and glycine (Fig. 3.38). Oxidation of GSH results in formation of a disulfide (GSSG) between the cysteine residues of two glu-

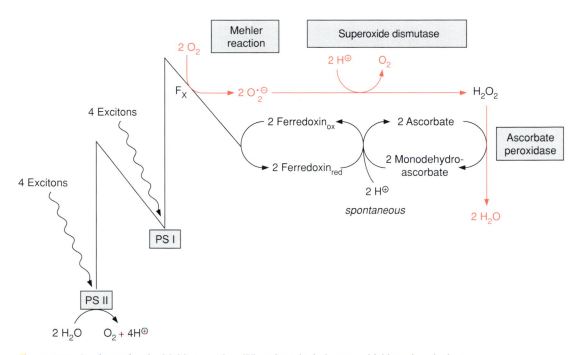

Figure 3.35 A scheme for the Mehler reaction. When ferredoxin becomes highly reduced, electrons are transferred by the Mehler reaction to oxygen and superoxide is formed. The elimination of this highly reactive radical involves reactions catalysed by superoxide dismutase and ascorbate peroxidase.

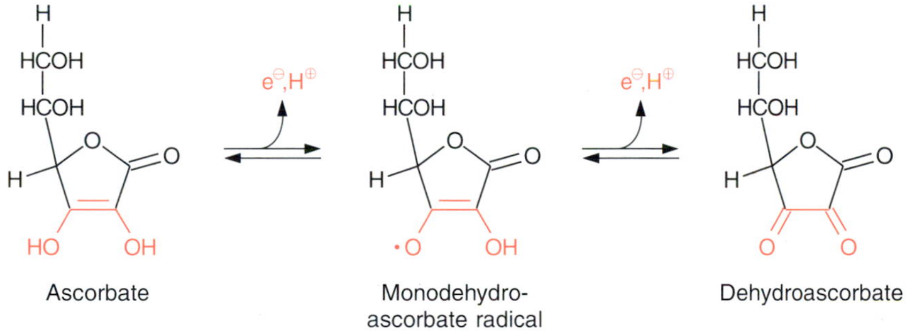

Figure 3.36 The oxidation of ascorbate proceeds via the formation of the monodehydroascorbate radical.

tathione molecules. Reduction of GSSG is catalysed by a *glutathione reductase* with NADPH as reductant (Fig. 3.37).

The major function of the Mehler ascorbate–peroxidase cycle is to dissipate excessive excitation energy of photosystem II as heat. The absorption of a total of eight excitons via PS I results in formation of two superoxide radicals and two molecules of reduced ferredoxin, the latter serving as a reductant for eliminating

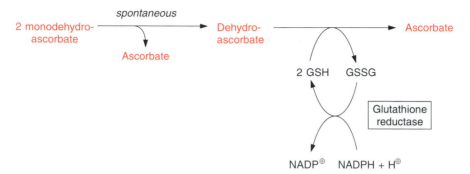

Figure 3.37 Dehydroascorbate can be reduced to form ascorbate.

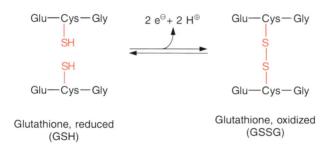

Figure 3.38 Redox reaction of glutathione.

H_2O_2 (Fig. 3.35). In a sense the transfer of electrons to oxygen by the Mehler reaction could be viewed as a reversal of the splitting of water by PS II. As will be discussed in the following section, the Mehler reaction occurs when ferredoxin is very highly reduced. The only gain from this reaction is the generation of a proton gradient from electron transport through photosystem II and the cyt-b_6/f complex. This proton gradient can be used for synthesis of ATP when ADP is present. But since there is usually a shortage of ADP under the conditions of the Mehler reaction, it mostly results in the formation of a high pH gradient. A feature common to the Mehler reaction and cyclic electron transport is that there is no net production of NADPH from reduction of NADP. For this reason electron transport via the Mehler reaction has been termed *pseudocyclic electron transport*.

Instead of ferredoxin, PS I can also reduce methylviologen. Methylviologen is used commercially as a herbicide, *paraquat* (Fig. 3.39). Its herbicidal effect is due to the reduction of oxygen to superoxide radicals by reduced paraquat. Additionally, paraquat competes with dehydroascorbate for the reducing equivalents provided by photosystem I. Therefore, in the presence of paraquat, ascorbate is no longer regenerated from dehydroascorbate and the ascorbate peroxidase

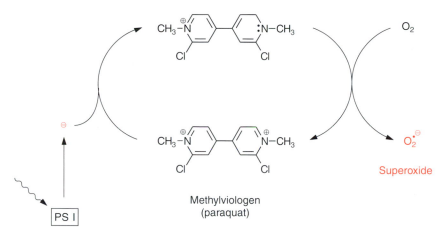

Figure 3.39 Methylviologen, a herbicide also named paraquat, is reduced by transfer of an electron from the excited PS I to form a radical substance. The latter transfers the electron to oxygen with formation of the chemically reactive superoxide radical. Paraquat is distributed as herbicide by ICI under the trade name Gramoxone.

reaction can no longer proceed. The increased production of superoxide and decreased detoxification of hydrogen peroxide in the presence of paraquat causes severe oxidative damage to mesophyll cells, noticeable by a bleaching of the leaves. In the past paraquat has been used to destroy marijuana fields.

3.10 Regulatory processes control the distribution of the captured excitons between the two photosystems

Non-cyclic photosynthetic electron transport through the two photosystems requires the even distribution of the captured excitons between them. As discussed in section 2.4, the excitons are transferred preferentially to that chromophore requiring the least energy for excitation. Photosystem I (P_{700}) requires less energy for excitation than photosystem II (P_{680}). Therefore, in an unrestricted competition between the two photosystems for the excitons, these would be directed mainly to PS I, and therefore distribution of the excitons between the two photosystems must be regulated. The spatial separation of PS I and PS II in the thylakoid membrane is an important element in this regulation.

In chloroplasts the thylakoid membranes are present in two different arrays, as *stacked* and *unstacked membranes*. The outer surface of the unstacked membranes has free access to the stromal space; these membranes are called *stromal lamellae* (Fig. 3.40). In the stacked membranes, the neighbouring thylakoid membranes are in direct contact with each other. These membrane stacks can be seen as grains (grana) in light microscopy and are therefore called *granal lamellae*.

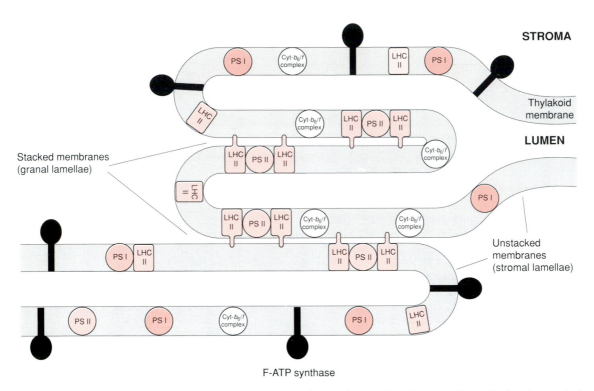

Figure 3.40 Distribution of photosynthetic protein complexes between the stacked and unstacked regions of thylakoid membranes. Stacking is caused by light harvesting complex II (LHC-II) (see Fig. 3.41).

ATP synthase and the PS I complex (including its light harvesting complexes, which will not be dealt with here) are located either in the stromal lamellae or in the outer membrane region of the granal lamellae. Therefore, these proteins have free access to ADP and NADP in the stroma. The PS II complex, on the other hand, is located mainly in the granalamellae. Peripheral LHC-II subunits attached to the PS II complex (section 2.4), contain a protein chain protruding from the membrane (Fig. 3.41), which binds electrostatically to the neighbouring membrane and causes stacking of granal lamellae. Besides PS II, the stacked membranes contain only cyt-b_6/f complexes. In this way the PS II complexes in the stacked membranes are separated spatially from the PS I complexes in the unstacked membranes. It is assumed that this prevents a spillover of excitons from PS II to PS I.

Stacking of membranes is a dynamic process, which can regulate both the spatial separation of the two photosystems and the spillover of excitons from PS II to PS I. If the excitation of PS II is greater than that of PS I, the result is an accumulation of plastohydroquinone, which PS I cannot oxidize rapidly enough via the cyt-b_6/f complex. Under these conditions a protein kinase is activated which catalyses

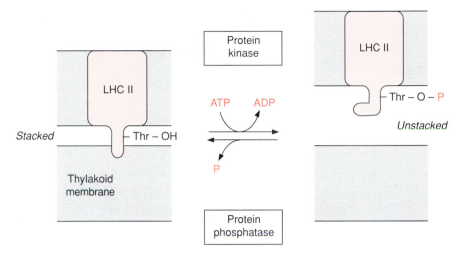

Figure 3.41 Conformation of the light harvesting complex II (LHC-II) is changed by phosphorylation of a threonine residue. The non-phosphorylated form of the LHC-II attaches to the neighbouring thylakoid membrane by a protein sequence protruding from the membrane and thus causes stacking of the membrane. After phosphorylation, LHC-II is no longer able to attach itself to the neighbouring thylakoid membrane. The phosphorylation is catalysed by a protein kinase and the dephosphorylation by a protein phosphatase.

phosphorylation of the hydroxyl group in threonine residues of peripheral LHC-II subunits (Fig. 3.41). As a result of this phosphorylation, the LHC-II protein loses its ability to anchor to the neighbouring membrane and this reduces the close connection between peripheral LHC-II and the smaller PS II complex. It is believed that the PS II complex with smaller antenna then migrates to regions of the stromal lamellae. The PS II complexes with fewer LHC-II in the stromal lamellae and, because of their smaller antenna, capture fewer excitons. It has been suggested that the single phosphorylated LHC-II subunits migrate to the stromal lamellae where they allow a spillover of excitons from LHC-II to PS I. In this way accumulation of reduced plastoquinone could decrease the excitation of PS II in favour of PS I. A protein phosphatase facilitates the reversal of this regulation. This regulatory process enables the plant to distribute the captured photons evenly between the two photosystems.

Excess light energy is eliminated as heat

Plants face the general problem that the energy of irradiating light can be much higher than the demand of photosynthetic metabolism for NADPH and ATP. This is, for instance, the case with very high light intensities when the metabolism cannot keep pace, or at low temperatures when the metabolism is slowed down because of decreased enzyme activities. Excess excitation of the photosystems could result in an excessive reduction of the components of the photosynthetic electron transport system.

Very high excitation of photosystem II, recognized by the accumulation of plastohydroquinone, results in damage to the photosynthetic apparatus, termed *photoinhibition*. The exact mechanism for this is not known. It is assumed that under these conditions the excited chlorophyll is converted increasingly to the triplet state. As dealt with in section 2.3, triplet chlorophyll can produce highly reactive and damaging singlet oxygen. The damaging effect of triplet chlorophyll can be demonstrated by placing a small amount of chlorophyll under the human skin. Illumination causes severe tissue damage. This is made use of in selective therapy for skin cancer.

Carotenoids (e.g. β-carotene, Fig. 2.9) are able to convert the triplet state of chlorophyll and also the singlet state of oxygen back to the corresponding ground states. An excited triplet carotenoid is formed which dissipates its energy as heat; thus carotenoids have an important protective function. This protective function of the carotenoids, however, may be unable to cope with excessive excitation of PS II, and the singlet oxygen then has a damaging effect on the PS II complex. The site of this damage could be the D_1 protein of the photosynthetic reaction centre

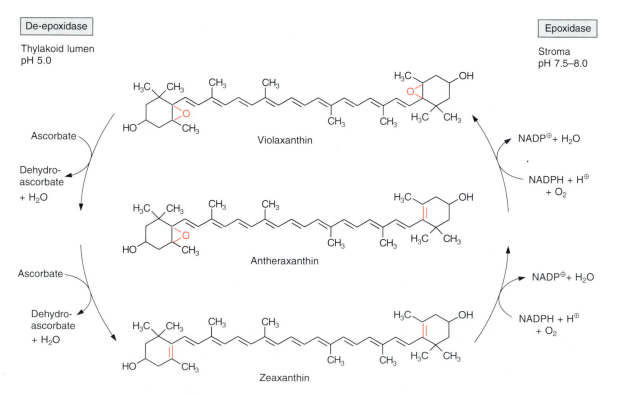

Figure 3.42 The zeaxanthin cycle (after Demmig-Adams and Adams, 1996).

in PS II, which even under normal photosynthetic conditions has a high turnover (section 3.6).

To protect the photosynthetic apparatus from light damage, plants possess mechanisms which dissipate the energy from excess exciton production as heat. This process is termed *non-photochemical quenching* of exciton energy. The mechanism of this quenching process is not exactly known. According to recent investigations, conversion of exciton energy to heat is enhanced by the presence of *zeaxanthin* (Fig. 3.42) in PS II. How this happens, and where in the photosynthetic apparatus, is not clear. Zeaxanthin is formed by the reduction of the diepoxide violaxanthin present in the antennae. The reduction proceeds with ascorbate as reductant and the monoepoxide antheraxanthin is formed as an intermediate. Zeaxanthin is reconverted to *violaxanthin* by epoxidation requiring NADPH and O_2 (Fig. 3.42). Formation of zeaxanthin by diepoxidase takes place on the luminal side of the thylakoid membrane at an optimum pH of 5.0, whereas the regeneration of violaxanthin by the epoxidase in the stroma proceeds at about pH 7.6. Therefore, the formation of zeaxanthin requires a high pH gradient across the thylakoid membrane. As discussed in connection with the Mehler reaction (section 3.9), a high pH gradient can be an indicator for the high excitation of photosystem I. It is assumed that when there is too much exciton energy, an increased pH gradient initiates zeaxanthin synthesis, dissipating excess exciton energy in the PS II complex as heat, thus protecting the plant from the adverse effect of excess light.

Further reading

Andersson, B. and Franzén, L. G. (1992). The two photosystems of oxygenic photosynthesis. In *Molecular mechanics in bioenergetics*, (ed. L. Ernster), pp. 103–20. Elsevier, Amsterdam.

Asada, K. (1994). Production and action of active oxygen species in photosynthetic tissues. In *Causes of photooxidative stress and amelioration of defence systems in plants*, (ed. C. H. Foyer and P. M. Mullineaux), pp. 77–104. CRC Press, Boca Raton, FL.

Baker, N. R. and Bowyer, J. R. (1994). *Photoinhibition of photosynthesis, from molecular mechanisms to the field*. Bios Scientific Publishers, Oxford.

Deisenhofer, J. and Michel, H. (1989). Nobel Lecture: The photosynthetic reaction centre from the purple bacterium *Rhodopseudomes viridis*. *EMBO Journal*, **8**, 2149–69.

Demmig-Adams, B. and Adams III, W. W. (1996). The role of the xanthophyll cycle carotenoids in the protection of photosynthesis. *Trends in Plant Science*, **1**, 21–6.

Ermler, U., Fritzsch, G., Buchanan, S. K., and Michel, H. (1994). Structure of the photosynthetic reaction centre from *Rhodobacter sphaeroides* at 2.65 Å resolution: cofactors and protein–cofactor interactions. *Structure*, **2**, 925–36.

Foyer, C. H. and Harbison, J. (1994). Oxygen metabolism and the regulation of photosynthetic electron transport. In *Causes of photooxidative stress and amelioration of defence systems in plants*, (ed. C. H. Foyer and P. M. Mullineaux), pp. 1–36. CRC Press, Boca Raton, FL.

Ghanotakis, D. F. and Yocum, C. F. (1990). Photosystem II and the oxygen-evolving complex. *Annual Review of Plant Physiology and Plant Molecular Biology*, **41**, 255–76.

Goldbeck, J. H. (1992). Structure and function of photosystem I. *Annual Review of Plant Physiology and Plant Molecular Biology*, **43**, 293–324.

Hauska, G. and Büttner, M. (1995). The cytochrome-b_6/f complexes. In *Bioelectrochemistry*, (ed. P. Gräber), Vol. 3, Bioenergetics. Birkhäuser Verlag, Basel.

Heller, B. A., Holten, D., and Kirmaier, C. (1995). Control of electron transfer between the L- and M-sides of photosynthetic reaction centres. *Science*, **269**, 940–5.

Horton, P., Ruban, A. V., and Walters, R. G. (1996). Regulation of light harvesting in green plants. *Annual Review of Plant Physiology and Plant Molecular Biology*, **47**, 655–84.

Krause, G. H. (1994). The role of oxygen in photoinhibition of photosynthesis. In *Causes of photooxidative stress and amelioration of defence systems in plants*, (ed. C. H. Foyer and P. M. Mullineaux), pp. 43–76. CRC Press, Boca Raton, FL.

Long, S. P., Humphries, S., and Falkowski, P. G. (1994). Photoinhibition of photosynthesis in nature. *Annual Review of Plant Physiology and Plant Molecular Biology*, 45, 633–62.

Park, Y-II, Wah, S. C., and Anderson, J. M. (1995). Light inactivation of functional photosystem II in leaves of peas grown in moderate light depends on photon exposure. *Planta*, **196**, 401–11.

Pfündel, E. and Bilger, W. (1994). Regulation and possible function of the violaxanthin cycle. *Photosynthesis Research*, **42**, 89–109.

Powles, S. B. and Holtum, J. A. M. (1994). *Herbicide resistance in plants*. Lewis Publishers, Boca Raton, FL.

Vermaas, W. (1993). Molecular-biological approaches to analyse photosystem II structure and function. *Annual Review of Physiology and Plant Molecular Biology*, **44**, 457–81.

chapter 4

ATP generation by photosynthesis

Chapter 3 discussed the transport of protons across a thylakoid membrane by photosynthetic electron transport and how, in this way, a proton gradient is generated. This chapter deals with how this proton gradient is utilized for the synthesis of ATP.

In 1954, in Berkeley, Daniel Arnon discovered that when suspended thylakoid membranes are illuminated, ATP is formed from ADP and inorganic phosphate. This process is called *photophosphorylation*. Further experiments showed that photophosphorylation is coupled to the generation of NADPH. This result was unexpected, as it was then generally believed that the synthesis of ATP in chloroplasts was driven, as in mitochondria, by electron transport from NADPH to oxygen. It soon became apparent, however, that the mechanism of photophosphorylation coupled to photosynthetic electron transport was very similar to that of ATP synthesis coupled to electron transport of mitochondria, termed *oxidative phosphorylation* (section 5.6).

In 1961 Peter Mitchell (Edinburgh) postulated in his *chemiosmotic hypothesis* that during electron transport-coupled ATP synthesis a proton gradient is formed and that it is the *proton motive force* of this gradient which drives the synthesis of ATP. At first this revolutionary hypothesis was strongly opposed by many workers in the field, but in the course of time experimental results of many researchers supported the chemiosmotic hypothesis which is now fully accepted. In 1978 Peter Mitchell was awarded the Nobel prize for Chemistry for this hypothesis.

4.1 A proton gradient serves as an energy-rich intermediate state during ATP synthesis

Let us first ask: how much energy is actually required in order to synthesize ATP?

The free energy for the synthesis of ATP from ADP and phosphate is calculated from the van't Hoff equation:

$$\Delta G = \Delta G^{0\prime} + RT \ln \frac{[ATP]}{[ADP][P]} \qquad (4.1)$$

The standard free energy for the synthesis of ATP is:

$$\Delta G^{0\prime} = + 30.5 \text{ kJ mol}^{-1}$$

The concentrations of ATP, ADP, and phosphate in the chloroplast stroma are very much dependent on metabolism. Typical concentrations are: ATP = 2.5×10^{-3} mol l^{-1}; ADP = 0.5×10^{-3} mol l^{-1}; P = 5×10^{-3} mol l^{-1}

When these values are introduced into equation 4.1 ($R = 8.32$ J (mol K)$^{-1}$, $T = 298$ K) the energy required for synthesis of ATP is evaluated as:

$$\Delta G = +47.8 \text{ kJ mol}^{-1}$$

This value is of course variable as it depends on the metabolic conditions. For further considerations an average value of 50 kJ mol^{-1} will be employed for ΔG_{ATP}.

The transport of protons across a membrane can have different effects. If the membrane is permeable to counter ions of the proton, e.g. a chloride ion (Fig. 4.1a), the charge of the proton will be compensated for, since each transported proton will pull a chloride ion across the membrane. This is how a proton concentration gradient can be generated. The free energy of the transport of protons from A to B is:

$$\Delta G = RT \ln \frac{[H^+]_B}{[H^+]_A} \qquad [J \text{ mol}^{-1}] \qquad (4.2)$$

If the membrane is impermeable to counter ions (Fig. 4.1b) a charge compensation for the transported proton is not possible. In this case the transfer of only a few protons across the membrane results in formation of a membrane potential $\Delta\Psi$ measured as voltage difference across the membrane. By convention, $\Delta\Psi$ is positive when a cation is transferred in the direction of the more positive region. Voltage and free energy are connected by the following equation:

$$\Delta G = m F \Delta\Psi \qquad (4.3)$$

(a) Membrane is permeable to counter ion

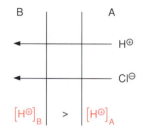

$\Delta\, pH$ $\left[H^{\oplus}\right]_B \;\; > \;\; \left[H^{\oplus}\right]_A$

(b) Membrane is impermeable to counter ion

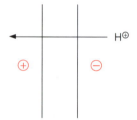

$\Delta\, \Psi$

Figure 4.1 (a) Transport of protons through a membrane, permeable to a counter ion such as chloride, results in the formation of a proton concentration gradient. (b) When the membrane is impermeable to a counter ion, proton transport results in the formation of a membrane potential.

where m is the charge of the ion (in the case of a proton, 1) and F the Faraday constant, $96\,480\ \mathrm{V^{-1}\,J\,mol^{-1}}$.

Proton transport across a biological membrane leads to the formation of a proton gradient and also of a membrane potential. The free energy for the transport of protons from A to B consists therefore of the sum of the free energies for the generation of the H^+ concentration gradient and the membrane potential:

$$\Delta G = RT \ln \frac{[H^+]_B}{[H^+]_A} + F\Delta\Psi \qquad (4.4)$$

In chloroplasts the energy stored in a proton gradient corresponds to the change in free energy during the flux of protons from the lumen into the stroma:

$$\Delta G = RT \ln \frac{[H^+]_S}{[H^+]_L} + F\Delta\Psi \qquad (4.5)$$

where S = stroma, L = lumen, Ψ = voltage difference stroma–lumen.

The conversion of the natural logarithm into the decadic logarithm yields:

$$\Delta G = 2.3\ RT \log \frac{[H^+]_S}{[H^+]_L} + F\Delta\Psi \qquad (4.6)$$

The logarithmic factor is the negative pH difference between lumen and stroma:

$$\log[H^+]_S - \log[H^+]_L = -\Delta pH$$

A rearrangement yields:

$$\Delta G = -2.3RT\Delta pH + F\Delta\Psi \qquad (4.7).$$

At 25°C:

$$2.3RT = 5700 \, \text{J mol}^{-1}$$

Thus:

$$\Delta G = -5700\Delta pH + F\Delta\Psi \qquad [\text{J mol}^{-1}] \qquad (4.8)$$

The expression $\Delta G/F$ is called the *proton motive force* (PMF), with units = volts:

$$\frac{\Delta G}{F} = \text{PMF} = -\frac{2.3RT}{F} \times \Delta pH + \Delta\Psi \qquad [\text{V}]. \qquad (4.9)$$

At 25°C:

$$\frac{2.3RT}{F} = 0.059 \, \text{V}$$

Thus:

$$\text{PMF} = -0.059\Delta pH + \Delta\Psi \qquad [\text{V}]. \qquad (4.10)$$

Equation 4.10 is of general significance for electron transport-coupled ATP synthesis. In mitochondrial oxidative phosphorylation the PMF is primarily the result of a membrane potential. In chloroplasts, on the other hand, the membrane potential does not contribute much to the PMF. Here the PMF is almost entirely due to the concentration gradient of protons across the thylakoid membrane. In illuminated chloroplasts one finds a ΔpH across the thylakoid membrane of about 3. Introducing this value into equation 4.10 yields:

$$\Delta G = -17 \, \text{kJ mol}^{-1}$$

A comparison of this value with ΔG for the formation of ATP (50 kJ mol^{-1}) shows that at least three protons are required for ATP synthesis from ADP and phosphate.

4.2 The electrochemical proton gradient can be dissipated by uncouplers

Photosynthetic electron transport from water to NADPH is coupled with photophosphorylation. Electron transport only occurs if ADP and phosphate are present as precursor substances for ATP synthesis. When an *uncoupler* is added, electron transport proceeds at a high rate in the absence of ADP; electron transport is then uncoupled from ATP synthesis. Therefore in the presence of an uncoupler ATP synthesis is abolished.

The chemiosmotic hypothesis explains the effect of uncouplers (Fig. 4.2).

Uncouplers, soluble in both water and lipids, permeate the lipid phase of a membrane by diffusion and transfer a proton or an alkali ion across the membrane, thus eliminating a proton concentration gradient or a membrane potential, respectively. In the presence of an uncoupler, due to the absence of the proton gradient, protons are transported by ATP synthase from the stroma to the thylakoid lumen at the expense of ATP which is hydrolysed to ADP and phosphate. This is the reason why uncouplers cause ATP hydrolysis (ATPase).

Figure 4.2a shows the effect of the uncoupler carbonylcyanide p-trifluoromethoxyphenylhydrazone (FCCP), a weak acid. FCCP diffuses in the undissociated (protonated) form from the compartment with a high proton concentration (on the left in Fig. 4.2a), through the membrane into the compartment with a low protein concentration, and dissociates there into a proton and the FCCP anion. The proton remains and the FCCP anion returns by diffusion to the other compartment, where it is protonated again. In this way the presence of FCCP at a concentration of only 7×10^{-8} mol l^{-1} results in a complete dissipation of the proton gradient. The substance SF 6847 (3.5-di-(*tert*-butyl)-4-hydroxybenzyldimalononitrile) (Fig. 4.3) has an even higher uncoupler effect. Uncouplers like FCCP or SF 6847, which transfer protons across a membrane, are called *protonophores*.

Besides the protonophores there is a second class of uncouplers, termed *ionophores*, which are able to transfer alkali cations across a membrane and thus dissipate a membrane potential. *Valinomycin*, an antibiotic from *Streptomyces*, is such an ionophore (Fig. 4.2b). Valinomycin is a cyclic molecule containing the sequence (L-lactate)–(L-valine)–(D-hydroxyisovalerate)–(D-valine) three times. Due to its hydrophobic outer surface, valinomycin is able to diffuse through a membrane. Oxygen atoms directed towards the inside of the valinomycin molecule form the binding site for dehydrated Rb$^+$ and K$^+$ ions. Na$^+$ ions are only very loosely bound. When K$^+$ ions are present, the addition of valinomycin results in the elimination of the membrane potential. The ionophore *gramicidin*, not dealt with here in detail, is also a polypeptide. Gramicidin incorporates into the membranes and forms a transmembrane ion channel by which both alkali cations and protons can diffuse through the membrane.

The chemiosmotic hypothesis was proved experimentally

In 1966 the American scientist André Jagendorf presented conclusive evidence for the validity of the chemiosmotic hypothesis in chloroplast photophosphorylation (Fig. 4.4). He incubated thylakoid membranes in an acidic medium of pH 4 in order to acidify the thylakoid lumen by unspecific uptake of protons. He added inorganic phosphate and ADP to the thylakoid suspension and then increased the pH of the medium to pH 8 by adding an alkaline buffer. This led to the sudden generation of a proton gradient of ΔpH = 4 and for a short time ATP was found to be synthesized. Since this experiment was carried out in the dark, it presented

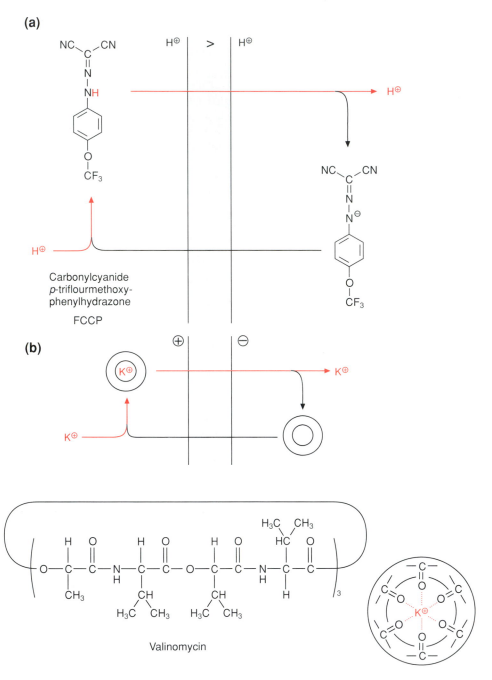

(a)

Carbonylcyanide
p-triflourmethoxy-
phenylhydrazone

FCCP

(b)

Valinomycin

Figure 4.2 The proton motive force of a proton gradient is eliminated by uncouplers. (a) The hydrophobicity of FCCP allows it to diffuse through a membrane in the protonated and also in the deprotonated form. This uncoupler, therefore, can dissipate a proton gradient by indirect proton transport. (b) Valinomycin, an antibiotic with a cyclic structure, folds to form a hydrophobic spherical molecule, which is able to bind K^+ ions in the interior. Loaded with K^+ ions, valinomycin can diffuse through a membrane. In this way valinomycin can eliminate a membrane potential by transferring K^+ ions across a membrane.

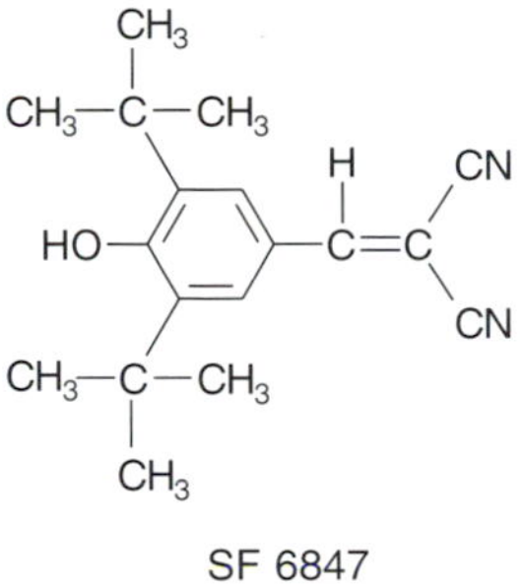

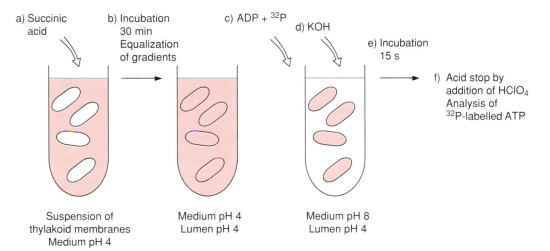

Figure 4.3 Di-(*tert*-butyl)-4-hydroxybenzyldimalononitrile (SF 6847) is an especially effective uncoupler. Only 10^{-9} mol l^{-1} of this substance result in the complete dissipation of a proton gradient across a membrane. This uncoupling is based in the permeation of the protonated and deprotonated molecule through the membrane, as shown in Fig. 4.2a for FCCP.

SF 6847

a) Succinic acid

b) Incubation 30 min Equalization of gradients

c) ADP + ^{32}P

d) KOH

e) Incubation 15 s

f) Acid stop by addition of HClO$_4$ Analysis of ^{32}P-labelled ATP

Suspension of thylakoid membranes Medium pH 4

Medium pH 4 Lumen pH 4

Medium pH 8 Lumen pH 4

Figure 4.4 Thylakoid membranes can synthesize ATP in the dark by an artificially formed proton gradient. In a suspension of thylakoid membranes the pH in the medium is lowered to 4.0 by addition of succinic acid (a). After incubation for about 30 minutes the pH in the thylakoid lumen is equilibrated with the pH of the medium due to a slow permeation of protons across the membrane (b). The next step is to add ADP and phosphate, the latter being radioactively labelled by the isotope ^{32}P (c). Then the pH in the medium is raised to 8.0 by adding KOH (d). In this way a pH gradient of 4.0 is generated between the thylakoid lumen and the medium, and this gradient drives the synthesis of ATP from ADP and phosphate. After a short time of reaction (e) the mixture is denatured by addition of perchloric acid, and the amount of radioactively labelled ATP formed in the deproteinized extract is determined. (After Jagendorf 1966.)

evidence that synthesis of ATP in chloroplasts can be driven without illumination, just by a pH gradient across a thylakoid membrane.

4.3 H$^+$-ATP synthases from bacteria, chloroplasts and mitochondria have a common basic structure

How is the energy of the proton gradient utilized to synthesize ATP? A proton-coupled ATP synthase (H$^+$–ATP synthase) is not unique to the chloroplasts. It was formed during an early stage of evolution and occurs in its basic form in bacteria, chloroplasts, and mitochondria. In bacteria this enzyme catalyses not only ATP synthesis driven by a proton gradient, but also (in a reversal of this reaction) the transport of protons against the concentration gradient at the expense of ATP. This was probably the original function of the enzyme. In some bacteria an ATPase homologous to the H^+-ATP synthase functions as an ATP-dependent Na^+ transporter.

Our present knowledge about the structure and function of the H^+-ATP synthase derives from investigations of mitochondria, chloroplasts, and bacteria. By 1960 progress in electron microscopy led to the detection of small particles which are attached by stalks to the inner membranes of mitochondria and the thylakoid membranes of chloroplasts. These particles occur only at the matrix or stromal side of the corresponding membranes. By adding urea Ephraim Racker and co-workers (Cornell University, USA), succeeded in removing these particles from mitochondrial membranes. The particles thus separated catalysed the hydrolysis of ATP to ADP and phosphate. Racker called them *F_1-ATPase*. Mordechai Avron from Rehovot in Israel showed that such particles from chloroplast membranes also have ATPase activity.

Vesicles containing F_1 particles could be prepared from the inner membrane of mitochondria. These membrane vesicles were able to carry out respiration coupled to ATP synthesis. As in intact mitochondria (section 5.6), the addition of uncouplers resulted in a high ATPase activity. The uncoupler-induced ATPase, as well as ATP synthesis carried out by these vesicles, was found to be inhibited by the antibiotic *oligomycin*. Mitochondrial vesicles where the F_1 particles had been removed, showed no ATPase activity but were highly permeable for protons. This proton permeability was eliminated by adding oligomycin. The ATPase activity of the removed F_1 particles, on the other hand, was not affected by oligomycin. These and other experiments showed that the H^+-ATP synthase of the mitochondria consists of two parts:

(1) a soluble factor 1 (F_1), which catalyses synthesis of ATP; and

(2) a membrane-bound factor enabling the flux of protons through the membrane to which oligomycin is bound.

Racker designated this factor F_o (O, Oligomycin) (Fig. 4.5). Basically the same result has also been found for H^+-ATP synthases of chloroplasts and bacteria, with the exception that the H^+-ATP synthase of chloroplasts is not inhibited by oligomycin. In spite of this, the membrane part of the chloroplastic ATP synthase is also designated as F_o. H^+-ATP synthases of chloroplasts, mitochondria, and bacteria, as well as the corresponding H^+ and Na^+ ATPases of bacteria are collectively termed *F-ATP synthases* or *F-ATPases*. The terms F_oF_1-ATP synthase and F_oF_1-ATPase are also used.

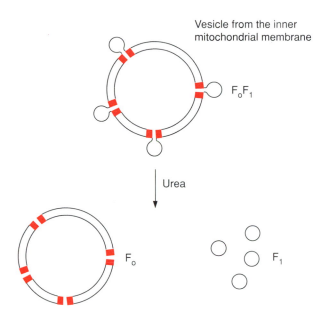

Vesicle from the inner
mitochondrial membrane

F_oF_1

Urea

F_o

F_1

F_o: no ATPase activity,
binds oligomycin

F_1: ATPase activity,
oligomycin insensitive

Figure 4.5 Vesicles prepared by ultrasonic treatment of mitochondria contain functionally intact H$^+$–ATP synthase. The soluble factor F_1 with ATPase function is removed by treatment with urea. The oligomycin binding factor F_o remains in the membrane.

F_1, after removal from the membrane, is a soluble oligomeric protein with the composition $\alpha_3\beta_3\gamma\delta\epsilon$ (Table 4.1). This composition has been found in chloroplasts, bacteria, and mitochondria.

F_o is a strongly hydrophobic protein complex which can only be removed from the membrane by detergents. Dicyclohexylcarbodiimide (DCCD) (Fig. 4.6) binds

Table 4.1 Components of the F-ATP synthase from chloroplasts

Subunits	Number in F_oF_1 molecule	Molecular mass (kDa)	Coded in
F_1:α	3	55	Plastid genome
β	3	54	Plastid genome
γ	1	36	Nucleus
δ	1	21	Nucleus
ϵ	1	15	Plastid genome
F_o: I	1	21	Nucleus
II	1	16	Plastid genome
III	12	8	Plastid genome
IV	1	27	Plastid genome

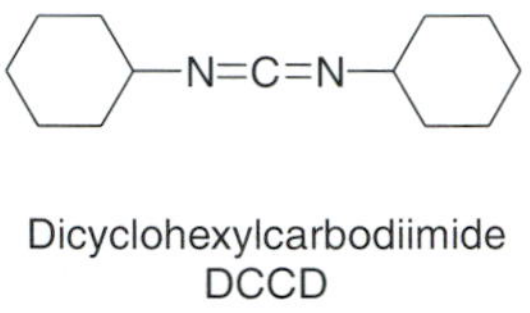

Dicyclohexylcarbodiimide
DCCD

Figure 4.6 An inhibitor of the F_o part of F-ATP synthase.

to the F_o embedded in the membrane, and thus closes the proton channel. In chloroplasts four different subunits have been detected as the main constituents of F_o, called I, II, III, and IV. Subunit III contains two transmembrane helices and a binding site for DCCD. About 10–12 subunits III form a tube (Fig. 4.7). Subunits I and II may be situated within this tube. Subunits γ, ϵ, and δ of F_1 probably form the stalk by which the F_o part is connected with the F_1 part.

Whereas the structure of the F_o part shown in Fig. 4.7 is still somewhat hypothetical, the structure of the F_1 part has been thoroughly investigated. The F_1 particles are so small that details of their structure are not visible on a single electron micrograph. However, details of the structure can be resolved if a very large number of F_1 images obtained by electron microscopy are subjected to *computer aided image analysis*. Figure 4.8 shows an averaged image of an F_1 particle from chloroplasts. In the side projection the stalk connecting the F part with the membrane can be recognized. In the vertical projection a hexagonal array is to be seen, corresponding to an alternating arrangement of α- and β-subunits. Investigations of the isolated F_1 protein showed that an F_oF_1 protein has three catalytic binding sites for ADP and ATP. One of these binding sites is occupied by very tightly

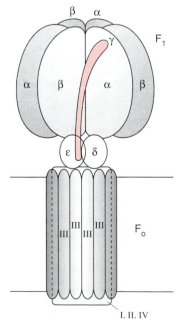

Figure 4.7 Scheme of the structure of an F-ATP synthase. The structure of the F_1 subunit concurs with the results of X-ray analysis dealt with in the text. The localization of subunits δ and ϵ is still uncertain, and the arrangement of subunits I, II, and III is hypothetical. The γ-subunit located in the centre of F_1 is positioned asymmetrically.

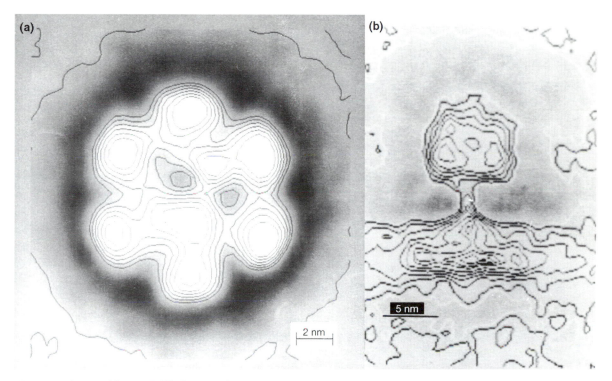

Figure 4.8 Averaged image of 483 electron micrographs of the F-ATP synthase from spinach chloroplasts. (a) Vertical projection of the F_1 part. A hexameric structure reflects the alternating $(\alpha\beta)$-subunits (by P. Graeber, Stuttgart). (b) Side projection, showing the stalk connecting the F_1 part with the membrane.

bound ATP, which is only released when energy is supplied from the proton gradient.

X-ray structure analysis of the F_1 part of ATP synthase yields an insight into the machinery of ATP synthesis

As recently as 1994 the group of John Walker in Cambridge (England) succeeded in analysing the three-dimensional structure of the F_1 part of ATP synthase. Crystals of F_1 from beef heart mitochondria were used for this analysis. Prior to crystallization the F_1 preparation was loaded with a mixture of ADP and an ATP analogue (5′-adenylyl imidodiphosphate, AMP-PNP). This ATP analogue differs from ATP in that the last two phosphate residues are connected by an N atom. It binds to the ATP binding site as ATP, but is not hydrolysed by ATPase. The structural analysis confirmed the alternating arrangement of the α- and β-subunits (Figs 4.7, 4.9). One α- and one β-subunit form a unit with a binding site for one adenine nucleotide. The β-subunit is primarily involved in catalysis of ATP synthesis. In the F_1 crystal investigated, one $(\alpha\beta)$-unit contained one ADP, the

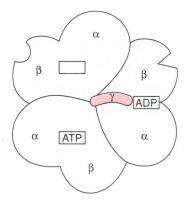

Figure 4.9 Scheme of the vertical projection of the F_1 part of the F-ATP synthase. The enzyme contains three nucleotide binding sites, each consisting of an α- and β-subunit. Each of the three β-subunits occurs in a different conformation. The γ-subunit in the centre, vertical to the viewer, is bent to the α- and β-subunit loaded with ADP. This representation corresponds to the results of X-ray structure analysis by Walker and co-workers mentioned in the text.

second the ATP analogue, whereas the third (αβ)-subunit was unoccupied. These differences in nucleotide binding were accompanied by differences in the conformation of the three β-subunits, shown in Fig. 4.9 in a schematic drawing. The γ-subunit is arranged asymmetrically; it protrudes through the centre of the F_1 part and is bent to the side of the (αβ)-unit loaded with ADP (Figs 4.7, 4.9). This asymmetry gives an insight into the function of the F_1 part of ATP synthase. Some general considerations about ATP synthase shall be made before dealing with this in more detail .

4.4 The synthesis of ATP is effected by a conformational change of the protein

For the reaction:

$$ATP + H_2O \rightarrow ADP + phosphate$$

the standard free energy is:

$$\Delta G^{0\prime} = -30.5 \, \text{kJ mol}^{-1}.$$

Because of its high free energy of hydrolysis, ATP is regarded as an energy-rich compound. It may be noted, however, that the standard value $\Delta G^{0\prime}$ has been determined for an aqueous solution of 1 mol l^{-1} ATP, ADP, and phosphate, corresponding to a water concentration of 55 mol l^{-1}. If the concentration of water were only 10^{-4} mol l^{-1}, the $\Delta G^{0\prime}$ for ATP hydrolysis would have a value of +2.2 kJ mol^{-1}. This means that at very low concentrations of water the reaction proceeds towards the synthesis of ATP. This example demonstrates that in the absence of water the synthesis of ATP does not require uptake of energy.

The catalytic site of an enzyme can form a reaction site where water is excluded. Catalytic sites are often located in a hydrophobic area of the enzyme protein in which the substrates are bound in the absence of water. Thus, with ADP and P tightly bound to the enzyme, synthesis of ATP could proceed spontaneously

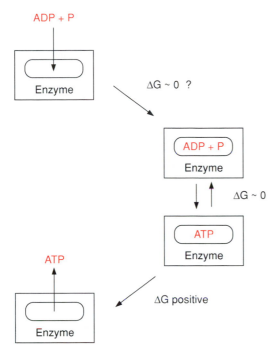

Figure 4.10 In the absence of H_2O, ATP synthesis can occur without the input of energy. In this case the energy required for ATP synthesis in an aqueous solution has to be spent on binding ADP and P and/or on the release of the newly formed ATP. From evidence available the latter case is more likely.

without requiring energy (Fig. 4.10). This has been proved for H^+-ATP synthase. Since the actual ATP synthesis does proceed without the uptake of energy, the amount of energy required to form ATP from ADP and phosphate in the aqueous phase has to be otherwise consumed, e.g. for the removal of the tightly bound newly synthesized ATP from the binding site. This could occur by an energy-dependent conformation change of the protein.

In 1977 Paul Boyer (USA) put forward the hypothesis that the *three catalytic sites of the F_1 protein alternate in their binding properties* (Fig. 4.11). One of the binding sites is present in the L form, which binds ADP and P loosely but is not catalytically active. A second binding site, T, binds ADP and ATP tightly and is catalytically active. The third binding site, O, is open, it binds ADP and ATP only very loosely and is catalytically inactive. According to this '*binding change*' *hypothesis* the synthesis of ATP proceeds in a cycle. First ADP and P are bound to the loose binding site, L. A conformation change of the F_1 protein converts site L into a binding site T, where ATP is synthesized from ADP and phosphate in the absence of water. The ATP formed remains tightly bound. Another conformation change converts the binding site T into an open binding site O, and the newly formed ATP is released. A crucial point of this hypothesis is that with the con-

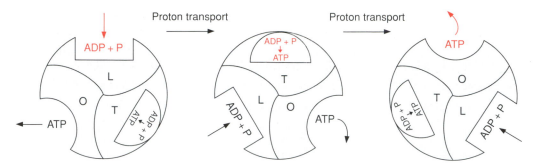

Figure 4.11 ATP synthesis by the binding change mechanism as proposed by Boyer. The central feature of this postulated mechanism is that synthesis of ATP proceeds in the F_1 complex by three nucleotide binding sites, which occur in three different conformations: conformation L binds ADP and P loosely, T binds ADP and P tightly and catalyses ATP formation; the ATP thus formed is tightly bound. The open form O, releases the newly formed ATP. The flux of protons through the F-ATP synthase, as driven by the proton motive force, results in a concerted conformational change of the three binding sites, probably as a rotation.

formational change of the F_1 protein, driven by the energy of the proton gradient, the conformation of each of the three catalytic sites is converted simultaneously into the next conformation (L → T, T → O, O → L).

The results of X-ray analysis, shown above, support the binding change hypothesis. The evaluated structure clearly shows that the three subunits of F_1, one free, one loaded with ADP, and one loaded with the ATP analogue AMP-PNP, have different conformations. A rotational movement could be deduced from the asymmetric arrangement of the γ-subunit, leaving it undecided which component actually rotates. It has been proposed that only the γ-subunit rotates, thereby changing the conformation of the three catalytic centres. The velocity of rotation of the F-ATP synthase in chloroplasts has been estimated to be about 100 revolutions per second.

Although there can no longer be any doubt that ATP synthesis is caused by a change of protein conformation, it is still not known by which mechanism the flux of protons through the F-ATP synthase causes this conformational change. Since there are F-ATPases which transport Na^+ ions instead of protons, it seems unlikely that the change of conformation is due to the protonation or deprotonation of acids or bases in the enzyme. The central question in bioenergetics about the mechanism by which photosynthesis or respiration is coupled to the generation of energy is thus still unsolved.

H^+-ATP synthase of chloroplasts is regulated by light

H^+-ATP synthase catalyses a reaction which is in principle reversible. In chloroplasts a pH gradient across the thylakoid membrane is only generated during illumination. In darkness therefore, due to the reversibility of ATP synthesis, one would expect that the ATP synthase then operates in the opposite direction by transporting protons into the thylakoid lumen at the expense of ATP. In order to avoid such a costly reversal, chloroplastic ATP synthase is subject to strict regulation. This is achieved in two ways. If the pH gradient across the thylakoid membrane decreases below a threshold value, the catalytic sites of the β-subunits are instantaneously switched off and they are switched on again when the pH gradient is restored upon illumination. The mechanism of this is not yet understood. Furthermore, chloroplastic ATP synthase is regulated by thiol modulation. By this process, described in detail in section 6.6, a disulfide bond in the γ-subunit of F_1 is reduced in the light, via reduced thioredoxin, by ferredoxin, to two SH groups. The reduction of the γ-subunit causes the activation of the catalytic centres in the β-subunits. In this way illumination turns F-ATP synthases on. Upon darkening, the two SH groups are reoxidized by oxygen from air to disulfide, and as a result of this the catalytic centres in the β-subunits are switched off. The simultaneous action of the two regulatory mechanisms allows an efficient control of ATP synthase in chloroplasts.

In photosynthetic electron transport the stoichiometry between the formation of NADPH and ATP is still a matter of debate

For a long time experts agreed that the synthesis of one molecule of ATP by F-ATP synthase requires the transport of three protons. Recently, however, experimental data from chloroplasts have been presented which allow the conclusion to be drawn that normally four protons are required for the synthesis of one ATP. It is still a matter of controversy to what extent the Q-cycle contributes to proton transport in chloroplasts. In linear (non-cyclic) electron transport, for each NADH formed without a Q-cycle probably four protons (PS II, $2H^+$; cyt-b_6/f complex, $2H^+$), and with a Q-cycle (cyt-b_6/f complex, $4H^+$) six protons are transported into the lumen (section 3.8). According to the previous notion, without a Q-cycle and an H^+/ATP ratio of 3 in non-cyclic electron transport for each molecule of NADPH, 1.33 molecules of ATP would be formed. However, with a Q-cycle operating and an H^+/ATP ratio of 4, in total 1.5 molecules of ATP would be formed for each molecule of NADPH. Thus the question concerning the stoichiometry of photophosphorylation is still not finally answered.

The V-ATPase is related to the F-ATP synthase

Vacuoles contain a proton-transporting *V-ATPase* (V = vacuole), which resembles the F-ATP synthase in its basic structure. It consists of several proteins embedded in the membrane, similar to the F_o part of the F-ATPase, to which a spherical part (like F_1) is attached by a stalk and protrudes into the cytosol The spherical part consists of 3A- and 3B-subunits, which are arranged alternately like the (αβ)-subunits of F-ATP synthase. F-ATP synthase and V-ATPase are derived from a common ancestor. The V-ATPase pumps two protons per molecule of ATP consumed and is able to generate proton concentrations of up to 1.4 mol l^{-1} within the vacuoles (section 8.5). Vacuolar membranes also contain an *H$^+$ pyrophosphatase* which, upon the hydrolysis of one molecule of pyrophosphate to phosphate, pumps one proton into the vacuole, but it does not reach such high proton gradients as the V-ATPase. H$^+$ pyrophosphatase probably only consists of a single protein with 13 transmembrane helices. The division of labour between H$^+$ pyrophosphatase and V-ATPase in transporting protons into the vacuole has not yet been resolved. Plasma membranes contain a proton transporting *P-ATPase* which will be dealt with in section 8.2.

Further reading

Abrahams, J. P., Leslie, A. G. W., Lutter, R., and Walker, J. E. (1994). Structure at 2.8 Å resolution of F$_1$-ATPase from bovine heart mitochondria. *Nature*, **370**, 621–8.

Boekema, E. J., van Heel, M., and Gräber, P. (1988). Structure of the ATP synthase from chloroplasts studied by electron microscopy and image processing. *Biochimica et Biophysica Acta*, **933**, 365–71.

Boyer, P. D. (1993). The binding change mechanism for ATP synthase—some probabilities and possibilities. *Biochimica et Biophysica Acta*, **1149**, 215–50.

Cramer, W. A. and Knaff, D. B. (1991). *Energy transduction in biological membranes*. Springer Verlag, New York.

Cross, R. (1992). The reaction mechanism of F$_o$F$_1$ synthases. In *Molecular mechanics in bioenergetics*, (ed. L. Ernster), pp. 317–30. Elsevier, Amsterdam.

Kobayashi, Y., Kaiser, W., and Heber, U. (1995). Bioenergetics of carbon assimilation in intact chloroplasts: Coupling of proton to electron transport at the ratio H$^+$/e = 3 is incompatible with H$^+$/ATP = 3 in ATP-synthesis. *Plant Cell Physiology*, **36**, 1629–37.

Leigh, R. A., Gordon-Weeks. R., Steele. S. H., and Korenkov, V. D. (1994). The H$^+$-pumping inorganic pyrophosphatase of the vacuolar membrane of higher plants. In *Membranes for experimental biology*, (ed. M. R. Blatt, R. A. Leigh, and D. Sanders), pp. 61–74. Cambridge University Press, Cambridge.

Lüttge, U. and Ratajczak, R. The physiology, biochemistry and molecular biology of the plant vacuolar ATP-ase. In *The plant vacuole*, (ed. R. A. Leigh and D. Sanders). Academic Press, in press.

Ort, D. R. and Oxborough, K. (1992). *In situ* regulation of the chloroplast coupling factor activity. *Annual Review of Plant Physiology and Plant Molecular Biology*, **43**, 269–91.

chapter 5

Mitochondria, the power stations of the cell

In the process of biological oxidation, substrates such as carbohydrates are oxidized to form water and CO_2. Biological oxidation can be seen as a reversal of the photosynthesis process. It evolved only after the oxygen in the atmosphere had been accumulated by photosynthesis. Both biological oxidation and photosynthesis serve the purpose of generating energy in the form of ATP. Biological oxidation involves a transport of electrons through a mitochondrial electron transport chain, which is, in part, similar to the photosynthetic electron transport dealt with in Chapter 3. This chapter will show that the machinery of mitochondrial electron transport is also constructed from modules. Of its three complexes, the middle one has the same basic structure as the cytochrome-b_6/f complex of the chloroplasts. Just as in photosynthesis, in mitochondrial oxidation electron transport and ATP synthesis are coupled to each other via the formation of a proton gradient. The synthesis of ATP proceeds by an F-ATP synthase, which has been described in Chapter 4.

5.1 Biological oxidation is preceded by substrate degradation

The total reaction of biological oxidation is equivalent to combustion. In contrast to combustion, however, biological oxidation proceeds in a sequence of partial reactions which allow the major part of the free energy to be utilized for synthesis of ATP.

   The principle of biological oxidation was formulated in 1932 by the Nobel prize winner Heinrich Wieland:

$$XH_2 + \tfrac{1}{2}O_2 \;\rightarrow\; X + H_2O.$$

First of all, hydrogen is removed from substrate XH_2 and afterwards oxidized to water. Thus during oxidation, carbohydrates $[CH_2O]_n$ are first degraded by reaction with water to form CO_2 and bound hydrogen $[H]$, and the latter is then oxidized to water:

$$[CH_2O] + H_2O \;\rightarrow\; CO_2 + 4[H]$$
$$4[H] + O_2 \;\rightarrow\; 2H_2O.$$

In 1934 Otto Warburg showed that the transfer of hydrogen from substrates to the site of oxidation occurs as bound hydrogen in the form of *NADH*. From results of studies with homogenates from pigeon muscle in 1937 Hans Krebs formulated the *citrate cycle* (also called Krebs cycle) as a mechanism for substrate degradation, yielding the reductive equivalents for the reduction of oxygen via biological oxidation. In 1953 he was awarded the Nobel prize for medicine for this discovery. The operation of the citrate cycle will be dealt with in detail in section 5.3.

5.2 Mitochondria are the sites of cell respiration

Microscopic studies of many different cells showed that they contain small granules similar in appearance to bacteria. At the beginning of this century the botanist C. Benda named these granules *mitochondria*, which means thread-like bodies. For a long time, however, the function of these mitochondria remained unclear.

   As early as 1913 Otto Warburg (Nobel prize winner for medicine, 1931) realized that cell respiration involves the function of granular cell constituents. He succeeded in isolating a protein from yeast which he termed 'Atmungsferment' (respiratory ferment), which catalyses the oxidation by oxygen. He also showed that iron atoms are involved in this catalysis. In 1925 David Keilin from Cambridge (England) discovered the cytochromes and their participation in cell respiration. Using a manual spectroscope he identified the cytochromes-a, -a_3, -b, and -c (Fig. 3.24). In 1928 Otto Warburg showed that his 'Atmungsferment' contained cytochrome-a_3. A further milestone in the clarification of cell respiration was reached in 1937 when Hermann Kalckar observed that the formation of ATP in aerobic systems depends on the consumption of oxygen. The interplay between cell respiration and ATP synthesis, named oxidative phosphorylation, was now apparent. In 1948 Eugene Kennedy and Albert Lehninger showed that mitochondria contain the enzymes of the citrate cycle and of *oxidative phosphorylation*. These findings demonstrated the function of the mitochondria as the *power stations of the cell* .

Mitochondria represent a separate metabolic compartment

Like plastids, mitochondria also form a separate metabolic compartment. The structure of the mitochondria is dealt with in section 1.4. Figure 5.1 provides an overview of mitochondrial metabolism. The degradation of substrates to CO_2 and hydrogen (the latter bound to the transport metabolite NADH) takes place in

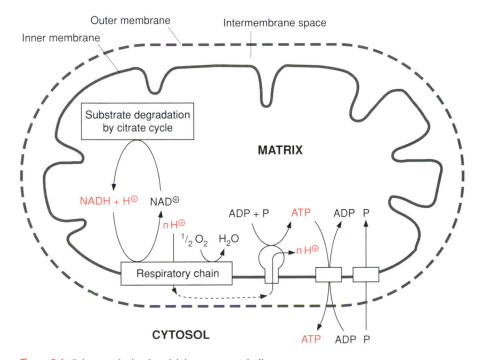

Figure 5.1 Scheme of mitochondrial energy metabolism.

the mitochondrial matrix. NADH thus formed diffuses through the matrix to the mitochondrial inner membrane and is oxidized there by the *respiratory chain*. The respiratory chain consists of a sequence of redox reactions by which electrons are transferred from NADH to oxygen. As in photosynthetic electron transport, in mitochondrial electron transport the released energy is used to generate a proton gradient, which in turn drives the synthesis of ATP. ATP thus formed is exported from the mitochondria and provides the energy required for cell metabolism. This is the universal function of mitochondria in all eukaryotic cells.

5.3 Degradation of substrates for biological oxidation takes place in the matrix compartment

Pyruvate, formed by glycolytic catabolism of glucose in the cytosol, is the starting compound for substrate degradation by the citrate cycle (Fig. 5.2). Pyruvate is first oxidized to acetate (in the form of acetyl coenzyme A), which is then degraded to CO_2 by the citrate cycle, yielding 10 reducing equivalents [H] to be oxidized by the respiratory chain to generate ATP. Figure 5.3 shows the reactions of the citrate cycle.

Pyruvate is oxidized by a multienzyme complex

Pyruvate oxidation is catalysed by the *pyruvate dehydrogenase complex*, a multi-enzyme complex located in the mitochondrial matrix. It consists of three different catalytic subunits, *pyruvate dehydrogenase*, *dihydrolipoyl transacetylase*, and *dihydrolipoyl dehydrogenase* (Fig. 5.4). The pyruvate dehydrogenase subunit contains *thiamine pyrophosphate* (TPP, Fig. 5.5a) as prosthetic group. The reactive group of TPP is the thiazole ring. Due to the presence of a positively charged N atom the thiazole ring contains an acidic C atom. After dissociation of a proton a carbanion is formed which is able to bind to the carbonyl group of the pyruvate. The positively charged N atom of the thiazole ring enhances the decarboxylation of the bound pyruvate to form hydroxethyl TPP (Fig. 5.4). The hydroxethyl group is then transferred to lipoic acid.

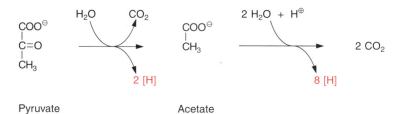

Pyruvate   Acetate

Figure 5.2 Overall reaction of the oxidation of pyruvate by mitochondria. The acetate is formed as acetyl coenzyme A. [H] means bound hydrogen in the form of NADH and $FADH_2$, respectively.

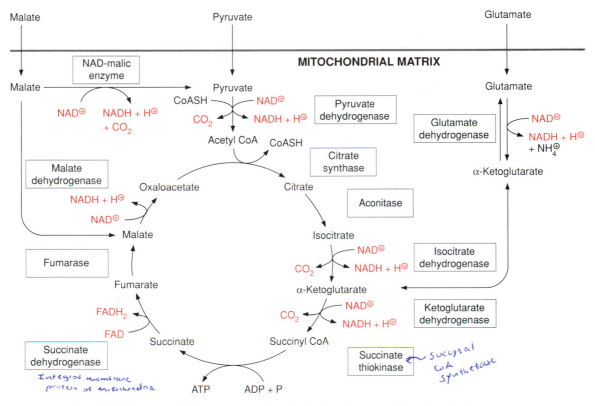

Figure 5.3 Scheme of the citrate cycle. The enzymes are localized in the mitochondrial matrix, with the exception of succinate dehydrogenase which is located in the inner mitochondrial membrane. As a special feature, plant mitochondria contain NAD-malic enzyme in the mitochondrial matrix. Therefore plant mitochondria are able to oxidize malate via the citrate cycle in the absence of pyruvate. Glutamate dehydrogenase enables mitochondria to oxidize glutamate.

   Lipoic acid is the prosthetic group of the dihydrolipoyl transacetylase subunit. It is covalently bound by its carboxyl group to a lysine residue of the enzyme protein via an amide bond (Fig. 5.5b). The lipoic acid residue is attached to the protein virtually by a long chain and so is able to react with the various reaction sites of the multienzyme complex. Lipoic acid contains two S atoms linked by a disulfide bond. When the hydroxyethyl residue is transferred to the lipoic acid residue, lipoic acid is reduced to dihydrolipoic acid and the hydroxyethyl residue is oxidized to an acetyl residue. The latter is attached to the dihydrolipoic acid by a thioester bond. This is how energy released during oxidation of the carbonyl group is conserved to generate an energy-rich thioester. The acetyl residue is now transferred by dihydrolipoyl transacetylase to the sulfhydryl group of coenzyme A (Fig. 5.5c) to form acetyl coenzyme A. *Acetyl coenzyme A (Acetyl CoA)*—also called active acetic acid—was discovered by Feodor Lynen from Munich (Nobel

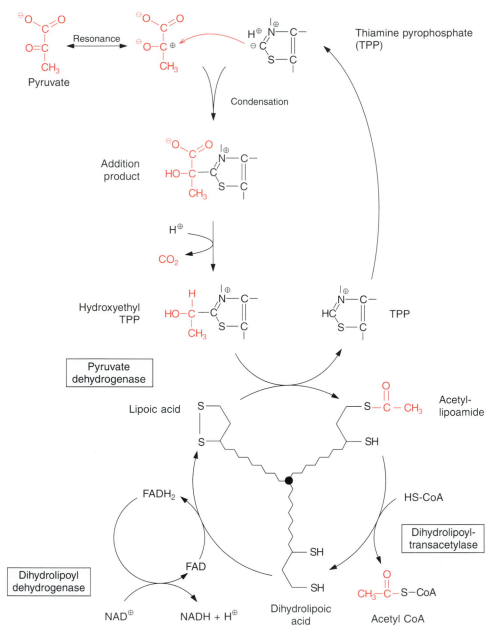

Figure 5.4 Oxidation of pyruvate by the pyruvate dehydrogenase complex, consisting of the subunits pyruvate dehydrogenase (with the prosthetic group thiamine pyrophosphate), dihydrolipoyl transacetylase (prosthetic group lipoic acid) and dihydrolipoyl dehydrogenase (prosthetic group FAD). The reactions of the cycle are described in the text.

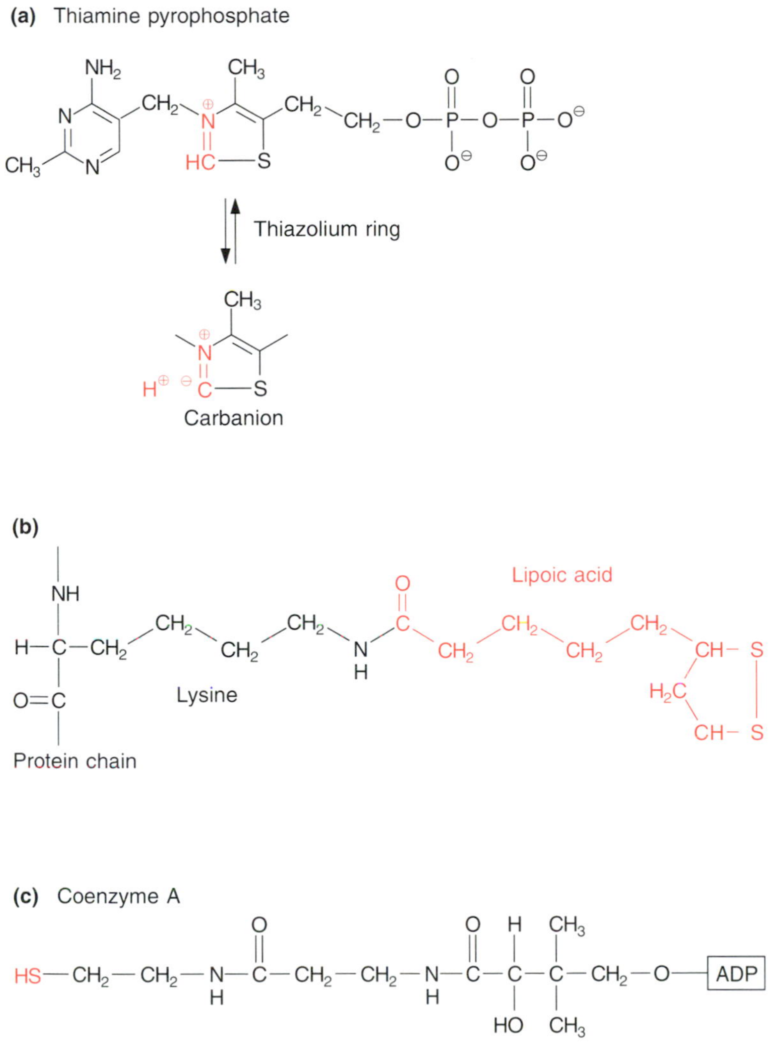

(a) Thiamine pyrophosphate

Thiazolium ring

Carbanion

(b)

Lysine

Lipoic acid

Protein chain

(c) Coenzyme A

Figure 5.5 Reaction partners of pyruvate oxidation: (a) thiamine pyrophosphate, (b) lipoic amide, and (c) coenzyme A.

prize for medicine, 1964). Dihydrolipoic acid is reoxidized to lipoic acid by dihydro-lipoyl dehydrogenase and NAD^+ is reduced to NADH via *FAD* (see Fig. 5.16). It should be noted that a pyruvate dehydrogenase complex is also found in the chloroplasts, but its function in lipid biosynthesis will be discussed in section 15.3.

Acetate is oxidized in the citrate cycle

Acetyl coenzyme A then enters the citrate cycle and condenses with oxaloacetate to form citrate (Fig. 5.6). This reaction is catalysed by the enzyme *citrate synthase*.

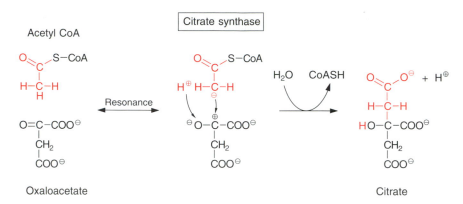

Figure 5.6 Condensation of acetyl CoA with oxaloacetate to form citrate.

The thioester group promotes the removal of a proton of the acetyl residue and the carbanion thus formed binds to the carbonyl carbon of oxaloacetate. Subsequent release of CoA-SH makes the reaction irreversible. The enzyme *aconitase* (Fig. 5.7) catalyses the reversible isomerization of citrate to isocitrate. In this reaction first water is released, the *cis*-aconitate thus formed remains bound to the enzyme and is converted to isocitrate by addition of water. Besides the mitochondrial aconitase, there is also an isoenzyme of aconitase in the cytosol of plant cells.

Oxidation of isocitrate to α-ketoglutarate by *NAD isocitrate dehydrogenase* (Fig. 5.8) results in the formation of NADH. Oxalosuccinate is formed as an intermediate, which, still tightly bound to the enzyme, is decarboxylated to α-ketoglutarate. This decarboxylation makes the oxidation of isocitrate irreversible. Besides NAD isocitrate dehydrogenase, some mitochondria also contain an NADP-dependent enzyme. NADP isocitrate dehydrogenases also occur in the chloroplast stroma and in the cytosol. The function of the latter enzyme will be dealt with in section 10.4.

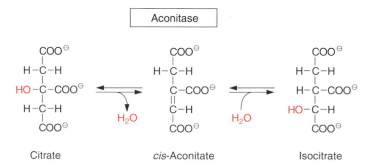

Figure 5.7 Isomerization of citrate to form isocitrate.

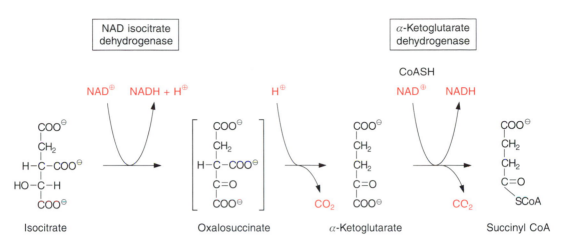

Figure 5.8 Oxidation of isocitrate to form succinyl CoA.

Oxidation of α-ketoglutarate to succinyl CoA (Fig. 5.8) is catalysed by the α-*ketoglutarate dehydrogenase multienzyme complex*, also involving thiamine pyrophosphate, lipoic acid, and FAD, analogously to the oxidation of pyruvate to acetyl CoA described above.

The thioester bond of the succinyl CoA is rich in energy. In the *succinate thiokinase* reaction the hydrolysis of this thioester is coupled to the formation of ATP (Fig. 5.9). The succinate formed is oxidized by *succinate dehydrogenase* to form fumarate. Succinate dehydrogenase is the only enzyme of the citrate cycle which is located not in the matrix but in the mitochondrial inner membrane, with its succinate binding site accessible from the matrix (section 5.5). Reducing equivalents derived from succinate oxidation are transferred to ubiquinone. Catalysed by *fumarase*, water reacts by *trans*-addition with the C–C double bond of fumarate to form L-malate. This is a reversible reaction (Fig. 5.9). Oxidation of malate by *malate dehydrogenase*, yielding oxaloacetate and NADH, is the final step in the citrate cycle (Fig. 5.9). The reaction equilibrium of this reversible reaction lies strongly towards the product malate:

$$\frac{[NADH][oxaloacetate]}{[NAD^+][malate]} = 2.8 \times 10^{-5} \, (pH \, 7).$$

Because of this equilibrium it is essential for the operation of the citrate cycle that the citrate synthase reaction is irreversible. In this way oxaloacetate can be withdrawn from the malate dehydrogenase equilibrium to react further in the cycle. Isoenzymes of malate dehydrogenase also occur outside the mitochondria. Both the cytosol and the peroxisomal matrix contain NAD–malate dehydrogenases, and the chloroplast stroma contains NADP–malate dehydrogenase. These enzymes will be dealt with in Chapter 7.

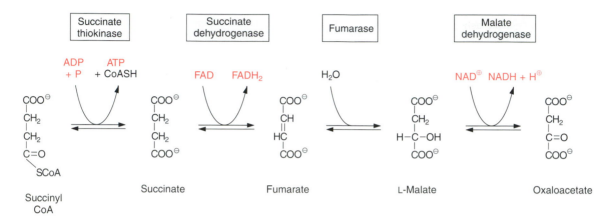

Figure 5.9 Conversion of succinyl CoA to form oxaloacetate.

A loss of intermediates of the citrate cycle is replenished by anaplerotic reactions

The citrate cycle can only proceed when the oxaloacetate required as acceptor for the acetyl residue is fully regenerated. Section 10.4 describes how citrate and α-ketoglutarate are withdrawn from the citrate cycle to synthesize the carbon skeletons of amino acids in the course of nitrate assimilation. It is necessary, there-fore, to replenish the loss of citrate cycle intermediates. In contrast to mitochondria from animal tissues, plant mitochondria are able to transport oxaloacetate into the chloroplasts via a specific translocator of the inner membrane (section 5.8). Therefore the citrate cycle can be replenished by the uptake of oxaloacetate which has been formed by phosphoenolpyruvate carboxylase in the cytosol (section 8.2). Oxaloacetate can also be delivered by oxidation of malate in the mitochondria. Malate is stored in the vacuole (sections 1.2, 8.2, 8.5) and is an important substrate for mitochondrial respiration. A special feature of plant mito-chondria is that malate is oxidized to pyruvate with the reduction of NAD and the release of CO_2 via *NAD–malic enzyme* in the matrix (Fig. 5.10). Thus an interplay

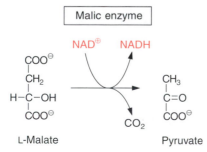

Figure 5.10 Oxidative decarboxylation of malate to form pyruvate.

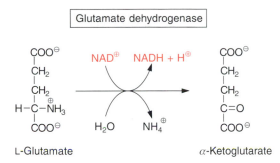

Figure 5.11 Oxidation of glutamate.

of malate dehydrogenase and NAD–malic enzyme allows citrate to be formed from malate without putting the complete citrate cycle into operation (Fig. 5.3). It may be noted that NADP-dependent malic enzyme is present in the chloroplasts, especially in C_4 plants (section 8.4).

Another important substrate of mitochondrial oxidation is glutamate, which is one of the main products of nitrate assimilation (section 10.1) and, beside sucrose, is the most highly concentrated organic compound in the cytosol of many plant cells. Glutamate oxidation, accompanied by formation of NADH, is catalysed by *glutamate dehydrogenase* located in the mitochondrial matrix (Fig. 5.11). This enzyme also reacts with NADP. NADP–glutamate dehydrogenase activity is also found in plastids, although its function there is not clear.

Glycine is the main substrate of respiration in the mitochondria from mesophyll cells of illuminated leaves. The oxidation of glycine as a partial reaction of the photorespiratory pathway will be dealt with in section 7.1.

5.4 How much energy can be gained by the oxidation of NADH?

Let us first ask ourselves how much energy is released during mitochondrial respiration, or to be more exact, how large is the difference in free energy in the mitochondrial redox processes? To answer this question one needs to know the differences of the potentials of the redox pairs involved. This potential difference can be calculated by the Nernst equation:

$$E = E^{0\prime} + \frac{RT}{nF} \ln \left[\frac{\text{oxidized substance}}{\text{reduced substance}} \right] \qquad (5.1)$$

where $E^{0\prime}$ is the standard potential at pH 7, 25 °C; R (gas constant) = 8.31 J K^{-1} mol^{-1}; T = 298 K; n is the number of electrons transferred; F (Faraday constant) = 96 480 J V^{-1} s mol^{-1}.

The standard potential for the redox pair NAD/NADH is:

$$E^{0\prime} = -0.320 \text{ V}.$$

Under certain metabolic conditions a NAD$^+$/NADH ratio of 3 was found in mitochondria from leaves. The introduction of this value into equation 5.1 yields:

$$E_{NAD^+/NADH} = -0.320 + \frac{RT}{2F}\ln 3 = -0.306\,\text{V}. \qquad (5.2)$$

The standard potential for the redox pair H_2O/O_2 is:

$$E^{0\prime} = +0.815\,\text{V}. \qquad ([H_2O]\text{ in water }55\,\text{mol}\,l^{-1})$$

The partial pressure of the oxygen in the air is introduced for the evaluation of the actual potential for $[O_2]$:

$$E_{H_2O/O_2} = 0.815 + \frac{RT}{2F}\ln\sqrt{pO_2}. \qquad (5.3)$$

The partial pressure of the oxygen in the air (pO_2) amounts to 0.2. Introducing this value into equation 5.3 yields:

$$E_{H_2O/O_2} = 0.805\,\text{V}.$$

The difference of the potentials amounts to:

$$\Delta E = E_{H_2O/O_2} - E_{NAD^+/NADH} = +1.13\,\text{V} \qquad (5.4)$$

The free energy (ΔG) is related to ΔE as follows:

$$\Delta G = -nF\Delta E. \qquad (5.5)$$

Two electrons are transferred in the reaction. The introduction of ΔE into equation 5.5 shows that the change of free energy during the oxidation of NADH by the respiratory chain amounts to:

$$\Delta G = -217\,\text{kJ mol}^{-1}.$$

How much energy is required for the formation of ATP? It has been calculated in section 4.1 that the synthesis of ATP under the metabolic conditions in the chloroplasts requires a change of free energy of $\Delta G \approx +50\,\text{kJ mol}^{-1}$. This value also applies approximately to the ATP provided by the mitochondria for the cytosol.

The free energy released with the oxidation of NADH would therefore be sufficient to generate four molecules of ATP, but in fact the amount of ATP formed by NADH oxidation is much lower (section 5.6).

5.5 The mitochondrial respiratory chain shares common features with the photosynthetic electron transport chain

The photosynthesis of cyanobacteria led to the accumulation of oxygen in the early atmosphere and thus formed the basis for the oxidative metabolism of mitochondria. Many cyanobacteria can satisfy their ATP demand both by photosynthesis and by oxidative metabolism. Cyanobacteria contain a photosynthetic electron transport chain which consists of three modules (complexes), namely photosystem II, the cyt-b_6/f complex, and photosystem I (Chapter 3, Fig. 5.12). These complexes are located in the inner membrane of cyanobacteria, where, however, there are also the enzymes of the respiratory electron transport chain. This respiratory chain consists once more of three modules, an NADH dehydrogenase complex, catalysing the oxidation of NADH, the same cyt-b_6/f complex which is also part of the photosynthetic electron transport chain and a cyt-a/a_3 complex, by which oxygen is reduced to water. Plastoquinone feeds the electrons into the cyt-b_6/f complex not only in photosynthesis (section 3.7) but also in the respiratory chain of the cyanobacteria. Likewise cytochrome-c mediates the electron transport from the cyt-b_6/f complex to photosystem II as well as to the cyt-a/a_3 complex. The relationship between photosynthetic and oxidative electron transport in cyanobacteria is obvious; both electron transport chains possess the same module as their middle part, the cyt-b_6/f complex. Section 3.7 described how in the cyt-b_6/f complex the energy released by electron transport is used to form a proton gradient. The function of the cyt-b_6/f complex in respiration and photosynthesis shows that the basic principle of energy conservation in photosynthetic and oxidative electron transport is the same.

The mitochondrial respiratory chain is analogous to the respiratory chain of cyanobacteria (Fig. 5.13), but with ubiquinone instead of plastoquinone as redox carrier and slightly different cytochromes. The mitochondria contain a cyt-b/c_1 complex instead of a cyt-b_6/f complex. Cyt-c and cyt-f both contain haem-c.

Figure 5.13 shows succinate dehydrogenase as an example of another electron

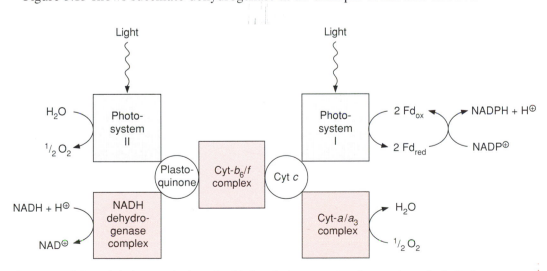

Figure 5.12 Scheme of photosynthetic and oxidative electron transport in cyanobacteria. In both electron transport chains the cytochrome-b_6/f complex functions as the central complex.

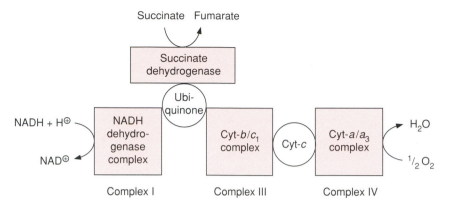

Figure 5.13 Scheme of mitochondrial electron transport. The respiratory chain consists of three complexes, the central cyt-b/c_1 complex corresponds to the cyt-b_6/f complex of cyanobacteria and chloroplasts.

acceptor of the mitochondrial respiratory chain. This enzyme (historically termed complex II) catalyses the oxidation of succinate to fumarate, a partial reaction of the citrate cycle (Fig. 5.9).

The complexes of the mitochondrial respiratory chain

The subdivision of the respiratory chain into several complexes goes back to the work of Youssef Hatefi, who in 1962, when working with beef heart mitochondria, succeeded in isolating four different complexes which he termed complexes I–IV. In complexes I, III, and IV electron transport is accompanied by a decrease in the redox potential (Fig. 5.14); the energy thus released is used to form a proton gradient.

The *NADH dehydrogenase complex* (complex I) (Fig. 5.15) feeds the respiratory chain with the electrons from NADH, formed from degradation of substrates in the matrix. The electrons are transferred to ubiquinone via a flavin adenine mononucleotide and several iron–sulfur centres. Complex I has the most complicated structure of all the mitochondrial electron transport complexes. Made up of more than 40 different subunits (of which eight are encoded in the mitochondria), it consists of one part which is embedded in the membrane (*membrane part*) and a *peripheral part* which protrudes into the matrix space. The peripheral part contains the binding site for NADH, a bound flavin mononucleotide (FMN) (Fig. 5.16) and at least three Fe–S centres (Fig. 3.26). The membrane part contains a further Fe–S centre and the binding site for ubiquinone. Despite intensive research, the exact path of the electron transport through complex I is not yet known. The electron transport is inhibited by a variety of poisons deriving from plants and bacteria, such as *rotenone* (which protects plants from being eaten by animals), the antibiotic *piericidin A*, and *amytal*, a barbiturate. The electron trans-

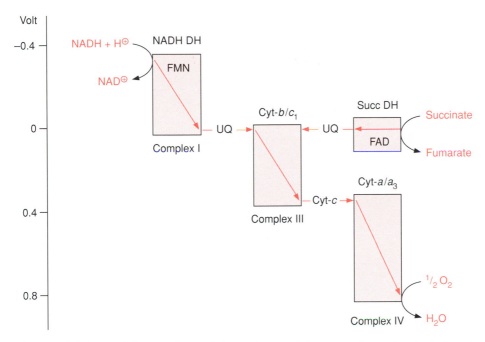

Figure 5.14 Scheme of the complexes of the respiratory chain arranged according to their redox potentials. DH, dehydrogenase; UQ, ubiquinone.

port catalysed by complex I is reversible. It is therefore possible for electrons to be transferred from ubiquinone to NAD, driven by the proton motive force of the proton gradient. In this way purple bacteria are provided with NADH by means of a homologous NADH dehydrogenase complex (see Fig. 3.1).

Succinate dehydrogenase (Fig. 5.9) contains a flavin adenine nucleotide (FAD, Fig. 5.16) as electron acceptor, several Fe–S centres (Fig. 3.26) as redox carriers and also one cytochrome-b, of which the function is not known. Electron transport by succinate dehydrogenase to ubiquinone proceeds with no major decrease in the redox potential, therefore no energy is gained in the electron transport from succinate to ubiquinone.

Ubiquinone reduced by the NADH dehydrogenase complex or succinate dehydrogenase, is oxidized by the *cyt-b/c_1 complex* (complex III) (Fig. 5.15). In mitochondria this complex consists of 9–11 subunits, of which just one (the cyt-b subunit) is encoded in the mitochondria. The cyt-b/c_1 complex is very similar in structure and function to the cyt-b_6/f complex of chloroplasts (section 3.7). Electrons are transferred by the cyt-b/c_1 complex to cyt-c which is bound to the outer surface of the inner membrane. Several antibiotics such as *antimycin A* and *myxothiazol* inhibit electron transport by the cyt-b/c_1 complex.

Due to its positive charge, reduced cyt-c diffuses along the negatively charged surface of the inner membrane to the *cyt-a/a_3 complex* (Fig. 5.15), also termed

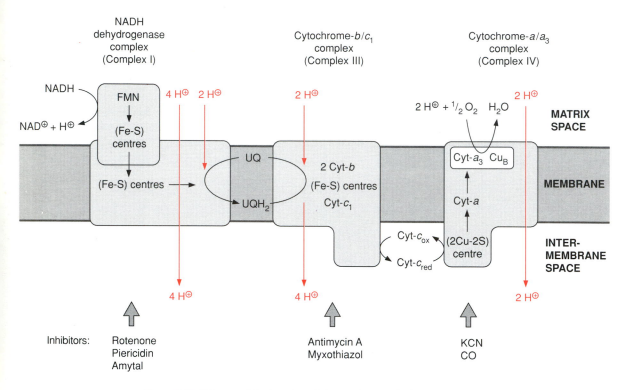

Figure 5.15 Scheme of the location of the respiratory chain complexes I, III, and IV in the mitochondrial inner membrane.

complex IV or cytochrome oxidase. The cyt-a/a_3 complex contains more than 10 different subunits, of which three are encoded in the mitochondria. Recently the three-dimensional structure of the cyt-a/a_3 complex has been resolved by X-ray structure analysis of these complexes from beef heart mitochondria and also from *Paracoccus denitrificans*. The complex has a large hydrophilic region, which protrudes into the intermembrane space and contains the binding site for cyt-c. In the oxidation of cyt-c the electrons are transferred to a *copper–sulfur cluster* containing two Cu atoms, called Cu_A. These two Cu atoms are linked by two S atoms of cysteine side chains (Fig. 5.17). This copper–sulfur cluster probably takes up one electron and transfers it via cyt-a to a *binuclear centre*, consisting of cyt-a_3 and a Cu atom (Cu_B), bound to histidine. This binuclear centre functions as a redox unit in which the Fe atom of the cyt-a_3, together with Cu_B, takes up two electrons:

$$[Fe^{3+}Cu^{2+}] + 2e^- \rightarrow [Fe^{2+}Cu_B^+].$$

In contrast to cyt-a and the other cytochromes of the respiratory chain, in cyt-a_3 the sixth coordination position of the Fe atom is not saturated by an amino acid of the protein (Fig. 5.18). This free coordination position, and also Cu_B, form the

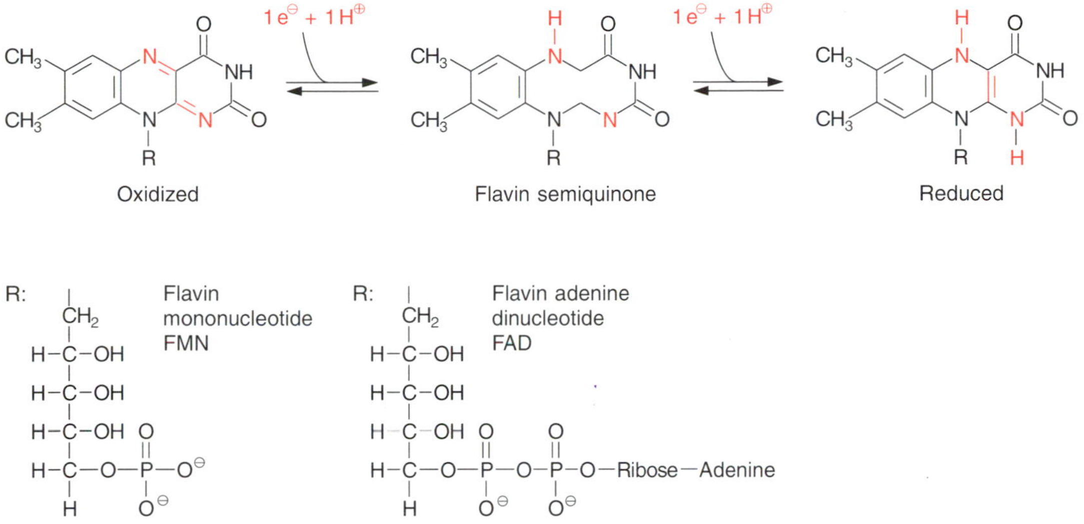

Figure 5.16 FMN, FAD, reduced and oxidized forms.

binding site for the oxygen molecule, which is reduced to water by the uptake of four electrons:

$$O_2 + 4e^- \rightarrow 2O^{2-} + 4H^+ \rightarrow 2H_2O.$$

Cu_B probably has an important function in electron-driven proton transport dealt with in the next section. Instead of O_2 also CO and CN^- can be very tightly bound to the free coordination position of the cyt-a_3, resulting in the inhibition of respiration. Therefore both carbon monoxide and prussic acid (HCN) are very potent poisons.

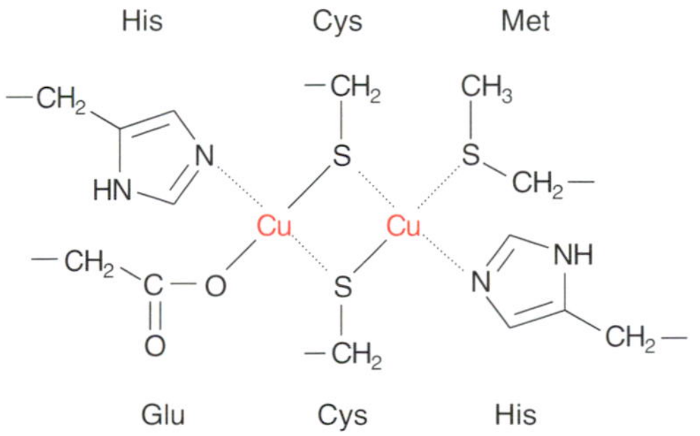

Figure 5.17 In a copper–sulfur cluster of the cytochrome-a/a_3 complex, termed Cu_A, a Cu^{2+} and a Cu^+ ion are linked by two cysteine residues of the protein and bound further to the protein by two histidine, one glutamate, and one methionine residue. Cu_A probably transfers one electron. The structure of this novel redox cluster was revealed by X-ray structural analysis of the cytochrome-a/a_3 complex carried out simultaneously in Frankfurt, Germany, (Iwata *et al.* 1995) and in Osaka, Japan (Tsukihara *et al.*

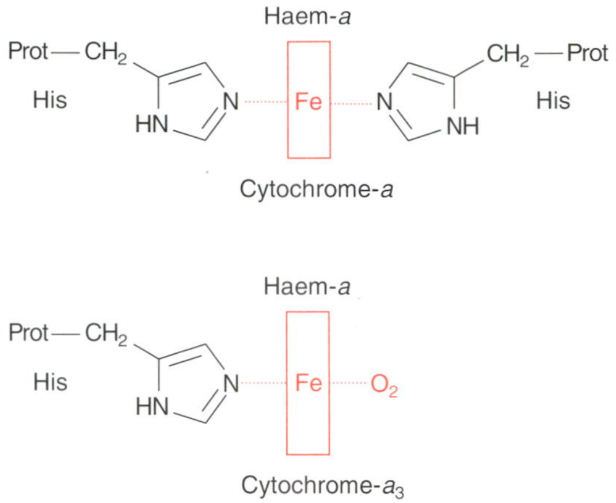

Figure 5.18 Axial ligands of the Fe atoms in the haem groups of cytochrome-*a* and -*a*$_3$. Of the six co-ordinate bonds formed by the Fe atom present in the haem, four are saturated by the N atoms present in the planar tetrapyrrole ring. Whereas in cytochrome-*a* the two remaining coordination positions of the central Fe atom bind to histidine residues of the protein, positioned at either side vertically to the plane of the tetrapyrrole, in cytochrome-*a*$_3$ one of these coordination positions is free and functions as binding site for the O$_2$ molecule.

5.6 Electron transport of the respiratory chain is coupled via proton transport to the synthesis of ATP

The electron transport of the respiratory chain is coupled to the formation of ATP. This is illustrated in the experiment of Fig. 5.19 in which the velocity of respiration in the mitochondrial suspension was determined by measuring the decrease in the oxygen concentration in the suspension medium. The addition of a substrate alone (e.g. malate) causes only a minor increase in respiration. The subsequent addition of a limited amount of ADP results in a considerable acceleration of respiration. After some time, however, respiration returns to the lower rate found prior to the addition of ADP, as the ADP has been converted to ATP. Respiration in the presence of ADP is called *active respiration*, while that after ADP is con-sumed, is called *controlled respiration*. As the ADP added to the mitochondria is completely converted into ATP, the amount of ATP formed with the oxidation of a certain substrate can be determined from the ratio of ADP added to oxygen consumed (ADP/O). An ADP/O of about 2.5 is determined for substrates oxidized in the mitochondria via formation of NADH (e.g. malate), and a ratio of about 1.6 for succinate, from which the redox equivalents are directly transferred to ubiquinone. The problem of ATP stoichiometry of respiration will be discussed at the end of this section.

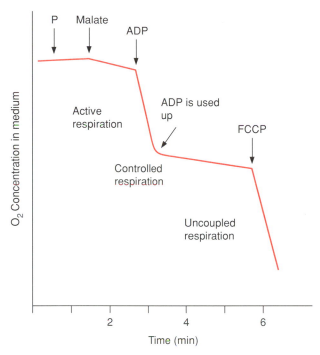

Figure 5.19 Measurement of oxygen consumption by isolated mitochondria. Phosphate and malate are added one after the other to mitochondria suspended in a buffered medium. Addition of ADP results in a high rate of respiration. The subsequent decrease of oxygen consumption indicates that the conversion of the added ADP into ATP is completed. Upon the addition of an uncoupler a high respiration rate is attained without ADP: the respiration is now uncoupled.

Like photosynthetic electron transport (Chapter 4), the electron transport of the respiratory chain is accompanied by the generation of a proton motive force (Fig. 5.15), which in turn drives the synthesis of ATP (Fig. 5.20). Therefore substances such as FCCP (Fig. 4.2) function as uncouplers of mitochondrial as well as of photosynthetic electron transport. Figure 5.19 shows that the addition of the uncoupler FCCP results in high stimulation of respiration. As discussed in section 4.2, the uncoupling function of the FCCP is due to a short circuit of protons across a membrane, resulting in the elimination of the proton gradient. The respiration is then uncoupled from ATP synthesis and the energy set free during electron transport is dissipated as heat.

Mitochondrial proton transport results in the formation of a membrane potential

Mitochondria, in contrast to chloroplasts, do not possess a closed thylakoid space for forming a proton gradient. Instead, in mitochondrial electron transport protons are transported from the matrix to the intermembrane space, which is con-

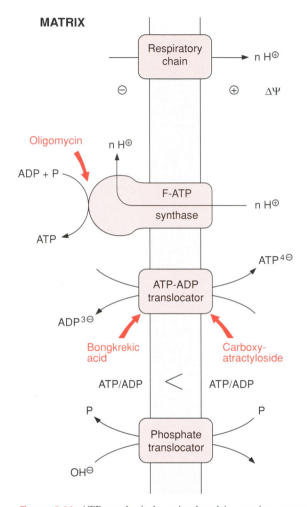

MATRIX

Figure 5.20 ATP synthesis by mitochondria requires an uptake of phosphate by the phosphate translocator in counter exchange for OH⁻ ions, and also an electrogenic exchange of ADP for ATP, as catalysed by the ATP–ADP translocator. Due to the membrane potential generated by electron transport of the respiratory chain, ADP is preferentially transported inward and ATP outward, and as a result of this the ATP : ADP ratio in the cytosol is higher than in the mitochondrial matrix. ATP–ADP transport is inhibited by carboxyactractyloside (binding from the outside) and bongkrekic acid (binding from the matrix side). F-ATP synthase of the mitochondria is inhibited by oligomycin.

nected to the cytosol by pores (formed by porins, Fig. 1.30). The formation in chloroplasts of a proton gradient of $\Delta pH = 3$ in the light results in a decrease of pH in the thylakoid lumen from about pH 7.5 to pH 4.5. If such a strong acidification were to occur in the cytosol it would have a grave effect on the activity of the cytosolic enzymes. In fact, during mitochondrial electron transport the pH gradient across the inner membrane is only about 0.2, as proton transport leads primarily to the formation of a membrane potential ($\Delta\Psi \approx 200$ mV).

Mitochondria are unable to generate a larger proton gradient as the inner membrane of the mitochondria is impermeable for anions, such as chloride. As shown in Fig. 4.1, a proton gradient can only be formed when the charge of the transported protons is compensated for by the diffusion of a counter anion.

Little is known about the mechanism of coupling between mitochondrial electron transport and transport of protons, despite intensive research for more than 30 years. Four protons are probably taken up during the transport of two electrons from the NADH dehydrogenase complex to ubiquinone on the matrix side and released again into the intermembrane space by the cyt-b/c_1 complex (Fig. 5.15). It is generally accepted that in mitochondria the cyt-b/c_1 complex catalyses a *Q-cycle* (Fig. 3.30) by which, when two electrons are transported, two additional protons are transported out of the matrix space into the intermembrane space. The cyt-a/a_3 complex transports at least two protons per two electrons. The mechanism of proton transport by the NADH dehydrogenase complex (complex I) and by the cyt-a/a_3 complex (complex IV) is not yet fully resolved. The three-dimensional structure of the cyt-a/a_3 complex determined recently indicates that the binuclear centre from cytochrome-a_3 and Cu_B is involved in this proton transport. If these stoichiometries are correct, altogether 10 protons would be transported during the oxidation of NADH and only six during the oxidation of succinate.

Mitochondrial ATP synthesis serves the energy demand of the cytosol

The energy of the proton gradient is used in the mitochondria for ATP synthesis by an F-ATP synthase (Fig. 5.20), which has the same basic structure as the F-ATP synthase of chloroplasts. However, there are differences regarding the inhibition by *oligomycin*, an antibiotic from *Streptomyces*. Whereas the mitochondrial F-ATP synthase is very strongly inhibited by oligomycin, due to the presence of an oligomycin binding protein, the chloroplast enzyme is insensitive towards this inhibitor. Despite these differences, the mechanism of ATP synthesis appears to be identical for both ATP synthases. Although the proton stoichiometry in mitochondrial ATP synthesis has not been resolved unequivocally, it is assumed that at least three protons are required for the synthesis of one molecule of ATP. Since the mechanism of the coupling between proton transport and ATP synthesis is still unclear (section 4.4) it is feasible that the proton stoichiometry is not a whole number and may be variable.

In contrast to chloroplasts, which synthesize ATP essentially for their own consumption, the ATP in mitochondria is synthesized mainly for export to the cytosol. This requires the uptake of ADP and phosphate from the cytosol into the mitochondria and the release of the ATP formed there. The uptake of phosphate proceeds by the *phosphate translocator* in a counter-exchange for OH ions, and the uptake of ADP and the release of ATP by the *ATP–ADP translocator* (Fig. 5.20). The mitochondrial ATP–ADP translocator is inhibited by *carboxy-*

atractyloside, a glucoside from the thistle *Atractylis gummifera*, and by *bonkrekic acid*, an antibiotic from the bacterium *Cocovenerans*, growing on coconuts. Both substances are deadly poisons.

The ATP–ADP translocator catalyses a strict *counter-exchange*; for each ATP transported out of the mitochondria an ADP is transported inwards. Since the transported ATP contains one negative charge more than the ADP, the transport is electrogenic. Due to the membrane potential generated by the proton transport of the respiratory chain, there is a preference for ADP to be taken up and ATP to be transported outwards. The result of this asymmetric transport of ADP and ATP is that the ADP:ATP ratio outside the mitochondria is much higher than in the matrix. In this way mitochondrial ATP synthesis maintains a high ATP:ADP ratio in the cytosol. With the exchange of ADP for ATP one negative charge is transferred from the matrix to the outside, which is in turn compensated for by the transport of a proton in the other direction. This is why protons from the proton gradient are not only required for ATP synthesis as such, but also for export of the synthesized ATP from the mitochondria.

Let us return to the stoichiometry between the transported protons and the ATP formation during respiration. It is customary to speak of three coupling sites of the respiratory chain, which correspond to the complexes I, III, and IV. Textbooks often state that when NADH is oxidized by the mitochondrial respiratory chain, one molecule of ATP is formed per coupling site, and as a result of this, the ADP/O quotient for oxidation of NADH amounts to three, and that for succinate to two. However, considerably lower values have been found in experiments with isolated mitochondria. The attempt was made to explain this discrepancy by assuming that owing to a proton leakage of the membrane the theoretical ADP/O values were not attained in the isolated mitochondria. It appears now that even in theory these whole numbers for ADP/O values are incorrect. Probably 10 protons are transported upon the oxidation of NADH. In the case that three protons are required for the synthesis of ATP and another one for its export from the mitochondria, the resulting ADP/O would be 2.5. Such values have in fact been obtained experimentally with isolated mitochondria.

At the beginning of this chapter the change in free energy during the oxidation of NADH was evaluated as -217 kJ mol^{-1} and for the synthesis of ATP at about $+50$ kJ mol^{-1}. An ADP/O of 2.5 for the respiration of NADH-dependent substrates means that only about 60 per cent of the free energy released during oxidation is used for synthesis of ATP.

5.7 Plant mitochondria have special metabolic functions

The function of the mitochondria as the power stations of the cell, dealt with so far, is relevant for all mitochondria, from unicellular organisms as well as from animals and plants. In plant cells performing photosynthesis, the role of the mitochondria as a supplier of energy is not restricted to the dark phase; the mitochondria provide the cytosol with ATP also during photosynthesis.

In addition, plant mitochondria fulfil special functions. The mitochondrial matrix contains enzymes for the oxidation of glycine to serine, an important step in the photorespiratory pathway (section 7.1):

$$2\,\text{glycine} + \text{NAD}^+ \rightarrow 1\,\text{serine} + \text{NADH} + \text{H}^+ + \text{CO}_2 + \text{NH}_4^+.$$

The NADH generated from glycine oxidation is the main fuel for mitochondrial ATP synthesis during photosynthesis. Another important role of plant mitochondria is the conversion of oxaloacetate and pyruvate to form citrate, a precursor for the synthesis of α-ketoglutarate. This pathway is important for providing the carbon skeletons for amino acid synthesis during nitrate assimilation (Fig. 10.11).

Mitochondria can oxidize surplus NADH without forming ATP

It is essential for a plant that the oxidation of glycine, formed by the photorespiratory cycle (section 7.1), and the production of citrate, required for the synthesis of carbon skeletons for the products of nitrate assimilation (section 10.4), can proceed even when the cell requires no ATP. Plant mitochondria have *overflow mechanisms*, which oxidize surplus NADH without synthesis of ATP (Fig. 5.21). The inner mitochondrial membrane contains an *alternative NADH dehydrogenase* which transfers electrons from NADH to ubiquinone, without coupling to proton

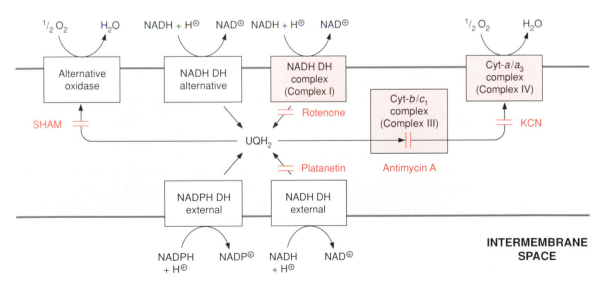

Figure 5.21 Beside rotenone-insensitive NADH dehydrogenase (NADH DH) there are three further dehydrogenases which transfer electrons to ubiquinone without accompanying proton transport. The external NADH dehydrogenase reacting with NADH of the intermembrane space is inhibited by platanetin. An alternative oxidase enables the oxidation of ubihydroquinone (UQH₂). This pathway is insensitive towards the inhibitors antimycin A and KCN, but it is inhibited by salicylhydroxamate (SHAM).

transport. This pathway is not inhibited by rotenone. However, oxidation of NADH via this rotenone insensitive pathway proceeds only when the NADH/NAD quotient in the matrix is exceptionally high.

Moreover, mitochondria possess an *alternative oxidase* by which electrons can be transferred directly from ubiquinone to oxygen; this pathway too is not coupled to proton transport. This alternative oxidation is insensitive to *antimycin A* and *KCN* (inhibitors of complex III and II, respectively), but is inhibited by *salicylhydroxamate* (SHAM). The alternative oxidation is catalysed by a membrane protein of 36 kDa with two transmembrane helices, containing an iron–sulfur centre, which mediates the oxidation of ubiquinone by oxygen.

Electron transport via the alternative oxidase can be understood as a short circuit. It only occurs when the mitochondrial ubiquinone pool is reduced to a very high degree. The alternative oxidase is activated by pyruvate, which ensures that it operates only when there are sufficient metabolites in the cell. When metabolites in the mitochondria are in excess, the interplay of the alternative NADH dehydrogenase and the alternative oxidase leads to their elimination by oxidation without accompanying ATP synthesis, and the oxidation energy is dissipated as heat. The capacity of the alternative oxidase in the mitochondria from different plant tissues is variable and also depends on the developmental state. An especially high alternative oxidase activity has been found in the spadix of the voodoo lily, *Saurum guttatum*, which uses the alternative oxidase to heat the spadix, producing a nasty smell like carrion or dung. This strong stench attracts insects from far and wide. The formation of the alternative oxidase is synchronized in these spadices with the beginning of flowering.

NADH and NADPH from the cytosol can be oxidized by the respiratory chain of plant mitochondria

In contrast to mitochondria from animal tissues, plant mitochondria can also oxidize cytosolic NADH and in some cases cytosolic NADPH. Oxidation of this external NADH and NADPH proceeds via two specific dehydrogenases of the inner membrane, of which the substrate binding site is directed towards the intermembrane space. The *external NADH dehydrogenase* is inhibited by *platanetin*, a flavone derivative from the buds of the platane tree. As in the case of succinate dehydrogenase, the electrons from external NADH and NADPH dehydrogenase are fed into the respiratory chain at the site of *ubiquinone* and therefore this electron transport is not inhibited by rotenone. As oxidation of external NADH and NADPH (like the oxidation of succinate) does not involve proton transport by complex I (Fig. 5.21), in the oxidation of external pyridine nucleotides the yield of ATP is lower than that in the oxidation of NADH provided from the matrix. Oxidation by external NADH dehydrogenase proceeds only when the cytosolic NAD pool is excessively reduced. Also the external NADH dehydrogenase may be regarded as part of an overflow mechanism which only comes into action when the NADH in the cytosol is reduced too much. As already discussed in section

3.10, in certain situations photosynthesis may produce a surplus of reducing power, which is hazardous for a cell. The plant cell has the capacity to eliminate excessive reducing power by making use of the external NADH dehydrogenase, the alternative dehydrogenase for internal NADH from the matrix, and the alternative oxidase mentioned earlier.

5.8 Compartmentalization of mitochondrial metabolism requires specific membrane translocators

The mitochondrial inner membrane is impermeable to metabolites. Specific translocators enable a specific transport of metabolites between the mitochondrial matrix and the cytosol in a counter-exchange mode (Fig. 5.22). The role of the

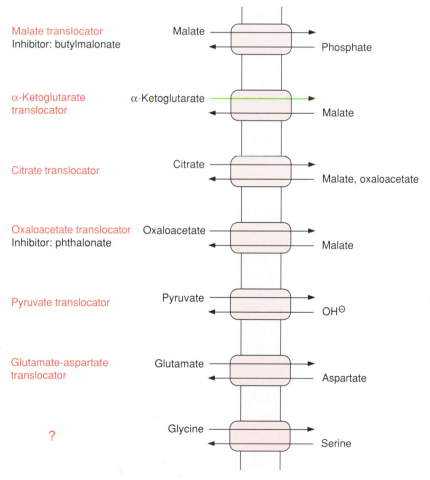

Figure 5.22 Important translocators of the inner mitochondrial membrane. The phosphate and the ATP–ADP translocator have been already shown in Fig. 5.20.

ATP–ADP and the phosphate translocators (Fig. 5.20) has already been dealt with in section 5.6. Malate and succinate are transported in counter-exchange for phosphate by a *dicarboxylate translocator*. This transport is inhibited by *butyl-malonate*. α-Ketoglutarate, citrate and oxaloacetate are transported in counter-exchange for malate. Glutamate is transported in counter-exchange for aspartate, and pyruvate in counter-exchange for OH^- ions. Although these translocators all occur in plant mitochondria, most of our present knowledge about them is based on studies with mitochondria from animal tissues. A comparison of the amino acid sequences of the ATP–ADP, phosphate, oxaloacetate, citrate, and glutamate–aspartate translocators shows a homology; the proteins of these translocators represent a family deriving from a common ancestor. As already mentioned in section 1.9, all these translocators are composed of two six transmembrane helices.

The malate–oxaloacetate translocator is a special feature of plant mitochondria and has an important function in the malate–oxaloacetate cycle described in section 7.3. It also transports citrate and is involved in providing the carbon skeletons for nitrate assimilation (Fig. 10.11). The oxaloacetate translocator, and to a lesser extent also the α-ketoglutarate translocator, are inhibited by the dicarboxylate phthalonate. The transport of glycine and serine, involved in the photorespiratory pathway (section 7.1), has not yet been characterized. Although final proof is still lacking, it is to be expected that this transport is mediated by one or two mito-chondrial translocators.

Further reading

Bachmann, G., Schulte, U., and Weiss, H. (1992). Mitochondrial ubiquinol–cytochrome-*c* oxidoreductase. In *Molecular mechanisms in bioenergetics*, (ed. L. Ernster), pp. 199–216. Elsevier, Amsterdam.

Calhoun, M. W., Thomas, J., and Gennis, R. B. (1994). The cytochrome oxidase superfamily of redox-driven proton pumps. *Trends in Biological Sciences*, **19**, 325–30.

Douce, R. and Neuburger, M. (1989). The uniqueness of plant mitochondria. *Annual Review of Plant Physiology and Plant Molecular Biology*, **40**, 371–414.

Friedrich, T., Weidner, U., Nehls, U., Fecke, W., Schneider, R., and Weiss, H. (1993). Minireview: Attempts to define distinct parts of NADH: ubiquinone oxidoreductase (Complex I). *Journal of Bioenergetics and Biomembranes*, **25**, 331–7.

Haltia, T. and Wikström, M. (1992). Cytochrome oxidase: notes on structure and mechanism. In *Molecular mechanisms in bioenergetics*, (ed. L. Ernster), pp. 217–40. Elsevier, Amsterdam.

Iwata, S., Ostermeier, C., Ludwig, B., and Michel, H. (1995). Structure at 2.8 Å resolution of cytochrome C oxidase from *Paracoccus denitrificans*. *Nature*, **376**, 660–9.

Klingenberg, M. (1989). Molecular aspects of the adenine nucleotide carrier from mito-chondria. *Archives of Biochemistry and Biophysics*, **270**, 1–14.

Krämer, R. and Palmieri, F. (1992). Metabolite carriers in mitochondria. In *Molecular mechanisms in bioenergetics*, (ed. L. Ernster), pp. 359–84. Elsevier, Amsterdam.

Krömer, S. (1995). Respiration during photosynthesis. *Annual Review of Plant Physiology Plant Molecular Biology*, **46**, 45–70.

Lambers, H. and van der Plaas, L. H. W. (1992). *Molecular, biochemical and physiological aspects of plant respiration.* SPB Academic Publishing, Den Haag.

McIntosh, L. (1994). Molecular biology of the alternative oxidase. *Plant Physiology*, **105**, 781–6.

Møller, I. M. (1986). Membrane-bound NAD(P)H dehydrogenases in higher plant cells. *Annual Review of Plant Physiology and Plant Molecular Biology*, **37**, 309–34.

Nicholls, D. G. and Ferguson, S. J. (1992). *Bioenergetics,* (2nd edn). Academic Press, London.

Tsukihara, T., Aoyama, H., Yamashita, E., Tomizaki, T., Yamaguchi, H., Shinzawa-Itoh, K. *et al.* (1995). Structure of metal sites of oxidized bovine heart cytochrome C oxidase at 2.8 Å. *Science*, **269**, 1069–74.

chapter 6

Photosynthetic CO_2 assimilation by the Calvin cycle

Chapters 3 and 4 showed how the electron transport chain and the ATP synthase of the thylakoid membrane use the energy from light to provide reducing equivalents in the form of NADPH, and chemical energy in the form of ATP. This chapter will describe how this NADPH and ATP are used for CO_2 assimilation.

6.1 CO_2 assimilation proceeds via the dark reaction of photosynthesis

It is relatively simple to isolate chloroplasts with an intact envelope from leaves (section 1.7). When these chloroplasts are added to an isotonic medium containing an osmotically active substance, a buffer, bicarbonate, and inorganic

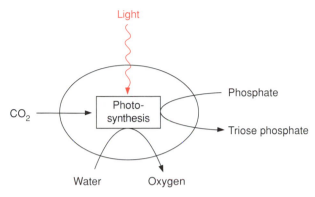

Figure 6.1 Basic scheme of photosynthesis by a chloroplast.

phosphate, and the light is switched on, one observes the generation of oxygen. Water is split to form oxygen by the light reaction described in Chapter 3 and the resulting reducing equivalents are used for CO$_2$ assimilation (Fig. 6.1). With intact chloroplasts there is no oxygen evolution in the absence of CO$_2$ or phosphate, demonstrating that the light reaction in intact chloroplasts is coupled to CO$_2$ assimilation and the product of this assimilation contains phosphate. The main assimilation product of the chloroplasts is *dihydroxyacetone phosphate*, a *triose phosphate*. Figure 6.2 shows that the synthesis of triose phosphate from CO$_2$ requires energy in the form of ATP and reducing equivalents in the form of NADPH, which have been provided by the *light reaction of photosynthesis*. The reaction chain for the formation of triose phosphate from CO$_2$, ATP, and NADPH is called the *dark reaction of photosynthesis*, as it requires no light *per se* and theoretically it should also be able to proceed in the dark. The fact is, however, that in leaves this reaction does not proceed during darkness, since some of the enzymes of the reaction chain, due to regulatory processes, are only active during illumination (section 6.6).

Between 1946 and 1953 Melvin Calvin and his collaborators Andrew Benson and James Bassham in Berkeley, California, resolved the mechanism of photosynthetic CO$_2$ assimilation. In 1961 Calvin was awarded the Nobel prize for

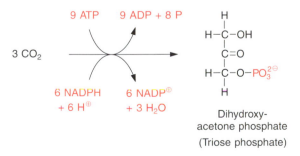

Figure 6.2 Overall reaction of photosynthetic CO$_2$ fixation.

Chemistry for this fundamental discovery. A prerequisite for the elucidation of the CO$_2$ fixation pathway was the discovery in 1940 of the radioactive carbon isotope ^{14}C which, as a by-product of nuclear reactors, was available in larger amounts in the USA after 1945. Calvin chose the green alga *Chlorella* for his investigations. He added radioactively labelled CO$_2$ to illuminated algal suspensions, killed the algae after a short incubation period by adding hot ethanol and, using paper chromatography, analysed the radioactively labelled products of CO$_2$ fixation. By successively shortening the incubation time, he was able to show that phosphoglycerate was formed as the first stable product of CO$_2$ fixation. More detailed studies revealed that CO$_2$ fixation proceeds by a cyclic process which has been named *Calvin cycle* after its discoverer. *The reductive pentose phosphate pathway* is another name which will be used in some sections of this book for the sake of systematics. This name derives from the fact that a reduction occurs and pentoses are formed in the cycle.

The Calvin cycle can be subdivided into three sections:

(1) the *carboxylation* of the C$_5$ sugar, ribulose 1,5-bisphosphate, leading to the formation of two molecules of 3-phosphoglycerate;

(2) the *reduction* of the 3-phosphoglycerate to triose phosphate; and

(3) the *regeneration* of the CO$_2$ acceptor ribulose 1,5-bisphosphate from triose phosphate (Fig. 6.3).

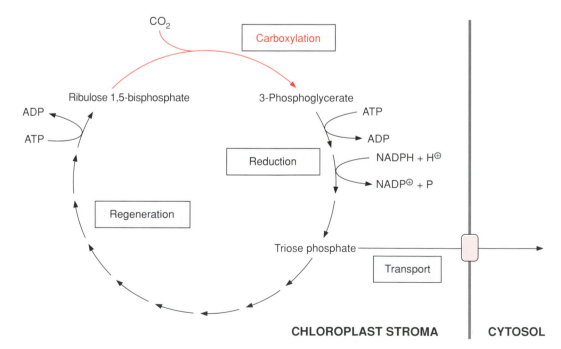

Figure 6.3 Overview of the basic reactions of the Calvin cycle.

As a product of photosynthesis, triose phosphate is exported from the chloroplasts by specific transport. However, most of the triose phosphate remains in the chloroplasts to regenerate ribulose 1,5-bisphosphate. These reactions will be dealt with in detail in the following sections.

6.2 Ribulose bisphosphate carboxylase catalyses the fixation of CO_2

The key reaction for photosynthetic CO_2 assimilation is the binding of atmospheric CO_2 to the acceptor ribulose 1,5-bisphosphate (RuBP) to form two molecules of 3-phosphoglycerate. The reaction is very exergonic ($\Delta G^{0\prime} = -35$ kJ mol^{-1}) and therefore virtually irreversible. It is catalysed by the enzyme ribulose bisphosphate carboxylase/oxygenase (abbreviated to *RubisCO*), so called because the same enzyme also catalyses a side-reaction in which the ribulose bisphosphate reacts with O_2 (Fig. 6.4).

Figure 6.5 shows the reaction sequence of the *carboxylase reaction*. Keto–enol isomerization of RuBP yields an enediol which reacts with CO_2 to form the intermediate 2-carboxy 3-ketoarabinitol 1,5-bisphosphate, which is cleaved to two molecules of 3-phosphoglycerate. In the *oxygenase reaction*, of which the mechanism is not yet known in detail, a peroxide is probably formed as an intermediate (Fig. 6.6). The products of the oxygenase reaction are 2-phosphoglycolate and 3-phosphoglycerate.

Ribulose bisphosphate carboxylase/oxygenase enables the fixation of atmospheric CO_2 for the formation of biomass. This enzyme is therefore a prerequisite for the existence of the present life on earth. In plants and cyanobacteria it consists of eight identical large subunits (molecular mass of 51 000–58 000 Da,

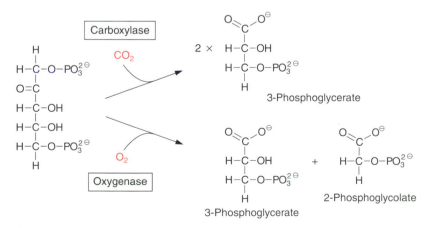

Figure 6.4 Ribulose bisphosphate carboxylase catalyses two reactions with RuBP: the carboxylation, which is the actual CO_2 fixation reaction, and also the oxygenation, an unavoidable side-reaction.

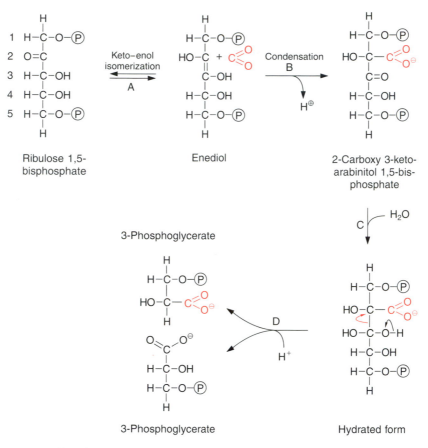

Figure 6.5 Reaction sequence in the carboxylation of RuBP by RubisCO. For the sake of simplicity -PO$_3^{2-}$ is symbolized as -P. An enediol, formed by keto–enol isomerization of the carbonyl group of the RuBP (A), allows the nucleophilic reaction of CO₂ with the C-2 atom of RuBP by which 2-carboxy-3-ketoarabinitol 1,5-bisphosphate (B) is formed. After hydration (C), the bond between C-2 and C-3 is cleaved and two molecules of 3-phosphoglycerate are formed (D).

depending on the species) and eight identical small subunits (molecular mass 12 000–18 000 Da). In plants the genetic information for the large subunit is encoded in the plastid genome and that for the small subunit in the nucleus. Each large subunit contains one catalytic centre. The function of the small subunits is not yet understood. It has been proposed that the eight small subunits stabilize the complex of the eight large subunits. The small subunit is apparently not essential for the process of CO₂ fixation *per se*. RubisCO occurs in some phototrophic purple bacteria as a dimer only of large subunits, and the catalytic properties of the corresponding bacterial enzymes are not basically different from those in plants. The one difference in these bacterial enzymes which consist of only two large subunits is that the ratio between oxygenase and carboxylase activity is

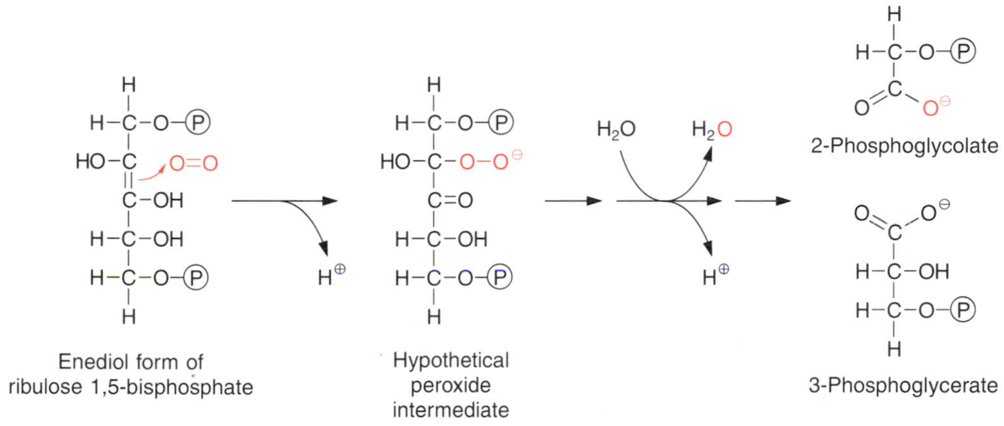

Figure 6.6 Part of the reaction sequence in the oxygenation of RuBP as catalysed by RubisCO. O_2 probably reacts in a similar way to CO_2 with the enediol of RuBP and thus forms a peroxide. In the subsequent cleavage of the O_2 adduct, one atom of the O_2 molecule is found in the water and the other in the carboxyl group of the 2-phosphoglycolate.

higher than in the plant enzyme which is made up of eight large and eight small subunits.

The oxygenation of ribulose bisphosphate: a costly side-reaction

Although the CO_2 concentration required for half-saturation of the enzyme ($K_M[CO_2]$) is much lower than that of O_2 ($K_M[O_2]$) (Table 6.1), the velocity of the oxygenase reaction is very high since the concentration of O_2 in air amounts to 21 per cent and that of CO_2 to only 0.035 per cent. Moreover, the CO_2 concentration in the gaseous space of the leaves can be considerably lower than the CO_2 concentration in the outside air. For these reasons, the ratio of oxygenation to carboxylation during photosynthesis of a leaf at 25 °C is mostly in the range of 1:4 to 1:2. This means that every third to fifth ribulose 1,5-bisphosphate molecule is consumed in the side-reaction. When the temperature rises, the CO_2/O_2 specificity (Table 6.1) decreases, and as a consequence the ratio of oxygenation to carboxylation increases.

It will be shown in Chapter 7 that recycling of the byproduct 2-phosphoglycolate is a very costly process for plants. This recycling process requires a metabolic chain with more than 10 enzymatic reactions distributed over three different organelles (chloroplasts, peroxisomes, mitochondria), and also very high energy consumption. Section 7.5 describes in detail that about one-third of the photons absorbed during photosynthesis of a leaf are consumed to reverse the consequences of oxygenation.

Apparently evolution has not been successful in eliminating this costly side-reaction of ribulose bisphosphate carboxylase. Between cyanobacteria and higher

Table 6.1 Kinetic properties of ribulose bisphosphate carboxylase/oxygenase (RubisCO) at 25 °C (data from Woodrow and Berry 1988)

Substrate concentrations at half saturation of the enzyme:

$K_M[CO_2]$:	9 µmol l^{-1}*
$K_M[O_2]$:	535 µmol l^{-1}*
$K_M[RuBP]$:	28 µmol l^{-1}

Maximal turnover:

kcat[CO$_2$]:	3.3 s^{-1}	(related to one subunit)
kcat[O$_2$]:	2.4 s^{-1}	

$$CO_2/O_2 \text{ specificity} = \left(\frac{k_{cat}[CO_2]}{K_M[CO_2]} \Big/ \frac{k_{cat}[O_2]}{K_M[O_2]} \right) = 82$$

* For comparison:
In equilibrium with air (0.035 per cent = 350 p.p.m. CO$_2$, 21 per cent O$_2$) the concentrations in water at 25 °C amount to:

CO$_2$:	11 µmol l^{-1}
O$_2$:	253 µmol l^{-1}

plants the ratio in the activities of carboxylase and oxygenase of RubisCO is increased by a factor of less than two. It seems as if we are confronted here with a case in which the evolutionary refinement of a key process of life has reached a limit set by the chemistry of the reaction. The reason is probably that early evolution of RubisCO occurred at a time when there was no oxygen in the atmosphere. A comparison of the RubisCO proteins from different organisms leads to the conclusion that RubisCO was already present at least 3.5 billion years ago, when the first chemolithotrophic bacteria evolved from primitive anaerobic bacteria. When, 1.5 billion years later, due to photosynthesis, oxygen appeared in the atmosphere in higher concentrations, the complexity of the RubisCO protein probably made it too difficult to change the catalytic centre to eliminate oxygenase activity. Experimental results support this notion. A large number of experiments, in which genetic engineering was employed to obtain site-specific mutations of the amino acid sequence in the region of the active centre of RubisCO, were unable to improve the ratio between the activities of carboxylation to oxygenation. The only way to specifically decrease the oxygenation reaction by genetic engineering may be a simultaneous exchange of several amino acids in the catalytic binding site of the enzyme, which evolution could not achieve. Section 7.7 will show how plants make a virtue of necessity and use the energy-consuming oxygenation to eliminate a surplus of NADPH and ATP produced by the light reaction.

Ribulose bisphosphate carboxylase/oxygenase: special features

The catalysis of the carboxylation of RuBP by RubisCO is very slow (Table 6.1): the turnover number for each subunit amounts to 3.3 s^{-1}. This means that at substrate saturation only about three molecules of CO_2 and RuBP are converted per second at a catalytic site of RubisCO. In comparison, the turnover numbers of dehydrogenases and carbonic anhydrase are in the order of 10^3 s^{-1}, and 10^5 s^{-1} respectively. Because of the extremely low turnover number of RubisCO, very large amounts of enzyme are necessary to catalyse the fluxes required for photo-synthesis. RubisCO can amount to 50 per cent of the total soluble proteins in leaves. The wide distribution of plants makes RubisCO by far the most abundant protein on earth. The concentration of the catalytic large subunits in the chloro-plast stroma is as high as 4–10×10^{-3} mol l^{-1}. A comparison of this value with the aqueous concentration of CO_2 in equilibrium with air (at 25 °C about 11×10^{-6} mol l^{-1}) shows the abnormal situation where the concentration of an enzyme is up to a thousand times higher than the concentration of the substrate CO_2 and at similar concentrations to the substrate RuBP.

Activation of ribulose bisphosphate carboxylase/oxygenase

All the large subunits of RubisCO contain a lysine in position 201 of their sequence of about 470 amino acids. RubisCO is only active when the ε-amino group of this lysine reacts with CO_2 to form a *carbamate* to which an Mg^{2+} ion is bound (Fig. 6.7). The activation is due to a change in the conformation of the large subunit. The active conformation is stabilized by the complex formation with Mg^{2+}. This carbamylation is a prerequisite for the activity of all known RubisCO proteins. It should be noted that the CO_2 bound as carbamate is different from the CO_2 which is a substrate of the carboxylation reaction of RubisCO.

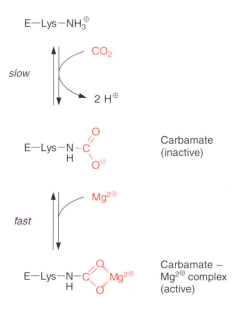

Figure 6.7 RubisCO is activated by the carbamyl-ation of a lysine residue.

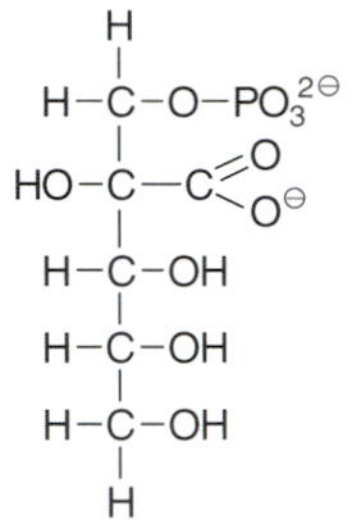

H
|
H—C—O—PO₃²⁻
| ⟋O
HO—C—C
| ⟍O⁻
H—C—OH
|
H—C—OH
|
H—C—OH
|
H

2-Carboxyarabinitol
1-phosphate

Figure 6.8 2-Carboxyarabinitol 1-phosphate, an inhibitor of RubisCO.

Chloroplasts contain the enzyme *RubisCO activase*, which facilitates carbamylation of RubisCO with consumption of ATP. The mechanism of this activation is not yet fully understood. The non-carbamylated, inactive form of RubisCO binds RuBP very tightly, preventing the carbamylation reaction and resulting in an inhibition of RubisCO. The activase probably releases RuBP, which is tightly bound to the inactive form of RubisCO, and thus promotes the carbamylation of the free enzyme.

RubisCO is inhibited by several hexose phosphates and by 3-phosphoglycerate, which all bind to the active site instead of RuBP. A very strong inhibitor is *2-carboxyarabinitol 1-phosphate* (CA1P) (Fig. 6.8). This substance has a very similar structure to 2-carboxy 3-ketoarabinitol 1,5-bisphosphate (Fig. 6.5) which is an intermediate of the carboxylation reaction. CA1P has a thousandfold higher affinity than RuBP for the RuBP binding site of RubisCO. In a number of species CA1P accumulates in the leaves during the night, blocking a large number of the binding sites of RubisCO and thus inactivating the enzyme. During the day CA1P is released by RubisCO activase and degraded by a specific phosphatase, which hydrolyses the phosphate residue from CA1P and thus eliminates the effect of the RubisCO inhibitor. The regulation of CA1P synthesis and degradation is not yet understood. Since CA1P is not formed in all plants, its role in the regulation of RubisCO is still a matter of debate.

6.3 The reduction of 3-phosphoglycerate yields triose phosphate

For the synthesis of dihydroxyacetone phosphate the carboxylation product 3-phosphoglycerate is phosphorylated by the enzyme *phosphoglycerate kinase* to 1,3-bisphosphoglycerate. In this reaction, with the consumption of ATP, a mixed anhydride is formed between the new phosphate residue and the carboxyl group (Fig. 6.9). As the free energy for the hydrolysis of this anhydride is similarly high to that of the phosphate anhydride in ATP, the phosphoglycerate kinase reaction is reversible. An isoenzyme of the chloroplastic phosphoglycerate kinase is also

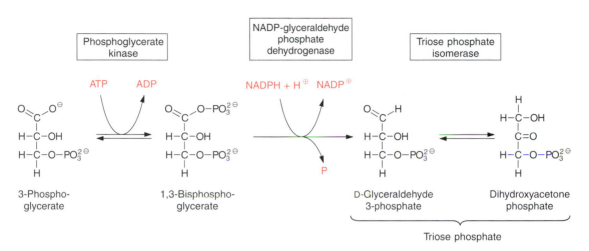

Figure 6.9 Conversion of 3-phosphoglycerate to triose phosphate.

involved in the glycolytic pathway proceeding in the cytosol and catalyses the formation there of ATP from ADP and 1,3-bisphosphoglycerate.

The reduction of 1,3-bisphosphoglycerate to D-glyceraldehyde 3-phosphate is catalysed by the enzyme *glyceraldehyde phosphate dehydrogenase* (Fig. 6.9). A thioester is formed as an intermediate of this reaction by the exchange of the phosphate residue bound to the carboxyl group for an SH-group of a cysteine residue in the active centre of the enzyme (Fig. 6.10). The free energy for the hydrolysis of this thioester is similarly high to that of the anhydride ('energy-rich bond'). When a thioester is reduced, a thio-semiacetal is formed which has a low free energy of hydrolysis.

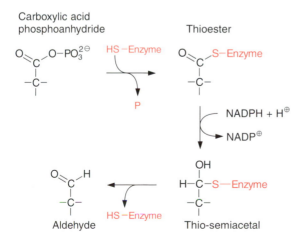

Figure 6.10 Reaction sequence catalysed by glyceraldehyde phosphate dehydrogenase. HS–enzyme symbolizes the sulfhydryl group of a cysteine residue in the active centre of the enzyme.

Through the catalysis of phosphoglycerate kinase and glyceraldehyde phosphate dehydrogenase the large difference in redox potentials between the aldehyde and the carboxylate in the reduction of 3-phosphoglycerate to glyceraldehyde phosphate is overcome by the consumption of ATP. This is a reversible reaction. A glyceraldehyde phosphate dehydrogenase in the cytosol catalyses the conversion of glyceraldehyde phosphate to 1,3-bisphosphoglycerate. In contrast to the cytosolic enzyme, which mainly catalyses the oxidation of glyceraldehyde phosphate with NAD^+ as hydrogen acceptor, the chloroplastic enzyme uses NADPH as a hydrogen donor.

This is an example of the different roles which the $NADH/NAD^+$ and $NADPH/NADP^+$ systems play in the metabolism of eukaryotic cells. Whereas the NADH system is specialized in collecting reducing equivalents to be oxidized for the synthesis of ATP, the NADPH system mainly gathers reducing equivalents to be donated to synthetic processes. Figuratively speaking, the NADH system has been compared with a *hydrogen low-pressure line* through which reducing equivalents are pumped off for oxidation to generate energy, and the NADPH system as a *hydrogen high-pressure line* through which reducing equivalents are pressed into synthesis processes. Usually the reduced/oxidized ratio is about a hundred times higher for the NADPH system than for the NADH system. The relatively high degree of reduction of the NADPH system in chloroplasts (about 50–60 per cent reduced) allows the very efficient reduction of 1,3-bisphosphoglycerate to glyceraldehyde-3-phosphate.

Triose phosphate isomerase catalyses the isomerization of glyceraldehyde phosphate to dihydroxyacetone phosphate. This conversion of an aldose into a ketose proceeds via an 1,2-enediol as intermediate and is basically similar to the reaction catalysed by ribose phosphate isomerase. The equilibrium of the reaction lies towards the ketone. Therefore triose phosphate as a collective term consists of about 96 per cent dihydroxyacetone phosphate and only 4 per cent glyceraldehyde phosphate.

6.4 Ribulose bisphosphate is regenerated from triose phosphate

From the fixation of three molecules of CO_2 in the Calvin cycle, six molecules of phosphoglycerate are formed and are converted to six molecules of triose phosphate (Fig. 6.11). Of these only one molecule of triose phosphate is the actual gain which is provided to the cell for various biosynthetic processes. The remaining five triose phosphates are needed to regenerate three molecules of ribulose bisphosphate so that the Calvin cycle can continue. Figure 6.12 shows the metabolic pathway for the conversion of the five triose phosphates (white boxes) to three pentose phosphates (red boxes).

The two trioses dihydroxyacetone phosphate and glyceraldehyde phosphate are

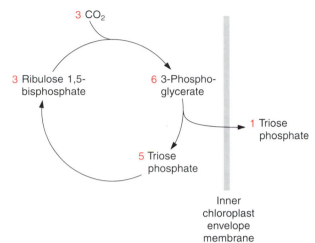

Figure 6.11 Five-sixths of the triose phosphate formed by photosynthesis are required for the regeneration of ribulose 1,5-bisphosphate. One molecule of triose phosphate represents the net product and can be utilized by the chloroplast or exported.

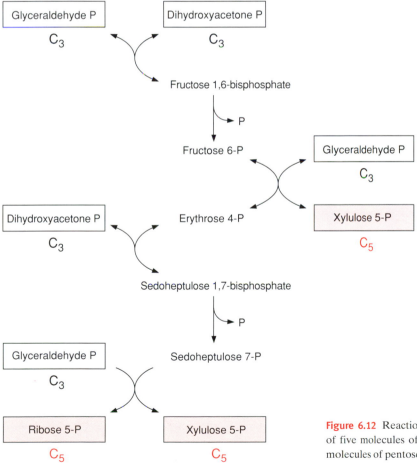

Figure 6.12 Reaction chain for the conversion of five molecules of triose phosphate to three molecules of pentose phosphate.

condensed in a reversible reaction to fructose 1,6-bisphosphate, as catalysed by the enzyme *aldolase* (Fig. 6.13). Figure 6.14 shows the reaction mechanism. As an intermediate of this reaction, a protonated Schiff base is formed between a lysine residue in the active centre of the enzyme and the keto group of the dihydroxyacetone phosphate. This Schiff base enhances the release of a proton from the C-3 position and enables a nucleophilic reaction with the C atom of the aldehyde group of glyceraldehyde phosphate. Fructose 1,6-bisphosphate is hydrolysed in an irreversible reaction to fructose 6-phosphate, as catalysed by *fructose 1,6 bisphosphatase* (Fig. 6.15).

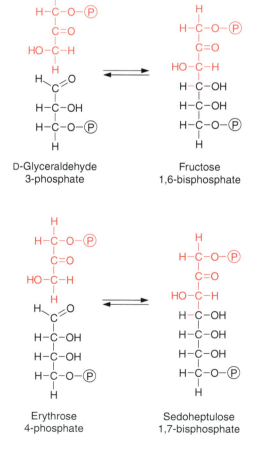

Figure 6.13 Aldolase catalyses the condensation of dihydroxyacetone phosphate with the aldoses glyceraldehyde 3-phosphate or erythrose 4-phosphate.

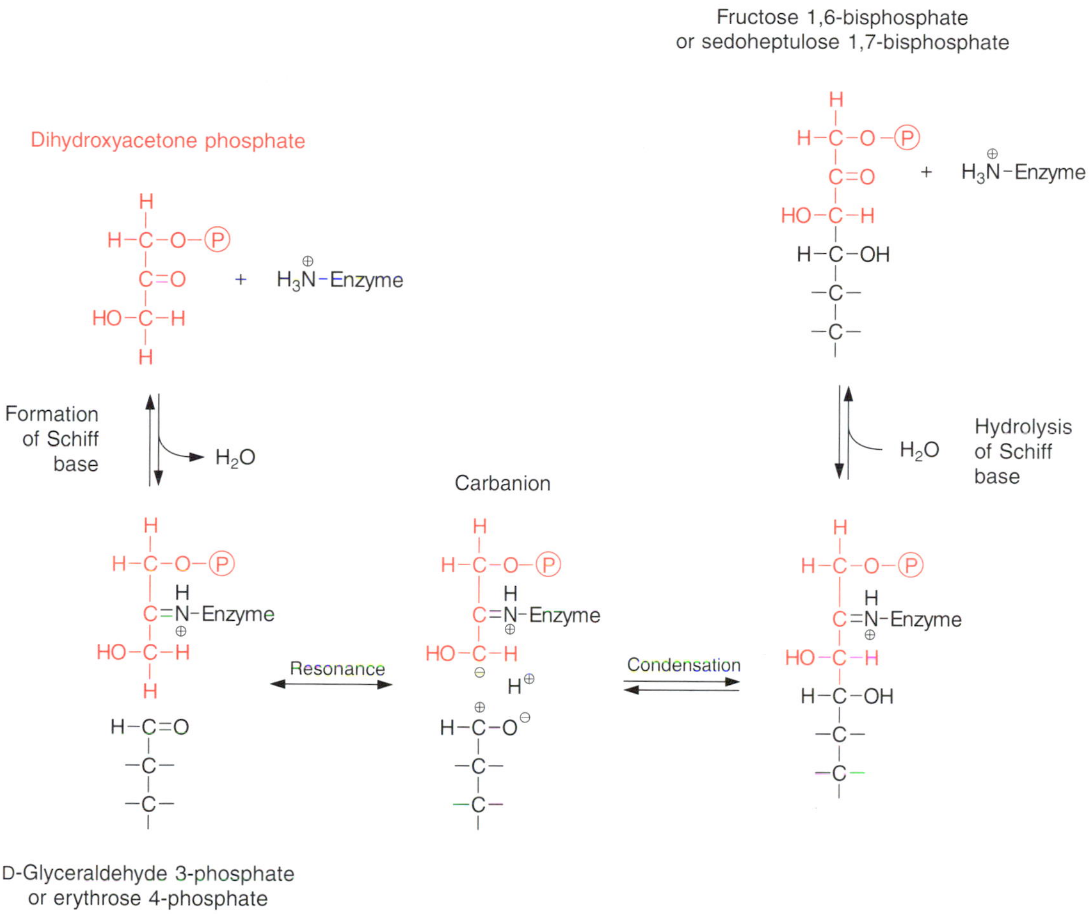

Dihydroxyacetone phosphate

Fructose 1,6-bisphosphate
or sedoheptulose 1,7-bisphosphate

Formation of Schiff base

Hydrolysis of Schiff base

H_2O

Carbanion

Resonance

Condensation

D-Glyceraldehyde 3-phosphate
or erythrose 4-phosphate

Figure 6.14 Pathway of the aldolase reaction. Dihydroxyacetone phosphate forms a Schiff base with the terminal amino group of a lysyl residue of the enzyme protein. The positive charge at the nitrogen favours the release of a proton at C-3, and thus a carbanion is formed. In one mesomeric form of the glyceraldehyde phosphate the C atom of the aldehyde group is positively charged. This enables condensation between this C atom and the negatively charged C-3 of the dihydroxyacetone phosphate. After condensation the Schiff base is cleaved again and fructose 1,6-bisphosphate is released. Sedoheptulose 1,7-bisphosphate is formed by the same enzyme from reaction with erythrose 4-phosphate. The aldolase reaction is reversible.

The enzyme *transketolase* transfers a carbohydrate residue with two carbon atoms from fructose 6-phosphate to glyceraldehyde 3-phosphate, yielding xylulose 5-phosphate and erythrose 4-phosphate in a reversible reaction (Fig. 6.16). In this reaction thiamine pyrophosphate (Fig. 5.5), already dealt with as a reaction partner of pyruvate oxidation (section 5.3), is involved as prosthetic group (Fig. 6.17).

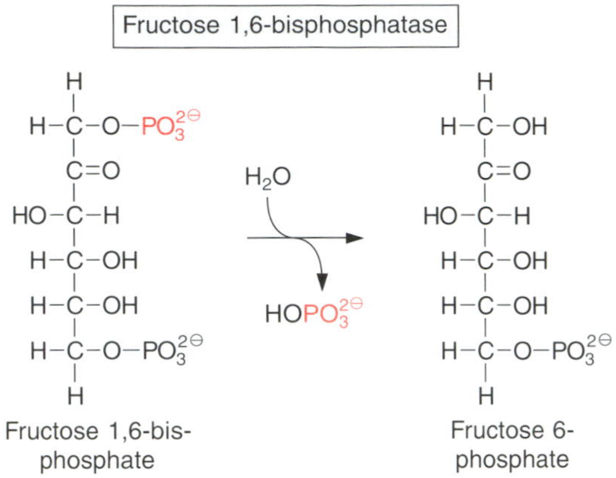

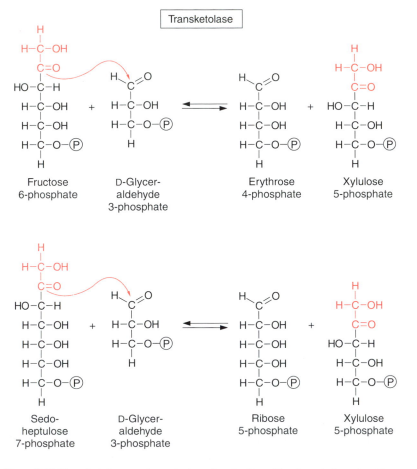

Figure 6.15 Hydrolysis of fructose 1,6-bisphosphate by fructose 1,6-bisphosphatase

Figure 6.16 Transketolase catalyses the transfer of a C_2 residue from ketoses to aldoses.

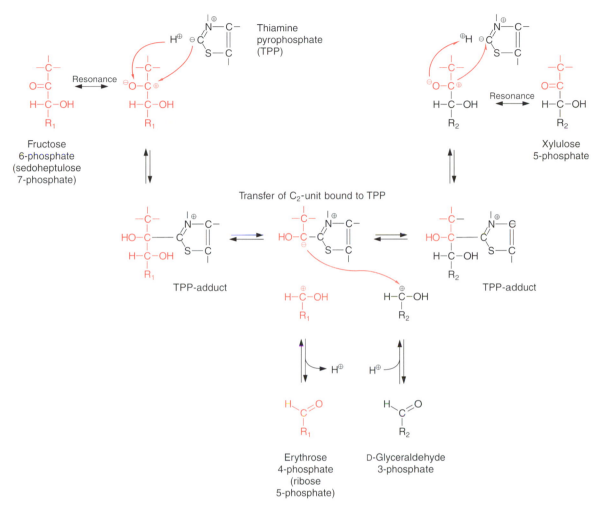

Figure 6.17 Mechanism of the transketolase reaction. The enzyme contains as a prosthetic group thiamine pyrophosphate with a thiazole ring as reactive component. The positive charge of the N atom in this ring enhances the release of a proton at the neighbouring C atom, resulting in a negatively charged C atom (carbanion), to which the partially positively charged C atom of the keto group is bound. The positively charged N atom of the thiazole favours the cleavage of the carbon chain, and the carbon atom in position 2 becomes a carbanion. The reaction mechanism is basically the same as that of the aldolase reaction in Fig. 6.14. The C_2 carbohydrate moiety bound to the thiazole is transferred to the C-1 position of the glyceraldehyde 3-phosphate

Once more, *aldolase* (Fig. 6.13) catalyses a condensation, this time of erythrose 4-phosphate with dihydroxyacetone phosphate to form sedoheptulose 1,7-bisphosphate. Subsequently the enzyme *sedoheptulose 1,7-bisphosphatase* catalyses the irreversible hydrolysis of sedoheptulose 1,7-bisphosphate. This reaction is similar to the hydrolysis of fructose 1,6-bisphosphate, although the two reactions are catalysed by different enzymes. Again a carbohydrate residue with two C atoms is transferred by *transketolase* from sedoheptulose 7-phosphate to dihydroxyacetone phosphate and this forms ribose 5-phosphate and xylulose 5-phosphate (Fig. 6.16).

The three pentose phosphates formed are then converted to ribulose 5-phosphate (Fig. 6.18). The conversion of xylulose 5-phosphate is catalysed by *ribulose phosphate epimerase*; this reaction proceeds via a keto–enol isomerization with a 2,3-enediol as intermediary product. The conversion of the aldose ribose 5-phosphate to the ketose ribulose 5-phosphate is catalysed by *ribose phosphate isomerase*, again via an enediol as intermediate, although in the 1,2-position. The three molecules of ribulose 5-phosphate formed in this way are converted to the CO$_2$ acceptor ribulose 1,5-bisphosphate with consumption of ATP by *phosphoribulose kinase* (Fig. 6.19). This kinase reaction is irreversible, since a phosphate is converted from the 'energy-rich' anhydride in the ATP to a phosphate ester with a low free energy of hydrolysis.

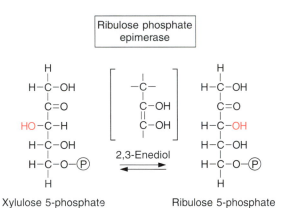

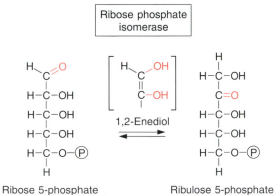

Figure 6.18 The conversion of xylulose 5-phosphate and ribose 5-phosphate into ribulose 5-phosphate. In both cases a *cis*-enediol is formed as intermediate.

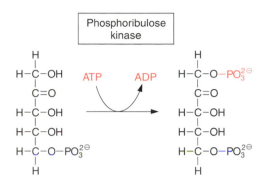

Figure 6.19 Phosphoribulose kinase catalyses the irreversible formation of ribulose 1,5-bisphosphate.

The scheme in Fig. 6.20 gives a summary of the various reactions of the Calvin cycle. There are four irreversible steps in the cycle: carboxylation, hydrolysis of fructose bisphosphate and sedoheptulose bisphosphate, and phosphorylation of ribulose 5-phosphate. Fixation of one molecule of CO_2 requires in total two molecules of NADPH and three molecules of ATP.

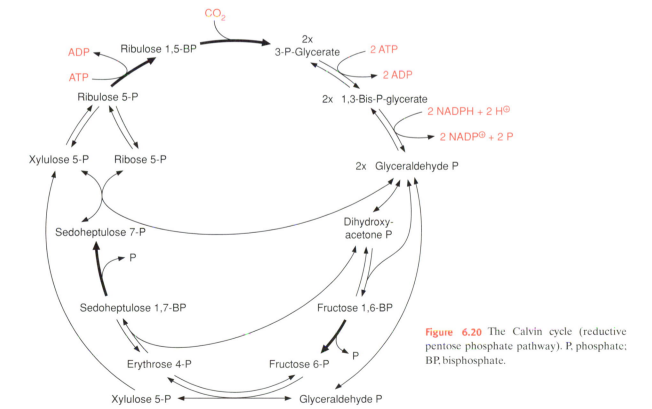

Figure 6.20 The Calvin cycle (reductive pentose phosphate pathway). P, phosphate; BP, bisphosphate.

6.5 Besides the reductive pentose phosphate pathway there is also an oxidative pentose phosphate pathway

Besides the reductive pentose phosphate pathway dealt with in the preceding section, the chloroplasts also contain the enzymes of an oxidative pentose phosphate pathway. This pathway, which occurs in both the plant and animal kingdoms, oxidizes a hexose phosphate to a pentose phosphate with the release of one molecule of CO_2. This pathway provides NADPH as 'high-pressure hydrogen' for biosynthetic processes. Figure 6.21 shows the pathway. Glucose 6-phosphate is first oxidized by *glucose 6-phosphate dehydrogenase* to 6-phosphogluconolactone (Fig. 6.22). This reaction is highly exergonic and therefore not reversible. 6-

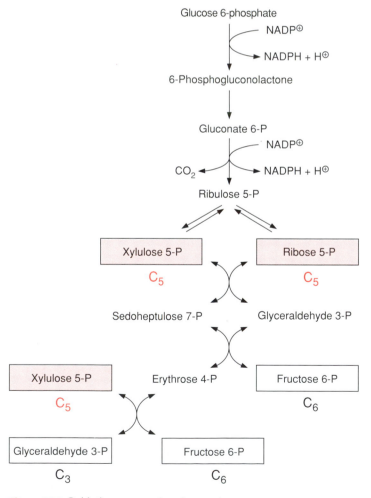

Figure 6.21 Oxidative pentose phosphate pathway.

Phosphogluconolactone, an internal ester, is hydrolysed by *lactonase*. The gluconate 6-phosphate thus formed is oxidized by the enzyme *gluconate 6-phosphate dehydrogenase* to ribulose 5-phosphate. In this reaction CO_2 is released and NADPH formed.

In the oxidative pathway xylulose 5-phosphate is formed from ribulose 5-phos-

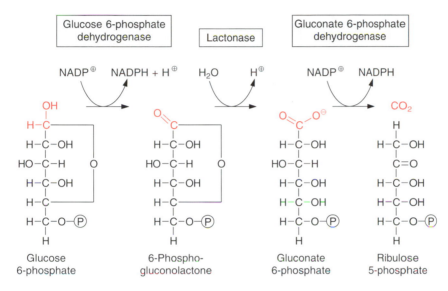

Figure 6.22 The two oxidation reactions of the pentose phosphate pathway.

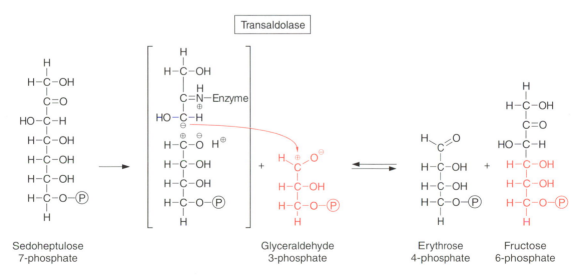

Figure 6.23 Transaldolase catalyses the transfer of the C_3 residue from a ketone to an aldehyde. The reaction is reversible.

phate by *ribulose phosphate epimerase* and from ribose 5-phosphate by *phosphoribose isomerase*. These two products are then converted by *transketolase* to sedoheptulose 7-phosphate and glyceraldehyde 3-phosphate. This reaction sequence is a reversal of the reductive pentose phosphate pathway. The further reaction sequence is a special feature of the oxidative pathway: *transaldolase* transfers a non-phosphorylated C₃ residue from sedoheptulose 7-phosphate to glyceraldehyde 3-phosphate, forming fructose 6-phosphate and erythrose 4-phosphate (Fig. 6.23). The reaction mechanism is basically the same as in the aldolase reaction (Fig. 6.13), with the only difference that after the cleavage of the C–C bond, the remaining C₃ residue continues to be bound to the enzyme via a Schiff base, until it is transferred. Erythrose 4-phosphate reacts with another xylulose 5-phosphate via *transketolase* to form glyceraldehyde 3-phosphate and fructose 6-phosphate. In this way two hexose phosphates and one triose phosphate are formed from three pentose phosphates:

$$3C_5\text{-P} \rightleftarrows 2C_6\text{-P} + 1C_3\text{-P}.$$

This reaction chain is reversible. It allows the cell to provide ribose 5-phosphate for nucleotide biosynthesis when no NADPH is required.

In the oxidative pathway, two molecules of NADPH are gained from the oxidation of glucose 6-phosphate with the release of one molecule CO_2, whereas in the reductive pathway the fixation of one molecule CO_2 requires not only two molecules of NADPH but also three molecules of ATP (Fig. 6.24). This expenditure of energy makes it possible for the reductive pentose phosphate pathway to proceed with a very high flux rate in the opposite direction to the oxidative pathway.

6.6 Reductive and oxidative pentose phosphate pathways are regulated

The enzymes of the reductive as well as of the oxidative pentose phosphate pathway are located in the chloroplast stroma (Fig. 6.24). A simultaneous operation of both metabolic pathways, in which one molecule of CO_2 is reduced to a carbohydrate residue at the expense of three ATP and two NADPH, which would then be reoxidized by the oxidative pathway to CO_2, yielding two molecules of NADPH, would represent a futile cycle in which three molecules of ATP were wasted in each turn. This is prevented by metabolic regulation which ensures that key enzymes of the reductive pentose phosphate pathway are only active during illumination and switched off in darkness, whereas the first enzyme of the oxidative pentose phosphate pathway is only active in the dark.

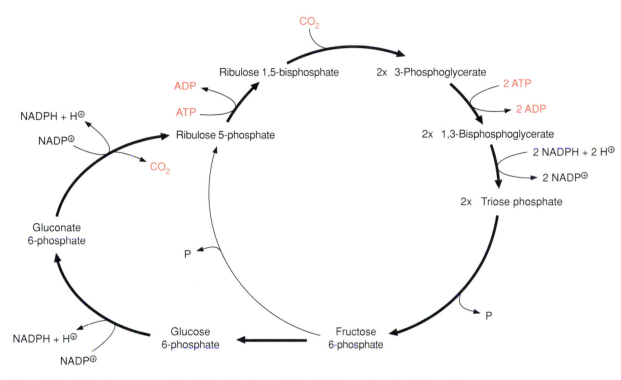

Figure 6.24 A simultaneous operation of the reductive and the oxidative pentose phosphate pathway would result in a futile, ATP-wasting cycle.

Reduced thioredoxins transmit the signal for 'illumination' to enzyme proteins

An important signal for the state 'illumination' is provided by photosynthetic electron transport as reducing equivalents in the form of reduced thioredoxin (Fig. 6.25). These reducing equivalents are transferred from ferredoxin to thioredoxin by the enzyme *ferredoxin–thioredoxin reductase.*

Thioredoxins form a family of small enzyme proteins, consisting of about 100 amino acids, which contain the sequence *Cys–Gly–Pro–Cys* in their catalytic centre, which is located at the periphery of the protein. Due to the neighbouring cysteine groups, the thioredoxin can be present in two redox states: the reduced thioredoxin with two SH-groups and the oxidized thioredoxin in which the two cysteines are linked by a disulfide (S–S) bridge.

Thioredoxins are found in all living organisms, from the archaebacteria to plants and animals. They function as *protein disulfide oxidoreductases*, in reducing disulfide bridges in target proteins to the SH form and reoxidizing them again to the S–S form. Furthermore, thioredoxins participate as redox carriers in the reduction of substances of low molecular weight. Its function as redox carrier for the re-

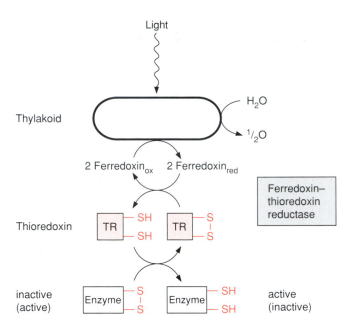

Figure 6.25 The light regulation of chloroplastic enzymes is mediated by reduced thioredoxin.

duction of ribonucleotides to deoxyribonucleotides led to the discovery of thioredoxin. Thioredoxin participates as a redox partner in the reduction of sulfate, a process occurring in plants and micro-organisms (section 12.1). An abundance of other processes are known in which thioredoxins play an essential role, ranging from the assembly of bacteriophages to hormone action or the blood-clotting process in animals. The reductive activation of seed proteins during germination is another example of a function of thioredoxins in plants. The involvement of thioredoxins in the light regulation of chloroplastic enzymes may be looked upon as a special function which they have attained in addition to their general metabolic functions.

The chloroplastic enzymes ribulose phosphate kinase, sedoheptulose 1,7-bisphosphatase, and the chloroplastic isoform of fructose 1,6-bisphosphatase are converted from an inactive state to an active state by reduced thioredoxin and are thus switched on by light. This also applies to other chloroplastic enzymes such as NADP–malate dehydrogenase (section 7.3) and F-ATPase (section 4.4). Reduced thioredoxin also converts NADP glyceraldehyde phosphate dehydrogenase contained in the chloroplasts from a less active to a more active state. On the other hand, reduced thioredoxin inactivates glucose 6-phosphate dehydrogenase, the first enzyme of the oxidative pentose phosphate pathway.

The thioredoxin-modulated activation of chloroplastic enzymes releases an inbuilt blockage

Important knowledge of the mechanism of thioredoxin action on the chloroplastic enzymes has been obtained from comparison with the corresponding isoenzymes from other cellular compartments. Isoenzymes of chloroplastic fructose 1,6-bis-phosphatase and malate dehydrogenase exist in the cytosol which are not regulated by thioredoxin. This also applies to F-ATPase in the mitochondria. A comparison of the amino acid sequences shows that at least in some cases the chloroplastic isoenzymes possess additional sections at the end or in an inner region of their sequence, containing two cysteine residues (Fig. 6.26). The SH-groups of these cysteine residues can be converted by oxidation to a disulfide and represent the substrate for the protein disulfide oxidoreductase activity of thioredoxin.

The isoenzymes, which are not regulated by thioredoxin and therefore do not contain these additional sequences, often have a higher activity than the isoenzymes regulated by thioredoxin. An exchange of the cysteine residues involved in the regulation for other amino acids by genetic engineering (Chapter 22) resulted in enzymes which were fully active also in the absence of reduced thioredoxin. Under oxidizing conditions the enzymes regulated by thioredoxin are forced by the formation of a disulfide bridge into a conformation in which the catalytic centre is inactivated. The reduction of this disulfide bridge by thioredoxin releases this blockage and the enzyme protein is converted to a relaxed conformation in which the catalytic centre is active.

The light activation discussed so far is not an all or nothing effect. It is due to a continuous change between the thioredoxin-mediated reduction of the enzyme protein and its simultaneous oxidation by oxygen. The degree of activation of the enzyme depends on the rate of reduction. This is due not only to the degree of the

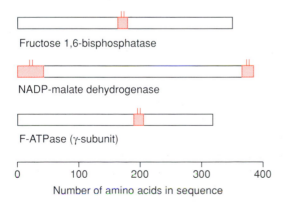

Fructose 1,6-bisphosphatase

NADP-malate dehydrogenase

F-ATPase (γ-subunit)

0   100   200   300   400

Number of amino acids in sequence

Figure 6.26 In contrast to the non-plastidic isoenzymes, several thioredoxin-modulated chloroplastic enzymes contain additional sections in the sequence (marked red) in which two cysteine residues are located (after Scheibe 1990).

reduction of thioredoxin (and thus to the degree of reduction of ferredoxin), but also to the presence of metabolites. Thus the reductive activation of fructose- and sedoheptulose bisphosphatases is enhanced by the corresponding bisphosphates. These effectors cause a decrease in the redox potential of the SH groups in the corresponding enzymes, which enhances the reduction of the disulfide group by thioredoxin. In this way the activity of these enzymes is increased when the concentration of their substrates rises. On the other hand, the reductive activation of NADP malate dehydrogenase is decreased by the presence of $NADP^+$. This has the effect that the enzyme is only active at a high NADPH/NADP ratio. In contrast, the oxidative activation of glucose 6-phosphate dehydrogenase is enhanced by NADP and this increases the activity of the oxidative pentose phosphate pathway when there is a demand for NADPH.

An abundance of further regulatory processes ensures that the various steps of the reductive pentose phosphate pathway are matched

An additional light regulation of the Calvin cycle is based on the effect of light-dependent changes of the proton and Mg^{2+} concentrations in the stroma on the activity of chloroplastic enzymes. When isolated chloroplasts are illuminated, the acidification of the thylakoid space (Chapter 3) is accompanied by an alkalization and an increase in the Mg^{2+} concentration in the stroma. During a dark–light transition the pH in the stroma may change from about pH 7.2 to 8.0. CO_2 fixation by isolated chloroplasts shows an optimum at about pH 8.0 with a sharp decline towards the acidic range. An almost identical pH dependence is found with the light-activated enzymes fructose 1,6-bisphosphatase and sedoheptulose 1,7-bisphosphatase. Moreover, the catalytic activity of both these enzymes is increased by the light-dependent increase in the Mg^{2+} concentration in the stroma. The light activation of these enzymes by the thioredoxin system, together with the increase in enzyme activity by light-induced changes of the pH and Mg^{2+} concentration in the stroma, of which each alone results in an extensive inactivation of the corresponding enzymes during darkness, is a very efficient system for switching these enzymes on and off, according to demand.

The activities of several stromal enzymes are also regulated by metabolite levels. The chloroplastic fructose 1,6-bisphosphatase and sedoheptulose 1,7-bisphosphatase are inhibited by their corresponding products, fructose 6-phosphate and sedoheptulose 7-phosphate, respectively. This can decrease the activity of these enzymes when their products accumulate. Ribulose phosphate kinase is inhibited by 3-phosphoglycerate and also by ADP. Inhibition by ADP is important for coordinating the two kinase reactions of the reductive pentose phosphate pathway. Whereas ribulose phosphate kinase catalyses an irreversible reaction, the phosphoglycerate kinase reaction is reversible. If both reactions were to compete for ATP in an unrestricted manner, in the case of a shortage of ATP, the irreversible phosphorylation of ribulose 5-phosphate would be at an advantage, resulting in an imbalance of the Calvin cycle. A decrease in the activity of ribulose phosphate kinase at an elevated level of ADP can prevent this.

Finally one observes a strong inhibition of fructose 1,6-bisphosphatase and of sedoheptulose 1,7-bisphosphatase by glycerate. As shown in section 7.1 glycerate is an intermediate in the recycling of phosphoglycolate formed by the oxygenase activity of RubisCO. This inhibition allows the accumulation of glycerate to slow down the regeneration of RuBP and its carboxylation, and thus also to lower the accompanying oxygenation resulting in the formation of glycolate as the precursor of glycerate.

Despite intensive research, the mechanism of the light regulation of RubisCO is still unclear. Experiments with whole leaves demonstrate that the degree of the activation of RubisCO correlates with the intensity of illumination and the rate of photosynthesis. The activation of RubisCO via the formation of the carbamate–Mg^{2+} complex (Fig. 6.7) might be increased by the light-dependent increase in the concentration of Mg^{2+} and the corresponding rise in the H^+ concentration in the

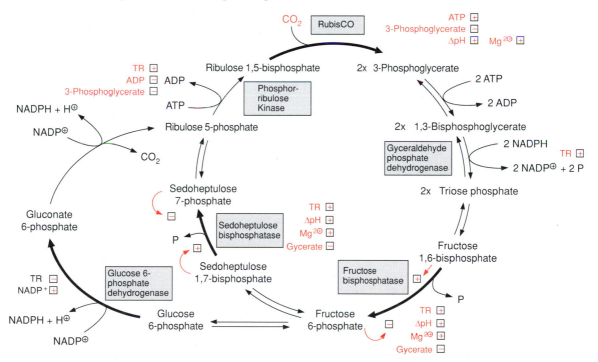

Figure 6.27 Regulation of the reductive and oxidative pentose phosphate pathways. Both pathways are represented in a simplified scheme. Only those enzymes for which regulation has been discussed are highlighted. ⊞ increase and ⊟ decrease indicates an activity caused by the factors marked in red, such as reduced thioredoxin (TR), light-dependent alkalization (ΔpH) and the increase in the Mg^{2+} concentration in the stroma and the presence of metabolites.

stroma. The activation state of RubisCO is probably adjusted via regulation of RubisCO activase. The activity of RubisCO activase is dependent on the ATP/ADP ratio. When there is rise in the ATP/ADP ratio in the stroma, the activity of the activase also rises. It is speculated that this is how the activity of RubisCO is adjusted to the delivery of ATP by the light reaction of photosynthesis. However, many observations suggest that this cannot be the only mechanism for a light regulation of RubisCO. It has been suggested that RubisCO activase is regulated by the light-dependent proton gradient across the thylakoid membrane. Furthermore, the activity of RubisCO is inhibited by its product 3-phosphoglycerate. In this way the activity of RubisCO can be decreased when its product accumulates.

Figure 6.27 shows a scheme of the various factors which influence the regulation of the enzymes of the reductive and oxidative pentose pathways. An abundance of regulatory processes ensures that the various steps of both reaction chains are adjusted to each other and to the demand of the cell.

Further reading

Andrews, T. J. and Lorimer, G. H. (1987). RubisCO: structure, mechanism and prospects for improvement. In *The biochemistry of plants*, Vol. 10 (ed. M. D. Hatch, N. K. Boardman), pp. 131–218. Academic Press, New York.

Buchanan, B. B. (1984). The ferredoxin/thioredoxin system: a key element in the regulatory function of light in photosynthesis. *BioScience*, **34**, 378–83.

Edwards, G. and Walker, D. A. (1983). *C₃, C₄ : Mechanisms and cellular and environmental regulation of photosynthesis*. Blackwell Scientific Publications, Oxford.

Furbank, R. T. and Taylor, W. C. (1995). Regulation of photosynthesis in C₃ and C₄ plants. *The Plant Cell*, **7**, 797–807.

Geiger, D. R. and Servaites, J. C. (1994). Diurnal regulation of photosynthetic carbon metabolism in C₃ plants. *Annual Review of Plant Physiology and Plant Molecular Biology*, **45**, 235–56.

Gutteridge, S. and Gatenby, A. (1995). RubisCO synthesis, assembly, mechanism and regulation. *The Plant Cell*, **7**, 809–19.

Portis, A. R. (1992). Regulation of ribulose 1,5-bisphosphate carboxylase/oxygenase activity. *Annual Review of Plant Physiology and Plant Molecular Biology*, **43**, 415–37.

Scheibe, R. (1990). Light/dark modulation: regulation of chloroplast metabolism in a new light. *Botanica Acta*, **103**, 327–34.

Schnarrenberger, C., Flechner, A., and Martin, W. (1995). Enzymatic evidence for a complete oxidative pentose pathway in chloroplasts and an incomplete pathway in the cytosol of spinach leaves. *Plant Physiology*, **108**, 609–14.

Stitt, M. (1997). Metabolic regulation of photosynthesis. In *Advances in Photosynthesis*, Vol. 5. *Environmental Stress and Photosynthesis*, (ed. N. Baker). Academic Press, 151–90.

Woodrow, I. E. and Berry, J. A. (1988). Enzymatic regulation of photosynthetic CO₂ fixation in C₃ plants. *Annual Review of Plant Physiology and Plant Molecular Biology*, **39**, 533–94.

chapter 7

Photorespiration

Section 6.1 described how large amounts of 2-phosphoglycolate are formed as a by-product during CO_2 fixation by RubisCO, due to the oxygenase activity of this enzyme. In the photorespiratory pathway, discovered in 1972 by the American scientist Edward Tolbert, the by-product 2-phosphoglycolate is recycled to ribulose 1,5-bisphosphate. The term photorespiration indicates that the process involves oxygen consumption occurring in the light which is accompanied by release of CO_2. Whereas in mitochondrial respiration (cell respiration; Chapter 5) the oxidation of substrates to CO_2 serves the purpose of producing ATP, in the case of photorespiration ATP is consumed.

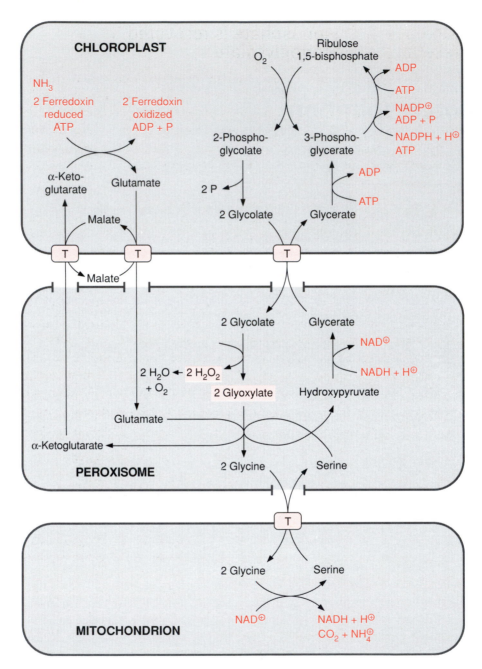

Figure 7.1 Compartmentalization of the photorespiration pathway. This scheme does not show the outer membranes of the chloroplasts and mitochondria, which are non-specifically permeable to metabolites, due to the presence of porins. T = translocator.

7.1 Ribulose 1,5-bisphosphate is recovered by recycling 2-phosphoglycolate

Figure 7.1 gives an overview of the reactions of the photorespiratory pathway and their localization. Recycling of 2-phosphoglycolate begins with the hydrolytic release of phosphate by *phosphoglycolate phosphatase* present in the chloroplast stroma (Fig. 7.2). The resultant glycolate leaves the chloroplasts by a specific translocator located in the inner envelope membrane and enters the peroxisomes via non-specific *pores* in the peroxisomal boundary membrane, probably formed by a *porin* (section 1.11).

In the peroxisomes the alcoholic group of glycolate is oxidized to a carbonyl group in an irreversible reaction catalysed by *glycolate oxidase*, resulting in the formation of glyoxylate. The reducing equivalents are transferred to molecular oxygen, forming H_2O_2 (Fig. 7.2). Like other H_2O_2-forming oxidases, glycolate oxidase contains a flavin mononucleotide cofactor (FMN, Fig. 5.16) as redox mediator between glycolate and oxygen. The H_2O_2 formed is converted to water and oxygen by the enzyme *catalase* present in the peroxisomes. Thus, in total, 0.5 mol of O_2 is consumed for the oxidation of 1 mol of glycolate to glyoxylate.

The glyoxylate formed is converted to the amino acid glycine by two different reactions proceeding in the peroxisome simultaneously at a 1 : 1 ratio. The enzyme *glutamate–glyoxylate aminotransferase* catalyses the transfer of an amino group from the donor glutamate to glyoxylate. This enzyme also reacts with alanine as amino donor. In the other reaction the enzyme *serine–glyoxylate aminotransferase* catalyses the transamination of glyoxylate by serine. These two aminotransferases,

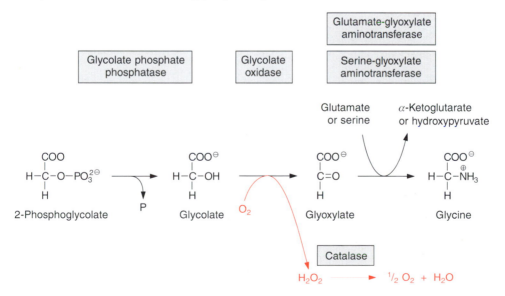

Figure 7.2 Reaction sequence for the conversion of 2-phosphoglycolate to glycine.

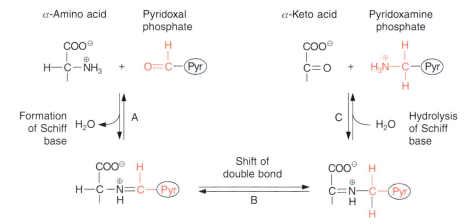

Pyridoxal phosphate   **Figure 7.3**

like other aminotransferases (e.g. glutamate–oxaloacetate aminotransferase, see section 8.4), contain bound pyridoxal phosphate with an aldehyde function as reactive group (Fig. 7.3). Figure 7.4 shows the reaction sequence.

The glycine formed leaves the peroxisomes via pores and is transported into the mitochondria. Although this transport has not yet been characterized in detail, one would expect it to proceed via a specific translocator. In the mitochondria two molecules of glycine are oxidized yielding one molecule of serine with release of CO_2 and NH_4^+ and a transfer of reducing equivalents to NAD^+ (Fig. 7.5). The oxidation of glycine is catalysed by the *glycine decarboxylase–serine hydroxy-methyltransferase complex*. This is a multienzyme complex, consisting of four different subunits (Fig. 7.7), which shows a great similarity to the pyruvate dehydrogenase complex described in section 5.3. The so-called *H-protein* with the

Figure 7.4 Reaction sequence of the aminotransferase reaction. The aldehyde group of pyridoxal phosphate forms a Schiff base with the α-amino group of the amino acid (in this case glutamate or serine) (A), which is subsequently converted to an isomeric form (B) by a base-catalysed movement of a proton. Hydrolysis of the isomeric Schiff base results in the formation of an α-ketoacid (α-keto-glutarate or hydroxypyruvate), and pyridoxamine remains (C). The amino group of this pyridoxamine then forms a Schiff base with another α-ketoacid (in this case glyoxylate), and glycine is formed by a reversal of the steps C, B, and A. Pyridoxal is thus regenerated and is available for the next reaction cycle.

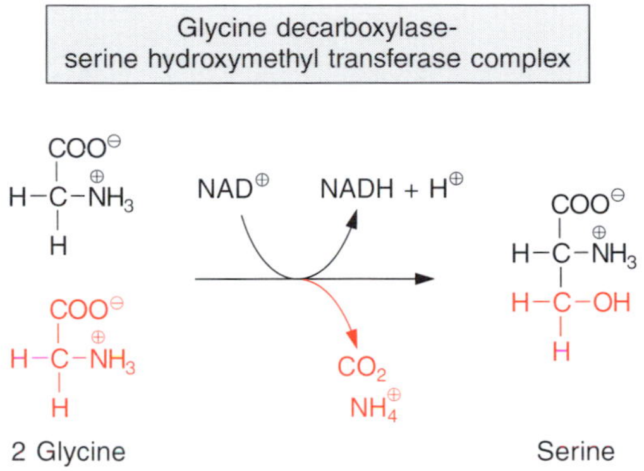

Figure 7.5 Overall reaction scheme for the conversion of two molecules of glycine to one molecule of serine as catalysed by the glycine decarboxylase complex.

prosthetic group lipoic acid amide (Fig. 5.5) represents the centre of the glycine decarboxylase complex. Around this centre are positioned the pyridoxal phosphate-containing *P-protein*, the *T-protein* with a tetrahydrofolate (Fig. 7.6) as a prosthetic group, and the *L-protein*, also named dihydrolipoate dehydrogenase, which is identical to the dihydrolipoate dehydrogenase of the pyruvate and α-

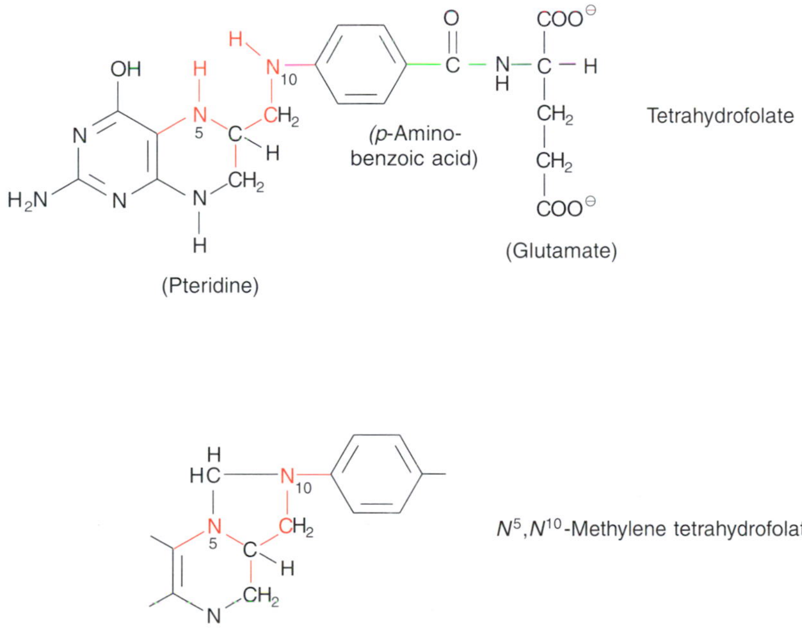

Figure 7.6 Tetrahydrofolate.

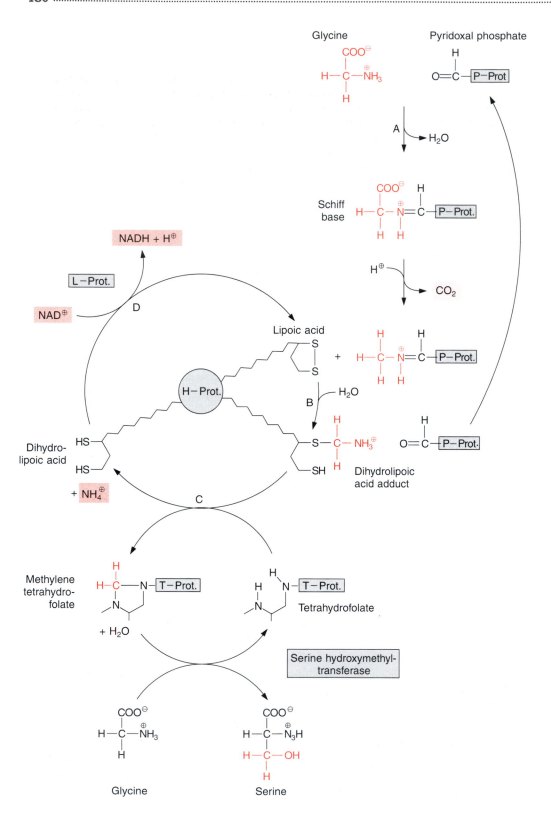

Glycine

Pyridoxal phosphate

Schiff base

Lipoic acid

Dihydro-lipoic acid

Dihydrolipoic acid adduct

Methylene tetrahydro-folate

Tetrahydrofolate

Serine hydroxymethyl-transferase

Glycine

Serine

Figure 7.7 Reaction sequence for the conversion of two molecules of glycine to one molecule of serine. The amino group of glycine first reacts with the aldehyde group of pyridoxal in the P-protein to form a Schiff base (A). The glycine residue is then decarboxylated and transferred from the P-protein to the lipoic acid residue of the H-protein (B). This is the actual oxidation step: the C_1 residue is oxidized to a formyl group and the lipoic acid residue is reduced to dihydrolipoic acid. The dihydrolipoic acid adduct then reacts with the T-protein, the C-residue is transferred to tetrahydrofolate, and the dihydrolipoic acid residue remains (C). The dihydrolipoic acid is reoxidized via the L-protein (dihydrolipoate dehydrogenase) to lipoic acid and the reducing equivalents are transferred to NAD (D). A new reaction cycle can then begin. The formyl residue bound to tetrahydrofolate is transferred to a second molecule of glycine by serine hydroxymethyl transferase and serine is formed (E).

ketoglutarate dehydrogenase complex (Figs. 5.4 and 5.8). Since the disulfide group of the lipoic acid amide in the H-protein is located at the end of a flexible polypeptide chain (see also Fig. 5.4), it is able to react with the three other subunits. Figure 7.7 shows the reaction sequence. The enzyme *serine hydroxymethyltransferase*, which is in close proximity to the glycine decarboxylase complex, catalyses the transfer of the formyl residue to another molecule of glycine to form serine.

The NADH produced in the mitochondrial matrix from glycine oxidation can be oxidized by the mitochondrial respiratory chain in order to generate ATP. Alternatively, these reducing equivalents can be exported from the mitochondria to other cell compartments, as will be discussed in section 7.3. The capacity for glycine oxidation in the mitochondria of green plant cells is very high. The glycine decarboxylase complex of the mitochondria can amount to 30–50 per cent of the total content of soluble proteins. In mitochondria of non-green plant cells, however, the proteins of glycine oxidation are present only in very low amounts or are absent.

Serine probably leaves the mitochondria via a specific translocator, possibly the same translocator by which glycine is taken up. After entering the peroxisomes through pores, serine is converted to hydroxypyruvate by the enzyme *serine–glyoxylate aminotransferase* mentioned above (Fig. 7.8). At the expense of NADH, hydroxypyruvate is reduced by *hydroxypyruvate reductase* to glycerate, which is released from the peroxisomes and imported into the chloroplasts.

The uptake of glycerate into the chloroplasts proceeds by the same translocator as that catalysing the release of glycolate from the chloroplasts (*glycolate–glycerate translocator*). This translocator facilitates a glycolate–glycerate counterexchange as well as a co-transport of just glycolate with a proton. In this way, the translocator enables the export of two molecules of glycolate from the chloroplasts in exchange for the import of one molecule of glycerate. Glycerate is converted to 3-phosphoglycerate, consuming ATP by *glycerate kinase* present in the chloroplast stroma. Finally, phosphoglycerate is reconverted to ribulose 1,5-bisphosphate via the reductive pentose phosphate pathway (section 6.3). These reactions complete the recycling of 2-phosphoglycolate.

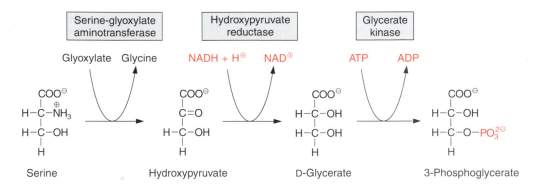

Figure 7.8 Reaction sequence for the conversion of serine to 3-phosphoglycerate.

7.2 The NH_4^+ released in the photorespiratory pathway is refixed in the chloroplasts

Nitrogen is an important plant nutrient. Nitrogen supply is often a limiting factor in plant growth. It is therefore necessary for the economy of plant metabolism that ammonium, which is released at very high rates in the photorespiratory pathway during glycine oxidation, is completely refixed. This refixation occurs in the chloroplasts. It is catalysed by the same enzymes as those participating in nitrate assimilation (Chapter 10). However, the rate of NH_4^+ refixation in the photorespiratory pathway is 5–10 times higher than the rate of NH_4^+ fixation in nitrate assimilation.

In a plant cell, chloroplasts and mitochondria are mostly in close proximity to each other. The NH_4^+ formed during oxidation of glycine passes through the inner membrane of the mitochondria and of the chloroplasts. Whether this passage occurs by simple diffusion or is facilitated by specific translocators or ion channels is still a matter of debate. The enzyme *glutamine synthetase*, present in the chloroplast stroma, catalyses the transfer of an ammonium ion to the δ-carboxyl group of glutamate (Fig. 7.9) to form glutamine. This reaction is driven by the conversion of one molecule of ATP to ADP and phosphate. In an intermediary step the δ-carboxyl group is activated by reaction with ATP to form a carboxy-phosphate anhydride. Glutamine synthetase has a high affinity for NH_4^+ and catalyses an essentially irreversible reaction. This enzyme has a key role in the fixation of NH_4^+ not only in plants, but also in bacteria and animals.

The nitrogen fixed as amide in glutamine is transferred by reductive amination to α-ketoglutarate (Fig. 7.9). In this reaction catalysed by *glutamate synthase*, two molecules of glutamate are formed. The reducing equivalents are provided by reduced ferredoxin, which is a product of photosynthetic electron transport (section 3.8). In green plant cells glutamate synthase is located exclusively in the chloroplasts.

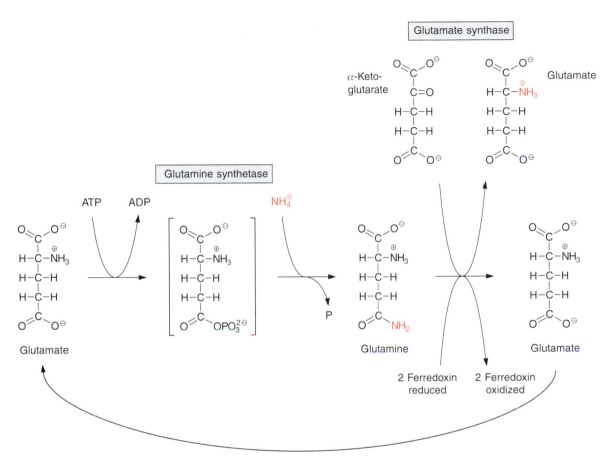

Figure 7.9 Reaction sequence for the fixation of ammonia with subsequent formation of glutamate from α-ketoglutarate.

Of the two glutamate molecules thus formed in the chloroplasts, one is exported by the *glutamate–malate translocator* in exchange for malate and, after entering the peroxisomes, is available as reaction partner for the transamination of glyoxy-late. α-Ketoglutarate formed in the peroxisomes is re-imported into the chloro-plasts by a *malate–α-ketoglutarate translocator*, again in counter-exchange for malate.

7.3 Peroxisomes have to be provided with reducing equivalents to reduce hydroxypyruvate

NADH is required as reductant for the conversion of hydroxypyruvate to glycerate in the peroxisomes. Since leaf peroxisomes have no metabolic pathway capable of delivering NADH at the very high rates required, peroxisomes are dependent on the supply of reducing equivalents from outside.

Reducing equivalents are taken up into the peroxisomes via a malate–oxaloacetate shuttle

The cytosolic NADH system of a leaf cell is oxidized to such an extent (NADH/NAD$^+$ = 10^{-3}), that the concentration of NADH in the cytosol is only about 10^{-6} mol l^{-1}. This very low concentration would not allow a diffusion gradient to build up, which is high enough to drive the necessary high diffusive fluxes of reducing equivalents in the form of NADH into the peroxisomes. Instead, the reducing equivalents are imported indirectly into the peroxisomes via the uptake of malate and the subsequent release of oxaloacetate (this is termed a malate–oxaloacetate shuttle) (Fig. 7.10).

Malate dehydrogenase (Fig. 5.9), which catalyses the oxidation of malate to

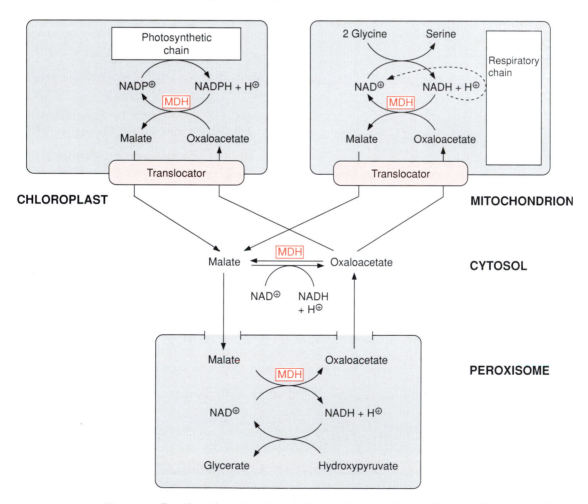

Figure 7.10 Reaction scheme for the transfer of reducing equivalents from the chloroplasts and the mitochondria to the peroxisomes; MDH, malate dehydrogenase.

oxaloacetate in a reversible reaction, has a key function in this shuttle. High malate dehydrogenase activity is found in the cytosol as well as in chloroplasts, mitochondria, and peroxisomes. The malate dehydrogenases in the various cell compartments are considered to be isoenzymes. They show some differences in their structure and are encoded by different, but homologous genes. Apparently, these are all related proteins, which have derived in the course of evolution from a common precursor. Whereas the NADH system is redox partner for malate dehydrogenases in the cytosol, mitochondria, and peroxisomes, the chloroplastic isoenzyme reacts with the NADPH system.

Mitochondria export reducing equivalents via a malate–oxaloacetate cycle

In contrast to mitochondria from animal tissues, where the inner membrane is impermeable to oxaloacetate, plant mitochondria contain in their inner membrane a specific *malate–oxaloacetate translocator*, which transports malate and oxaloacetate in a counter-exchange. Since the activity of malate dehydrogenase in the mitochondrial matrix is very high, NADH formed in mitochondria during glycine oxidation can be captured to reduce oxaloacetate to form malate for export by the *malate-oxaloacetate shuttle*. This shuttle has a high capacity. As can be seen from Fig. 7.1, the amount of NADH generated in the mitochondria from glycine oxidation is equal to that of NADH required for the reduction of hydroxypyruvate in the peroxisomes. If all the oxaloacetate formed in the peroxisomes should reach the mitochondria, the NADH generated from glycine oxidation would be totally consumed in the formation of malate and would no longer be available to support ATP synthesis by the respiratory chain. However, mitochondrial ATP synthesis is required during photosynthesis to supply energy to the cytosol of mesophyll cells. In fact, mitochondria deliver only about half the reducing equivalents required for peroxisomal hydroxypyruvate reduction, with the remaining portion being provided by the chloroplasts (Fig. 7.10). Thus only about half of the NADH formed during glycine oxidation is captured by the malate–oxaloacetate shuttle for export, and the remaining NADH is oxidized by the respiratory chain for synthesis of ATP.

A 'malate valve' controls the export of reducing equivalents from the chloroplasts

Chloroplasts are also able to export reducing equivalents by a malate–oxaloacetate cycle. Malate and oxaloacetate are transported across the inner envelope membrane via a specific translocator, probably in a counter-exchange. In spite of the high activity of the chloroplastic malate–oxaloacetate shuttle, a high gradient exists between the chloroplastic and cytosolic redox systems: the NADPH/NADP$^+$ ratio in chloroplasts is more than 100 times higher than the corresponding

NADH/NAD$^+$ ratio in the cytosol. Whereas malate dehydrogenases usually catalyse a reversible equilibrium reaction, the reduction of oxaloacetate by chloroplastic malate dehydrogenase is virtually irreversible and does not reach equilibrium. This is due to a regulation of chloroplastic malate dehydrogenase.

Section 6.6 described how chloroplastic malate dehydrogenase is activated by thioredoxin and therefore only active in the light. In addition to this, increasing concentrations of NADP$^+$ inhibit the reductive activation of the enzyme by *thioredoxin*. NADP$^+$ increases the redox potential of the regulatory SH-groups of malate dehydrogenase, and, as a result of this, the reductive activation of the enzyme by thioredoxin is lowered.

Thus a decrease in the NADP$^+$ concentration, which corresponds to an increase in the reduction of the NADPH/NADP$^+$ system, switches on chloroplastic malate dehydrogenase. This allows the enzyme to function like a *valve*, through which excessive reducing equivalents can be released by the chloroplasts to prevent harmful over-reduction of the redox carriers of the photosynthetic electron transport chain. At the same time, this valve makes it possible for the chloroplasts to provide reducing equivalents for the reduction of hydroxypyruvate in the peroxisomes and also for other processes, e.g. nitrate reduction in the cytosol (section 10.1).

An alternative way of exporting reducing equivalents from chloroplasts to the cytosol is the triose phosphate–3-phosphoglycerate shuttle (Fig. 7.11). By this shuttle, ATP is delivered together with NADH.

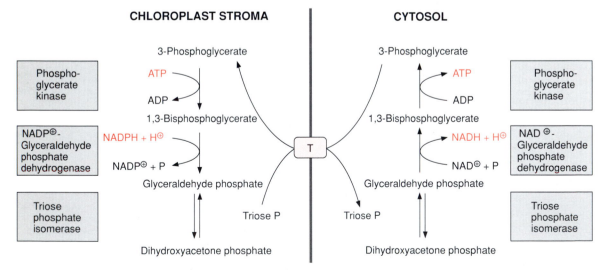

Figure 7.11 Triose phosphate–3-phosphoglycerate shuttle operating between the chloroplast stroma and the cytosol. In the chloroplast stroma triose phosphate is formed from 3-phosphoglycerate at the expense of NADPH and ATP. Triose phosphate is transported by the triose phosphate–phosphate translocator across the inner envelope membrane in exchange for 3-phosphoglycerate. In the cytosol triose phosphate is reconverted to 3-phosphoglycerate with the generation of NADH and ATP.

7.4 The peroxisomal matrix is a special compartment for the metabolism of toxic products

The question arises, why, besides the chloroplasts, are two other organelles involved in the process for recycling 2-phosphoglycolate? That mitochondria are the site of the conversion of glycine to serine appears to have the advantage that the respiratory chain can utilize the resultant NADH for synthesis of ATP. During the conversion of glycolate to glycine two toxic intermediates are formed: glyoxylate and H_2O_2. In isolated chloroplasts photosynthesis is completely inhibited by the addition of low concentrations of H_2O_2 or glyoxylate. The inhibitory effect of H_2O_2 is due to the oxidation of SH-groups in thioredoxin-activated enzymes of the reductive pentose phosphate pathway, resulting in their inactivation. Glyoxylate, a very reactive carbonyl compound, also has a strong inhibitory effect on thioredoxin-activated enzymes by reacting also with their SH-groups. Glyoxylate also inhibits RubisCO. Compartmentalization of the conversion of glycolate to glycine in the peroxisomes serves the purpose of eliminating the toxic intermediate products glyoxylate and H_2O_2 at the site of their formation so that they do not invade other cell compartments.

How is such a compartmentalization possible? Compartmentalization of metabolic processes in other cell compartments, e.g. the chloroplast stroma or the mitochondrial matrix, is achieved by separating membranes, which contain specific translocators for the passage of certain metabolites but are impermeable to metabolic intermediates present in these different compartments. This principle, however, does not apply to the compartmentalization of glycolate oxidation, since membranes are normally quite permeable to H_2O_2 as well as to glyoxylate. For this reason a boundary membrane would be unable to prevent these substances escaping from the peroxisomes.

The very efficient compartmentalization of the conversion of glycolate to glycine and of serine to glycerate in the peroxisomes is due to specific properties of the peroxisomal matrix. When the boundary membrane of chloroplasts or mitochondria is disrupted, e.g. by suspending the organelles for a short time in pure water to cause an osmotic shock, the proteins of the stroma or the matrix, respectively, which are all soluble, are released from these disrupted organelles. However, in peroxisomes, after disruption of the boundary membrane the peroxisomal matrix proteins remain aggregated in the form of particles of a size similar to peroxisomes, and the compartmentalization of the peroxisomal reactions is maintained. Glyoxylate, H_2O_2, and hydroxypyruvate, intermediates of peroxisomal metabolism, are not released from these particles in the course of glycolate oxidation. Apparently, the enzymes of the photorespiratory pathway are arranged in the peroxisomal matrix in the form of a multienzyme complex, by which the product of one enzymatic reaction binds immediately to the enzyme of the following reaction and is therefore not released.

This process, termed '*metabolite channelling*', probably occurs not only in the peroxisomal matrix but may also occur in a similar orderly manner in other metabolic pathways, e.g. the Calvin cycle in the chloroplast stroma (Chapter 6). It seems to be a special feature of the peroxisomes, however, that such an enzyme complex remains intact after disruption of the boundary membrane. This may be a protective function to avoid the escape of glycolate oxidase after eventual damage to the peroxisomal membrane. An escape of glycolate oxidase would result in glycolate being oxidized outside the peroxisomes and poisoning of the cell due to the accumulation of the products glyoxylate and H_2O_2 in the cytosol.

For any glyoxylate and hydroxypyruvate leaking out of the peroxisomes in spite of metabolite channelling, rescue enzymes are present in the cytosol which use NADPH to convert glyoxylate to glycolate (*NADPH-glyoxylate reductase*) and hydroxypyruvate into glycerate (*NADPH-hydroxypyruvate reductase*).

7.5 How high are the costs of the ribulose bisphosphate oxygenase reaction for the plant?

On the basis of the metabolic schemes in Figs. 6.20 and 7.1 the expenditure in ATP and NADPH (respectively the equivalent of two reduced ferredoxins) for oxygenation and carboxylation of RuBP by RubisCO is listed in Table 7.1. The data

Table 7.1 Expenditure of ATP and NADPH for carboxylation of ribulose 1,5-bisphosphate in comparison to the corresponding expenditure for oxygenation

		ATP	NADPH or 2 reduced ferredoxin
Carboxylation:			
Fixation of 1 mol CO_2:			
1 CO_2	$\rightarrow$ 0.33 triose phosphate	**3**	**2**
Oxygenation:			
2 ribulose 1,5-bisphosphate	+ 2 O_2 $\rightarrow$ 2 3-phosphoglycerate + 2 2-phosphoglycolate		
2 2-phosphoglycolate	$\rightarrow$ 3-phosphoglycerate + 1 CO_2	2	1
1 CO_2	$\rightarrow$ 0.33 triose phosphate	3	2
3 3-phosphoglycerate	$\rightarrow$ 3 triose phosphate	3	3
3.33 triose phosphate	$\rightarrow$ 2 ribulose 1,5-bisphosphate	2	
		Σ 10	6
Oxygenation by 1 mol O_2:		5	3

Table 7.2 Additional consumption for RuBP oxygenation as related to the consumption for CO_2 fixation

Carboxylation/oxygenation ratio	Additional consumption	
	ATP	NADPH
2	83%	75%
4	42%	38%

illustrate that the consumption of ATP and NADPH, required to compensate the consequences of oxygenation, is much higher than the ATP and NADPH expenditure for carboxylation. Whereas in CO_2 fixation the conversion of CO_2 to triose phosphate requires three molecules of ATP and two of NADPH, the oxygenation of RuBP costs in total five molecules of ATP and three molecules of NADPH per molecule of O_2. Table 7.2 shows the additional expenditure in ATP and NADPH at various ratios of carboxylation to oxygenation. In the leaf, where the carboxylation/oxygenation ratio is usually between two and four, the additional expenditure of NADPH and ATP to compensate for the oxygenation is more than 50 per cent of the corresponding expenditure for CO_2 fixation. Thus the oxygenase side-reaction of RubisCO costs the plant more than one-third of the captured photons.

7.6 There is no net CO_2 fixation at the compensation point

At a carboxylation/oxygenation ratio of 1/2 there is no net CO_2 fixation, as the amount of CO_2 fixed by carboxylation is equal to the amount of CO_2 released by the photorespiratory pathway as a result of oxygenation. One can simulate this situation experimentally by illuminating a plant in a closed chamber. Due to photosynthesis, the CO_2 concentration decreases until it reaches a concentration at which the fixation of CO_2 and the release of CO_2 are counterbalanced. This state is termed the *compensation point*. Although the release of CO_2 is caused not only by the photorespiratory pathway but also by other reactions, e.g. the citrate cycle in mitochondria, the latter sources of CO_2 release are negligible compared with the photorespiratory pathway. For the plants dealt with so far, designated C_3 *plants* (this term is derived from the fact that the first carboxylation product is the C_3 compound 3-phosphoglycerate), the CO_2 concentration in air at the compensation point, depending on the species and temperature, is in the range of 35–70 p.p.m., which corresponds to 10–20 per cent of the CO_2 concentration in the atmosphere. In the aqueous phase, where RubisCO is present, this corresponds at

$25\,°C$ to a CO_2 concentration of $1\text{–}2 \times 10^{-6}$ mol l^{-1}. For C_4 *plants*, dealt with in section 8.4, the CO_2 concentration at the compensation point is only about 5 p.p.m. How these plants manage to have such a low compensation point in comparison with C_3 plants will be discussed in detail in section 8.4.

With a plant kept in a closed system, it is possible to decrease the CO_2 concentration to a value below the compensation point by trapping CO_2 with KOH. In this case, on illumination, oxygenation by RubisCO and the accompanying photorespiratory pathway results in a net release of CO_2 at the expense of the plant matter, which is degraded to produce carbohydrates to allow the regeneration of ribulose 1,5-bisphosphate. In such a situation illumination of a plant causes its consumption.

7.7 The photorespiration pathway, although energy-consuming, may also have a useful function for the plant

Due to the high ATP and NADPH consumption during photorespiration, photosynthetic metabolism proceeds at full speed at the compensation point, although there is no net CO_2 fixation. Such a situation arises when, in leaves exposed to full light, the stomata are closed because of water shortage (section 8.1) and therefore CO_2 cannot be taken up. An over-reduction and an over-energization of the photosynthetic electron transport carriers can cause severe damage to them (section 3.10). The plant utilizes the energy-consuming photorespiratory pathway to eliminate ATP and NADPH which have been produced by light reactions, but which cannot be used for CO_2 assimilation. Photorespiration, the unavoidable side-reaction of photosynthesis, is thus utilized by the plant for its protection. It is feasible, therefore, that lowering the oxygenase reaction of RubisCO by molecular engineering (Chapter 22), as attempted by many researchers, though still without success, may lead not only to the plant using energy more efficiently, but at the same time also may increase its vulnerability towards excessive illumination or shortage of water (see Chapter 8).

Further reading

Douce, R., Macherel D., and Neuburger M. (1994). The glycine decarboxylase system in higher plant mitochondria—structure, function and biogenesis. *Biochemical Society Transactions*, **22**, 184–8.

Heldt, H. W. and Flügge, U. I. (1992). Metabolite transport in plant cells. In *Plant organelles*, (ed. A. K. Tobin), pp. 21–47. Cambridge University Press, Cambridge.

Husic, D. W., Husic, H. D., and Tolbert, N. E. (1987). The oxidative photosynthetic carbon cycle or C_2 cycle. *CRC Critical Reviews in Plant Sciences*, **1**, 45–100.

Oliver, D. J. (1994). The glycine decarboxylase complex from plant mitochondria. *Annual Review of Plant Physiology and Plant Molecular Biology*, **45**, 323–37.

Reumann, S., Maier, E., Benz, R., and Heldt, H. W. (1995). The membrane of leaf peroxisomes contains a porin-like channel. *Journal of Biological Chemistry*, **270**, 17559–65.

Tolbert, N. E. (1980). Microbodies—peroxisomes and glyoxysomes. In *The biochemistry of plants*, (ed. P. K. Stumpf and E. E. Conn), Vol. 1, pp. 359–88. Academic Press, London.

chapter 8

Photosynthesis and water consumption

This chapter deals with the fact that photosynthesis is unavoidably linked with a substantial loss of water and therefore is often limited by the lack of it. Biochemical mechanisms will be described which enable certain plants living in hot and dry habitats to reduce their need of water.

8.1 The entry of CO_2 into the leaf is accompanied by an escape of water vapour from the leaves

Since CO_2 assimilation is linked with a high water demand, plants require an ample water supply for their growth. A normal plant growing in temperate climates requires 700–1300 mol H_2O for the fixation of 1 mol CO_2. In this assessment the water consumption for photosynthetic water oxidation is negligible in quantitative terms. Water demand is dictated by the fact that water escaping from the leaves in the form of water vapour has to be replenished by water taken up through the roots. Thus during photosynthesis there is a steady flow of water, termed the *transpiration stream*, from the roots via the xylem vessels into the leaves.

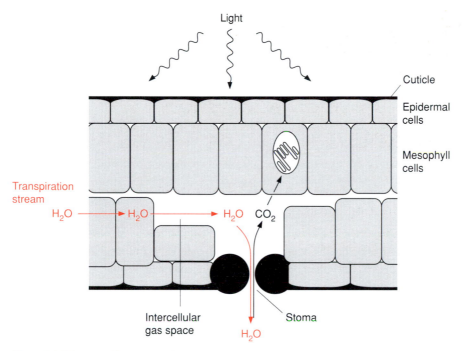

Figure 8.1 Diagram of a cross-section of a leaf. The stomata are often located on the lower surface of the leaf. CO_2 diffuses through the stomata into the intercellular air space and thus reaches the mesophyll cells carrying out photosynthesis. Water escapes from the cells into the atmosphere by diffusion in the form of water vapour. This scheme is simplified. In reality a leaf is formed mostly from several cell layers.

The loss of water during photosynthesis is unavoidable, as the uptake of CO_2 into the leaves requires openings in the leaf surface, termed *stomata*. The stomata open up to allow the diffusion of CO_2 from the atmosphere into the air space within the leaf, but at the same time water vapour escapes through the open stomata (Fig. 8.1). As the water vapour concentration in the air space of a leaf in equilibrium with the cell water (31 000 p.p.m., 25 °C) is higher by two orders of magnitude than the CO_2 concentration in air (350 p.p.m.), the escape of a very high amount of water vapour during the influx of CO_2 is inevitable. To minimize this water loss from the leaves, opening of the stomata is regulated. Opening and closing of the stomata are caused by biochemical processes and will be dealt with in the next section.

Even when the water supply is adequate, plants open their stomata just enough to provide CO_2 for photosynthesis. During water shortage, plants prevent dehydration by closing their stomata partially or completely, which results in a slowing down or even cessation of CO_2 assimilation. Therefore water shortage is often a decisive factor in limiting plant growth, especially in the warmer and drier regions of our planet, where a large number of plants have evolved a strategy for decreasing water loss during photosynthesis. In CO_2 fixation the first product is 3-phosphoglycerate, a compound with three carbon atoms, hence the name C_3 *plants* (see section 6.2). Some plants save water by first producing the C_4 compound oxaloacetate to fix CO_2 and are therefore named C_4 *plants*.

8.2 Stomata regulate the gas exchange of a leaf

Stomata are formed by two *guard cells*, which are surrounded in some plants by subsidiary cells. Figures 8.2 and 8.3 show an open and a closed stomatal pore. The pore is opened by the increase in osmotic pressure in the guard cells, resulting in the uptake of water. The corresponding increase in the cell volume 'puffs up' the guard cells and the pore opens.

In order to study the mechanism of the opening process, the guard cells must be isolated. Biochemical and physiological studies are difficult, as the guard cells are very small and can be isolated with only low yields. Although guard cells can be regarded as one of the most thoroughly investigated plant cells, our knowledge of the mechanism of stomatal closure is still limited.

Malate plays an important role in guard cell metabolism

The increase in osmotic pressure in guard cells during stomatal opening is mainly due to an accumulation of *potassium salts*. The corresponding anions are usually *malate* but, depending on the species, sometimes *chloride*. Figure 8.4 shows a scheme of the metabolic reactions occurring during the opening process with malate as the main anion. An H^+–*P-ATPase* pumps protons across the plasma membrane into the extracellular compartment. The H^+–P-ATPase, entirely different from the F-ATPase and V-ATPase (sections 4.3, 4.4), is of the same type as the Na^+–K^+-ATPase in animal cells. An aspartyl residue of the P-ATPase protein is phosphorylated during the transport process. The potential difference generated by the H^+–P-ATPase drives the influx of K^+ ions into the guard cells via a K^+ *channel*. This channel is voltage regulated (section 1.10) and allows only an inwardly directed flux as the channel is only open at a negative voltage. For this reason it is called a K^+ *inward channel*. Most of the K^+ ions taken up into the cell are transported into the vacuole, but the mechanism of this transport is still not known. Probably a vacuolar H^+-ATPase (*V-ATPase*, see section 4.4) is involved,

Figure 8.2 Scanning electron micrograph of stomata from the lower epidermis of hazel leaves in (a) the closed and (b) the open state. (By R. S. Harrison-Murray and C. M. Clay, Wellesbourne.) (c) Traverse section of a pair of guard cells from a tobacco leaf. The large central vacuole and the gap between the two guard cells can be seen. (By D. G. Robinson, Göttingen.)

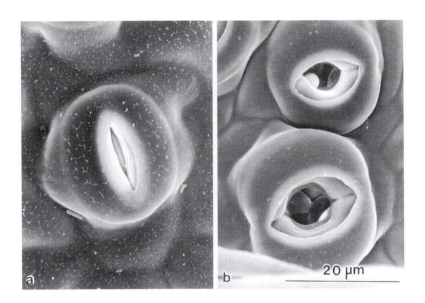

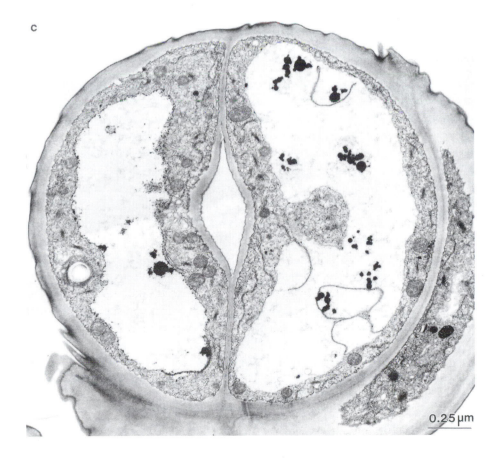

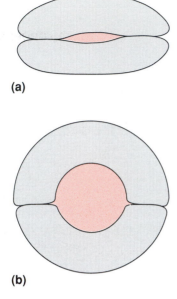

(a)

(b)

Figure 8.3 Diagram of a stoma formed from the two guard cells; (a) closed and (b) open state.

pumping protons into the vacuoles which could then be exchanged for K^+ ions via a vacuolar potassium channel.

Accumulation of cations in the vacuole leads to the formation of a potential difference across the vacuolar membrane, driving the influx of malate via a channel specific for organic anions. Malate is provided by glycolytic degradation of the starch stored in the chloroplasts. As described in section 9.1, this degradation yields triose phosphate, which is released from the chloroplasts to the cytosol in exchange for inorganic phosphate via the *triose phosphate–phosphate translocator* (section 1.9) and is subsequently converted to phosphoenolpyruvate (see Fig. 10.11). Phosphoenolpyruvate reacts with HCO_3^- to form oxaloacetate in a reaction catalysed by the enzyme *phosphoenolpyruvate carboxylase* (Fig. 8.5), in which the high energy enol ester bond is cleaved, making the reaction irreversible. The oxaloacetate formed is transported via a specific translocator to the chloroplasts and reduced to malate via *NADP–malate dehydrogenase* (Fig. 8.4). The malate is then released to the cytosol, probably by the same translocator as that which transports oxaloacetate.

During stomatal closure most of the malate is released from the guard cells. As the guard cells contain only very low activities of RubisCO, they are unable to fix CO_2 in significant amounts. Starch is regenerated from glucose, which is taken up into the guard cells. The chloroplast glucose 6-phosphate–phosphate translocator of guard cells differs from the triose phosphate–phosphate translocator of mesophyll cells by transporting glucose 6-phosphate as well as triose phosphate, 3-phosphoglycerate, and phosphate. Such a glucose 6-phosphate transport is also found in plastids from non-green tissues, such as roots (section 13.3).

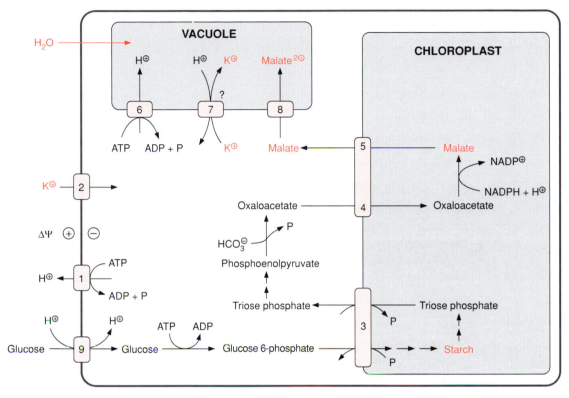

Figure 8.4 Diagram of the processes in operation during the opening of stomata with malate as the main anion. The proton transport by H^+–P-ATPase (1) of the plasmalemma of the guard cell results in an increase in the proton potential and the depolarization of this membrane. The voltage-dependent K^+ inward channel (2) is opened by this depolarization and the proton potential drives the influx of potassium ions through this channel. Simultaneously, starch degradation occurs in the chloroplasts, yielding triose phosphate which is then released from the chloroplasts via the triose phosphate–phosphate translocator (3) and converted in the cytosol to oxaloacetate. Oxaloacetate is transported into the chloroplasts (4) and converted to malate by reduction. This malate is transported from the chloroplast to the cytosol, possibly via the same translocator responsible for the influx of oxaloacetate (5). The mechanism of the accumulation of potassium malate in the guard cell vacuoles is not yet known in detail. Protons are transported into the vacuole (6) probably by an H^+-ATPase, and these protons are exchanged for potassium ions (7). The electric potential difference formed in this way drives the influx of malate ions via a malate channel (8). The accumulation of potassium malate increases the osmotic potential in the vacuole and results in an influx of water. For resynthesis of starch, glucose is taken up into the guard cells via H^+ symport (9), where it is converted in the cytosol to glucose 6-phosphate which is then transported into the chloroplast via a glucose 6-phosphate–phosphate translocator

Our knowledge of the regulation of stomatal opening is still fragmentary

Several parameters are known which influence the stomatal opening. The opening is regulated by light via a blue light receptor (section 19.6). An important factor is the CO_2 concentration in the intercellular air space, although the nature of the

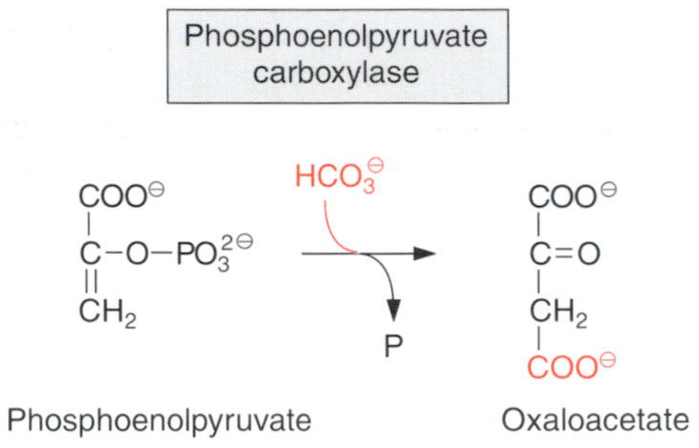

Phosphoenolpyruvate
carboxylase

$HCO_3^{\ominus}$

$COO^{\ominus}$
|
$C{-}O{-}PO_3^{2\ominus}$
‖
CH_2

P

$COO^{\ominus}$
|
$C{=}O$
|
CH_2
|
$COO^{\ominus}$

Phosphoenolpyruvate   Oxaloacetate   **Figure 8.5** Phosphoenolpyruvate carboxylase.

CO_2 sensor is not known. At micromolar concentrations *abscisic acid* (ABA) (section 19.4) causes the closure of the stomata. This effect of ABA is dependent on the intercellular CO_2 concentration and the water status of the plant. It has been suggested that the binding of ABA to a membrane receptor causes an increase in the cellular concentration of free Ca^{2+} ions via an *inositol triphosphate cascade* (section 19.7) which, in turn, inactivates K^+ inward channels. The efflux of anions via regulated anion channels from the guard cells causes depolarization of the plasmalemma and, by this, opening of *K^+ outward channels* (section 1.10). The resulting release of K^+, malate^{2-} and Cl^- ions from the guard cells lowers the osmotic pressure, leading to a decrease in the guard cell volume and hence to a closure of the stomata. The introduction of the patch clamp technique (section 1.10) has brought important insights into the role of specific ion channels in the stomatal opening process. The signal transduction chain governing the regulation of the stomatal aperture is still unknown.

8.3 The diffusive flux of CO_2 into a plant cell

The movement of CO_2 from the atmosphere to the catalytic centre of RubisCO—through the stomata, the intercellular air space, across the plasmalemma, the chloroplast envelope, and the chloroplast stroma—proceeds by diffusion.

According to a simple derivation of the Fick law, the diffusive flux, *I*, over a certain distance is:

$$I = \frac{\Delta C}{R}$$

where *I* is defined as the amount of a substance diffusing per unit of time and surface area; ΔC, the diffusion gradient, is the difference of concentrations between start and end-point; and *R* is the diffusion resistance. *R* of CO_2 is 10^4 times larger in water than in air.

In Fig. 8.6a a model illustrates the diffusive flux of CO_2 into a leaf of a C_3 plant with a limited water supply. The control of the aperture of the stomata leads to a stomatal diffusion resistance, by which a diffusion gradient of 100 p.p.m. is main-

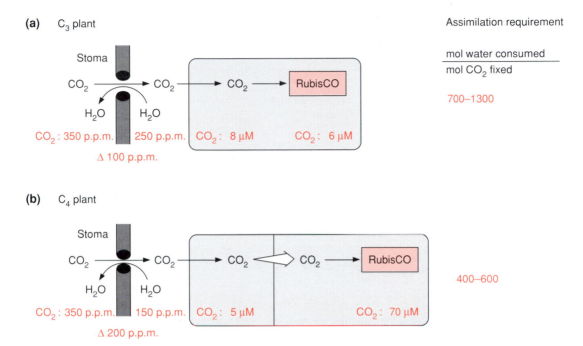

Figure 8.6 Reaction diagrams for the uptake of CO_2 in C_3 and C_4 plants. This scheme shows typical stomatal resistances for C_3 and C_4 plants. The values for the CO_2 concentration in the vicinity of RubisCO are taken from von Caemmerer and Evans (1991)(C_3 plants) and Hatch (1992) (C_4 plants).

tained. The resultant CO_2 concentration of 250 p.p.m. in the intercellular air space is in equilibrium with the CO_2 concentration in an aqueous solution of 8×10^{-6} mol l^{-1} (8 μM). In water saturated with air containing 350 p.p.m. CO_2 the equilibrium concentration of the dissolved CO_2 is 11.5 μM at 25°C.

Since the chloroplasts are positioned at the inner surface of the mesophyll cells (see Fig. 1.1), within the mesophyll cell the major distance for the diffusion of CO_2 to the reaction site of RubisCO is the passage through the chloroplast stroma. To facilitate this diffusive flux, the stroma contains high activities of *carbonic anhydrase*. This enzyme allows the CO_2 entering the chloroplast stroma, after crossing the envelope, to equilibrate with HCO_3^- (Fig. 8.7). At pH 8.0, 1 μM CO_2 is in equilibrium with 50 μM HCO_3^- (25°C). Thus, in the presence of carbonic anhydrase the gradient for the diffusive movement of HCO_3^- is 50 times higher than that of CO_2. As the diffusion resistance for HCO_3^- is only about 20 per cent higher than that of CO_2, the diffusive flux of HCO_3^- in the presence of carbonic anhydrase is about 40 times higher than that of CO_2. Due to the presence of carbonic anhydrase in the stroma, the diffusive flux of CO_2 from the intercellular air space to the site of RubisCO in the stroma results in a decrease in CO_2 concentration of only about 2 μM. At the site of RubisCO a CO_2 concentration of about 6 μM has

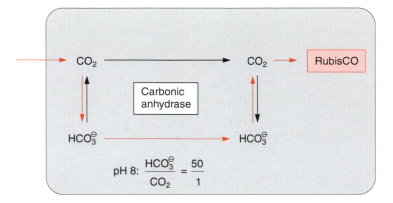

Figure 8.7 Carbonic anhydrase catalyses the rapid equilibration of CO_2 with HCO_3^- and thus increases the diffusion gradient and hence the diffusive flux of the inorganic carbon across the chloroplast stroma. The example is based on the assumption that the pH value is 8.0. Dissociation constant $[HCO_3^-][H^+]/[CO_2] = 5 \times 10^{-7}\,M^{-1}$.

been measured. In equilibrium with air the O_2 concentration at the carboxylation site is 250 µM. This results in a ratio of carboxylation/oxygenation of about 2.5.

Let us turn our attention again to Fig. 8.6. Since CO_2 and O_2 are competitors for the active site of RubisCO, and the CO_2 concentration in the atmosphere is very low as compared with the O_2 concentration, the concentration decrease of CO_2 during the diffusive flux from the atmosphere to the active site of carboxylation is still a limiting factor for efficient CO_2 fixation by RubisCO. Naturally, the stomatal resistance could be decreased by increasing the aperture of the stomata, e.g. by a factor of two. In this case, with still the same diffusive flux, the CO_2 concentration in the intercellular air space would be increased from 250 to 300 p.p.m., and the ratio of carboxylation to oxygenation by RubisCO increased accordingly. The price, however, for such a reduction of the stomatal diffusion resistance would be a doubling of the water loss. Since the diffusive efflux of water vapour from the leaves is proportional to the diffusion gradient, air humidity is also a decisive factor governing water loss. These considerations illustrate the important function of stomata for the gaseous exchange of the leaves. The regulation of the stomatal aperture determines how high the rate of CO_2 assimilation may be, without the plant losing too much of the essential water.

8.4 C$_4$ plants perform CO_2 assimilation with less water consumption

In equilibrium with fluid water, the density of water vapour increases exponentially with the temperature. A temperature increase from 20° to 30°C leads to almost a doubling of water vapour density. Therefore at high temperatures the problem of water loss during CO_2 assimilation becomes very serious for plants. C_4 plants found a way to decrease this water loss considerably. At around 25 °C these plants use only 400–600 mol H_2O for the fixation of 1 mol CO_2, which is just about half the water consumption of C_3 plants, and this difference is even greater at higher temperatures. C_4 plants grow mostly in hot areas which are often also dry; they include important crop plants such as maize, sugar cane, and millet. The principle by which these C_4 plants save water can be demonstrated by comparing the models of C_3 and C_4 plants in Fig. 8.6. By doubling the stomatal resistance prevailing in C_3 plants, the C_4 plant can decrease the diffusive efflux of water vapour by 50 per cent.

In order to maintain the same diffusive flux of CO_2 as in C_3 plants at this increased stomatal resistance, in accordance with the Fick law the diffusion gradient has to be increased by a factor of two. This means that at 350 p.p.m. CO_2 in the air, the CO_2 concentration in the intercellular air space would be only 150 p.p.m., which is in equilibrium with 5 μM CO_2 in water. At such low CO_2 concentrations C_3 plants would be approaching the compensation point (section 7.6) and therefore the rate of net CO_2 fixation by RubisCO would be very low.

Under these conditions the crucial factor for the maintenance of CO_2 assimilation in C_4 plants is the presence of a pumping mechanism which elevates the concentration of CO_2 at the carboxylation site from 5 μM to about 70 μM. This pumping requires two compartments and the input of energy. However, the energy costs may be recovered, since with this high CO_2 concentration at the carboxylation site, the oxygenase reaction is eliminated to a great extent and the loss of energy connected with the photorespiratory pathway is largely decreased (section 7.5). For this reason C_4 metabolism does not necessarily imply a higher energy demand; in fact, at higher temperatures C_4 photosynthesis is more efficient than C_3 photosynthesis. This is due to the fact that the oxygenase activity of RubisCO increases more rapidly than the carboxylase activity with increasing temperature.

The discovery of C_4 metabolism was stimulated by an unexplained experimental result. After Melvin Calvin and Andrew Benson had established that 3-phosphoglycerate is the primary product of CO_2 assimilation by plants, Hugo Kortschak and colleagues studied the incorporation of radioactively labelled CO_2 during photosynthesis of sugar-cane leaves at a sugar-cane research institute in Hawaii. The result was surprising. The primary fixation product was not, as expected, 3-phosphoglycerate, but the C_4 compounds *malate* and *aspartate*. This result questioned whether the then fully accepted Calvin cycle was universally valid for CO_2 assimilation. Perhaps Kortschak was reluctant to raise these doubts and his results remained unpublished for almost 10 years. It is interesting to note that during this time and without knowing these results, Yuri Karpilov in the Soviet Union observed similar radioactive CO_2 fixation into C_4 compounds during photosynthesis in maize.

Following the publication of these puzzling results Hal Hatch and Roger Slack in Australia set out to solve the riddle by systematic studies. They found that the incorporation of CO_2 in malate was a reaction preceding the CO_2 fixation by the Calvin cycle and that this first carboxylation reaction was part of a CO_2 concentration mechanism, the function of which was elucidated by the two researchers by 1970. Earlier, this process was known as the Hatch–Slack pathway. However, they used the term *C_4 dicarboxylic acid pathway* of photosynthesis and this was later abbreviated to the C_4 pathway or C_4 photosynthesis.

The CO_2 pump in C_4 plants

The requirement of two different compartments for pumping CO_2 from a low to a high concentration is reflected in the leaf anatomy of C_4 plants. The leaves of all C_4 plants show a so-called *Kranz anatomy* (Fig. 8.8). The vascular bundles, containing sieve tubes and xylem vessels, are surrounded by a sheath of cells (*bundle sheath cells*), and these are encircled by *mesophyll cells*. The latter are in contact with the intercellular air space of the leaves. At the beginning of this century the German botanist Gustav Haberland described in his textbook *Physiologische Pflanzenanatomie* (*Physiological plant anatomy*) that the assimilatory cells in several plants, including sugar cane and millet, are arranged in what he termed a Kranz-type mode. With remarkable foresight, he suggested that this

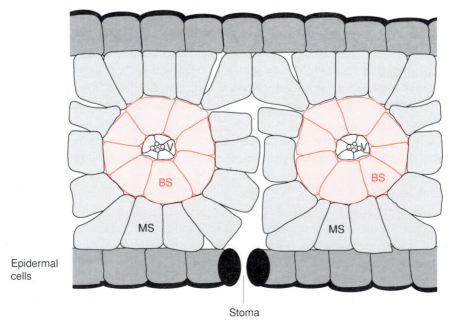

Figure 8.8 Characteristic leaf anatomy of a C_4 plant. Schematic diagram. V, Vascular bundle; BS, bundle sheath cells; MS, mesophyll cells.

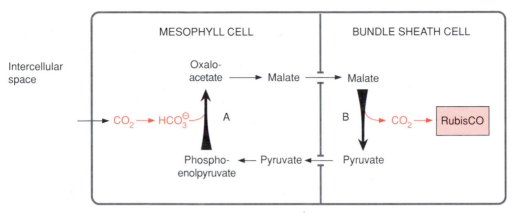

Figure 8.9 Principle of C$_4$ metabolism.

special anatomy may indicate a division of labour between the chloroplasts of the mesophyll and bundle sheath cells.

Mesophyll and bundle sheath cells are separated by a cell wall, which in some instances contains a *suberin layer* which is probably gas impermeable. Suberin is a polymer of phenolic compounds which are impregnated with wax (section 18.3). The border between the mesophyll and bundle sheath cells is penetrated by a large number of *plasmodesmata* (section 1.1). These plasmodesmata enable the diffusive flux of metabolites between the mesophyll and bundle sheath cells.

The CO$_2$ pumping of C$_4$ metabolism does not rely on the specific function of a membrane transporter but is due to a prefixation of CO$_2$, after conversion to HCO$_3^-$ by reaction with phosphoenolpyruvate to form oxaloacetate in the mesophyll cells. After the conversion of this oxaloacetate to malate, the malate diffuses through the plasmodesmata into the bundle sheath cells, where CO$_2$ is released as a substrate for RubisCO. Figure 8.9 shows a simplified scheme of this process. The formation of the CO$_2$ gradient between the two compartments by this pumping process is due to the fact that the prefixation of CO$_2$ and its subsequent release are catalysed by two different reactions, each of which is virtually irreversible. As a crucial feature of C$_4$ metabolism, RubisCO is located exclusively in the bundle sheath chloroplasts.

The reaction of HCO$_3^-$ with phosphoenolpyruvate is catalysed by the enzyme *phosphoenolpyruvate carboxylase*. This enzyme has already been discussed when dealing with the metabolism of guard cells (Figs 8.4, 8.5). The reaction is strongly exergonic and irreversible. As the enzyme has a very high affinity for HCO$_3^-$, micromolar concentrations of HCO$_3^-$ are fixed very efficiently. The formation of HCO$_3^-$ from CO$_2$ is facilitated by carbonic anhydrase present in the cytosol of the mesophyll cells.

The release of CO$_2$ in the bundle sheath cells occurs in three different ways (Fig. 8.10). In most C$_4$ species decarboxylation of malate with an accompanying

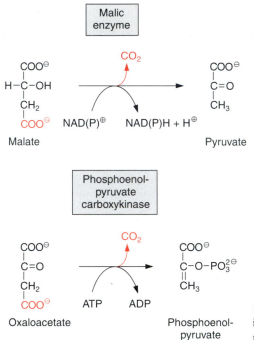

Figure 8.10 Reactions by which CO_2 pre-fixed in C_4 metabolism in mesophyll cells can be released in bundle sheath cells.

oxidation to pyruvate is catalysed by *malic enzyme*. In one group of these species, termed *NADP–malic enzyme type* plants, release of CO_2 occurs in the bundle sheath chloroplasts and oxidation of malate to pyruvate is coupled with the reduction of $NADP^+$. In other plants, termed *NAD–malic enzyme type*, decarboxylation takes place in the mitochondria of the bundle sheath cells and is accompanied by the reduction of NAD^+. In the *phosphoenolpyruvate carboxykinase type* plants, oxaloacetate is decarboxylated in the cytosol of the bundle sheath cells. ATP is required for this reaction and phosphoenolpyruvate is formed as a product.

The following deals in more detail with the metabolism and its compartmentalization in the three different types of C_4 plants.

C_4 metabolism of the NADP–malic enzyme type plants

This group includes the important crop plants maize and sugar cane. Figure 8.11 shows the reaction chain and its compartmentalization. The oxaloacetate arising from the carboxylation of phosphoenolpyruvate is transported via a specific translocator into the chloroplasts where it is reduced by NADP–malate dehydrogenase to malate, which is subsequently transported into the cytosol. (The reduction of oxaloacetate in the chloroplasts has been dealt with in section 7.3 in connection with photorespiratory metabolism). Malate diffuses via plasmo-

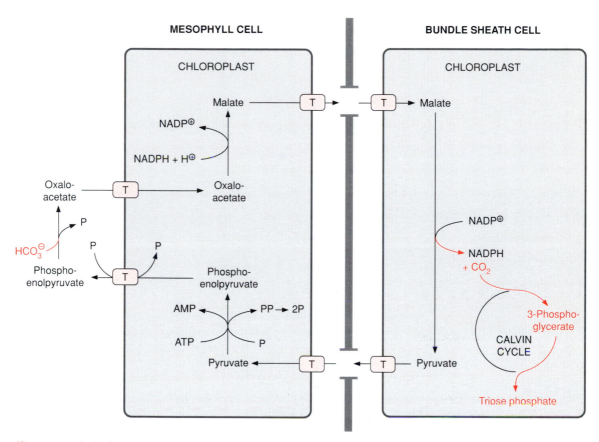

Figure 8.11 Mechanism for concentrating CO$_2$ in plants of the C$_4$ NADP–malic enzyme type (as an example, maize). In the cytosol of the mesophyll cells HCO$_3^-$ is fixed by reaction with phospho-enolpyruvate, the oxaloacetate formed is reduced in the chloroplast to form malate. After leaving the chloroplasts malate diffuses into the bundle sheath cells where it is oxidatively decarboxylated leading to the formation of pyruvate, CO$_2$, and NADPH. The pyruvate formed is phosphorylated to phospho-enolpyruvate in the chloroplasts of the mesophyll cells. The transport across the chloroplast membranes proceeds by specific translocators. The diffusive flux between the mesophyll and the bundle sheath cells proceeds through plasmodesmata. The transport of oxaloacetate into the mesophyll chloroplasts and the subsequent release of malate from the chloroplasts is probably facilitated by the same translocator (T).

desmata from the mesophyll to the bundle sheath cells. The diffusive flux of malate between the two cells requires a diffusion gradient of about 2×10^{-3} mol l^{-1}. The malic enzyme present in the bundle sheath cells causes the release of CO$_2$, which is fixed there by RubisCO.

Pyruvate is exported by a specific translocator from the bundle sheath chloroplasts, diffuses through the plasmodesmata into the mesophyll cells where it is transported by another specific translocator into the chloroplasts. The enzyme *pyruvate phosphate dikinase* in the mesophyll chloroplasts converts pyruvate to

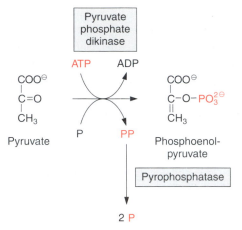

Reaction mechanism:

E−His + ATP + P ⇌ E−His−P + AMP + PP

E−His−P + Pyr ⇌ E−His + PEP

Figure 8.12 Pyruvate phosphate dikinase. In this reaction one phosphate residue is transferred from ATP to inorganic phosphate, resulting in the formation of pyrophosphate, and a second phosphate residue is transferred to a histidine residue at the catalytical site of the enzyme. In this way a phosphoric amide (-H-N-PO_3^{2-}) is formed as an intermediate, and this phosphate residue is then transferred to pyruvate, resulting in the formation of phosphoenolpyruvate.

phosphoenolpyruvate by a rather unusual reaction (Fig. 8.12). The name dikinase means an enzyme that catalyses a twofold phosphorylation. In a reversible reaction one phosphate residue is transferred from ATP to pyruvate and a second one to phosphate, converting it to pyrophosphate. A *pyrophosphatase* present in the chloroplast stroma immediately hydrolyses the newly formed pyrophosphate and thus makes this reaction irreversible. In exchange for inorganic phosphate the phosphoenolpyruvate is exported from the chloroplasts via a phosphoenol-pyruvate–phosphate translocator.

With the high CO_2 gradient between bundle sheath and mesophyll cells, the question arises why does most of the CO_2 not leak out before it is fixed by RubisCO. As the bundle sheath chloroplasts, in contrast to those from mesophyll cells (see Fig. 8.7), do not contain carbonic anhydrase, the diffusion of CO_2 through the stroma of bundle sheath cells proceeds more slowly than in the mesophyll cells. The suberin layer between the cells of some plants probably prevents the leakage of CO_2 through the cell wall, and in this case there would only be a diffusive loss through plasmodesmata. The portion of CO_2 concentrated in the bundle sheath cells that is lost by diffusion back to the mesophyll cells, is estimated at 10–30 per cent in different species.

In maize leaves the chloroplasts from mesophyll cells differ in their structure from those of bundle sheath cells. Mesophyll chloroplasts have many grana, whereas bundle sheath chloroplasts contain mainly stroma lamellae, with only very few grana stacks and little photosystem II activity (Chapter 3.10). The major function of the bundle sheath chloroplasts is to provide ATP by cyclic photophosphorylation via photosystem I. NADPH required for the reductive pentose phosphate pathway (Calvin cycle) is provided mainly by the linear electron transport in the mesophyll cells. This NADPH is delivered in part via the oxidative decarboxylation of malate (by NADP–malic enzyme), but this reducing power is actually provided by the mesophyll cells for the reduction of oxaloacetate. The rest of the NADPH required is transferred along with ATP from the mesophyll chloroplasts to the bundle sheath chloroplasts by a *triose phosphate–3-phosphoglycerate shuttle* via the triose phosphate–phosphate translocators of the inner envelope membranes of the corresponding chloroplasts (Fig. 8.13).

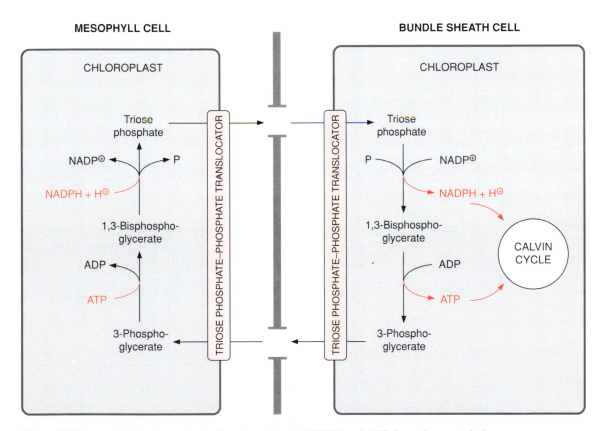

Figure 8.13 C$_4$ metabolism in maize. Indirect transfer of NADPH and ATP from the mesophyll chloroplast to the bundle sheath chloroplast via a triose phosphate–3-phosphoglycerate shuttle. In the chloroplasts of mesophyll cells, 3-phosphoglycerate is reduced to triose phosphate at the expense of ATP and NADPH. In the bundle sheath chloroplasts triose phosphate is reconverted to phosphoglycerate leading to the formation of NADPH and ATP. Transport across the chloroplast membranes proceeds by counter-exchange via triose phosphate–phosphate translocators.

C₄ metabolism of the NAD–malic enzyme type

The metabolism of the NAD–malic enzyme type, shown in the metabolic scheme of Fig. 8.14, is found in a large number of species, including millet. Here the oxaloacetate formed by phosphoenolpyruvate carboxylase is converted to aspartate by transamination via *glutamate–aspartate aminotransferase*. Since the

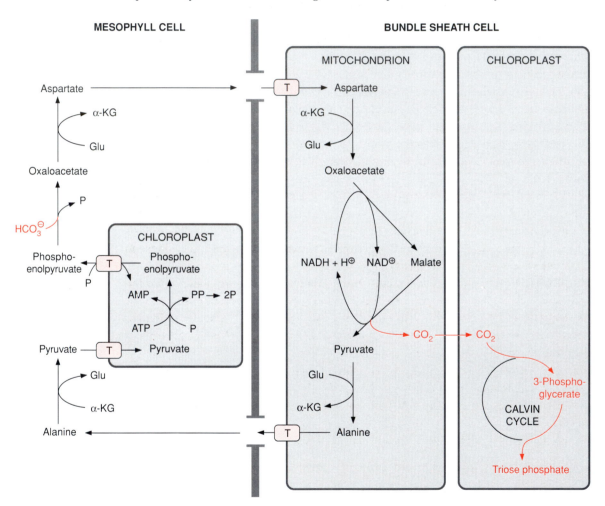

Figure 8.14 Scheme for the CO_2 concentrating mechanism in plants of the C_4 NAD–malic enzyme type. In contrast to C_4 metabolism described in Fig. 8.11, oxaloacetate is transaminated in the cytosol to aspartate. After diffusion through the plasmodesmata, aspartate is transported into the mitochondria of the bundle sheath cells by a specific translocator and is reconverted there to oxaloacetate and the latter reduced to malate. In the mitochondria malate is oxidized by NAD–malic enzyme, giving CO_2 and pyruvate. Pyruvate is transaminated to alanine which then diffuses into the mesophyll cells. The CO_2 released from the mitochondria diffuses into the bundle sheath chloroplasts, which are in close contact with the mitochondria and this CO_2 serves as substrate for RubisCO. Abbreviations: Glu, glutamate; α-KG, α-ketoglutamate; P, phosphate; PP pyrophosphate; T, translocator.

oxaloacetate concentration in the cell is below 0.1×10^{-3} mol l^{-1}, oxaloacetate cannot form a high enough diffusion gradient for the necessary diffusive flux into the bundle sheath cells. Because of the high concentration of glutamate in a cell, the transamination of oxaloacetate yields aspartate concentrations in a range between 5×10^{-3} and 10×10^{-3} mol l^{-1}. This is the reason why aspartate is very suitable for supporting a diffusive flux between the mesophyll and bundle sheath cells.

After diffusing into the bundle sheath cells, aspartate is transported by a translocator into the mitochondria. An isoenzyme of glutamate–aspartate amino-transferase present in the mitochondria catalyses the conversion of aspartate to oxaloacetate, which is then transformed by *NAD–malate dehydrogenase* to malate. This malate is decarboxylated by *NAD–malic enzyme* to pyruvate and the NAD formed during the malate dehydrogenase reaction is reduced to NADH again. CO$_2$ thus released in the mitochondria then diffuses into the chloroplasts where it is available for assimilation via RubisCO. The pyruvate formed in the mitochondria is converted via *alanine–glutamate aminotransferase* to alanine and this alanine probably leaves the mitochondria via a specific translocator. Alanine diffuses into the mesophyll cells, where it is transformed to pyruvate by an isoenzyme of the aminotransferase mentioned above. Pyruvate is transported into the chloroplasts where it is converted to phosphoenolpyruvate by pyruvate phosphate dikinase in the same way as in the chloroplasts of the NADP–malic enzyme type.

All NADH formed by malic enzyme in the mitochondria is sequestered for the reduction of oxaloacetate and thus there are no reducing equivalents left to be oxidized by the respiratory chain (Fig. 8.14). To enable mitochondrial oxidative phosphorylation to form ATP, some of the oxaloacetate formed in the mesophyll cells by phosphoenolpyruvate carboxylase is reduced in the mesophyll chloro-plasts to malate, as in the NADP–malic enzyme type metabolism. This malate dif-fuses into the bundle sheath cells, is taken up by the mitochondria, and is oxidized there by malic enzyme to yield NADH. ATP is generated from oxidation of this NADH by the respiratory chain. This pathway also operates in the phospho-enolpyruvate carboxykinase type metabolism, described below.

C$_4$ metabolism of the phosphoenolpyruvate carboxykinase type

This type of metabolism is found in several of the fast-growing tropical grasses used as forage crops. Figure 8.15 shows a scheme of the metabolism. As in the NAD–malic enzyme type, oxaloacetate is converted in the mesophyll cells to aspartate and the latter diffuses into the bundle sheath cells, where the oxalo-acetate is regenerated via an aminotransferase in the cytosol. In the cytosol the oxaloacetate is converted to phosphoenolpyruvate at the expense of ATP via phosphoenolpyruvate carboxykinase. The CO$_2$ released in this reaction diffuses into the chloroplasts and the remaining phosphoenolpyruvate diffuses back into

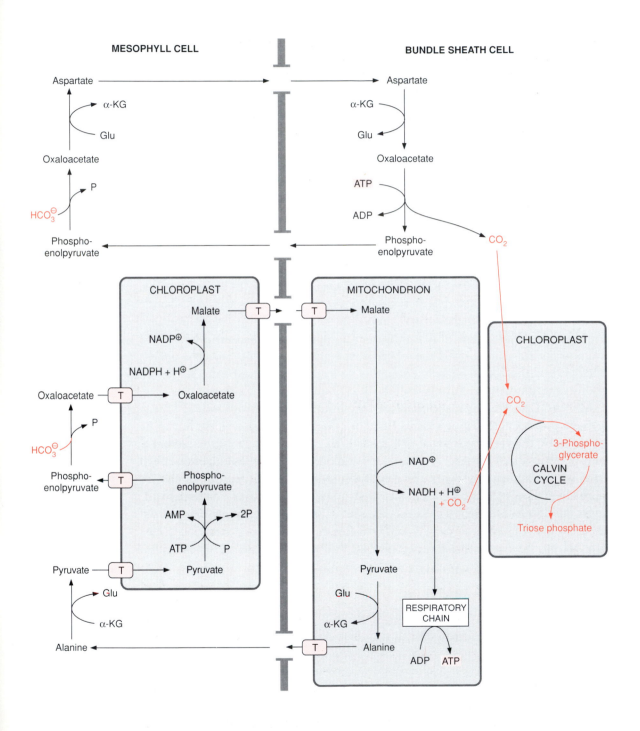

Figure 8.15 Scheme of the CO$_2$ concentrating mechanism in plants of the C$_4$ phosphoenolpyruvate carboxykinase type. In contrast to C$_4$ metabolism described in Fig. 8.14, oxaloacetate is formed from aspartate in the cytosol of the bundle sheath cells, and is then decarboxylated to phosphoenolpyruvate and CO$_2$ via the enzyme phosphoenolpyruvate carboxykinase. Phosphoenolpyruvate diffuses back into the mesophyll cells. Simultaneously, as in Fig. 8.11, some malate formed in the mesophyll cells diffuses into the bundle sheath cells and is oxidized there by an NAD–malic enzyme in the mitochondria. The NADH thus formed serves as a substrate for the formation of ATP by mitochondrial oxidative phosphorylation. This ATP is transported to the cytosol to be used for phosphoenolpyruvate carboxykinase reaction. The CO$_2$ released in the mitochondria, together with the CO$_2$ released by phosphoenolpyruvate carboxykinase in the cytosol, serves as a substrate for the RubisCO in the bundle sheath chloroplasts. T, translocator.

the mesophyll cells. In this C$_4$ type the ATP demand for pumping CO$_2$ into the bundle sheath compartment is due to ATP consumption by the *phosphoenolpyruvate carboxykinase* reaction (Fig. 8.10). The mitochondria provide the ATP required for phosphoenolpyruvate carboxykinase reaction by oxidizing malate via NAD–malic enzyme. This malate originates from mesophyll cells as in the NADP–malic enzyme type. Thus in the C$_4$ phosphoenolpyruvate carboxykinase type plants a minor portion of the CO$_2$ is released in the mitochondria and the remainder in the cytosol.

Enzymes of C$_4$ metabolism are regulated by light

Phosphoenolpyruvate carboxylase (PEP carboxylase), the key enzyme of C$_4$ metabolism, is highly regulated. In a darkened leaf this enzyme is present in a state of low activity. In this state the affinity of the enzyme to its substrate phosphoenolpyruvate is very low and it is inhibited by low concentrations of malate. Therefore, during the dark phase the enzyme in the leaf is practically inactive. Upon illumination of the leaf a serine protein kinase (see also Figs 9.18, 10.9) is activated which phosphorylates the hydroxyl group of a serine residue in PEP carboxylase. The enzyme can be inactivated again by hydrolysis of the phosphate group by a protein serine phosphatase. The active phosphorylated enzyme is also inhibited by malate, but in this case very much higher concentrations of malate are required for its inhibition than for the non-phosphorylated inactive enzyme. The rate of the irreversible carboxylation of phosphoenolpyruvate can be adjusted in such a way through a feedback inhibition by malate that a certain malate level is maintained in the mesophyll cell. It is not yet known which signal turns on the activity of the protein kinase when the leaves are illuminated.

NADP–malate dehydrogenase is activated by light via reduction by thioredoxin as described in section 6.6.

Pyruvate phosphate dikinase (Fig. 8.12) is also subject to dark/light regulation. It is inactivated in the dark by phosphorylation of a threonine residue. This phosphorylation is rather unusual, as it requires ADP rather than ATP as phosphate donor. The enzyme is activated in the light by the phosphorolytic cleavage of the threonine phosphate group. Thus, the regulation of pyruvate phosphate dikinase proceeds in a completely different way from the regulation of PEP carboxylase. The signal chain from illumination to the dephosphorylation of the pyruvate phosphate dikinase is not yet known.

Products of C_4 metabolism can be identified by mass spectrometry

Measuring the distribution of the ^{12}C and the ^{13}C isotopes in a photosynthetic product can show whether it has been formed by C_3 or C_4 metabolism. ^{12}C and ^{13}C occur as natural carbon isotopes in the CO_2 of the atmosphere (98.89 per cent and 1.11 per cent, respectively). RubisCO reacts with $^{12}CO_2$ more rapidly than with $^{13}CO_2$. This is due to a kinetic isotope effect. For this reason the $^{13}C/^{12}C$ ratio is lower in the products of C_3 photosynthesis than in the atmosphere. The $^{13}C/^{12}C$ ratio can be determined by mass spectrometry and is expressed as $\delta^{13}C$ value.

$$\delta^{13}C[\text{‰}] = \left(\frac{^{13}C/^{12}C \text{ in sample}}{^{13}C/^{12}C \text{ in standard}} - 1 \right) \times 1000.$$

As a standard one uses the distribution of the two isotopes in a defined limestone. In products of C_3 photosynthesis $\delta^{13}C$ values of -28‰ are found. In the PEP carboxylase reaction, the preference for ^{12}C over ^{13}C is less pronounced. Since in C_4 plants practically all the CO_2, which had been prefixed by PEP carboxylase, reacts further in a gastight compartment with RubisCO, the photosynthesis of C_4 plants has a $\delta^{13}C$ value in the range of -14‰ only. Therefore it is possible to determine by mass spectrometric analysis of the ^{13}C to ^{12}C ratio, whether, for instance, sucrose has been formed by sugar beet (C_3 metabolism) or by sugar cane (C_4 metabolism).

C_4 plants include important crop plants but also many of the worst weeds

In C_4 metabolism ATP is consumed to concentrate the CO_2 in the bundle sheath cells. This avoids the loss of energy incurred by photorespiration in C_3 plants. The ratio of oxygenation versus carboxylation by RubisCO increases with temperature (section 6.2). At a low temperature, with resultant low photorespiratory activity, the C_3 plants are at an advantage. Under these circumstances C_4 plants have no advantage and very few C_4 plants occur as wild plants in a temperate climate. However, at temperatures of about 25°C or above, the C_4 plants are at an advantage as, under these conditions, the energy consumption for C_4 photosynthesis (measured as quantum requirement of CO_2 fixation) is lower than in C_3 plants. As indicated above, this is due largely to additional photorespiration resulting from a high increase in the oxygenase reaction of RubisCO. A further advantage of C_4 plants is that, because of the high CO_2 concentration in the bundle sheath chloroplasts, they need less RubisCO. Since RubisCO is a main protein of leaves, C_4 plants require less nitrogen for growth than C_3 plants. Last, but not least, C_4 plants require less water. In warmer climates these advantages make C_4 plants very suitable as crop plants. It has been estimated that about 30 per cent of global photosynthesis by terrestrial plants comes from C_4 plants, growing mainly in savannahs, the largest habitats on the globe. From the 12 most rapidly growing crop or pasture plants, 11 are C_4 plants. One disadvantage, however, is that C_4 crop plants such as maize, millet, and sugar cane are very sensitive to chilling and this restricts them to the warmer areas. Especially persistent weeds are members of the C_4 plants, including 8 of the 10 worst weed species world-wide, e.g. Bermuda grass, jungle rice, and Congo grass.

8.5 Crassulacean acid metabolism makes it possible for plants to survive even during a very severe water shortage

Many plants growing in very dry and often hot habitats have developed a strategy not only for surviving periods of severe water shortage, but also for carrying out photosynthesis under such conditions. The ornamental plant *Kalanchoe* and many cacti (e.g. *Opuntia*) are examples of such plants, as are plants which grow as epiphytes in tropical rain forests, including a number of orchids. These plants solve the problem of water loss during photosynthesis by opening their stomata only at night-time when it is cool and air humidity is comparatively high. CO_2 taken up during the night through the open stomata is fixed in an acid which is stored until the following day, when it releases the CO_2 and feeds it into the Calvin cycle, which can now proceed while the stomata are closed. Figure 8.16 shows the basic scheme of this process. Note the strong similarity of this scheme to the basic scheme of C_4 metabolism in Fig. 8.9. The scheme in Fig. 8.16 only differs from that of C_4 metabolism in that carboxylation and decarboxylation are separated in time instead of being separated spatially. As this metabolism has first been elucidated in *Crassulaceae* and involves the storage of an acid, it has been named *crassulacean acid metabolism* (abbreviated *CAM*). Important CAM crop plants are pineapples and also the agave, sisal, yielding natural fibres.

The first observations concerning CAM were made at the beginning of the nineteenth century. In 1804 the French scientist de Saussure observed that upon illumination and in the absence of CO_2, branches of the cactus *Opuntia* produced oxygen. He concluded that these plants consumed their own matter to produce CO_2, which was then used for CO_2 assimilation. An English gentleman, Benjamin Heyne, noticed in his garden in India that the leaves of the then very popular ornamental plant *Bryophyllum calycinum*, had a herby taste in the afternoon, whereas in the morning the taste was as acid as sorrel. He found this observation

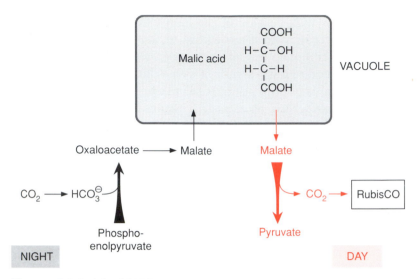

Figure 8.16 Principle of CAM.

so remarkable that, after his return to England in 1813, he communicated it in a letter to the Linnean Society.

CO_2 fixed during the night is stored in the form of malic acid

Nocturnal fixation of CO_2 is brought about by reaction with phosphoenolpyruvate, as catalysed by phosophoenolpyruvate carboxylase, in the same way as in C_4 metabolism and in the metabolism of guard cells. Figure 8.17 shows a scheme of CAM. Starch located in the chloroplasts is degraded to triose phosphate (section 9.1) which is then exported via the triose phosphate–phosphate translocator (section 1.9) and is converted to phosphoenolpyruvate in the cytosol. In some CAM plants the carbohydrates required for the regeneration of phosphoenolpyruvate are stored as soluble sugars, such as sucrose (section 9.2) or fructans (section 9.5).

The oxaloacetate formed from the prefixation of CO_2 is reduced in the cytosol to malate via NAD–malate dehydrogenase. The NADH required for this is provided by the oxidation of triose phosphate in the cytosol. Malate is pumped at the expense of energy into the vacuoles. As already described for the guard cells (section 8.2), the energy-dependent step in this pumping process is the transport of protons by the H^+-ATPase (V-ATPase, section 4.4) located in the vacuolar membrane. In contrast to the guard cells, these transported protons are not exchanged for potassium ions. The malate which, by uptake through a malate channel driven by the proton potential, has accumulated in the vacuoles, is stored there as malic acid and makes the vacuolar content very acidic (about pH 3) during the night. The two carboxyl groups of malic acid have pK values of 3.4 and 5.1, respectively. Thus at pH 3 malic acid is largely undisocciated and the osmotic

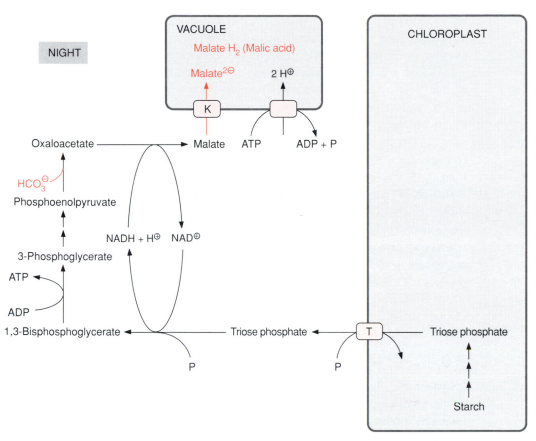

Figure 8.17 CAM during the night. Degradation of starch in the chloroplasts provides triose phosphate which is converted along with the generation of NADH and ATP to phosphoenolpyruvate, the acceptor for HCO_3^-. The oxaloacetate is reduced in the cytosol to malate. An H^+-ATPase (V-ATPase) in the vacuolar membrane drives the accumulation of malate anions in the vacuole, where they are stored as malic acid.

pressure deriving from the accumulation of malic acid is only about one-third of the osmotic pressure produced by the accumulation of potassium malate ($2K^+$ + Mal^{2-}) in the guard cells. In other words, at a certain osmotic pressure, three times as much malate can be stored as malic acid than as potassium malate. In order to gain a high storage capacity most CAM plants have unusually large vacuoles and are succulent. The ATP required for CAM is formed by mitochondrial oxidative phosphorylation from the oxidation of malate.

Photosynthesis proceeds with closed stomata

The malate stored in the vacuoles during the night is released during the day by a regulated efflux through the malate channel. It is still not known how this efflux, and also the malate flux from the guard cell vacuoles, is regulated. In CAM, as in C_4 metabolism, different plants release CO_2 in various ways: via *NADP–malic enzyme*, *NAD–malic enzyme*, or *phosphoenolpyruvate carboxykinase*.

CAM of the *NADP–malic enzyme type* is described in Fig. 8.18. A specific translocator takes up malate into the chloroplasts where it is decarboxylated to give pyruvate, NADPH, and CO_2. The latter reacts as substrate with RubisCO and the pyruvate is converted via pyruvate phosphate dikinase to phosphoenolpyruvate (see also Figs 8.11, 8.12, 8.14, 8.15). Since plastids are normally unable to convert phosphoenolpyruvate to 3-phosphoglycerate (still to be investigated for CAM chloroplasts) the phosphoenolpyruvate is exported in exchange for 3-phosphoglycerate (probably catalysed by two different translocators) as shown in Fig. 8.18. As in C_4 plants, CAM chloroplasts contain, in addition to a triose phosphate–phosphate translocator (transporting in a counter-exchange triose phosphate, phosphate, and 3-phosphoglycerate), a phosphoenolpyruvate–phosphate translocator (catalysing a counter-exchange for phosphate). The 3-phosphoglycerate taken up into the chloroplasts is fed into the Calvin cycle. The triose phosphate thus formed is used primarily for resynthesis of the starch consumed during the previous night. Only a small surplus of triose phosphate remains and this is the actual gain of CAM photosynthesis.

Since photosynthesis proceeds with closed stomata, the water loss during CAM is very low. In CAM photosynthesis the water requirement for CO_2 assimilation (compare Fig. 8.6) amounts to only 5–10 per cent of the water needed for photosynthesis of C_3 plants. Since the storage capacity for malate is limited, the daily increase in biomass in CAM plants is usually very low. Thus the growth rate for plants relying solely on CAM is limited.

Quite frequently plants use CAM as a strategy for surviving extended dry periods. Some plants perform normal C_3 photosynthesis when water is available, but switch to CAM during drought or salt stress by inducing the corresponding enzymes. It is possible to determine by mass spectrometric analysis of the $^{13}C/^{12}C$ ratio (section 8.4) whether a facultative CAM plant performs C_3 metabolism or CAM. During extreme drought cacti can survive for a long time without even opening their stomata during the night. Under these conditions they can conserve carbon by refixing respiratory CO_2 by CAM photosynthesis.

C_4 metabolism and CAM have been developed several times during evolution

One finds C_4 and CAM plants in many unrelated families of monocot and dicot plants. This shows that on several occasions C_4 metabolism and CAM have both evolved independently from C_3 precursors. As the structural elements and the

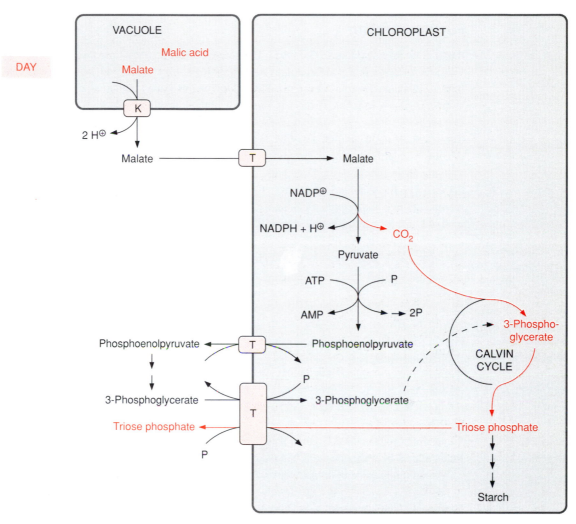

Figure 8.18 CAM during the day. Malate and the accompanying protons are released from the vacuole by a mechanism which is not yet known in detail. In the example given, malate is oxidized in the chloroplasts to pyruvate, yielding CO_2 for the CO_2 fixation by RubisCO. The pyruvate is converted via pyruvate phosphate dikinase to phosphoenolpyruvate which is probably converted in the cytosol to 3-phosphoglycerate. After transport into the chloroplasts, 3-phosphoglycerate is converted to triose phosphate which is used mainly for the regeneration of starch. The transport of phosphoenolpyruvate, 3-phosphoglycerate, triose phosphate, and phosphate proceeds via the triose phosphate translocator. T, translocator; K, channel.

enzymes of C_4 and CAM plants are also present in C_3 plants (e.g. in the guard cells of stomata), the conversion of C_3 plants to C_4 and CAM plants seems to involve relatively simple evolutionary processes.

Further reading

Badger, M. R. and Price, D. G. (1995). The role of carbonic anhydrase in photosynthesis. *Annual Review of Plant Physiology and Plant Molecular Biology*, **45**, 369–92.

Barkla, B. J. and Pantoja, O. (1996). Physiology of ion transport across the tonoplast of higher plants. *Annual Review of Plant Physiology and Plant Molecular Biology*, **47**, 159–84.

Chollet, R., Vidal, J., and O'Leary, M. H. (1996). Phospho*enol*pyruvate carboxylase: a ubiquitous, highly regulated enzyme in plants. *Annual Review of Plant Physiology and Plant Molecular Biology*, **47**, 273–98.

Edwards, G. and Walker, D. A. (1983). C_3, C_4: *Mechanisms and cellular and environmental regulation of photosynthesis*. Blackwell Scientific Publications, Oxford.

Evans, J. R. and von Caemmerer, S. (1996). Carbon dioxide diffusion inside leaves. *Plant Physiology*, **110**, 339–402.

Farquhar, G. D., Ehleringer, J. R., and Hubick, K. T. (1989). Carbon isotope discrimination and photosynthesis. *Annual Review of Plant Physiology and Plant Molecular Biology*, **40**, 503–37.

Hatch, M. D. (1987). C_4 Photosynthesis: a unique blend of modified biochemistry, anatomy and ultrastructure. *Biochimica et Biophysica Acta*, **895**, 81–106.

Hatch, M. D. (1992). C_4 Photosynthesis: an unlikely process full of surprises. *Plant Cell Physiology*, **33**, 333–42.

Kramer, P. J. and Boyer, J. S. (1995). *Water relations of plants and soils*. Academic Press, New York.

Leegood, R. C. and Osmond, C. B. (1990). The flux of metabolites in C_4 and CAM plants. In *Plant physiology, biochemistry and molecular biology*, (ed. D. T. Dennis and D. H. Turpin), pp. 284–308. Longman Scientific and Technical, Burnt Mill, Essex.

Mansfield, T. A., Hetherington, A. M., and Atkinson, C. J. (1990). Some current aspects of stomatal physiology. *Annual Review of Plant Physiology and Plant Molecular Biology*, **41**, 55–75.

Nelson, T. and Langdale, J. A. (1992). Developmental genetics of C_4-photosynthesis. *Annual Review of Plant Physiology and Plant Molecular Biology*, **43**, 25–47.

Von Caemmerer, S. and Evans, J. R. (1991). Determination of the average partial pressure of CO_2 in the leaves of several C_3 plants. *Australian Journal of Plant Physiology*, **18**, 287–305.

chapter 9

Polysaccharides

In higher plants, photosynthesis in the leaves provides substrates, such as carbohydrates, for the various heterotrophic plant tissues, e.g. the roots. Substrates delivered from the leaves are oxidized in the root cells by the large number of mitochondria present there. The ATP thus generated is required for driving the ion pumps of the roots by which mineral nutrients are taken up from the surrounding soil. Therefore respiratory metabolism of the roots, supported by photosynthesis of the leaves, is essential for plants. The plant dies when the roots are not sufficiently aerated and there is not enough oxygen available for respiration.

Various plant parts are supplied with carbohydrates mostly in the form of the disaccharide sucrose, but in some plants also as tri- and tetrasaccharides or sugar alcohols. Since the synthesis of carbohydrates by photosynthesis only occurs during the day, these carbohydrates have to be stored in the leaves to ensure their continued supply to the rest of the plant during the night or during unfavourable weather conditions. Moreover, plants need to build up carbohydrate stores to tide them over the winter or dry periods, and also as a reserve in seeds for the growth of the following generation. For this purpose carbohydrates are stored primarily in the form of high molecular weight polysaccharides, in particular as starch or fructans, but also as low molecular weight oligosaccharides.

Starch and sucrose are the main products of CO_2 assimilation in many plants

In most crop plants, e.g. cereals, potato, sugar beet, and rapeseed, carbohydrates are stored in the leaves as *starch* and exported as *sucrose* to other parts of the plants such as the roots or growing seeds. CO_2 assimilation in the chloroplasts yields phosphate which is transported by the *triose phosphate–phosphate translocator* (section 1.9) in counter-exchange for phosphate into the cytosol, where it is converted to sucrose, accompanied by the release of inorganic phosphate (Fig. 9.1). The return of this phosphate is essential, since if there were a phosphate deficiency in the chloroplasts, photosynthesis would come to a stop. Part of the triose phosphate generated by photosynthesis is converted in the chloroplasts to starch, serving primarily as a reserve for the following night period.

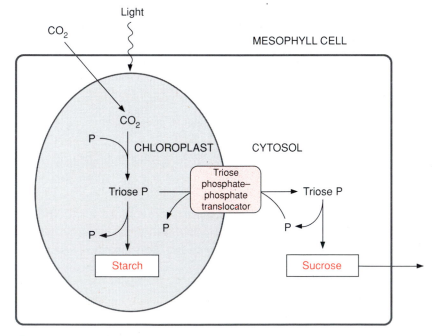

Figure 9.1 Triose phosphate, the product of photosynthetic CO_2 fixation, is either converted in the chloroplasts to starch or, after transport out of the chloroplasts, transformed to sucrose and in the latter form exported from the mesophyll cells.

9.1 Large quantities of carbohydrate can be stored as starch in the cell

Glucose is a relatively unstable compound since its aldehyde group can be oxidized spontaneously to a carboxyl group. Therefore glucose is not suitable as a carbohydrate storage compound. Moreover, for osmotic reasons, the cell has a limited storage capacity for monosaccharides. By polymerization of glucose to the osmotically inert starch, large quantities of glucose molecules can be deposited in a cell without effecting an increase in the osmotic pressure of the cell sap. This may be illustrated by an example: at the end of the day the starch content in potato leaves may amount to 10^{-4} mol glucose units per mg of chlorophyll. If this amount were dissolved as free glucose in the aqueous phase of the mesophyll cell, this would yield a glucose concentration of 0.25 mol l^{-1}. Such an accumulation of glucose would result in an increase by more than 50 per cent in the osmotic pressure of the cell sap.

The glucose molecules in starch are primarily connected by ($\alpha1\rightarrow4$) glycosidic linkages (Fig. 9.2). These linkages protect the aldehyde groups of the glucose molecules against oxidation; only the first glucose molecule, coloured red in Fig. 9.2, is unprotected. In this way long glucose chains are formed which can be branched by ($\alpha1\rightarrow6$) glycosidic linkages. Branched starch molecules contain many terminal glucose residues at which the starch molecule can be enlarged.

In plants the formation of starch is restricted to plastids, namely *chloroplasts* in leaves and green fruits, and *leucoplasts* in heterotrophic tissues. Starch is deposited in the plastids in the form of *starch granules* (Fig. 9.3). The starch granules in a leaf are very large at the end of the day and are usually degraded extensively during the following night. This starch is called *transitory starch*. In contrast,

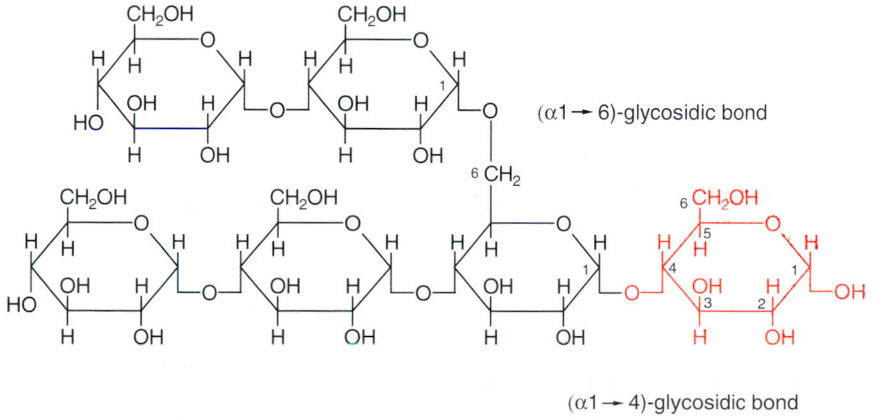

Figure 9.2 The glucose molecules in starch are connected by ($\alpha1\rightarrow4$) and ($\alpha1\rightarrow6$) glycosidic bonds to form a polyglucan. Only the glucose residue coloured red contains a reducing group.

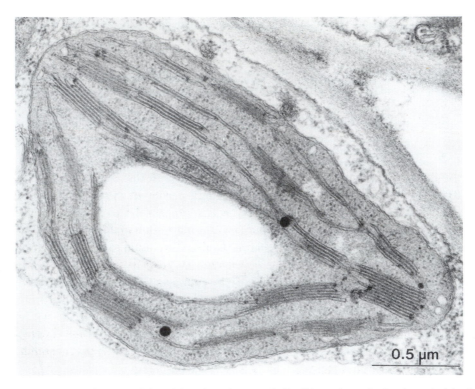

Figure 9.3 Transitory starch in a chloroplast of a mesophyll cell in a tobacco leaf at the end of the day. The starch granule in chloroplasts appears as a large white spot. (By D. G. Robinson, Göttingen.)

the starch in storage organs, e.g. seeds or tubers, is deposited for longer time periods, and therefore is called *reserve starch*. In cereals the reserve starch often represents 65–75 per cent and in potato tubers even 80 per cent of the dry weight.

Starch granules consist primarily of *amylopectin, amylose,* and, in some cases, *phytoglycogen* (Table 9.1). Starch granules contain the enzymes for starch synthesis and degradation. These enzymes are present in several isoforms, some of which are bound to the starch granules while others are soluble. Amylose consists mainly of unbranched chains of about 1000 glucose molecules. Amylopectin, with 10^4–10^5 glucose molecules, is much larger than amylose and has a branching point at every 20–25 glucose residues (Fig. 9.4). The exact structure of the starch granule is not yet known. It appears to have a semi-crystalline structure in which the amylopectin molecules are probably arranged in layers (Fig. 9.5). The reducing glucose (coloured red in Figs 9.2 and 9.4) is directed towards the inside, and glucose residues at the ends of the branches (coloured black) towards the outside. Neighbouring branches of amylopectin form double helices which are packed in a crystalline array. Amylose, on the other hand, is probably present in an amorphous form. Some starch granules contain phytoglycogen, a particularly highly branched starch.

Table 9.1 Constituents of plant starch

	Number of glucose residues	Number of glucose residues per branching	Absorption maximum of the glucan iodine complex
Amylose	10^3		660 nm
Amylopectin	10^4–10^5	20–25	530–550 nm
Phytoglycogen	10^5	10–15	430–450 nm

A starch granule usually contains 20–30 per cent amylose, 70–80 per cent amylopectin, and, in some cases, up to 20 per cent phytoglycogen. Wrinkled peas, which Gregor Mendel used in his classical breeding experiments, have an amylase content of up to 80 per cent. In the so-called waxy maize mutants the starch granules consist almost entirely of amylopectin. On the other hand, the starch of the maize variety amylomaize consists of 50 per cent amylose. Recently transgenic potato plants have been generated which contain only amylopectin in their tubers. A uniform starch content in potato tubers is of importance for the use of starch as a multi-purpose raw material in the chemical industry.

Amylose, amylopectin, and phytoglycogen form blue- to violet-coloured complexes with iodine molecules (Table 9.1). This makes it very easy to detect starch in a leaf by a simple iodine test.

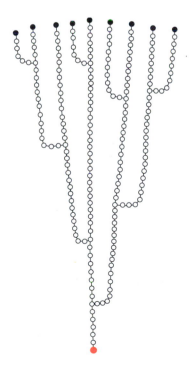

Figure 9.4 The polyglucan chains in amylopectin contain a branch point at every 20–25 glucose residues. Neighbouring chains are arranged in an ordered structure. The glucose residue, coloured red at the beginning of the chain, contains a reducing group. The groups coloured black at the end of the branches are the acceptors for the addition of further glucose residues by starch synthase.

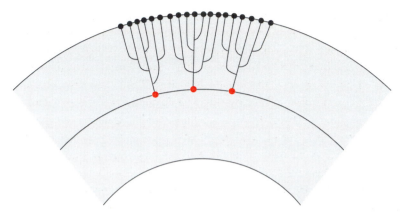

Figure 9.5 In a starch granule the amylopectin molecules are arranged in layers. Compare with Fig. 9.4.

Starch is synthesized via ADP-glucose

Fructose 6-phosphate, an intermediate of the Calvin cycle, is the precursor for starch synthesis in chloroplasts (Fig. 9.6). Fructose 6-phosphate is converted by hexose phosphate isomerase to glucose 6-phosphate, and a *cis*-enediol is formed as an intermediate of this reaction. Phosphoglucomutase transfers the phosphate residue from the 6-position of glucose to the 1-position. Activation of glucose 1-phosphate by reaction with ATP to ADP-glucose, accompanied by release of

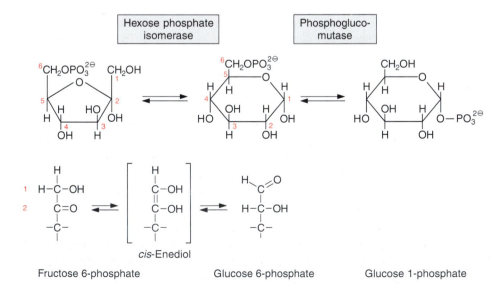

Figure 9.6 Conversion of fructose 6-phosphate to glucose 1-phosphate. In the hexose phosphate isomerase reaction a *cis*-enediol is formed as an intermediary product.

pyrophosphate, is a crucial step for starch synthesis. This reaction, catalysed by the enzyme *ADP-glucose pyrophosphorylase* (Fig. 9.7), is reversible. The high activity of pyrophosphatase in the chloroplast stroma, however, ensures that the pyrophosphate formed is immediately hydrolysed to phosphate and thus withdrawn from equilibrium. Therefore the formation of ADP-glucose is an irreversible process and very suitable for regulating starch synthesis. The American biochemist Jack Preiss, who has studied the properties of ADP-glucose pyrophosphorylase in detail, found that this enzyme is allosterically activated by 3-phosphoglycerate and inhibited by phosphate. The significance of this regulation will be dealt with at the end of this section. The glucose residue is transferred by *starch synthase* from ADP-glucose to the OH-group in the 4-position of the terminal glucose molecule in the polysaccharide chain of starch (Fig. 9.7).

Branches are formed by a *branching enzyme*. At certain chain lengths the polysaccharide chain is cleaved at the ($\alpha 1 \rightarrow 4$) glycosidic bond (Fig. 9.8) and the chain fragment thus separated is connected via a newly formed ($\alpha 1 \rightarrow 6$) bond to a neighbouring chain. These chains are elongated further by starch synthase until a new branch develops. In the course of starch synthesis branches are also cleaved again by a *debranching enzyme*, which will be dealt with at a later point. It is assumed that the activities of the branching and the debranching enzymes determine the degree of branching in starch. The wrinkled peas with the high amylose content already mentioned are the result of a decrease in the activity of the branching enzyme in these plants, leading on the whole to a lowered starch content.

Degradation of starch proceeds in two different ways

Degradation of starch proceeds in two basically different reactions (Fig. 9.9). *Amylases* catalyse a hydrolytic cleavage of ($\alpha 1 \rightarrow 4$) glycosidic bonds. Different amylases attack the starch molecule at different sites (Fig. 9.10). *Exoamylases* hydrolyse starch at the end of the molecules. β-*Amylase* is an important amylase which splits off two glucose residues in the form of the disaccharide maltose from the end of the starch molecule (Fig. 9.11). The enzyme is named after its product, β-maltose, in which the OH-group in the 1-position is present in the β-configuration. Amylases that hydrolyse starch in the interior of the glucan chain (*endoamylases*) produce cleavage products in which the OH-group in the 1-position is in α-configuration, and are therefore called α-*amylases*. ($\alpha 1 \rightarrow 6$) Glycosidic bonds at the branch points are hydrolysed by *debranching enzymes*.

Phosphorylases (Fig. 9.9) cleave ($\alpha 1 \rightarrow 4$) bonds phosphorolytically, resulting in the formation of glucose 1-phosphate. The energy of the glycosidic bond is used here to form a phosphate ester, making this phosphorolytic starch degradation more economical than starch hydrolysis by amylases.

Mobilization of chloroplast transitory starch involves both phosphorolytic and hydrolytic degradation. The phosphorylases alone are not capable of degrading

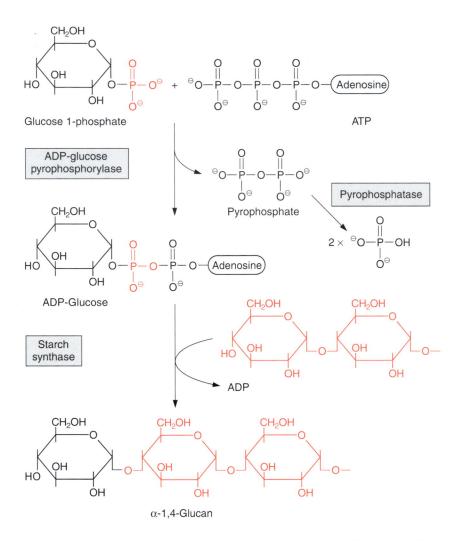

Figure 9.7 Biosynthesis of starch. Glucose 1-phosphate reacts with ATP to give ADP-glucose. The pyrophosphate formed is hydrolysed by pyrophosphatase and in this way the formation of ADP-glucose becomes irreversible. The glucose activated by ADP is transferred by starch synthase to a terminal glucose residue in the glucan chain.

large, branched starch molecules. Degradation of transitory starch begins with α-amylases cleaving large starch molecules into smaller fragments, which are then degraded further not only by amylases but also by phosphorylases. Phosphorylases also occur in the cytosol. It is feasible that small oligosaccharides, products of hydrolytic starch degradation, are released from the chloroplast to the cytosol to be degraded completely by the phosphorylases present there. Our knowledge about this is still incomplete.

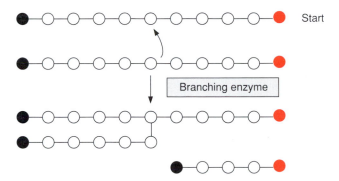

Figure 9.8 In a polyglucan chain an ($\alpha 1 \rightarrow 4$) linkage is cleaved by the branching enzyme and the disconnected chain fragment is linked to a neighbouring chain by an ($\alpha 1 \rightarrow 6$) glycosidic bond.

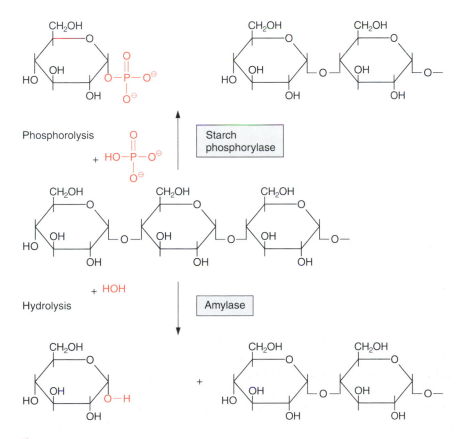

Figure 9.9 The ($\alpha 1 \rightarrow 4$) linkage in a starch molecule can be cleaved by phosphorolysis or by hydrolysis.

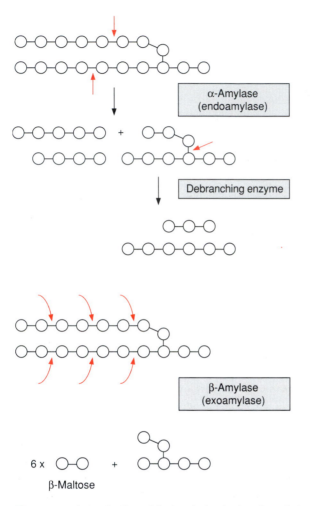

Figure 9.10 (α1→4) Glycosidic bonds in the interior of the starch molecule are hydrolysed by α-amylases. The debranching enzyme hydrolyses (α1→6) linkages. β-Amylases release the disaccharide β-maltose by hydrolysis of (α1→4) bonds successively from the end of the starch molecules.

Surplus photosynthesis products can be stored temporarily in chloroplasts by starch synthesis

Figure 9.12 gives an outline of the synthesis and degradation of transitory starch in chloroplasts. Regulation of ADP-glucose pyrophosphorylase by *3-phosphoglycerate* (3-PGA) and *phosphate* (P) mentioned earlier enables the flux of carbohydrates into starch to be regulated. The activity of the enzyme is governed by the 3-PGA/P concentration ratio. 3-PGA is a major metabolite in the chloroplast stroma. Due to the equilibrium of the reactions catalysed by phosphoglycerate kinase and glyceraldehyde phosphate dehydrogenase (section 6.3) the stromal

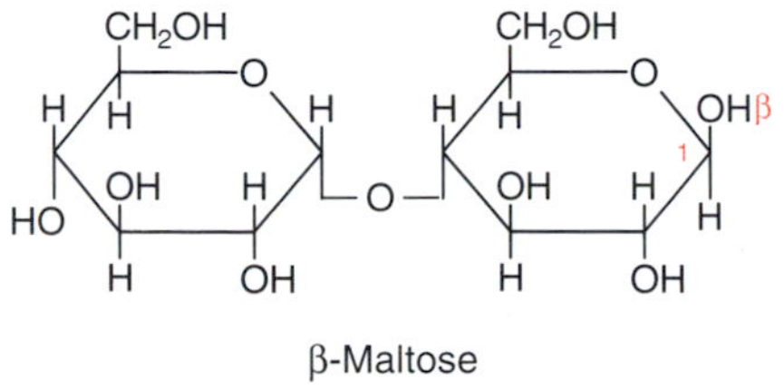

Figure 9.11 In the disaccharide, β-maltose, the OH-group in position 1 is in the β-configuration.

β-Maltose

PGA concentration is much higher than that of triose phosphate. Since in the chloroplast stroma the total amount of phosphate and phosphorylated intermediates of the Calvin cycle is kept virtually constant by the counter-exchange of the triose phosphate–phosphate translocator (section 1.9), an increase in the concentration of 3-PGA results in a decrease in the concentration of phosphate. The 3-PGA/P ratio is therefore a very sensitive indicator of the metabolite status in the chloroplast stroma. When a decrease in sucrose synthesis brings about a decrease in phosphate liberation in the cytosol, the chloroplasts suffer from phosphate deficiency which limits their photosynthesis (Fig. 9.1). In such a situation the PGA/P quotient increases, leading in turn to an increase in starch synthesis, by which phosphate is released thus allowing photosynthesis to continue. Starch functions here as a buffer. Assimilates that are not utilized for synthesis of sucrose or other substances are deposited temporarily in the chloroplasts as transitory starch. Moreover, starch synthesis is programmed in such a way (by a mechanism largely unknown) that starch is deposited each day for use during the following night.

So far very little is known about the regulation of transitory starch degradation. It is probably stimulated by an increase in the stromal phosphate concentration, but the mechanism for this is still unclear. An increase in the stromal phosphate concentration indicates a shortage of substrates. Glucose arising from the hydrolytic degradation of starch is released from the chloroplasts via a *glucose translocator* (Fig. 9.12). Glucose 1-phosphate, derived from phosphorolytic starch degradation, is converted in a reversal of the starch synthesis pathway to fructose 6-phosphate, and the latter to fructose 1,6-bisphosphate by fructose 6-phosphate kinase. Triose phosphate formed from fructose 1,6-bisphosphate by aldolase is released from the chloroplasts via the *triose phosphate–phosphate translocator*. Part of the triose phosphate is oxidized within the chloroplasts to 3-phosphoglycerate and is subsequently exported also via the triose phosphate–phosphate translocator. This translocator, as well as the glucose translocator, is thus involved in the mobilization of the chloroplastic transitory starch.

9.2 Sucrose synthesis takes place in the cytosol

The synthesis of sucrose, a disaccharide of glucose and fructose (Fig. 9.13), takes place in the cytosol of mesophyll cells. As in starch synthesis, the glucose residue is

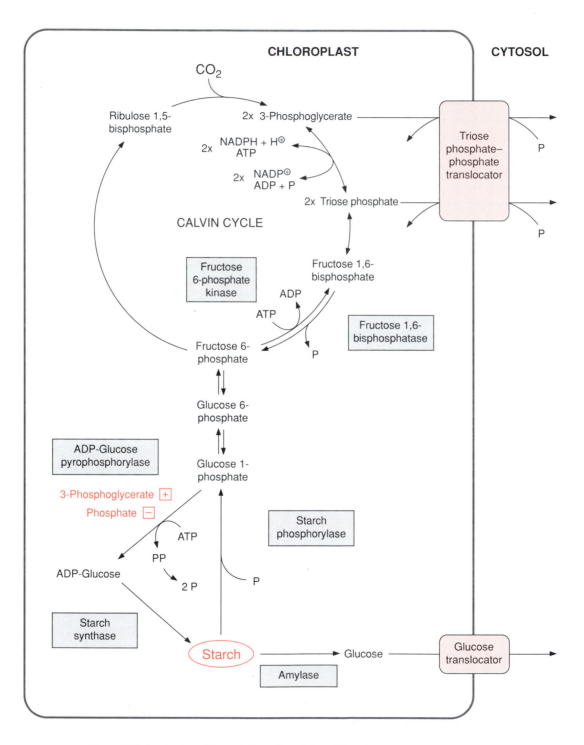

Figure 9.12 Synthesis and degradation of starch in a chloroplast.

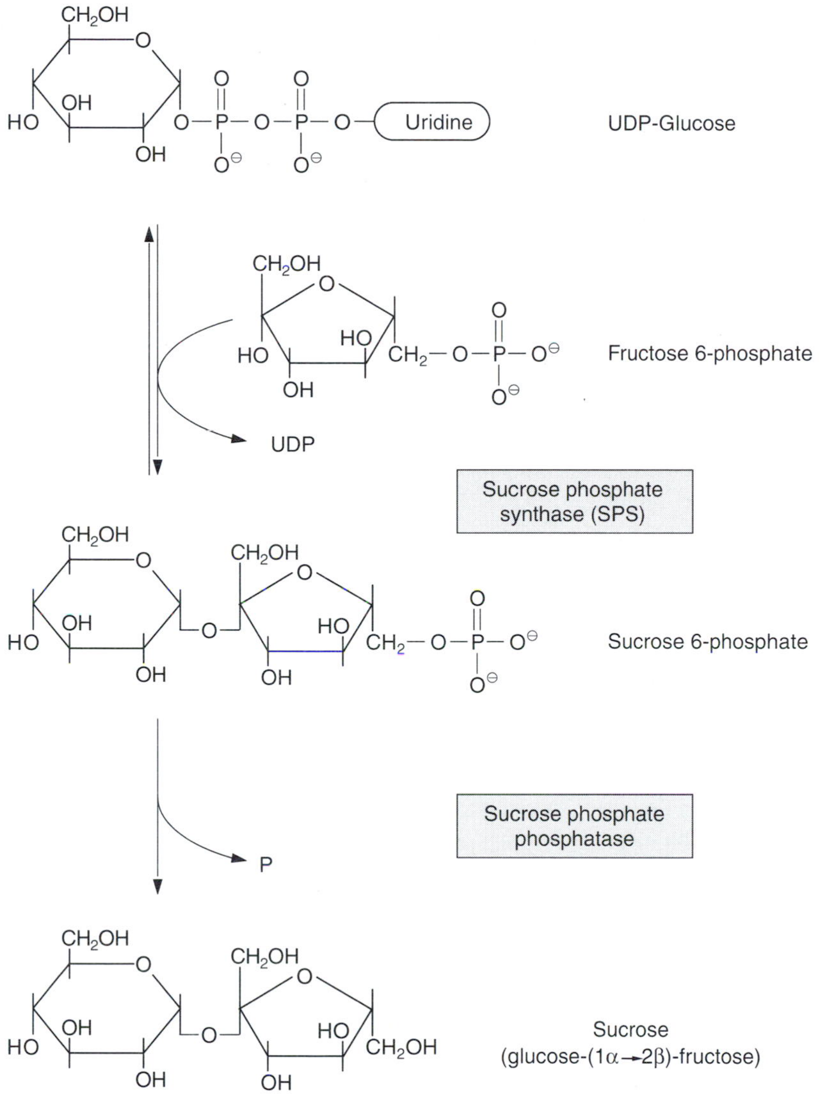

Figure 9.13 Synthesis of sucrose. The glucose activated by UDP is transferred to fructose 6-phosphate. The total reaction becomes irreversible by hydrolysis of the formed sucrose 6-phosphate.

activated as nucleoside diphosphate–glucose, although in this case via *UDP-glucose pyrophosphorylase*:

$$\text{UDP-glucose pyrophosphorylase}$$
$$\text{glucose 1-phosphate} + \text{UTP} \rightleftarrows \text{UDP-glucose} + \text{PP}$$

In contrast to the chloroplast stroma, the cytosol of mesophyll cells does not contain pyrophosphatase to withdraw pyrophosphate from the equilibrium, and therefore the UDP-glucose pyrophosphorylase reaction is reversible. *Sucrose phosphate synthase* (SPS, Fig. 9.13) catalyses the transfer of the glucose residue from UDP-glucose to fructose 6-phosphate, forming sucrose 6-phosphate. *Sucrose phosphate phosphatase* hydrolyses sucrose 6-phosphate, thus withdrawing it from the sucrose phosphate synthase reaction equilibrium. Therefore, the overall reaction of sucrose synthesis is an irreversible process.

Besides sucrose phosphate synthase, plants also contain a *sucrose synthase*:

$$\text{Sucrose synthase}$$
$$\text{sucrose} + \text{UDP} \rightleftarrows \text{UDP-glucose} + \text{fructose}$$

This reaction is reversible. It is not involved in sucrose synthesis but in the utilization of sucrose by catalysing the formation of UDP-glucose and fructose from UDP and sucrose. This enzyme occurs primarily in non-photosynthetic tissues. It is involved in sucrose breakdown for the synthesis of starch in amyloplasts of storage tissue such as potato tubers (section 13.3). It also plays a role in the synthesis of cellulose and callose, where the sucrose synthase, otherwise soluble, is membrane bound (section 9.6).

9.3 Utilization of triose phosphate formed during photosynthesis is strictly regulated

As already shown in Fig. 6.11, five-sixths of the triose phosphate generated in the Calvin cycle is required for the regeneration of the CO_2 acceptor ribulose bisphosphate. Therefore a maximum of one-sixth of the triose phosphate formed is available for export from the chloroplasts. In fact, due to photorespiration (Chapter 7) the portion of available triose phosphate is only about one-eighth of the triose phosphate formed in the chloroplasts. If more triose phosphate were withdrawn from the Calvin cycle, the CO_2 acceptor ribulose bisphosphate could no longer be regenerated and the Calvin cycle would collapse. Therefore it is crucial for the functioning of the Calvin cycle that the withdrawal of triose phosphate does not exceed this limit. On the other hand, photosynthesis of chloroplasts can only proceed if its product triose phosphate is utilized, e.g. for synthesis of sucrose, and phosphate is released. A phosphate deficiency would result in a decrease or even a total stop in photosynthesis. Thus it is important for a plant that when the rate of photosynthesis is increased, e.g. by strong sunlight, the increase in the formation of assimilation products is matched by a corresponding increase in their utilization.

Therefore the utilization of the triose phosphate generated by photosynthesis should be regulated in such a way that as much as possible is utilized without exceeding the set limit and so ensuring the regeneration of the CO_2 acceptor ribulose bisphosphate.

Fructose 1,6-bisphosphatase functions as an entry valve for the pathway of sucrose synthesis

In mesophyll cells sucrose synthesis is normally the main consumer of triose phosphate generated by CO_2 fixation. The withdrawal of triose phosphate from chloroplasts for the synthesis of sucrose is not regulated at the translocation step. Due to the hydrolysis of fructose 1,6-bisphosphate and sucrose 6-phosphate the total reaction of sucrose synthesis (Fig. 9.14) is an irreversible process which, owing to the high enzymatic activities, has a high synthetic capacity. Sucrose synthesis has to be strictly regulated to ensure that not more than the permitted amount of triose phosphate (see above) is withdrawn from the Calvin cycle.

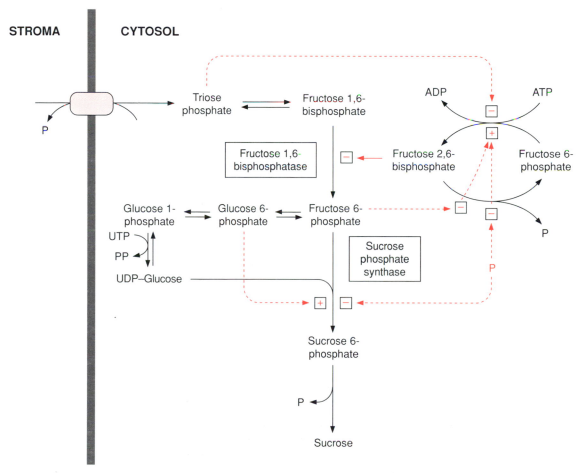

Figure 9.14 Conversion of triose phosphate into sucrose. The dashed red lines represent the regulation of single reactions by metabolites; $\boxed{-}$ means inhibition; $\boxed{+}$, activation. The effect of the regulatory substance fructose 2,6-bisphosphate is explained in detail in Fig. 9.15.

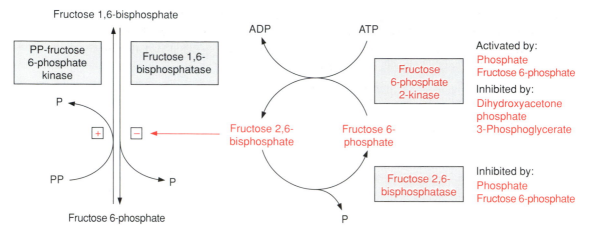

Figure 9.15 Fructose 1,6-bisphosphatase represents the entry valve for the conversion of the CO_2 assimilate into sucrose. The enzyme is inhibited by the regulatory substance fructose 2,6-bisphosphate (Fru2,6BP). The pyrophosphate-dependent fructose 6-phosphate kinase, which synthesizes fructose 1,6-bisphosphate from fructose 6-phosphate, with the consumption of pyrophosphate, is only active in the presence of the regulatory substance Fru2,6BP. The concentration of this Fru2,6BP is adjusted by continuous synthesis and degradation. The enzymes catalysing Fru2,6BP synthesis and degradation are regulated by metabolites. In this way the presence of triose phosphate and 3-phosphoglycerate decreases the concentration of Fru2,6BP and thus increases the activity of fructose 1,6-bisphosphatase.

The first irreversible step of sucrose synthesis is catalysed by the *cytosolic fructose 1,6-bisphosphatase*. This reaction is an important control point and is the entry valve where triose phosphate is withdrawn for the synthesis of sucrose. Figure 9.15 shows how this valve functions. An important role is played here by *fructose 2,6-bisphosphate* (Fru2,6BP), a regulatory substance, which differs from the metabolic fructose 1,6-bisphosphate only in the positioning of one phosphate group (Fig. 9.16).

Fru2,6BP was discovered to be a potent activator of ATP-dependent fructose 6-kinase and an inhibitor of fructose 1,6-bisphosphatase in liver. Later it became apparent that Fru2,6BP has a general function in controlling glycolysis and gluconeogenesis in animals, plants, and fungi. It is a powerful regulator of cytosolic fruc-

Fructose 1,6-bisphosphate, a metabolite

Fructose 2,6-bisphosphate, a regulatory substance

Figure 9.16 The regulatory substance fructose 2,6-bisphosphate differs from the metabolite fructose 1,6-bisphosphate only in the position of one phosphate group.

tose 1,6-bisphosphatase in mesophyll cells and just micromolar concentrations of Fru2,6BP, as may occur in the cytosol of mesophyll cells, result in a large decrease in the affinity of the enzyme towards its substrate, fructose 1,6-bisphosphate. On the other hand, Fru2,6BP activates a *pyrophosphate-dependent fructose 6-phosphate kinase* found in the cytosol of plant cells. Without Fru2,6BP this enzyme is inactive. The pyrophosphate-dependent fructose 6-phosphate kinase can utilize pyrophosphate which is formed in the UDP-glucose pyrophosphorylase reaction.

Fru2,6BP is formed from fructose 6-phosphate by a specific kinase (*fructose 6-phosphate 2-kinase*) and is degraded hydrolytically by a specific phosphatase (*fructose 2,6-bisphosphatase*) to fructose 6-phosphate again. The level of the regulatory substance Fru2,6BP is adjusted by regulation of the relative rates of synthesis and degradation. Triose phosphate and 3-phosphoglycerate inhibit the synthesis of Fru2,6BP, whereas fructose 6-phosphate and phosphate stimulate the synthesis and decrease the hydrolysis. In this way an increase in the triose phosphate concentration results in a decrease in the level of Fru2,6BP and thus in an increased affinity of the cytosolic fructose 1,6-bisphosphate towards its substrate, fructose 1,6-bisphosphate. Moreover, due to the equilibrium catalysed by cytosolic aldolase, an increase in the triose phosphate concentration results in an increase in the concentration of fructose 1,6-bisphosphate. The simultaneous increase in substrate concentration and substrate affinity has the effect that the rate of sucrose synthesis increases in a sigmoidal fashion with rising triose phosphate concentrations (Fig. 9.17). In this way the rate of sucrose synthesis can be adjusted effectively to the supply of triose phosphate.

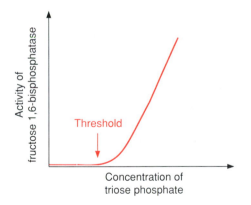

Figure 9.17 Cytosolic fructose 1,6-bisphosphatase acts as an entry valve to adjust the synthesis of sucrose to the supply of triose phosphate. Increasing triose phosphate leads, via aldolase, to an increase in the substrate fructose 1,6-bisphosphate (Fig. 9.14) and, in parallel (Fig. 9.15), to a decrease in the concentration of the regulatory substance fructose 2,6-bisphosphate. As a consequence of these two synergistic effects, the activity of fructose 1,6-bisphosphatase increases with the concentration of triose phosphate in a sigmoidal manner. The enzyme becomes active only after the triose phosphate concentration reaches a threshold concentration, and then responds in its activity to increasing triose phosphate concentrations.

The principle of regulation can be compared with an overflow valve. A certain threshold concentration of triose phosphate has to be exceeded for an appreciable metabolite flux via fructose 1,6-bisphosphatase to occur. This ensures that the triose phosphate level in chloroplasts does not decrease below the minimum level required for the Calvin cycle to function. When this threshold is overstepped, a further increase in triose phosphate results in a large increase in enzyme activity, whereby the surplus triose phosphate can be channelled very efficiently into sucrose synthesis.

Cytosolic fructose 1,6-bisphosphatase not only adjusts its activity, as shown above, to the substrate supply, but also to the demand for its product. With an increase in fructose 6-phosphate, the level of the regulatory substance Fru2,6BP is increased by stimulation of fructose 6-phosphate 2-kinase and simultaneous inhibition of fructose 2,6-bisphosphatase, whereby the activity of cytosolic fructose 1,6-bisphosphatase is reduced (Fig. 9.15).

Sucrose phosphate synthase is regulated not only by metabolites but also by covalent modification

Sucrose phosphate synthase (SPS) (Fig. 9.14) is also subject to strict metabolic control. This enzyme is activated by glucose 6-phosphate and inhibited by phosphate. Due to hexose phosphate isomerase, the activator glucose 6-phosphate is in equilibrium with fructose 6-phosphate. In this equilibrium the concentration of glucose 6-phosphate greatly exceeds the concentration of the substrate fructose 6-phosphate. In equilibrium, therefore, the change in the concentration of the substrate results in a much larger change in the concentration of the activator. In this way the activity of the enzyme is adjusted effectively to the supply of the substrate.

Moreover, the activity of sucrose phosphate synthase is altered by a covalent modification of the enzyme. The enzyme contains at least one serine residue, of which the OH-group is phosphorylated by a special protein kinase, termed *sucrose phosphate synthase kinase* (SPS kinase) and is dephosphorylated by the corresponding *SPS phosphatase* (Fig. 9.18). The SPS phosphatase is inhibited by *okadaic acid*, an inhibitor of protein phosphatases of the so-called 2A-type (not dealt with in more detail here).

The phosphorylated form of sucrose phosphate synthase is less active than the dephosphorylated form. The activity of the enzyme is adjusted by the relative rates of its phosphorylation and dephosphorylation. Illumination of a leaf increases the activity of SPS phosphatase and thus the sucrose phosphate synthase is converted into the more active form. The mechanism for this is still not known. It is debated that the decrease of SPS phosphatase activity observed during darkness is due to a lowered rate of the synthesis of the SPS phosphatase. The synthesis and hydrolysis of the regulatory substance Fru2,6BP is probably also subject to control by diurnal factors.

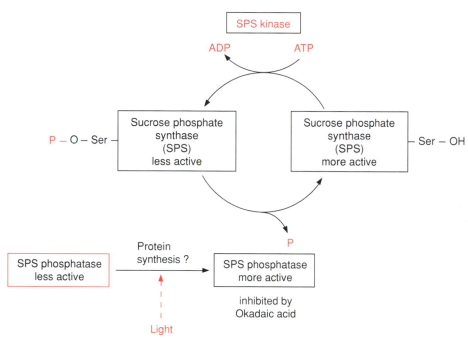

Figure 9.18 Sucrose phosphate synthase (abbreviated SPS) is converted to a less active form by phosphorylation of a serine residue via a special protein kinase (SPS kinase). The hydrolysis of the phosphate residue by SPS phosphatase results in an increase in the activity. The activity of SPS phosphatase is increased by illumination, probably via a *de novo* synthesis of the enzyme protein. (After Huber and Huber 1996.)

Partitioning of assimilates between sucrose and starch is due to the interplay of several regulatory mechanisms

The preceding section discussed various regulatory processes involved in the regulation of sucrose synthesis. Metabolites acting as enzyme inhibitors or activators can adjust the rate of sucrose synthesis immediately to the prevailing metabolic conditions in the cell. Such an immediate response is called *fine control*. The covalent modification of enzymes, influenced by diurnal factors and probably also by phytohormones (Chapter 19), results in a *general regulation of metabolism* according to the metabolic demand of the plant. This includes partitioning of assimilates between sucrose, starch, and amino acids (Chapter 10). Thus, slowing down sucrose synthesis, which results in an increase in triose phosphate and also of 3-phosphoglycerate, can lead to an increase in the rate of starch synthesis (Fig. 9.12). As already discussed, during the day a large part of the photoassimilates is deposited temporarily in the chloroplasts of leaves as transitory starch, to be converted during the following night to sucrose and delivered to other parts of the plant. However, in some plants, such as barley, large quantities of the photoassimilates are stored as sucrose in the leaves during the day. Therefore during darkness the rate of sucrose synthesis varies in the leaves of different plants.

9.4 In some plants assimilates from the leaves are exported as sugar alcohols or oligosaccharides of the raffinose family

Sucrose is not the main transport form in all plants for the translocation of assimilates from the leaves to other parts of the plant. In some plants photoassimilates are translocated as sugar alcohols. In the *Rosaceae* (these include orchard trees in temperate regions) the assimilates are translocated from the leaves in the form of *sorbitol* (Fig. 9.19). Other plants, such as squash, several deciduous trees, (lime, hazelnut, elm), and olive trees, translocate in their sieve tubes oligosaccharides of the *raffinose family*, in which sucrose is linked by a glycosidic bond to one or more galactose molecules (Fig. 9.20). Oligosaccharides of the raffinose family include *raffinose* with one, *stachyose* with two, and *verbascose* with three galactose residues. These also function as storage compounds and, for example in pea and bean seeds, make up 5–15 per cent of the dry matter. Humans do not have the enzymes that catalyse the hydrolysis of α-galactosides and are therefore unable to digest oligosaccharides of the raffinose family. When these sugars are ingested they are decomposed in the last section of the intestines by anaerobic bacteria to form digestive gases.

The galactose required for raffinose synthesis is formed by epimerization of UDP-glucose (Fig. 9.21). *UDP-glucose epimerase* catalyses the oxidation of the OH-group in position 4 of the glucose molecule by NAD, which is tightly bound to the enzyme. A subsequent reduction results in the formation of glucose as well as galactose residues in an epimerase equilibrium. The galactose residue is transferred by a transferase to the cyclic alcohol myo-inositol producing galactinol. Myo-inositol–galactosyl transferases catalyse the transfer of the galactose residue from galactinol to sucrose, to form raffinose, and correspondingly also stachyose and verbascose:

$$\text{sucrose} + \text{galactinol} \rightarrow \text{raffinose} + \text{myo-inositol}$$
$$\text{raffinose} + \text{galactinol} \rightarrow \text{stachyose} + \text{myo-inositol}$$
$$\text{stachyose} + \text{galactinol} \rightarrow \text{verbascose} + \text{myo-inositol}.$$

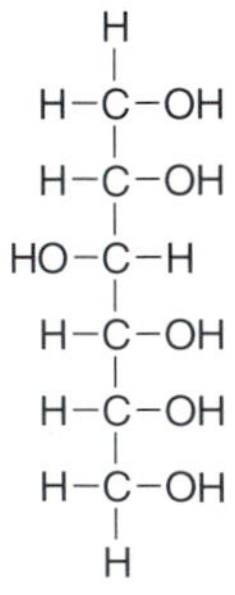

Figure 9.19 In some plants assimilated CO_2 is exported from the leaves in the form of the sugar alcohol sorbitol.

Raffinose

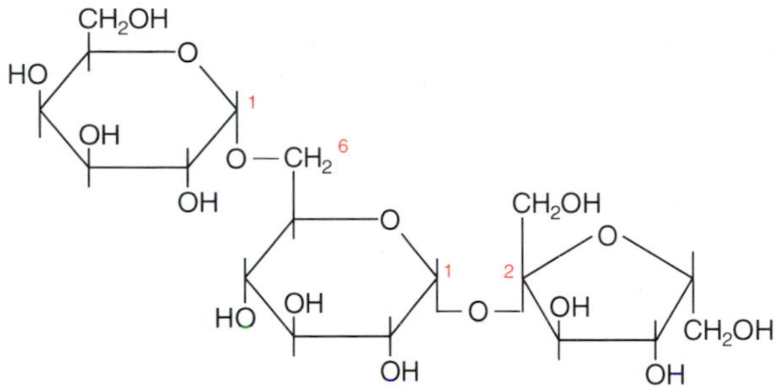

Gal-(1α→6)-Glc-(1α→2β)-Fru

Stachyose

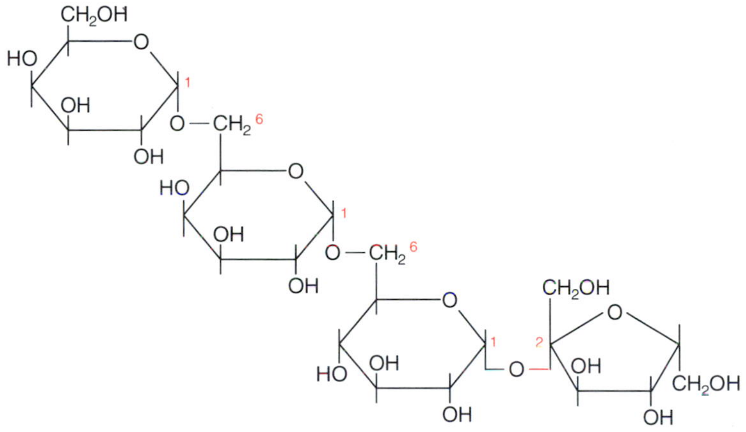

Gal-(1α→6)-Gal-(1α→6)-Glc-(1α→2β)-Fru

Verbascose

Gal-(1α→6)-Gal-(1α→6)-Gal-(1α→6)-Glc-(1α→2β)-Fru

Figure 9.20 In the oligosaccharides of the raffinose family, one to three galactose residues are linked to the glucose residue of sucrose in position 6. Abbreviations: Gal, galactose; Glc, glucose; Fru, fructose.

9.5 Fructans are deposited as storage substances in the vacuole

In addition to starch, many plants use *fructans* as carbohydrate storage compounds. Whereas starch is an insoluble polyglucose formed in the plastids, fructans are soluble polyfructoses which are synthesized and stored in the vacuole. They were first found in the tubers of ornamental flowers such as dahlias. Wheat, barley, and many other grasses from temperate climates, for example, store fruc-

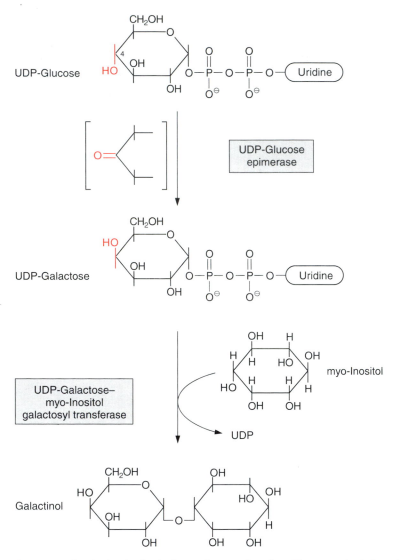

Figure 9.21 Synthesis of galactinol as an intermediate in raffinose synthesis from UDP-glucose and myo-inositol. The epimerization of UDP-glucose to UDP-galactose proceeds via the formation of a keto group as intermediate in position 4.

tans, often in the leaves and stems. Fructans are also the major carbohydrate found in onions and, like the raffinose sugars, cannot be digested by humans. Because of their sweet taste, fructans are used as natural, calorie-free sweeteners. Fructans are also used in the food industry as replacement of fat.

A sucrose molecule to which additional fructose molecules are linked by a glyco-sidic bond is the precursor for the polysaccharide chain of fructans. The basic structure of a fructan in which sucrose is linked with one additional fructose

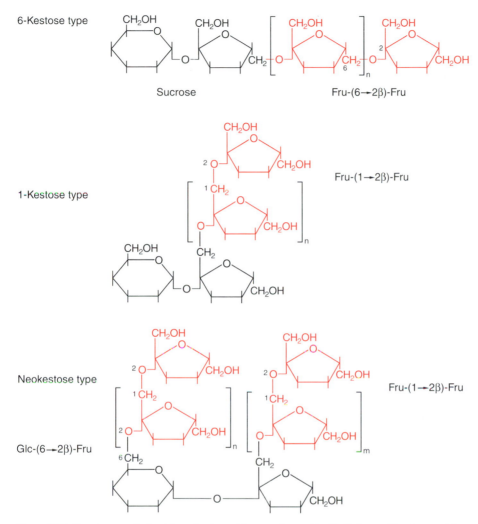

6-Kestose type

Sucrose   Fru-(6→2β)-Fru

1-Kestose type

Fru-(1→2β)-Fru

Neokestose type

Glc-(6→2β)-Fru

Fru-(1→2β)-Fru

Figure 9.22 Fructans are derived from kestoses. They are formed by the linkage of fructose residues to a sucrose molecule. In fructans of the 6-kestose type $n = 200$ fructose residues, and in those of the 1-kestose type $n < 50$. In fructans of the neokestose type both n and $m < 10$.

molecule to a trisaccharide is called *kestose*. Figure 9.22 shows three major types of fructans.

In fructans of the *6-kestose type* the fructose residue of sucrose in position 6 is glycosidically linked with another fructose in the 2β-position. Chains of different lengths (10–200 fructose residues) are formed by (6→2β) linkages with further fructose residues. These fructans are also called *levan-type fructans* and are often found in grasses.

The fructose residues in fructans of the *1-kestose type* are linked to the sucrose molecule and to each other by (1→2β) glycosidic linkages. These fructans, also called *inulin-type fructans*, consist of up to 50 fructose molecules. Inulin is found in dahlia tubers.

In fructans of the *neokestose type* two polyfructose chains are connected to sucrose, one as in 1-kestose via (1→2β) glycosidic linkage with the fructose moiety, and the other in (6→2β) glycosidic linkage with the glucose moiety of the sucrose molecule. The neokestose type are the smallest fructans, with only 5–10 fructose molecules. Branched fructans, in which the fructose molecules are connected by both (1→β2) and (6→β2) glycosidic linkages, are to be found in wheat and barley and are called *graminanes*.

Although fructans appear to have an important function in the metabolism of many plants, our knowledge of their function and metabolism is still fragmentary. Fructan synthesis occurs in the vacuoles and sucrose is the precursor for this synthesis. The fructose moiety of a sucrose molecule is transferred by a *sucrose–sucrose fructosyl transferase* to a second sucrose molecule, resulting in the formation of a 1-kestose with a glucose molecule remaining (Fig. 9.23a). Additional fructose residues are transferred not from another sucrose molecule but from another kestose molecule for the elongation of the kestose chain (Fig. 9.23b). The enzyme *fructan–fructan 1-fructosyl transferase* preferentially transfers the fructose residue from a trisaccharide to a larger size kestose. Correspondingly the formation of 6-kestoses is catalysed by a *fructan–fructan 6-fructosyl transferase*.

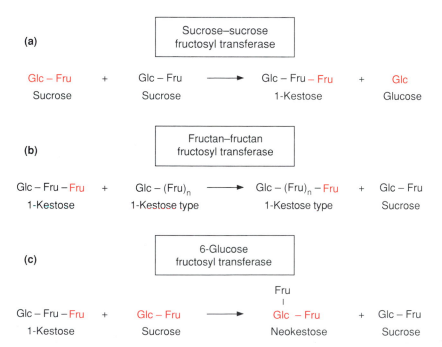

Figure 9.23 Sucrose is the precursor for synthesis of kestoses. Three important reactions of the kestose biosynthesis pathway proceeding in the vacuole are shown.

For the formation of neokestoses a fructose residue is transferred via a 6-glucose fructosyl transferase from a 1-kestose to the glucose moiety of sucrose (Fig. 9.23c). The trisaccharide thus formed is a precursor for further chain elongation as shown in Fig. 9.23b.

The degradation of fructans proceeds by the successive hydrolysis of fructose residues from the end of the fructan chain by exohydrolytic enzymes.

In many grasses fructans are accumulated for a certain time period in the leaves and in the stems. They can comprise up to 30 per cent of the dry matter. Often carbohydrates are accumulated as fructans before the onset of flowering as a reserve for rapid seed growth after pollination of the flowers. Plants in marginal habitats, where periods of positive CO_2 balance are succeeded by periods in which adequate photosynthesis is not possible, use fructans as a reserve to survive unfavourable conditions. Thus in many plants fructans are formed when plants are subjected to water or cold stress.

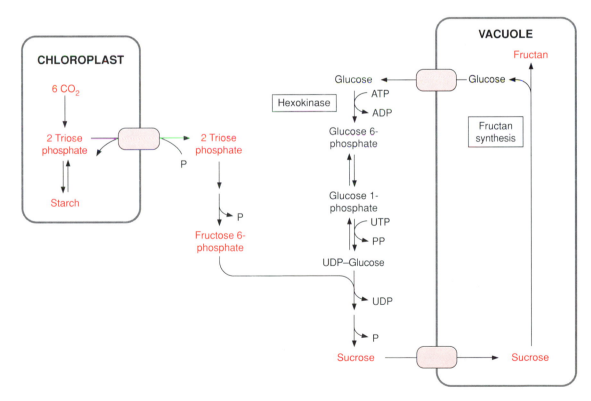

Figure 9.24 The conversion of assimilated CO_2 to fructan. Fructose 6-phosphate, which is provided as product of photosynthesis in the cytosol, is first converted to sucrose. The glucose required for this is formed as a by-product in the synthesis of fructan in the vacuole (see Fig. 9.23). Phosphorylation is catalysed by a hexokinase present in the cytosol. The entry of sucrose into the vacuole and the release of glucose are facilitated by different translocators.

Plants that accumulate fructans usually also store sucrose and starch in their leaves. Fructan is then an additional store. Figure 9.24 shows a simplified scheme of the formation of fructans as alternative storage compounds in a leaf. For the synthesis of fructans, sucrose is first synthesized in the cytosol (for details see Fig. 9.14). The UDP-glucose required is formed from the glucose molecule which is released from the vacuole in the course of fructan synthesis and phosphorylated by hexokinase. Thus the conversion of fructose 6-phosphate, generated by photosynthesis, into fructan needs altogether 2 ATP equivalents per molecule, which is twice the amount necessary for the synthesis of starch in plastids. One reason for this additional ATP expenditure is because pyrophosphatase, which renders the ADP-glucose pyrophosphorylase reaction irreversible in the plastids (Fig. 9.7), is not present in the cytosol.

The large size of the leaf vacuoles, comprising 80 per cent of the total leaf volume, offers the plant a very advantageous storage capacity for carbohydrates in the form of fructan. Thus in a leaf, on top of the diurnal carbohydrate stores in the form of transitory starch and sucrose, an additional carbohydrate reserve can be maintained for purposes such as rapid seed production or the endurance of unfavourable growth conditions.

9.6 Cellulose is synthesized by enzymes located in the plasma membrane

Cellulose, an important cell constituent (section 1.1), is a glucan in which the glucose residues are linked by ($\beta 1 \rightarrow 4$) glycosidic bonds forming a very long chain (Fig. 9.25). The synthesis of cellulose is catalysed by *cellulose synthase* located in the plasma membrane. The glucose molecules required are delivered as *UDP-glucose* from the cytosol of the cell and the newly synthesized cellulose chain is secreted into the extracellular compartment (Fig. 9.26). It has been shown in cotton-producing cells—a useful system for studying cellulose synthesis—that UDP-glucose is supplied by a membrane-bound *sucrose synthase* (section 9.2) from sucrose provided by the cytosol. The UDP-glucose formed is transferred directly to cellulose synthase. Cellulose never occurs in single chains but always in a crystalline array of many chains called a *microfibril* (section 1.1). It is assumed that due to the many neighbouring cellulose synthases in the membrane, all the β-1,4-glucan chains of a microfibril are synthesized simultaneously and form a microfibril spontaneously.

Synthesis of callose is often induced by wounding

Callose is a β-*1,3-glucan* (Fig. 9.25) with a long, unbranched helical chain. Callose by itself forms very compact structures and functions as a *universal isolating material* in the plant. In response to wounding of a cell, large amounts of callose

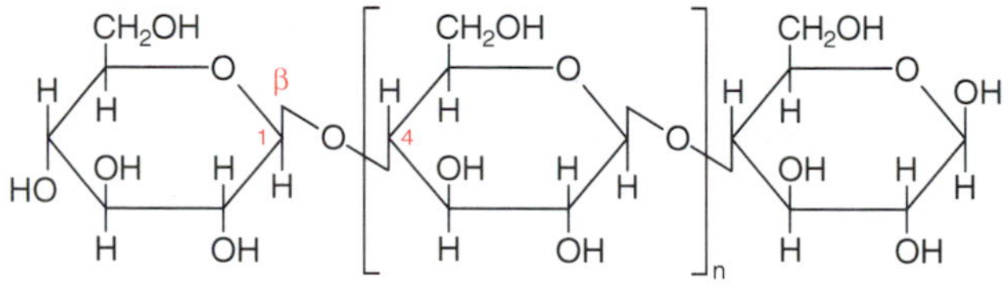

β-1,4-Glucan: cellulose

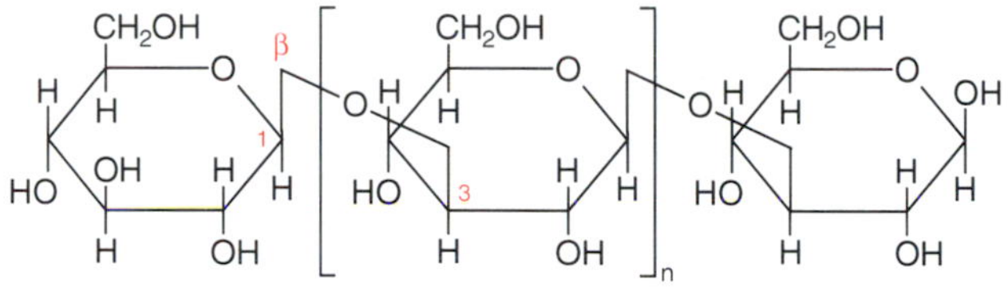

β-1,3-Glucan: callose

Figure 9.25 Cellulose and callose.

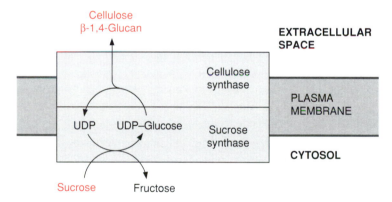

Figure 9.26 Synthesis of β-1,4-glucan chains by a membrane-bound cellulose synthase. The UDP–glucose required is formed from sucrose by a membrane-bound sucrose synthase.

can be synthesized at the plasma membrane very rapidly. It is currently believed that its synthesis proceeds like the synthesis of cellulose (shown in Fig. 9.26). Membrane-bound sucrose synthase also seems to provide UDP-glucose for callose synthesis. It has been suggested that callose and cellulose are synthesized by the same enzyme, but definite proof is still lacking. Callose synthesis is stimulated by an increase in the cytosolic calcium concentration. Wounding, leading to calcium influx and increasing the cytosolic calcium concentration, thus induces the synthesis of callose as an isolating material. Plasmodesmata of damaged cells are closed by callose formation in order to prevent damage to other cells of the symplast (section 1.1). Moreover, callose serves as a filling material to close defective sieve tubes (section 13.2).

Further reading

Beck, E. and Ziegler, P. (1989). Biosynthesis and degradation of starch in higher plants. *Annual Review of Plant Physiology and Plant Molecular Biology*, **40**, 95–117.

Brown, R. M., Saxena, I. M., and Kuklicka, K. (1996). Cellulose biosynthesis in higher plants. *Trends in Plant Science*, **1**, 149–56.

Delmer, D. P. and Amor, Y. (1995). Cellulose biosynthesis. *The Plant Cell*, **7**, 787–1000.

Huber, S. C. and Huber, J. L. (1996). Role and regulation of sucrose-phosphate synthase. *Annual Review of Plant Physiology and Plant Molecular Biology*, **47**, 431–44.

Martin, C. and Smith, A. M. (1995). Starch biosynthesis. *The Plant Cell*, **7**, 971–85.

Mendel, G. (1865). Versuche über Pflanzen-Hybriden Verh. *Naturforscher-Verein Brünn*, **4**, 3–47.

Pollock, C. J. and Cairns, A. J. (1991). Fructan metabolism in grasses and cereals. *Annual Review of Plant Physiology and Plant Molecular Biology*, **42**, 77–101.

Preiss, J. (1991). Biology and molecular biology of starch synthesis and its regulation. In *Oxford Surveys of Plant Molecular and Cell Biology*, (ed. B. J. Miflin), Vol. 7, pp. 59–114.

Smith, A. S., Denyer, K., and Martin, C. R. (1995). What controls the amount and structure of starch in storage organs? *Plant Physiology*, **107**, 673–7.

Stitt, M. (1990). Fructose-2,6-bisphosphate as a regulatory molecule in plants. *Annual Review of Plant Physiology and Plant Molecular Biology*, **41**, 153–85.

Stitt, M. (1997). Metabolic regulation of photosynthesis. In *Advances in photosynthesis*, Vol. 5. *Environmental stress and photosynthesis*, (ed. N. Baker). Academic Press, New York, 151–90.

Zimmermann, M. H. and Ziegler, H. (1975). List of sugars and sugar alcohols in sieve-tube exudates. In *Encyclopedia of Plant Physiology*, (ed. M. H. Zimmermann and J. A. Milburn), Vol. 1, pp. 480–503. Springer Verlag, Heidelberg.

chapter 10

Nitrate assimilation

Living matter contains a large amount of nitrogen incorporated in proteins, nucleic acids, and many other biomolecules. This organic nitrogen is present in oxidation state $-III$ (as in NH$_3$). During autotrophic growth the nitrogen demand for the formation of cellular matter is met by inorganic nitrogen in two alternative ways:

(1) fixation of molecular nitrogen (N$_2$) from air

(2) assimilation of the nitrate or ammonia contained in water or soil.

Only some bacteria, including cyanobacteria, are able to fix N$_2$ from air. Some plants enter a symbiosis with N$_2$-fixing bacteria, which supply them with organic bound nitrogen. Chapter 11 deals with this in detail.

However, about 99 per cent of the organic nitrogen in the biosphere is derived from the assimilation of nitrate. NH$_4^+$ is formed as end-product of the degradation of organic matter, primarily by the metabolism of animals and bacteria, and is oxidized to nitrate by nitrifying bacteria in the soil. Thus a continuous cycle exists between the nitrate in the soil and the organic nitrogen in the plants growing on it. NH$_4^+$ accumulates only in poorly aerated soils with insufficient drainage, where, due to lack of oxygen, nitrifying bacteria cannot grow. Mass animal production can lead to a high ammonia input into the soil, from manure as well as from the air. If available, many plants can also utilize NH_4^+ instead of nitrate as a nitrogen source.

10.1 The reduction of nitrate to NH_3 proceeds in two partial reactions

Nitrate is assimilated in the leaves and also in the roots. In most fully grown herbaceous plants nitrate assimilation occurs primarily in the leaves, although nitrate assimilation in the roots often plays a major role at an early growth stage of these plants. In contrast, many woody plants (trees, shrubs), and also legumes such as soybean, assimilate nitrate mainly in the roots.

Roots contain an energy-dependent uptake system with a high affinity for nitrate (Fig. 10.1). The efficiency of the uptake system enables plant growth when the external nitrogen concentration is as low as 10×10^{-6} mol l^{-1}. According to recent results, the uptake of nitrate proceeds via a *symport* with *two protons*. The ATP required for the formation of the proton gradient is mostly provided by *mitochondrial respiration*. When inhibitors or uncouplers of respiration abolish mitochondrial ATP synthesis in the roots, nitrate uptake normally comes to a stop. The transport of nitrate is inhibited by NH_4^+ ions.

The nitrate taken up into the root cells can be stored there temporarily in the vacuole. As dealt with in detail in section 10.2, nitrate is reduced to NH_4^+ in the epidermal and cortical cells of the root. This NH_4^+ is mainly used for the synthesis of glutamine and asparagine (collectively named amide in Fig. 10.1). These two amino acids can be transported to the leaves by the *transpiration stream* in the *xylem vessels*. However, when the capacity for nitrate assimilation in the roots is exhausted, nitrate is released from the roots into the xylem vessels and is carried by the transpiration stream to the leaves. There it is taken up into the mesophyll cells, probably also by proton symport. Large quantities of nitrate can be stored in a leaf by uptake into the vacuole. Sometimes this vacuolar store is emptied by nitrate assimilation during the day and replenished during the night. Thus in spinach leaves, for instance, the highest nitrate content is found in the early morning.

The nitrate in the mesophyll cells is reduced to nitrite by nitrate reductase present in the cytosol and then to NH_4^+ by nitrite reductase in the chloroplasts (Fig. 10.1).

Figure 10.1 Nitrate assimilation in the roots and leaves of a plant. Nitrate is taken up from the soil by the root. It can be stored in the vacuoles of the root cells or assimilated in the cells of the root epidermis and the cortex. Surplus nitrate is carried via the transpiration stream in the xylem vessels to the mesophyll cells, where nitrate can be temporarily stored in the vacuole. Nitrate is reduced to nitrite in the cytosol and then nitrite is reduced further in the chloroplasts to NH_4^+, from which amino acids are formed.

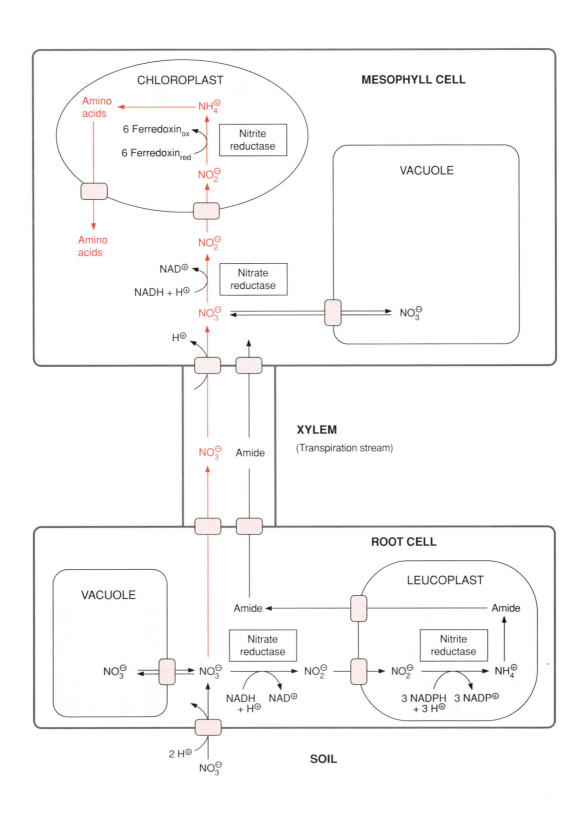

Nitrate is reduced to nitrite in the cytosol

Nitrate reduction uses mostly NADH as reductant, although some plants contain a nitrate reductase reacting with NADH as well as with NADPH. The *nitrate reductase* of higher plants consists of two identical subunits. The molecular mass of the subunits varies from 99 to 104 kDa, depending on the species. Each subunit contains an electron transport chain (Fig. 10.2), consisting of one *flavin adenine dinucleotide* (FAD) molecule, one *haem* of the cytochrome-*b* type (cyt-b_{557}), and one cofactor containing molybdenum (Fig. 10.3). This cofactor is a pteridine with a side chain to which the molybdenum is attached by two sulfur bonds, and is called *molybdenum cofactor*, abbreviated *MoCo*. Bound to the cofactor, the Mo probably changes between oxidation states +IV and +VI. The three redox carriers of nitrate reductase are each covalently bound to the subunit of the enzyme. The protein chain of the subunit can be cleaved by limited proteolysis into three domains, each of which contains only one of the redox carriers. These separated domains, as well as the holoenzyme, are able to catalyse, via their redox carriers, electron transport to artificial electron acceptors, e.g. from NADPH to Fe^{3+} ions via the FAD domain or from methylviologen (Fig. 3.39) to nitrate via the MoCo domain. Nitrate reductase also reduces chlorate (ClO_3^-) to chlorite (ClO_2^-). The latter is a very strong oxidant and therefore highly toxic to plant cells. Years ago chlorate was used as an inexpensive non-selective herbicide for keeping railway tracks free of vegetation.

The reduction of nitrite to ammonia proceeds in the plastids

The reduction of nitrite to ammonia requires the uptake of six electrons. This reaction is catalysed by one enzyme only, *nitrite reductase* (Fig. 10.4), which is located

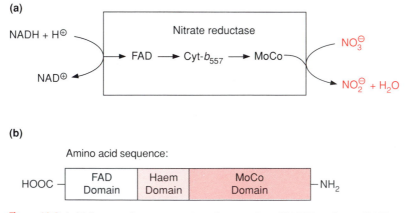

Figure 10.2 (a.) Nitrate reductase transfers electrons from NADH to nitrate. (b) The enzyme contains three domains where FAD, haem, and the molybdenum cofactor (MoCo) are bound.

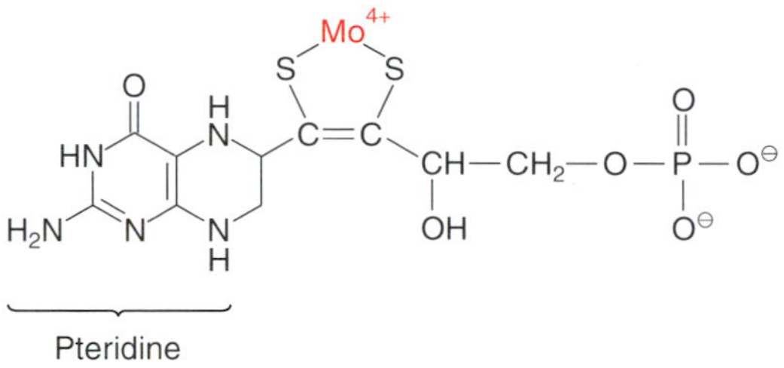

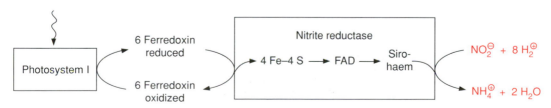

Pteridine

Figure 10.3 The molybdenum cofactor (MoCo).

exclusively in plastids. This enzyme utilizes reduced ferredoxin as electron donor, which is supplied by photosystem I as a product of photosynthetic electron transport (Fig. 3.31). To a much lesser extent, the ferredoxin required for nitrite reduction in a leaf can also be provided during darkness via reduction by NADPH, which is generated by the oxidative pentose phosphate pathway (Figs 6.21, 10.8).

Nitrite reductase contains a covalently bound *4Fe–4S cluster* (Fig. 3.26), one molecule of FAD, and one sirohaem. *Sirohaem* (Fig. 10.5) is a cyclic tetrapyrrole with one Fe atom in the centre. Its structure is different from that of haem as it contains additional acetyl and propionyl residues deriving from pyrrole synthesis (section 10.5).

The 4Fe–4S cluster, FAD, and sirohaem form an electron transport chain by which electrons are transferred from ferredoxin to nitrite. Nitrite reductase has a very high affinity for nitrite. The capacity for nitrite reduction in the chloroplasts is much greater than that for nitrate reduction in the cytosol. Therefore all nitrite formed by nitrate reductase can be totally converted into ammonia. This is important since nitrite is toxic to the cell. It forms diazo compounds with amino groups of nucleobases (R–NH₂), which are converted into alcohols with the release of nitrogen:

$$R–NH_2 + NO_2^- \rightarrow [R–N=N–OH + OH^-] \rightarrow R–OH + N_2 + OH^-$$

Thus, for instance, cytosine can be converted into uracil. This reaction can lead to mutations in nucleic acids. The very efficient reduction of nitrite by chloroplastic nitrite reductase prevents nitrite from accumulating in the cell.

Figure 10.4 Nitrite reductase in chloroplasts transfers electrons from ferredoxin to nitrite. Reduction of ferredoxin by photosystem I is shown in Fig. 3.16.

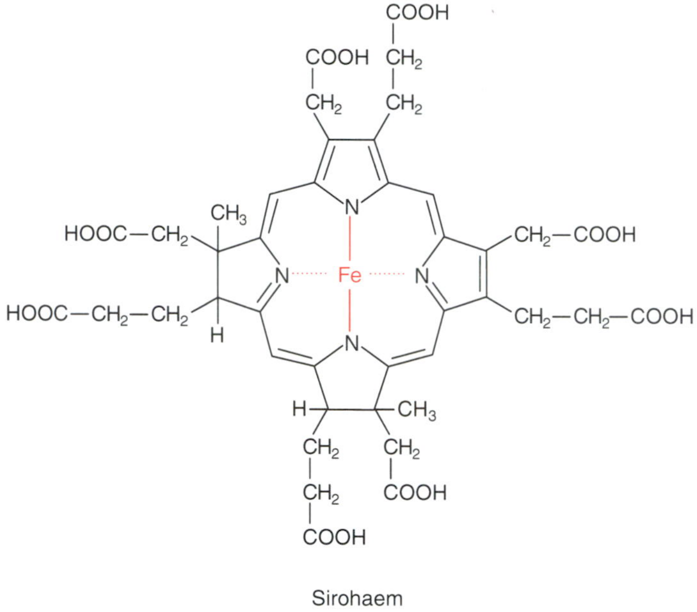

Sirohaem

Figure 10.5

The fixation of NH_4^+ proceeds in the same way as in photorespiration

Glutamine synthetase in the chloroplasts transfers the newly formed NH_4^+ at the expense of ATP to glutamate, forming glutamine (Fig. 10.6). The activity of gluta-mine synthetase and its affinity for NH_4^+ ($K_m \approx 5 \times 10^{-6}$ mol l^{-1}) are so high that the NH_4^+ produced by nitrite reductase is taken up completely. The same reaction also fixes the NH_4^+ released during photorespiration (see Fig. 7.9). Because of the high rate of photorespiration, the amount of NH_4^+ produced by the oxidation of glycine is about 5–10 times higher than the amount of NH_4^+ generated by nitrate assimilation. Thus only a minor proportion of glutamine synthesis in the leaves is actually involved in nitrate assimilation. Leaves also contain an isoenzyme of glutamine synthetase in their cytosol.

Glufosinate (Fig. 10.7), a substrate analogue of glutamate, inhibits glutamine synthesis. Plants in which the addition of glufosinate has inhibited the synthesis of glutamine, accumulate toxic levels of ammonia and die. Ammonium glufosinate is distributed as a herbicide (section 3.6) under the trade name Basta (AgrEvo). It has the advantage that it is degraded rapidly in the soil, leaving behind no toxic degradation products. Recently glufosinate-resistant crop plants have been generated by genetic engineering, enabling the use of glufosinate as a selective herbicide for eliminating weeds in growing cultures (section 22.6).

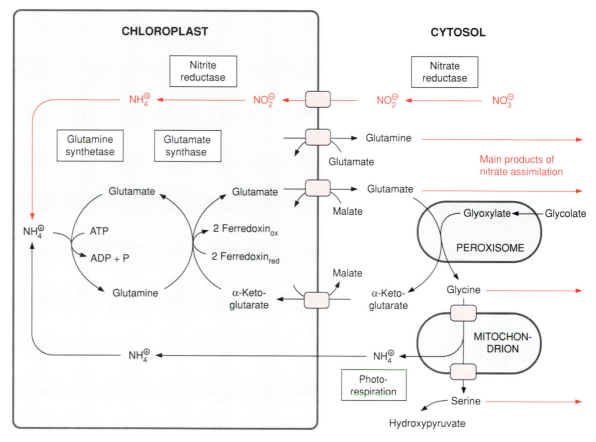

Figure 10.6 Compartmentalization of partial reactions of nitrate assimilation and the photorespiratory pathway in mesophyll cells. NH$_4^+$ formed in the photorespiratory pathway is coloured black and NH$_4^+$ formed by nitrate assimilation coloured red. The main products of nitrate assimilation are marked with a red arrow.

The glutamine formed in the chloroplasts is converted to two molecules of glutamate, via *glutamate synthase* (also named glutamine–oxoglutarate amino transferase, abbreviated GOGAT), by reaction with α-ketoglutarate (see also Fig. 7.9) with ferredoxin as the reductant. Some chloroplasts also contain an NADPH-dependent glutamate synthase. Glutamate synthases are inhibited by the substrate analogue *azaserine* (Fig. 10.7), which is toxic to plants.

α-Ketoglutarate, required for the glutamate synthase reaction, is transported into the chloroplasts by a specific translocator in counter-exchange for malate, and the glutamate formed is transported out of the chloroplasts into the cytosol by another translocator, also in exchange for malate (Fig. 10.6). A further translocator in the chloroplast envelope transports glutamine in counter-exchange for glutamate, enabling the export of glutamine from the chloroplasts.

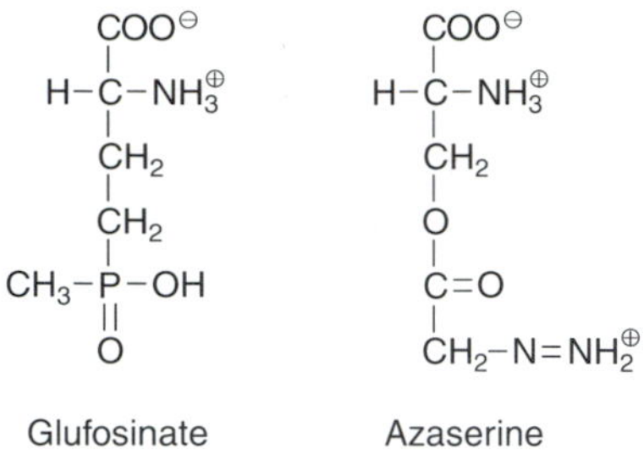

```
        COO⊖                 COO⊖
         |                    |
    H-C-NH₃⊕             H-C-NH₃⊕
         |                    |
        CH₂                  CH₂
         |                    |
        CH₂                   O
         |                    |
  CH₃-P-OH                   C=O
         ||                   |
         O              CH₂-N=NH₂⊕

     Glufosinate           Azaserine
```

Figure 10.7 Glufosinate (also named phosphinotricine) is a substrate analogue of glutamate and a strong inhibitor of glutamine synthetase. Ammonium glufosinate is a herbicide (Basta, AgrEvo). Azaserine is a substrate analogue of glutamate and an inhibitor of glutamate synthase.

10.2 Nitrate assimilation also takes place in the roots

As already mentioned, nitrate assimilation occurs in part, and in some species even mainly, in the roots. NH_4^+ taken up from the soil is normally fixed in the roots. The reduction of nitrate and of nitrite as well as the fixation of NH_4^+ proceeds in the root cells in an analogous way to the mesophyll cells. However, in the root cells the necessary reducing equivalents are supplied exclusively by oxidation of carbohydrates. The reduction of nitrite and the subsequent fixation of NH_4^+ (Fig. 10.8) occur in the leucoplasts, a differentiated form of plastid (section 1.3).

The oxidative pentose phosphate pathway provides reducing equivalents for nitrite reduction in leucoplasts

The reducing equivalents required for the reduction of nitrite and for the formation of glutamate are provided in leucoplasts by oxidation of glucose 6-phosphate via the oxidative pentose phosphate pathway dealt with in section 6.5 (Fig. 10.8). The uptake of glucose 6-phosphate proceeds in counter-exchange for triose phosphate. The *glucose 6-phosphate–phosphate translocator* of leucoplasts differs from the triose phosphate–phosphate translocator of chloroplasts in transporting glucose 6-phosphate in addition to phosphate, triose phosphate, and 3-phosphoglycerate. In the oxidative pentose phosphate pathway, three molecules of glucose 6-phosphate are converted into three molecules of ribulose 5-phosphate with the release of three molecules of CO_2, yielding six molecules of NADPH. The subsequent reactions yield one molecule of triose phosphate and two molecules of fructose 6-phosphate; the latter are reconverted into glucose 6-phosphate via phosphohexose isomerase. In the cytosol glucose 6-phosphate is regenerated from two molecules of triose phosphate via aldolase, cytosolic fructose 1,6-bisphosphatase, and phosphohexose isomerase. In this way glucose 6-phosphate can be completely oxidized to CO_2 in order to produce NADPH.

As in chloroplasts, nitrite reduction in leucoplasts also requires reduced ferredoxin as reductant. In the leucoplasts ferredoxin is reduced by NADPH. The ATP required for glutamine synthesis in the leucoplasts can be generated by the mito-

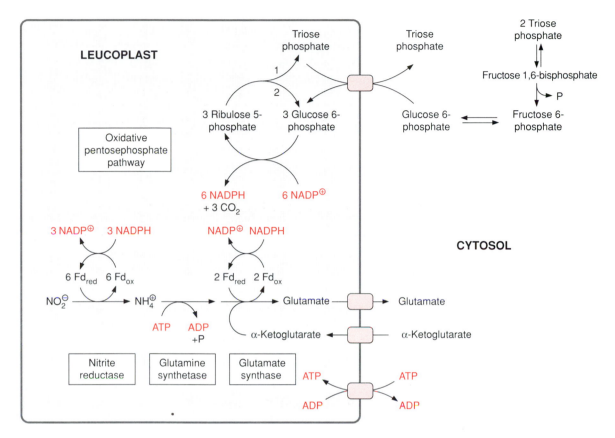

Figure 10.8 The oxidative pentose phosphate pathway provides the reducing equivalents for nitrite reduction in plastids from non-green tissues (leucoplasts). In some plastids glucose 1-phosphate is transported in counter-exchange for triose phosphate. Fd, ferredoxin.

chondria and transported into the leucoplasts by a plastidic ATP translocator in counter-exchange for ADP. The glutamate synthase of the leucoplasts also uses reduced ferredoxin as redox partner. NADPH generated by the oxidative phosphate pentose pathway reduces ferredoxin and thus serves as reductant for nitrite reduction, as well as for NH_4^+ fixation in the leucoplasts of root cells. Some leucoplasts also contain a glutamate synthase which utilizes NADPH or NADH directly as reductant. Nitrate reduction in the roots provides the shoot with organic nitrogen compounds mostly as *glutamine* and *asparagine* via the transpiration stream in the xylem vessels.

10.3 Nitrate assimilation is strictly controlled

During photosynthesis, CO_2 assimilation and nitrate assimilation have to be matched to each other. Nitrate assimilation can only progress when CO_2 assimilation provides the carbon skeletons for the amino acids. Moreover, nitrate assimilation must be regulated in such a way that the production of amino acids does not exceed demand. Finally, it is important that nitrate reduction does not proceed faster than nitrite reduction, since otherwise toxic levels of nitrite (section 10.1) would accumulate in the cells. Under certain conditions such a dangerous accumulation of nitrite can occur in roots, when excessive moisture makes the soil anaerobic. Flooded roots are able to discharge nitrite into water, avoiding the build-up of toxic levels of nitrite, but this escape route is not open to leaves, making the strict control of nitrate reduction there especially important.

The NADH required for nitrate reduction in the cytosol can also be provided during darkness, e.g. by glycolytic degradation of glucose. However, reduction of nitrite and fixation of NH_4^+ in the chloroplasts depends largely on photosynthesis providing reducing equivalents and ATP. The oxidative pentose phosphate pathway can provide only very limited amounts of reducing equivalents in the dark. Therefore, during darkness nitrate reduction in the leaves has to be slowed down or even switched off to prevent an accumulation of nitrite. This illustrates how essential it is for a plant to regulate the activity of nitrate reductase, the entry step of nitrate assimilation.

The synthesis of the nitrate reductase protein is regulated at the level of gene expression

Nitrate reductase is an exceptionally short-lived protein. Its half-life is only a few hours. The rate of *de novo* synthesis of this enzyme is therefore very high. Thus by regulating its synthesis, the activity of nitrate reductase can be altered within hours.

Various factors control the synthesis of the enzyme at the level of gene expression. Nitrate and light stimulate its synthesis. The question as to whether a phytochrome or a blue light receptor (Chapter 19) is involved in this effect of light, has not been fully resolved. The synthesis of the nitrate reductase protein is stimulated by glucose and other carbohydrates and is inhibited by glutamine and other amino acids (Fig. 10.9), but it is still not known which of these metabolites directly affects the expression of the gene. Sensors seem to be present in the cell which adjust the capacity of nitrate reductase both to the demand for amino acids and to the supply of carbon skeletons from CO_2 assimilation for its synthesis via regulation of gene expression.

Nitrate reductase is also regulated by reversible covalent modification

The regulation of *de novo* synthesis of nitrate reductase (NR) allows regulation of the enzyme activity within a time span of hours. This would not be sufficient to prevent an accumulation of nitrite in the plants during darkening or sudden shading of the plant. Rapid inactivation of nitrate reductase in the time span of minutes

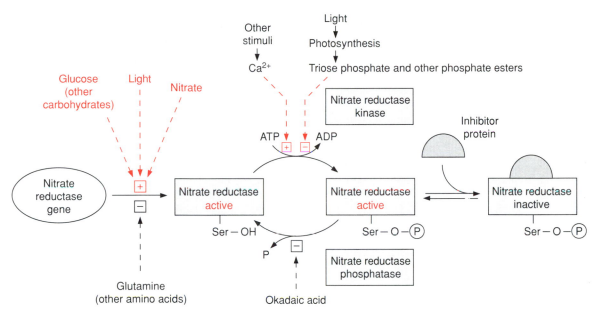

Figure 10.9 Regulation of nitrate reductase (NR). Synthesis of the NR protein is stimulated by carbohydrates (perhaps glucose or its metabolic products) and light $\boxed{+}$, and inhibited by glutamine or other amino acids $\boxed{-}$. The newly formed NR protein is degraded within a time span of a few hours. Nitrate reductase is inhibited by phosphorylation of a serine residue and the subsequent interaction with an inhibitor protein. After hydrolytic liberation of the phosphate residue by a protein phosphatase, the inhibitor is dissociated and nitrate reductase regains its full activity. The activity of the nitrate reductase kinase is inhibited in the light, yielding nitrate reductase active in the light. Okadaic acid, an inhibitor of protein phosphatases, prevents the activation of nitrate reductase. (After Huber *et al.* 1996.)

occurs via *phosphorylation of the nitrate reductase protein* (Fig. 10.9). Upon darkening, a *serine* residue, which is located in the nitrate reductase protein between the haem and the MoCo domain, is phosphorylated by a specific protein kinase termed *nitrate reductase kinase*. This protein kinase is stimulated by Ca^{2+} ions and inhibited by dihydroxyacetone phosphate and glucose 6-phosphate. The inhibition of nitrate reductase is caused by the binding of an *inhibitor protein* to the phosphorylated serine residue, resulting in an inhibition of electron transport between cyt-b_{557} and the MoCo domain (Fig. 10.2). The inhibitor protein belongs to a family of ubiquitous proteins, designated as *14-3-3 proteins*, which are highly conserved and occur in yeast, plants, and animals. The 14-3-3 proteins bind to a large variety of proteins and alter their activity. Present results indicate that all the 14-3-3 proteins bind to a specific sequence of five amino acids containing one phosphoserine. When the corresponding serine in the nitrate reductase protein is replaced with aspartate by site-directed mutagenesis, this altered nitrate reductase is not inhibited by nitrate reductase kinase any more. A specific protein phosphatase catalyses the hydrolysis of the phosphoserine, and as a consequence the inhibitor protein loses its function and the enzyme is active again. *Okadaic acid* inhibits the protein phosphatase and thus the reactivation of nitrate reductase. It is still not clear which are the effectors signalling the state 'illumination' to the nitrate reductase kinase. The inhibition of nitrate reductase kinase by triose phosphate and other phosphate esters may ensure that nitrate reductase is only active when CO_2 fixation is operating for delivery of the carbon skeletons for amino acid synthesis.

The mechanism shown here for the reversible inhibition of nitrate reduction by phosphorylation of serine residues of the enzyme protein by special protein kinases and protein phosphatases is remarkably similar to the regulation of sucrose phosphate synthase dealt with in the previous chapter (Fig. 9.18). Upon darkening, both enzymes are inactivated by phosphorylation, which in the case of nitrate reductase requires also a binding of an inhibitor protein. Both enzymes are reactivated by protein phosphatases, which are inhibited by okadaic acid. Although many details are still unknown, it now seems clear that the basic mechanisms for the rapid light regulation of sucrose phosphate synthase and nitrate reductase are similar. It may be noted, however, that the protein kinases and phosphatases involved in the regulation of both enzymes are probably not identical.

10.4 The end-product of nitrate assimilation is a whole spectrum of amino acids

As described in Chapter 13, the carbohydrates formed as the product of CO_2 assimilation are transported from the leaves via the sieve tubes to various parts of the plants only in defined transport forms, such as sucrose, sugar alcohols (e.g. sorbitol), or raffinoses, depending on the species. There are no such special transport forms for the products of nitrate assimilation. All amino acids present in the mesophyll cells are exported via the sieve tubes. Therefore the sum of amino acids can be regarded as the final product of nitrate assimilation. Synthesis of these amino acids takes place mainly in the chloroplasts. The pattern of the amino acids synthesized varies largely, depending on the species and the metabolic conditions. In most cases *glutamate* and *glutamine* represent the major portion of the synthesized amino acids. Glutamate is exported from the chloroplasts in exchange for malate, and glutamine in exchange for glutamate (Fig. 10.6). Also *serine* and *glycine*, which are formed as intermediate products in the photorespiratory cycle, may represent a considerable portion of the total amino acids present in the mesophyll cells. Large amounts of *alanine* are often formed in C_4 plants.

CO_2 assimilation provides the carbon skeletons required to synthesize the end-products of nitrate assimilation

CO_2 assimilation provides the carbon skeletons required for the synthesis of the various amino acids. Figure 10.10 gives an overview of the origin of the carbon skeletons of various amino acids.

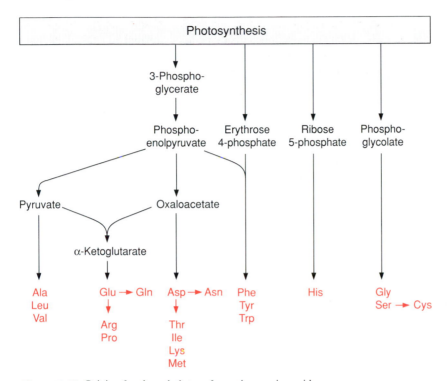

Figure 10.10 Origin of carbon skeletons for various amino acids.

3-Phosphoglycerate is the most important carbon precursor for synthesis of amino acids. It is generated in the Calvin cycle and exported from the chloroplasts to the cytosol by the triose phosphate–phosphate translocator in exchange for phosphate (Fig. 10.11). 3-Phosphoglycerate is converted in the cytosol by *phosphoglycerate mutase* and *enolase* to phosphoenolpyruvate (PEP). From phosphoenolpyruvate two pathways branch off, the reaction via pyruvate kinase leading to *pyruvate,* and via PEP carboxylase to *oxaloacetate.* Moreover, phosphoenolpyruvate together with erythrose 4-phosphate is the precursor for the synthesis of aromatic amino acids via the *shikimate pathway*, dealt with later in this chapter.

The PEP carboxylase reaction has already been dealt with in conjunction with metabolism of stomatal cells (section 8.2) and C₄ metabolism and CAM (sections 8.4 and 8.5). Oxaloacetate formed by PEP carboxylase has two functions in nitrate assimilation:

1. It is converted by transamination to aspartate which is precursor for synthesis of five other amino acids (asparagine, threonine, isoleucine, lysine, and methionine).

2. Together with pyruvate it is the precursor for the formation of α-ketoglutarate, which is converted by transamination to glutamate, being the precursor of three other amino acids (glutamine, arginine, and proline).

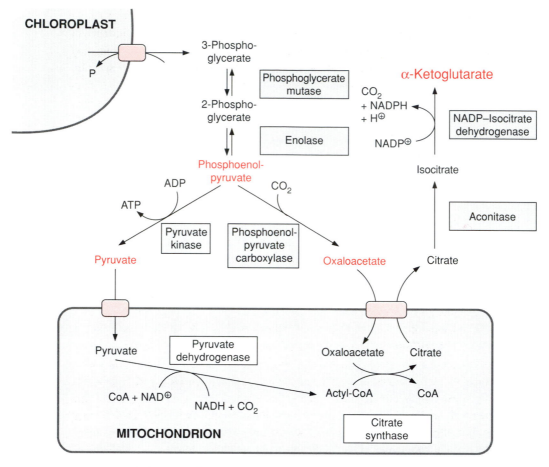

Figure 10.11 Carbon skeletons for the synthesis of amino acids are provided by CO_2 assimilation. Important precursors for amino acid synthesis are coloured red.

Glycolate formed by photorespiration is precursor for the formation of glycine and serine (see Fig. 7.1), and from the latter cysteine is formed (Chapter 12). In non-green cells serine and glycine can also be formed from 3-phosphoglycerate. Details of this are not dealt with here. Ribose 5-phosphate is a precursor for the synthesis of histidine. This pathway has not yet been fully resolved in plants.

The synthesis of glutamate requires the participation of mitochondrial metabolism

Figure 10.6 shows that glutamate is formed from α-ketoglutarate, which can be provided by a partial sequence of the mitochondrial citrate cycle (Fig. 10.11). Pyruvate and oxoloacetate are transported from the cytosol to the mitochondria by specific translocators. Pyruvate is oxidized by *pyruvate dehydrogenase* (see

Fig. 5.4), and the acetyl CoA thus generated condenses with oxaloacetate to form citrate (see Fig. 5.6). This citrate can be converted in the mitochondria via *aconitase* (Fig. 5.7), oxidized further by *NAD–isocitrate dehydrogenase* (Fig. 5.8) and the resultant α-ketoglucerate can be transported into the cytosol by a specific translocator. But often a major part of the citrate produced in the mitochondria is exported to the cytosol and converted there to α-ketoglucerate by cytosolic isoenzymes of aconitase and NADP–isocitrate dehydrogenase. In this case, only a short partial sequence of the citrate cycle is involved in the synthesis of α-ketoglutarate from pyruvate and oxaloacetate. Citrate is released from the mitochondria by a specific translocator in exchange for oxaloacetate.

Biosynthesis of proline and arginine

Glutamate is a precursor for synthesis of *proline* (Fig. 10.12). Its δ-carboxylic group is first converted by a *glutamate kinase* to an energy-rich phosphoric acid anhydride and is then reduced by NADH to an aldehyde. The accompanying hydrolysis of the energy-rich phosphate drives the reaction, resembling the reduction of 3-phosphoglycerate to glyceraldehyde 3-phosphate in the Calvin cycle. A ring is formed by the condensation of the carbonyl group with the α-amino group. Reduction by NADPH results in formation of proline.

Besides its role as a protein constituent, proline has a special function as a *protective substance against desiccation damage* in leaves. When exposed to dryness or to a high salt content in the soil (both leading to water stress), many plants accumulate very high amounts of proline in their leaves, in some cases several times the sum of all the other amino acids. It is assumed that the accumulation of proline during water stress is due to the induction of the synthesis of *pyrroline 5-carboxylate reductase*.

Proline protects a plant against desiccation because, in contrast to inorganic salts, it has no inhibitory effect on enzymes even at very high concentrations. Therefore proline is classified as a *compatible substance*. Other compatible substances, formed in certain plants in response to water stress, are sugar alcohols such as *mannitol* (Fig. 10.13) and *betains*, consisting of amino acids, such as proline, glycine, and alanine, of which the amino groups are methylated. The latter are termed proline, glycine, and alanine betains. The accumulation of such compatible substances, especially in the cytosol, chloroplasts, and mitochondria, minimizes in these compartments the water loss due to dryness or a high salt content of the soil.

In the first step of the synthesis of *arginine*, the α-amino group of glutamate is acetylated by reaction with acetyl CoA and is thus protected. Subsequently the δ-carboxylic group is phosphorylated and reduced to a semi-aldehyde in basically the same reaction as in proline synthesis. Here the α-amino group is protected and the formation of a ring is not possible. By transamination with glutamate the aldehyde group is converted into an amino group, and after cleavage of the acetyl

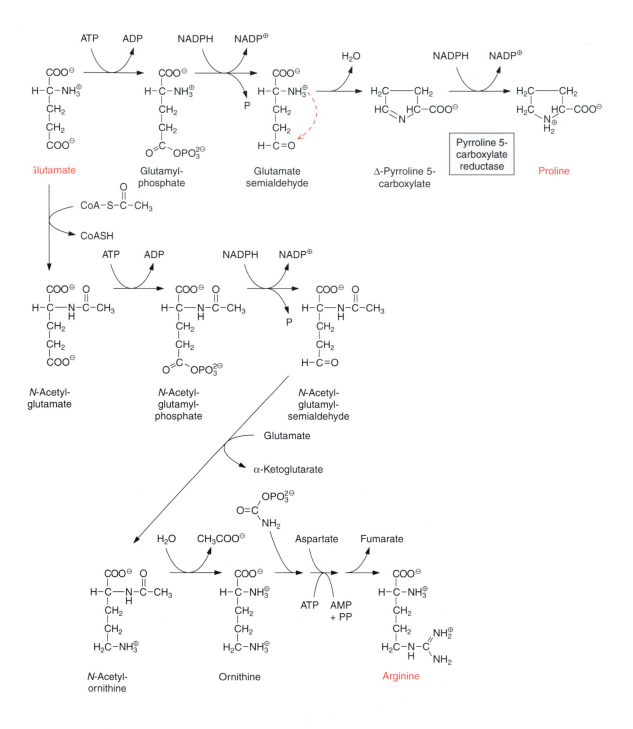

Figure 10.12 Pathway for the formation of amino acids from glutamate.

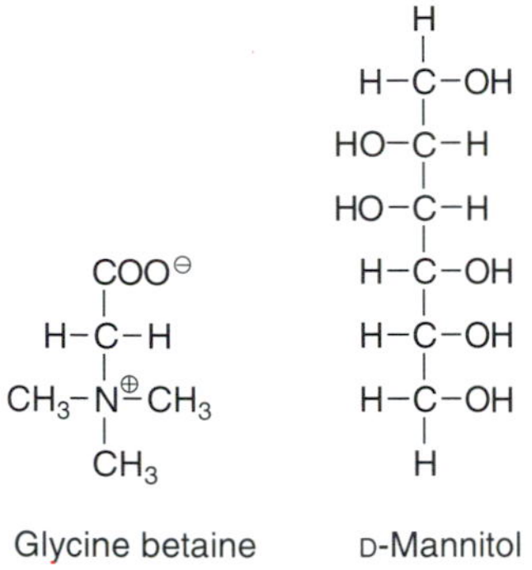

```
                                    H
                                    |
                              H−C−OH
                                    |
                           HO−C−H
                                    |
                           HO−C−H
                                    |
        COO⊖                 H−C−OH
         |                          |
      H−C−H                 H−C−OH
         |                          |
   CH₃−N⊕CH₃              H−C−OH
         |                          |
        CH₃                   H

  Glycine betaine        D-Mannitol
```

Figure 10.13 Two compatible substances which, like proline, are accumulated in plants as protective agents against desiccation and high salt contents in the soil.

residue ornithine is formed. The conversion of ornithine to arginine (shown only in Fig. 10.12 summarily) proceeds in the same way as in the urea cycle of animals, by condensation with carbamoyl phosphate to citrulline. An amino group is transferred from aspartate to citrulline resulting in the formation of arginine and fumarate.

Aspartate is the precursor of five amino acids

Aspartate is formed from oxoloacetate by transamination with glutamate by glutamate–oxaloacetate aminotransferase (Fig. 10.14). The synthesis of *asparagine* from aspartate requires a transitory phosphorylation of the terminal carboxylic group by ATP, as in the synthesis of glutamine. In contrast to glutamine synthesis, however, it is not NH_4^+ but the amide group of glutamine which usually serves as amino donor in asparagine synthesis. Therefore, the energy expenditure for the amidation of aspartate is twice as high as for the amidation of glutamate. Asparagine is formed to a large extent in the roots (section 10.2). Synthesis of asparagine in the leaves often plays only a minor role.

For the synthesis of *lysine, isoleucine, threonine,* and *methionine* the first two steps are basically the same as for proline synthesis: after phosphorylation by a kinase, the γ-carboxylic group is reduced to a semi-aldehyde. For the synthesis of lysine (not shown in detail in Fig. 10.14) the semi-aldehyde condenses with pyruvate and, in a sequence of six reactions, involving reduction by NADPH and transamination by glutamate, *meso*-2,6-diaminopimelate is formed and from this lysine arises by decarboxylation.

For the synthesis of *threonine* the semi-aldehyde is further reduced to homoserine. After phosphorylation of the hydroxyl group by homoserine kinase, threonine is formed by isomerization of the hydroxyl group, accompanied by the removal of phosphate. The synthesis of isoleucine from threonine will be dealt with in the following paragraph and the synthesis of methionine in conjunction with sulfur metabolism in Chapter 12.

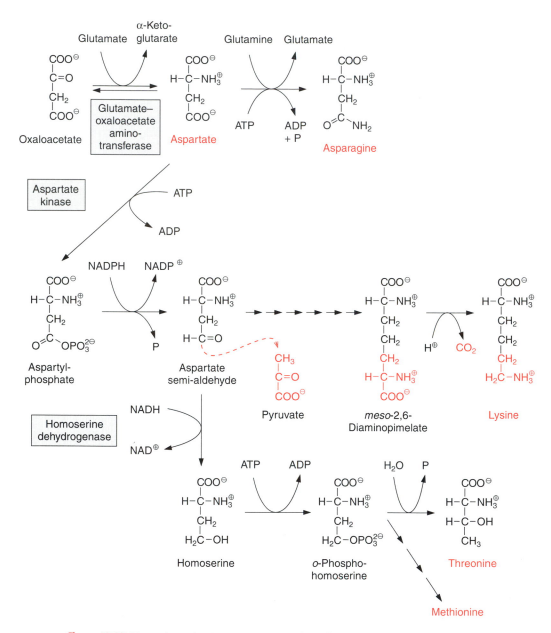

Figure 10.14 The pathway for the formation of amino acids from aspartate.

Synthesis of amino acids from aspartate is subject to strong feedback control by its end-products (Fig. 10.15). Aspartate kinase, the entrance valve for the synthetic pathways, is inhibited by threonine as well as lysine. In addition, the reactions of aspartate semi-aldehyde at the branch points of both synthetic pathways are inhibited by the corresponding end-products.

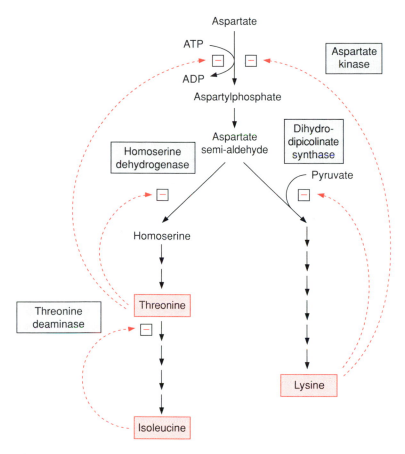

Figure 10.15 Feedback inhibition by end-products regulates the first enzyme for the synthesis of amino acids from aspartate according to demand. $\boxed{-}$ indicates inhibition.

Acetolactate synthase participates in the synthesis of hydrophobic amino acids

Pyruvate can be converted by transamination to *alanine* (Fig. 10.16A). This reaction plays a special role in C_4 metabolism (see Figs 8.14, 8.15).

Synthesis of *valine* and *leucine* begins with the formation of acetolactate from two molecules of pyruvate. *Acetolactate synthase*, catalysing this reaction, contains *thiamin pyrophosphate* (TPP) as its prosthetic group. The reaction of TPP with pyruvate yields hydroxyethyl-TPP and CO_2, in the same way as in the pyruvate dehydrogenase reaction (see Fig. 5.4). The hydroxyethyl residue is transferred to a second molecule of pyruvate and thus *acetolactate* is formed. Its reduction and rearrangement and the release of water yield α-ketoisovalerate and a subsequent transamination by glutamate produces valine.

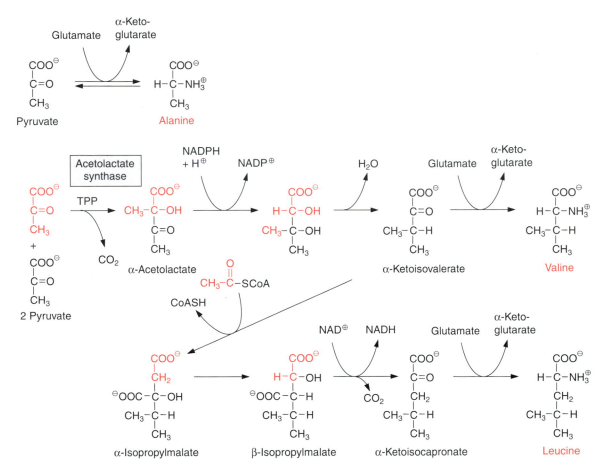

Figure 10.16A Pathways for the formation of amino acids from pyruvate.

The formation of *leucine* from α-ketoisovalerate proceeds in basically the same reaction sequences as for the formation of glutamate from oxaloacetate, shown in Fig. 5.3. First acetyl CoA condenses with α-ketoisovalerate (analogous to the formation of citrate), the product α-isopropylmalate isomerizes (analogous to isocitrate formation), and the β-isopropylmalate thus formed is oxidized by NAD^+ with the release of CO_2 to α-ketoisocapronate (analogous to the synthesis of α-ketoglutarate by isocitrate dehydrogenase). Finally, in analogy to the synthesis of glutamate, α-ketoisocapronate is transformed by transamination to leucine.

For the synthesis of *isoleucine* from threonine the latter is first converted by a deaminase to α-ketobutyrate (Fig. 10.16B). Acetolactate synthase condenses α-ketobutyrate with pyruvate, in an reaction analogous to the synthesis of aceto-lactate from two molecules of pyruvate (Fig. 10.16A). The further reactions in the

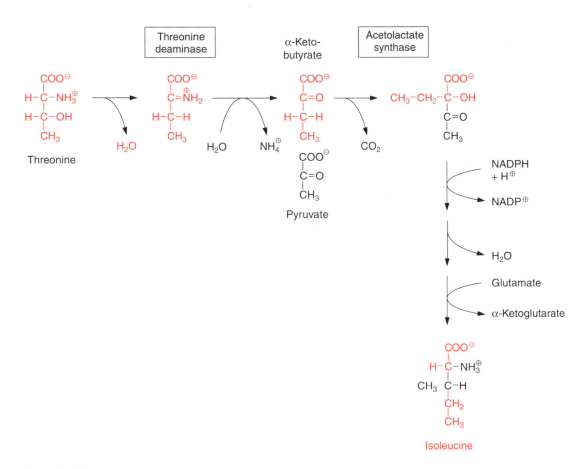

Figure 10.16B Pathways for the formation of amino acids from pyruvate.

synthesis of isoleucine correspond to the reaction sequence in the synthesis of valine.

The synthesis of leucine, valine, and isoleucine is also subject to feedback control by the end-products. Isopropylmalate synthase is inhibited by leucine (Fig. 10.17) and threonine deaminase by isoleucine (Fig. 10.15). The first enzyme, *acetolactate synthase* (ALS) is inhibited by valine and leucine. Sulfonyl ureas (e.g. chlorsulfurone) and imidazolinones (e.g. imazethapyr) (Fig. 10.18) are very strong inhibitors of ALS, where they bind to the pyruvate binding site. A concentration as low as 10^{-9} mol l^{-1} of chlorsulfurone is sufficient to inhibit ALS by 50 per cent. Since the pathway for the formation of valine, leucine, and isoleucine is present only in plants and micro-organisms, the inhibitors mentioned above are suitable for destroying plants specifically and are therefore used as *herbicides* (section 3.6). Chlorsulfurone (trade name Glean, Du Pont) is used as a selective herbicide

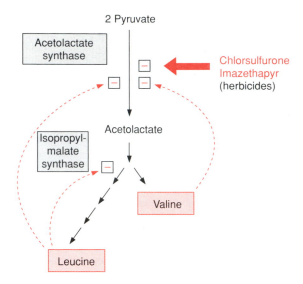

Figure 10.17 Synthesis of valine and leucine is adjusted to demand by the inhibitory effect of both amino acids on acetolactate synthase and the inhibition of isopropylmalate synthase by leucine. The herbicides chlorsulfurone and imazethapyr inhibit acetolactate synthase. $\boxed{-}$ indicates inhibition.

in the cultivation of cereals, and imazethapyr (Pursuit, American Cyanamide Co.) for protecting soybeans. From the application of these herbicides mutants of maize, soybean, rapeseed, and wheat have evolved naturally, which are resistant against sulfonyl ureas or imidazolinones, or even against both herbicides. In each

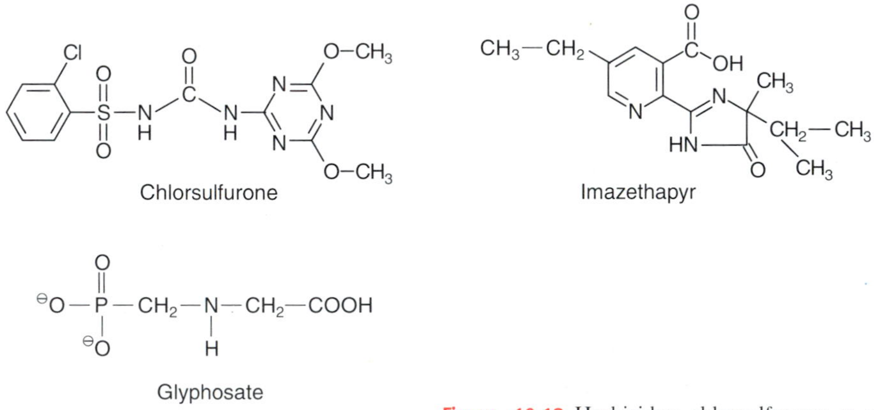

Chlorsulfurone

Imazethapyr

Glyphosate

Figure 10.18 Herbicides: chlorsulfurone, a sulfonyl urea (trade name Glean, Du Pont), and imazethapyr, an imidazolinone (trade name Pursuit, ACC), inhibit acetolactate synthase (Fig. 10.16A). Glyphosate (trade name Roundup, Monsanto) inhibits EPSP synthase (Fig. 10.19).

case a mutation was found in the gene for acetolactate synthase, making the enzyme insensitive towards the herbicides without affecting its enzyme activity. By crossing these mutants with other lines, herbicide-resistant varieties have been bred and are, in part, already commercially cultivated.

Aromatic amino acids are synthesized via the shikimate pathway

Precursors for the formation of aromatic amino acids are erythrose 4-phosphate and phosphoenolpyruvate. These two compounds condense to form cyclic dehydrochinate accompanied by liberation of both phosphate groups (Fig. 10.19). Following the removal of water and reduction of the carbonyl group, *shikimate* is formed. After protection of the 3'-hydroxyl group by phosphorylation, the 5'-hydroxyl group of the shikimate reacts with phosphoenolpyruvate to give the enolether 5'-enolpyruvylshikimate 3-phosphate (EPSP) and chorismate is formed from this by removal of phosphate. Chorismate represents a branch point for two biosynthetic pathways:

(1) Tryptophan is formed via four reactions which are not dealt with in detail here.

(2) Prephenate is formed by a rearrangement in which the side chain is transferred to the 1'-position of the ring, and arogenate is formed after transamination of the keto group. Removal of water results in the formation of the third double bond and *phenylalanine* is formed by decarboxylation. Oxidation of arogenate by NAD, accompanied by a decarboxylation, results in formation of *tyrosine*. According to recent results the enzymes of the shikimate pathway are located exclusively in the plastids.

The synthesis of aromatic amino acids is also controlled by the end-products at several steps in the pathway (Fig. 10.20).

Glyphosate acts as a herbicide

Glyphosate (Fig. 10.18), a structural analogue of phosphoenolpyruvate, is a very strong inhibitor of EPSP synthase. Glyphosate inhibits specifically the synthesis of aromatic amino acids but has only a low effect on other phosphoenolpyruvate metabolizing enzymes (e.g. pyruvate kinase or PEP carboxykinase). Interruption of the shikimate pathway by glyphosate has a lethal effect on plants. Since the shikimate pathway is not present in animals, glyphosate (under the trade name Roundup, Monsanto) is used as a herbicide (section 3.6). Due to its simple structure, glyphosate is degraded very rapidly and completely by bacteria present in the soil. Glyphosate is the herbicide with the highest sales worldwide. Recently genetic engineering has been successful in creating glyophosate-resistant crop plants (section 22.6).

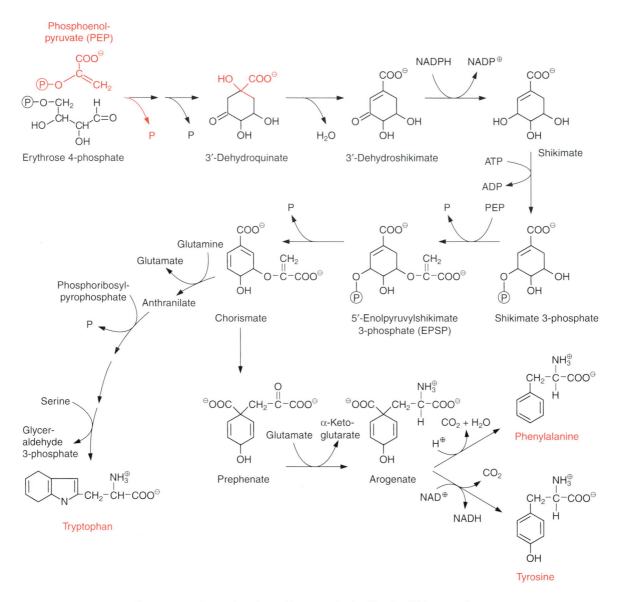

Figure 10.19 Aromatic amino acids are synthesized by the shikimate pathway.

A large proportion of the total plant matter can be formed by the shikimate pathway

The function of the shikimate pathway is not restricted only to the generation of amino acids for protein biosynthesis. It also provides precursors for a large variety of other substances (Fig. 10.21) formed by plants in large quantities, in particular

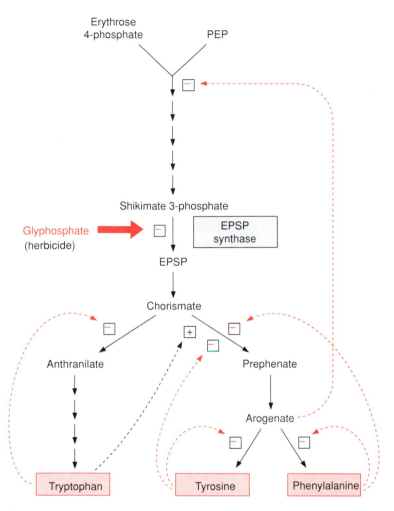

Figure 10.20   Several steps in the synthesis of aromatic amino acids are regulated by product feedback inhibition, thus adjusting the rate of synthesis to demand. Tryptophan stimulates the synthesis of tyrosine and phenylalanine +. The herbicide glyphosate (Fig. 10.19) inhibits EPSP synthase −.

phenylpropanoids such as *flavonoids* and *lignin* (Chapter 18). As the sum of these products can amount to a high proportion of the total cellular matter, in some plants up to 50 per cent of the dry matter, the shikimate pathway can be regarded as one of the main biosynthetic pathways of plants.

10.5 Glutamate is the precursor for synthesis of chlorophylls and cytochromes

Chlorophyll amounts to 1–2 per cent of the dry matter of leaves. Its synthesis proceeds in the plastids. As already shown in Fig. 2.4, chlorophyll consists of a

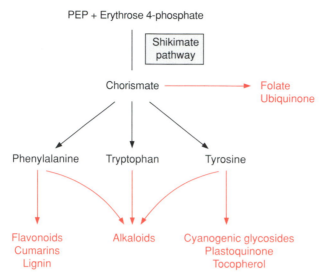

PEP + Erythrose 4-phosphate

Shikimate
pathway

Chorismate ⟶ Folate
Ubiquinone

Phenylalanine   Tryptophan   Tyrosine

Flavonoids   Alkaloids   Cyanogenic glycosides
Cumarins   Plastoquinone
Lignin   Tocopherol

Figure 10.21 Several secondary metabolites are synthesized via the shikimate pathway.

tetrapyrrole ring with *magnesium* as central atom and with a *phytol side chain* as a membrane anchor. *Haem*, likewise a tetrapyrrole, but with *iron* as central atom, is a constituent of cytochromes and catalase.

Porphobilinogen, a precursor for the synthesis of tetrapyrroles, is formed by condensation of two molecules of δ-*aminolevulinate*. δ-Aminolevulinate is synthesized in animals, yeast, and some bacteria from succinyl CoA and glycine, accompanied by liberation of CoASH and CO_2. In plastids, cyanobacteria, and many eubacteria, however, δ-aminolevulinate is formed by reduction of *glutamate*. As already dealt with in section 6.3, the difference in redox potentials between a carboxylate and an aldehyde is so high that a reduction of a carboxyl group by NADPH is only possible when this carboxyl group has been previously activated, e.g. as a thioester (Fig. 6.10) or as a mixed phosphoric acid anhydride (Fig. 10.12). In the plastidic δ-aminolevulinate synthesis glutamate is activated in a very unusual way by a covalent linkage to a *transfer RNA* (tRNA) (Fig. 10.22). This tRNA for glutamate is encoded in the plastids and is involved there in the synthesis of δ-aminolevulinate as well as in protein biosynthesis. As in protein biosynthesis (see Fig. 21.1), the linkage of the carboxyl group of glutamate to tRNA is accompanied by consumption of ATP. During reduction of glutamyl-tRNA by *glutamyl-tRNA reductase*, tRNA is liberated and in this way the reaction becomes irreversible. The *glutamate 1-semi-aldehyde* thus formed is converted into δ-aminolevulinate by an amino transferase containing pyridoxal phosphate as a prosthetic group. This reaction proceeds according to the same mechanism as the aminotransferase reaction shown in Fig. 7.4, with the only difference that here the amino group as amino donor, and the keto group as amino acceptor, are present in the same molecule.

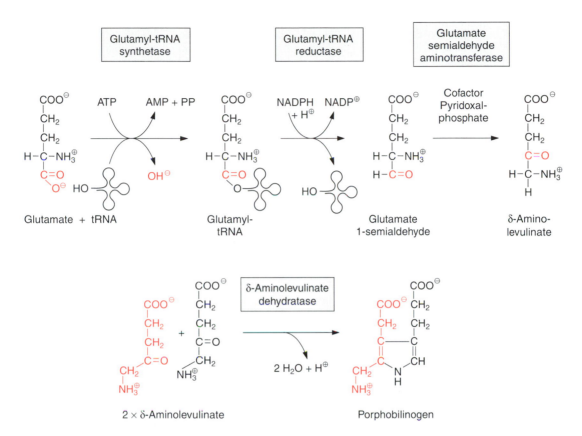

Figure 10.22 In chloroplasts glutamate is the precursor for the synthesis of δ-aminolevulinate, which is condensed to porphobilinogen.

Two molecules of δ-aminolevulinate condense to form porphobilinogen (Fig. 10.22). The open-chain tetrapyrrole hydroxymethylbilan is formed from four molecules of porphobilinogen, via *porphobilinogen deaminase* (Fig. 10.23). The enzyme contains a dipyrrole as cofactor, which it produces itself. Uroporphyrinogen III is formed after the exchange of the two side chains on ring d and by closure of the ring, and protoporphyrin IX by reaction with decarboxylases and oxidases (not shown in detail). Mg^{2+} is incorporated into the tetrapyrrole ring by magnesium chelatase and the resultant Mg-protoporphyrin IX is converted by three more enzymes to protochlorophyllide. The tetrapyrrole ring of protochlorophyllide contains the same number of double bonds as protoporphyrin IX. The reduction of one double bond in ring d by NADPH yields chlorophyllide. *Protochlorophyllide reductase*, which catalyses this reaction, is only active when protochlorophyllide is activated by absorption of light. The transfer of a phytyl chain, activated by pyrophosphate via a prenyl transferase named *chlorophyll synthetase* (see section 17.7), completes the synthesis of the chlorophyll.

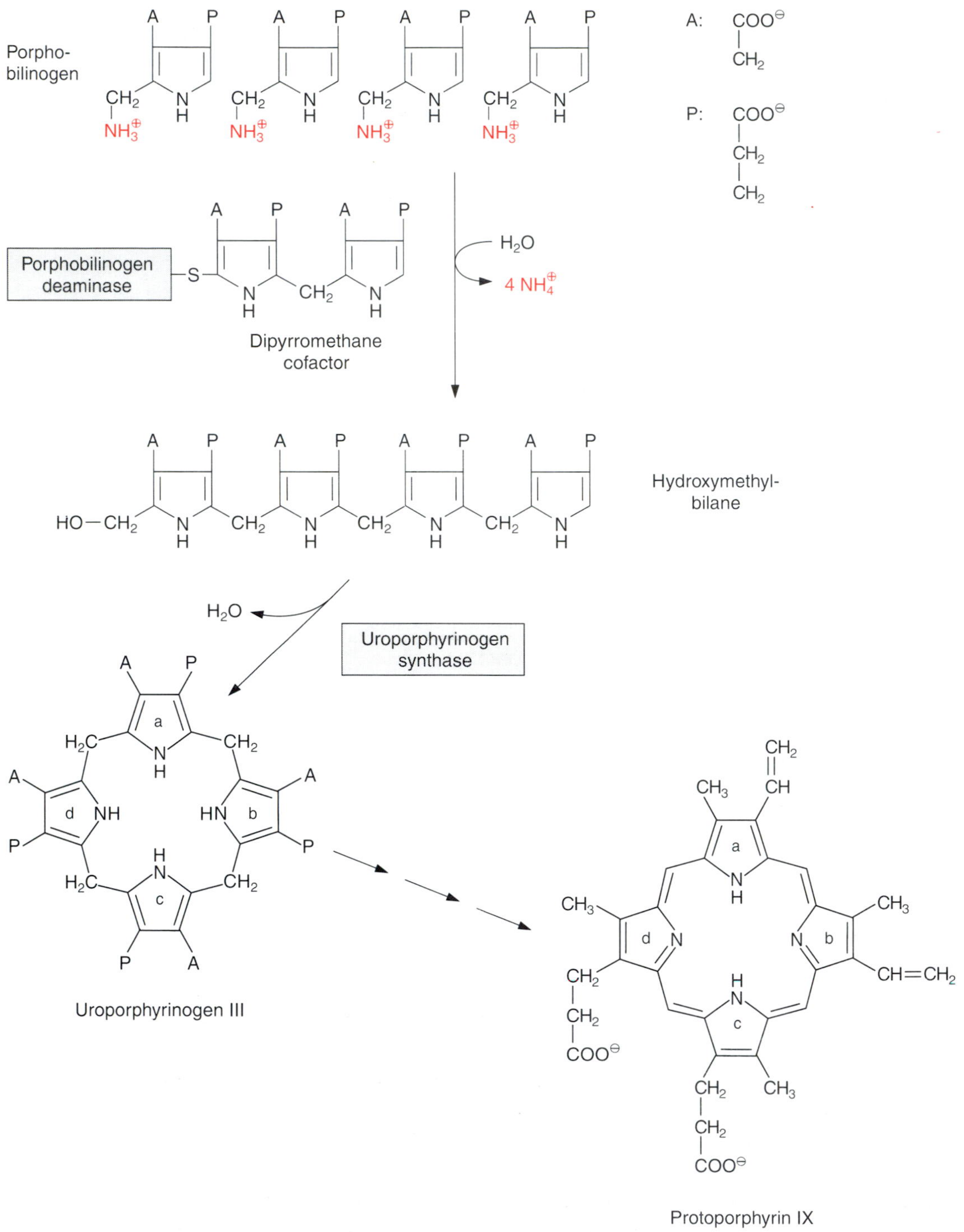

Figure 10.23 Protoporphyrin synthesis.

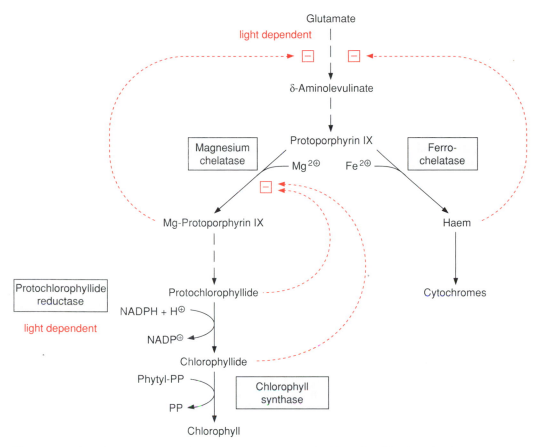

Figure 10.24 Overview of the synthesis of chlorophyll and haem in chloroplasts. The dashed red lines symbolize inhibition of enzymes by products of the biosynthesis chain.

Because of the light dependence of protochlorophyllide reductase, the shoot only starts greening when it reaches the light. Also, synthesis of the chlorophyll binding proteins of the light harvesting complexes is light dependent. The only exceptions are some gymnosperms, e.g. pine, in which protochlorophyllide reduction as well as the synthesis of chlorophyll binding proteins also progresses during darkness. Porphyrins are photo-oxidized in light, which may lead to photochemical cell damage. It is therefore important that intermediates of chlorophyll biosynthesis do not accumulate. To prevent this, the synthesis of δ-aminolevulinate is light dependent, but the mechanism of this regulation is not yet understood. Moreover, δ-aminolevulinate synthesis is subject to feedback inhibition by Mg-protoporphyrin IX and by haem. The end-products, protochlorophyllide and chlorophyllide inhibit magnesium chelatase. Regulation of chlorophyll synthesis is summarized in Fig. 10.24.

Protophorphyrin is also the precursor for haem synthesis

Incorporation of an iron atom into protoporphyrin IX by a ferro-chelatase results in the formation of haem. By assembling the haem with apoproteins, chloroplasts are able to synthesize their own cytochromes. Porphyrin biosynthesis in the plastids also delivers precursors for the synthesis of mitochondrial *cytochromes* and other haem binding proteins located outside the plastids. It is assumed that porphyrins are released from the chloroplasts and taken up into mitochondria and other cell compartments. Thus chloroplasts have a central function in the synthesis of tetrapyrrole compounds in a plant.

Further reading

Crawford, N. M. (1995). Nitrate: nutrient and signal for plant growth. *The Plant Cell*, **7**, 859–68.

Galili, G. (1995). Regulation of lysine and threonine synthesis. *The Plant Cell*, **7**, 899–906.

Hendrich, W. and Bereza, B. (1993). Spectroscopic characterisation of protochlorophyllide and its transformation. *Photosynthetica*, **28**, 1–16.

Hermann, K. M. (1995). The shikimate pathway: early steps in the biosynthesis of aromatic compounds. *The Plant Cell*, **7**, 907–19.

Huber, S. C., Bachmann, M., and Huber, J. L. (1996). Post-translational regulation of nitrate reductase activity: a role for Ca^{++} and 14-3-3 proteins. *Trends in Plant Science*, **1**, 432–8.

Huppe, H. C. and Turpin, D. H. (1994). Integration of carbon and nitrogen metabolism in plant and algal cells. *Annual Review of Plant Physiology and Plant Molecular Biology*, **45**, 577–607.

Kaiser, W. M. and Huber, S. C. (1994). Posttranslational regulation of nitrate reductase in higher plants. *Plant Physiology*, **106**, 817–21.

Kumar, A. M., Schaub, U., Söll, D., and Ujwal, M. L. (1996). Glutamyl-transfer RNA: at the crossroad between chlorophyll and protein biosynthesis. *Trends in Plant Science*, **1**, 371–6.

Lam, H.-M., Coshigano, K., Schultz, C., Melo-Oliveira, R., Tjadjen, G., Oliveira J., *et al.* (1995). Use of *Arabidopsis* mutants and genes to study amino acid biosynthesis. *The Plant Cell*, **7**, 887–98.

Miflin, B. J. and Lea P. J. (1990). Intermediary nitrogen metabolism. *The Biochemistry of Plants*, Vol. 16. Academic Press, New York.

Radwanski, E. R. and Last, R. L. (1995). Tryptophan biosynthesis and metabolism: biochemical and molecular genetics. *The Plant Cell*, **7**, 921–34.

Singh, B. K. and Shaner, D. L. (1995). Biosynthesis of branched chain amino acids: from test tube to field. *The Plant Cell*, **7**, 935–44.

Solomonson, L. P. and Barber, M. J. (1990). Assimilatory nitrate reductase. Functional properties and regulation. *Annual Review of Plant Physiology and Plant Molecular Biology*, **41**, 225–53.

v. Wettstein, D., Gough, S., and Kannagara, C. G. (1995). Chlorophyll biosynthesis. *The Plant Cell*, **7**, 1039–57.

chapter 11

Nitrogen fixation

In a closed ecological system the nitrate required for plant growth is derived from the degradation of biomass. In contrast to other plant nutrients, e.g. phosphate or sulfate, nitrate cannot be delivered by the weathering of rocks. Smaller amounts of nitrate are generated by lightning and carried into the soil by rain water (in temperate areas about 5 kg N ha^{-1} year^{-1}). Due to the effects of civilization (car traffic, mass animal production, etc.) the amount of nitrate, other nitrous oxides, and ammonia carried into the soil by rain can be in the range of 15–70 kg N ha^{-1} year^{-1}. Fertilizers are essential for agricultural production to compensate for the nitrogen lost by the withdrawal of harvest products. For the cultivation of maize, for instance, about 200 kg N ha^{-1} year^{-1} have to be added as fertilizers in the form of nitrate or ammonia. Ammonia, the primary product for the synthesis of nitrate fertilizer, is produced from nitrogen and hydrogen by the Haber–Bosch process:

$$3H_2 + N_2 \rightarrow 2NH_3 \qquad (\Delta G^{0\prime} = -33.5 \text{ kJ mol}^{-1}).$$

The synthesis requires temperatures of 400–500°C and a pressure of several hundred atmospheres and thus involves very high energy costs. The synthesis of nitrogen fertilizer costs about one-third of the total energy expenditure for the cultivation of maize.

Only certain bacteria, including cyanobacteria, are able to form ammonia from nitrogen in air. A number of plants live in symbiosis with N$_2$-fixing bacteria, in which the plants provide the bacteria with metabolites for their nutrition and in turn are supplied with organic nitrogen by the bacteria. The symbiosis of legumes with nodule-forming bacteria (rhizobia) is widespread. Legumes, a large order with about 12 000 species, include soybean, lentil, pea, clover, and lupins. Of all the legumes investigated so far, 90 per cent have been shown to form a symbiosis with rhizobia. In temperate climates the cultivation of legumes can lead to a fixation of 100–400 kg N_2 ha^{-1} yr^{-1}. Therefore legumes are important as green manure; in crop rotation they are an inexpensive alternative to artificial fertilizers. The symbiosis of the water fern *Azolla* with the cyanobacterium *Anabaena* supplies rice fields with nitrogen. N_2-fixing actinomycetes of the genus *Frankia* form a symbiosis with woody plants such as the alder or the Australian casuarina. The latter is notable as a pioneer plant on nitrogen-deficient soils.

11.1 Legumes form a symbiosis with nodule-forming bacteria

Initially it was thought that the nodules of legumes (Fig. 11.1) were caused by a plant disease, until their function in N_2 fixation was recognized by H. Hellriegel and H. Wilfarth in 1888. They found that beans containing these nodules were able to grow without nitrogen fertilizer.

The nodule-forming bacteria (*Rhizobiaceae*) are subdivided into the three genera *Rhizobium*, *Bradyrhizobium*, and *Azorhizobium*. Species of the genus *Rhizobium* form nodules with species of, for example, pea and lucerne, *Bradyrhizobium* with soybeans, and *Azorhizobium* with the tropical legume *Sesbania*. The *Rhizobiaceae* are strictly aerobic Gram-negative rods, which live in the soil and grow heterotrophically in the presence of organic compounds. Some species are also able to grow autotrophically in the presence of H_2, although at a low growth rate.

The uptake of rhizobia into the host plant is a *controlled infection*. The molecular basis of specificity and recognition is still only partially known. The rhizobia form species-specific nodulation factors (*nod factors*). These are lipo-oligosaccharides which acquire a high structural specificity by acylation, acetylation, and sulfonation. They are like a security key with many notches and open the house of the specific host with which the rhizobia associate. The 'key' induces the root hair of the host to curl and the root cortex cells to divide, forming the primary nodule meristem. The calcium-binding protein rhizadhesine mediates the binding of the rhizobia to the developing root hair. After the root hair has been invaded by the rhizobia, an *infection thread* forms (Fig. 11.2), which extends into the cortex of the roots, forms branches there, and infects the primary nodule meristem. The accumulation of infected cells forms a *nodule*. The morphogenesis of the nodule is of similarly high complexity to that of any other plant organ such as the root or shoot. The nodules are connected with the root via vascular tissues which supply them with substrate. The rapidly multiplying rhizobia develop into *bacteroids*. The

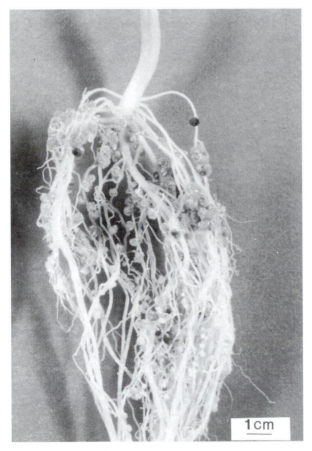

Figure 11.1 Root system of *Phaseolus vulgaris* (bean) with a dense formation of nodules after infection with *Rhizobium etli*. (By P. Vinuesa-Fleischmann and D. Werner, Marburg.)

volume of these bacteroids can be ten times the volume of rhizobia. In many cases several of these bacteroids are surrounded by a symbiosome membrane (also named peribacteroid membrane) and are thus separated from the cytoplasm of the host cell in a *symbiosome* (Fig. 11.3). In other legumes only a single bacteroid is found in a symbiosome. The peribacteroid membrane is synthesized by the plant host.

Rhizobia possess a respiratory chain with a basic structure corresponding to that of the mitochondrial respiratory chain (see Fig. 5.15). An additional electron transport path is formed during differentiation of the rhizobia to bacteroids. In *Bradyrhizobium japonicum* this path branches at the cyt-bc_1 complex of the respiratory chain and conducts electrons to another terminal oxidase, enabling an increased respiratory rate. This part of the respiratory chain is encoded by symbiosis-specific genes.

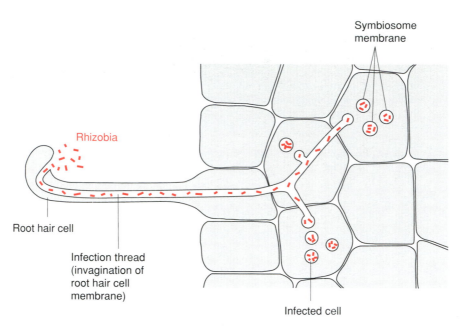

Symbiosome
membrane

Rhizobia

Root hair cell

Infection thread
(invagination of
root hair cell
membrane)

Infected cell

Figure 11.2 Controlled infection of a host cell by rhizobia is induced by an interaction with the root hairs. The rhizobia induce the formation of an infection thread, which is formed by invagination of the root hair cell wall and protrudes into the cells of the root cortex. In this way the rhizobia invaginate into the host cell where they are separated by a symbiosome membrane from the cytosol of the host cells. The rhizobia grow and differentiate into large bacteroids.

The formation of nodules is due to a regulated interplay of the expression of specific bacterial and plant genes

Rhizobia capable of entering a symbiosis contain a large number of genes which are switched off in the free-living bacteria and are activated only after an interaction with the host. The bacterial genes for proteins required for N_2 fixation are named *nif* and *fix* genes, and those for formation of root nodules, *nod* genes. In all *Rhizobium* species that are able to enter a symbiosis with legumes the *nif* and the *nod* genes are located on a very long plasmid of about 200 000 base pairs, whereas in *Bradyrhizobium* those genes are located on the chromosome.

The host plant signals its readiness for the formation of nodules (e.g. upon nitrate deficiency in the soil) by excreting several *flavonoids* (section 18.5) as signal compounds. These flavonoids bind to a bacterial protein which is encoded by a constitutive (which means expressed at all times) *nod* gene. The protein, to which the flavonoid is bound, activates the transcription of the other *nod* genes. The proteins encoded by these *nod* genes are involved in the synthesis of the nod factors mentioned above. Four so-called 'general' nodulation genes are present in all rhizobia. In addition, more than 20 other *nod* genes are known which are also responsible for the host's specificity.

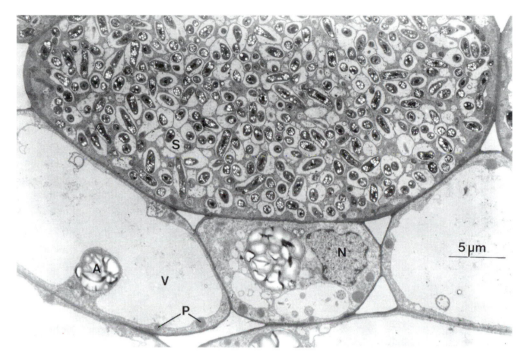

Figure 11.3 Electron microscopic cross-section through a nodule of *Glycine max* cv. Caloria (soybean) infected with *Bradyrhizobium japonicum*. The upper large infected cell shows intact symbiosomes (S) with one or two bacteroids per symbiosome. In the lower section three non-infected cells with nucleus (N), central vacuole (V), amyloplasts (A), and peroxisomes (P) are to be seen. (By E. Mörschel and D. Werner, Marburg.)

The nodule-specific proteins, which are synthesized by the host plant in the course of nodule formation, are named *nodulins*. These nodulins include the enzymes of carbohydrate degradation, the citrate cycle, the synthesis of glutamine and asparagine, and, if applicable, also of ureide synthesis. Nodules also include a malate translocator on the symbiosome membrane (see following section), and last, but not least, leghaemoglobin (section 11.2). The plant genes encoding these proteins are named nodulin genes. Early nodulins are involved in the process of infection and in forming the nodules. The expression of the corresponding genes is induced in part by signal substances released from the rhizobia. Late nodulins are synthesized only after the formation of the nodules. In many cases, nodulins are isoforms of proteins found in other plant tissues.

Metabolic products are exchanged between bacteroids and host cells

The main substrate provided by the host cells to the bacteroids is *malate* (Fig. 11.4), formed from sucrose, which is delivered by the sieve tubes. The sucrose is metabolized

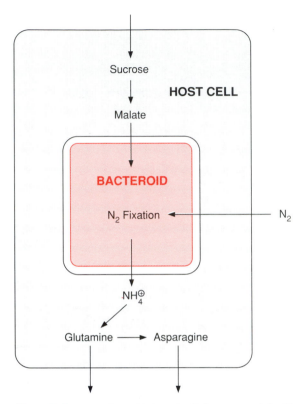

Figure 11.4 Metabolism of infected cells in a root nodule. Glutamine and asparagine are formed as the main products of N_2 fixation (see also Fig. 11.5).

by sucrose synthase (Fig. 13.5), degraded by glycolysis to phosphoenolpyruvate, which is carboxylated to oxaloacetate (see Fig. 10.11), and the latter is reduced to malate. Nodule cells contain high activities of phosphoenolpyruvate carboxylase. NH_4^+ is delivered as a product of N_2 fixation via a specific channel on the symbiosome membrane to the host cell, where it is subsequently converted mainly into *glutamine* (Fig. 7.9) and *asparagine* (Fig. 10.14) and then transported via the xylem vessels to the other parts of the plant.

Some nodules, e.g. those of soybean, export the fixed nitrogen as ureides (urea degradation products), especially *allantoin* and *allantoic acid* (Fig. 11.5). These compounds have a particularly high nitrogen to carbon ratio. The formation of ureides in the host cells requires a complicated synthetic pathway. Inosine monophosphate is synthesized via the pathway of purine synthesis and is then degraded via xanthine and ureic acid to the ureides mentioned above.

Malate taken up into the bacteroids is oxidized by the citrate cycle (Fig. 5.3). The reducing equivalents thus generated are the fuel for fixation of N_2.

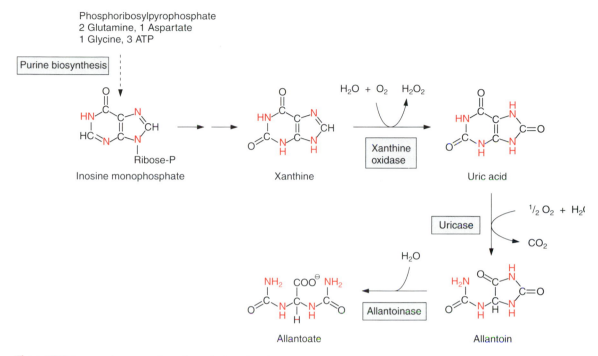

Phosphoribosylpyrophosphate
2 Glutamine, 1 Aspartate
1 Glycine, 3 ATP

Figure 11.5 In many legumes allantoin and allantoic acid are formed as products of N_2 fixation and are delivered via the roots to the xylem. Their formation proceeds first via inosine monophosphate by the purine synthesis pathway. Inosine monophosphate is oxidized to xanthine and then further to uric acid. Allantoin and allantoic acid are formed by hydrolysis and opening of the ring.

Dinitrogenase reductase delivers electrons for the dinitrogenase reaction

Nitrogen fixation is catalysed by the *nitrogenase complex*, a highly complex system with dinitrogenase reductase and dinitrogenase as the main components (Fig. 11.6). This complex is highly conserved and is present in the cytoplasm of the bacteroids. From NADH formed in the citrate cycle, electrons are transferred via soluble ferredoxin to *dinitrogenase reductase*. The latter is a one-electron carrier, consisting of two identical subunits, which together form a *4Fe–4S cluster* (see Fig. 3.26) and contain two binding sites for ATP. After reduction of dinitrogenase reductase, two molecules of ATP bind to it, resulting in a conformational change of the protein by which the redox potential of the 4Fe–4S cluster is raised from -0.25 to -0.40 V. Following the transfer of an electron to dinitrogenase, the two ATP molecules bound to the protein are hydrolysed to ADP and phosphate, and then released from the protein. As a result of this the conformation with the lower redox potential is restored and the enzyme is again ready to take up one electron from ferredoxin. Thus with the consumption of two molecules of ATP, one electron is transferred from NADH to dinitrogenase by dinitrogenase reductase.

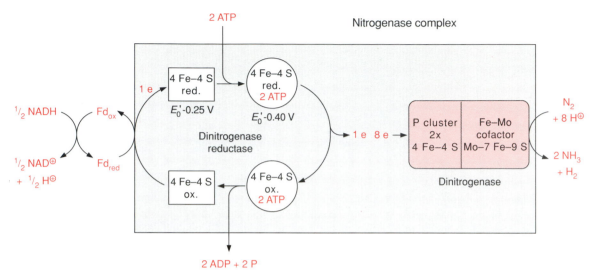

Figure 11.6 The nitrogenase complex consists of the dinitrogenase reductase and the dinitrogenase. Their structure and function are described in the text. The reduction of one molecule of N_2 is accompanied by the reduction of at least two protons to form molecular hydrogen.

N_2 as well as H^+ is reduced by dinitrogenase

Dinitrogenase is an $\alpha_2\beta_2$ tetramer. The α- and β-subunits have a similar size and are similarly folded. The tetramer contains two catalytic centres, probably re-acting independently of each other, and each contains a so-called *P cluster*, con-sisting of two *4Fe–4S clusters* and an iron–molybdenum cofactor (*FeMoCo*). FeMoCo is a large redox centre made up of Fe_4S_3 and Fe_3MoS_3, which are linked to each other via three inorganic sulfide groups (Fig. 11.7). A further constituent of the cofactor is *homocitrate*, which is linked via oxygen atoms of the hydroxyl and carboxyl group to molybdenum. Another ligand of molybdenum is the imida-zole ring of a histidine residue of the protein. The function of the Mo atom is still unclear. Alternative dinitrogenases are known, in which Mo is replaced by vana-dium or iron, but these dinitrogenases are much more unstable than the dinitro-genase containing FeMoCo. The Mo atom possibly causes a more favourable geometry and electron structure of the centre. It is not yet known how the nitrogen reacts with the iron–molybdenum cofactor. One possibility would be that the N_2 molecule is bound in the cavity of the FeMoCo centre (Fig. 11.7) and that the electrons required for N_2 fixation are transferred by the P cluster to the FeMoCo centre.

Dinitrogenase is able to reduce other substrates beside N_2, e.g. protons, which are reduced to molecular hydrogen:

$$2H^+ + 2e^- \xrightarrow{\text{Dinitrogenase}} H_2$$

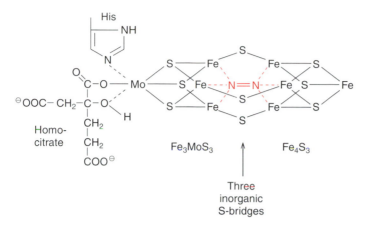

Figure 11.17 The iron–molybdenum cofactor consists of the fragments Fe_4S_3 and $MoFe_3S_3$ which are linked to each other by three inorganic sulfide bridges. In addition to this, the molybdenum is ligated with homocitrate and the histidine side chain of the protein. The cofactor binds one N_2 molecule and reduces it to two molecules of NH_3 by successive uptake of electrons. The position where N_2 is bound in the cofactor has not yet been proven experimentally. (After Karlin 1993.)

During N_2 fixation at least one molecule of hydrogen is formed per N_2 reduced:

$$8H^+ + 8e^- + N_2 \xrightarrow{\text{Dinitrogenase}} 2NH_3 + H_2$$

Thus the balance of N_2 fixation is:

$$N_2 + 4NADPH + 4H^+ + 16ATP \rightarrow 2NH_3 + H_2 + 4NADP^+ + 16ADP + 16P$$

In the presence of sufficient concentrations of acetylene only this is reduced and ethylene is formed:

$$HC{\equiv}CH + 2e^- + 2H^+ \xrightarrow{\text{Dinitrogenase}} H_2C{=}CH_2$$

This reaction is used to measure the activity of dinitrogenase.

The reason why H_2 is evolved during N_2 fixation is not known. It may be part of the catalytic mechanism or a side reaction or a reaction to protect the active centre against the inhibitory action of oxygen. The formation of molecular hydrogen during N_2 fixation can be detected in a clover field.

Many bacteroids, however, possess hydrogenases by which H_2 is reoxidized by electron transport:

$$2H_2 + O_2 \xrightarrow{\text{Hydrogenase}} 2H_2O$$

It is questionable, however, whether this reaction is coupled in the bacteroids to the generation of ATP.

11.2 N_2 fixation can proceed only at very low oxygen concentrations

Dinitrogenase is extremely sensitive to oxygen. Therefore N_2 fixation can proceed only at very low oxygen concentrations. Since N_2 fixation depends on the uptake of nitrogen from air, how is the enzyme protected against oxygen present in air?

The answer is that oxygen, which has diffused together with nitrogen into the nodules, is consumed by the *respiratory chain* contained in the bacteroid membrane. Due to the very high affinity of the bacteroid cytochrome-a/a_3 complex, respiration is still possible with an oxygen concentration of only 10^{-9} mol l^{-1}. As described above, a total of 16 molecules of ATP are required for the fixation of one molecule of N_2. Upon oxidation of one molecule of NADH about 2.5 molecules of ATP are generated by the mitochondrial respiratory chain (section 5.6). In the bacterial respiratory chain, which normally has a lower degree of coupling than that of mitochondria, only about two molecules of ATP might be formed per NADH oxidized. Thus about four molecules of O_2 have to be consumed for the formation of 16 molecules of ATP (Fig. 11.8). When the bacteroids possess a hydrogenase, due to the oxidation of H_2 formed during N_2 fixation, oxygen consumption is further increased by half an oxygen molecule. Thus during N_2 fixation, for each N_2 molecule at least four O_2 molecules are consumed by bacteroid respiration ($O_2/N_2 \geqslant 4$). In contrast to this, the O_2/N_2 ratio in air is about 0.25. This comparison shows that air required for N_2 fixation contains, in relation to nitrogen, far too little oxygen.

The outer layer of the nodules, consisting of intercellular spaces filled with water, is a considerable *diffusion barrier* for the entry of air. The diffusive resistance is so large that bacteroid respiration is limited by the uptake of oxygen. This leads to the astonishing situation that N_2 fixation is limited by the influx of O_2 for formation of the required ATP. Experiments by the Australian Fraser Bergersen have presented evidence for this; in soybean nodules, doubling the O_2 content in air (with a corresponding decrease of the N_2 content) resulted in a doubling of the rate of N_2 fixation. But, because of the O_2 sensitivity of the dinitrogenase, a further increase in O_2 resulted in a steep decline in N_2 fixation.

Since the bacterial respiratory chain is located in the membrane and dinitrogenase in the interior of the bacteroids, O_2 is kept at a safe distance from dinitrogenase. The high diffusive resistance for O_2, which, as shown in the experiment, can even limit N_2 fixation, ensures that even at low temperatures, in which N_2 fixation and the bacterial respiration are slowed down, oxygen is kept away from the nitrogenase complex.

Cells infected by rhizobia form *leghaemoglobin*, which is very similar to the myoglobin of animals but has a tenfold higher affinity for oxygen. The oxygen concentration required for half-saturation of leghaemoglobin amounts to only

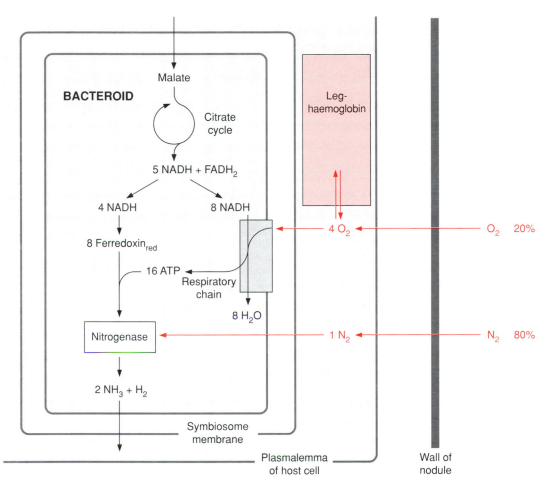

Figure 11.8 N_2 fixation by bacteroids. The total oxidation of malate by the citrate cycle yields five NADH and one $FADH_2$ (see Fig. 5.3). The formation of two NH_3 from N_2 and the accompanying reduction of $2H^+$ to H_2 requires 16 molecules of ATP. Generation of this ATP by the respiratory chain localized in the bacteroid membrane requires the oxidation of about eight molecules of NADH. Thus for each molecule of N_2 fixed four molecules of O_2 are consumed by the oxidation of NADH in the respiratory chain of the bacteroid membrane.

$10–20 \times 10^{-9}$ mol l⁻¹. Leghaemoglobin is located outside the symbiosome membrane in the cytosol of the host cell and is present there in unusually high concentrations (3×10^{-3} mol l⁻¹ in soybeans). Leghaemoglobin can amount to 40 per cent of the total soluble protein of the nodules and gives these a pink colour. Leghaemoglobin probably serves as an O_2 *buffer* to ensure continuous electron transport in the bacteroids at the very low prevailing O_2 concentration in the nodules.

11.3 The energy costs for the utilization of N_2 as a nitrogen source are much higher than for the utilization of NO_3^-

As shown in Fig. 11.8, at least six molecules of NADH are consumed in the formation of NH_4^+ from N_2. Assimilation of nitrate, in contrast, requires only four NAD(P)H equivalents for the formation of NH_4^+ (Fig. 10.1). Therefore it is much more economical for plants, which are able to fix N_2 with the help of their symbionts, to satisfy their nitrogen demand by nitrate assimilation. This is the reason why the formation of nodules is regulated. Nodules are only formed when the soil is nitrate deficient. The advantage of this symbiosis is that legumes can grow in soils with very low nitrogen content, where other plants have no chance.

Further reading

Dixon, R. O. D. and Wheeler, C. T. (1986). *Nitrogen fixation in plants*. Blackie Press, London.

Geurts, R. and Franssen, H. (1996). Signal transduction in *Rhizobium*-induced nodule formation. *Plant Physiology*, **112**, 447–54.

Karlin, K. D. (1993). Metalloenzymes, structural motifs and inorganic models. *Science*, **261**, 701–8.

Mylona, P., Pawlowski, K., and Bisseling, T. (1995). Symbiotic nitrogen fixation. *The Plant Cell*, **7**, 869–85.

Sanchez, F., Padilla, J. E., Pérez, H., and Lara, M. (1991). Control of nodulin genes in root-nodule development and metabolism. *Annual Review of Plant Physiology and Plant Molecular Biology*, **42**, 507–28.

Tyerman, S. D., Whitehead, L. F., and Day, D. A. (1995). A channel-like transporter for NH_4^+ on the symbiotic interface of N_2-fixing plants. *Nature*, **378**, 629–32.

Werner, D. (1992). *Symbiosis of plants and microbes*. Chapman & Hall, London.

chapter 12

Sulfate assimilation

Sulfate is an essential constituent of living matter. In the oxidation state $-II$, it is contained in the two amino acids cysteine and methionine, in the detoxifying agent glutathione, in various iron–sulfur redox clusters, and in thioredoxins. Plants, bacteria, and fungi are able to synthesize these substances by assimilating sulfate taken up from the environment. Animal metabolism is dependent on nutrients to supply amino acids containing sulfur. Therefore sulfate assimilation of plants is a prerequisite for animal life just like the carbon and nitrate assimilation discussed already.

Whereas the plant uses nitrate only in its reduced form, sulfur in the form of sulfate is also an essential plant constituent. Sulfate is contained in sulfolipids which comprise about 5 per cent of the lipids of the thylakoid membrane (Chapter 15). In sulfolipids sulfur is attached as sulfonic acid via a C–S bond to a carbohydrate residue of the lipid. The biosynthesis of this sulfonic acid group is, to a great extent, unknown.

12.1 Sulfate assimilation proceeds by photosynthesis

Sulfate assimilation in plants occurs primarily in the chloroplasts and is then a part of photosynthesis. However, the rate of sulfate assimilation is relatively low, amounting to only about 5 per cent of the rate of nitrate assimilation and only 0.1–0.2 per cent of the rate of CO_2 assimilation. The activities of the enzymes involved in sulfate assimilation are minute, making it very difficult to elucidate the reactions involved. Therefore our knowledge about sulfate assimilation is still fragmentary. It may be noted that enzymes of sulfate metabolism occur also in the cytosol and in mitochondria.

Sulfate assimilation shows parallels to but also differences from nitrogen assimilation

Plants take up *sulfate* via a specific translocator of the roots, in a similar manner to that described for nitrate in Chapter 10. The transpiration stream in the xylem vessels carries the sulfate to the leaves where it is taken up by a specific translocator, probably a symport with three protons, into the mesophyll cells (Fig. 12.1). Surplus sulfate is transported to the vacuole and is deposited there.

The basic scheme for sulfate assimilation in the mesophyll cells corresponds to that of nitrate assimilation. Sulfate is reduced to *sulfite* by the uptake of two electrons and then by the uptake of another six electrons, to *hydrogen sulfide*:

$$SO_4^{2-} + 2e^- + 2H^+ \rightarrow SO_3^{2-} + H_2O$$
$$SO_3^{2-} + 6e^- + 8H^+ \rightarrow H_2S + 3H_2O$$

Whereas the NH_3 formed during nitrite reduction is fixed by the formation of the amino acid glutamine (Fig. 10.6), the hydrogen sulfide formed during sulfite reduction is fixed to form the amino acid cysteine. A distinguishing difference between nitrate assimilation and sulfate assimilation is that the latter requires a much higher input of energy. This is shown in an overview in Fig. 12.1. The reduction of sulfate to sulfite, which in contrast to nitrate reduction occurs in the chloroplasts, requires in total the cleavage of three energy-rich phosphate anhydride bonds, and the fixation of the hydrogen sulfide into cysteine, another two. Thus the ATP consumption of sulfate assimilation is five times higher than that of nitrate assimilation. Let us now look at the partial reactions.

Sulfate is activated prior to reduction

Sulfate is taken up into the chloroplasts via the *triose phosphate–phosphate translocator* (section 1.9) in counter-exchange for phosphate. Sulfate cannot be directly reduced in the chloroplasts, because the redox potential of the substrate pair SO_3^{2-}/SO_4^{2-} ($\Delta E^{0\prime} = -517$ mV) is too high. No reductant is available in the chloroplasts which could reduce SO_4^{2-} to SO_3^{2-} in one reaction step. To reduce the difference in redox potentials, sulfate is activated prior to reduction.

As shown in Fig. 12.2, activation of sulfate proceeds via the formation of an anhydride bond with the phosphate residue of AMP. Sulfate is exchanged by the enzyme *ATP sulfurylase* for a pyrophosphate residue of ATP, and AMP sulfate (APS) is thus formed. Since the free enthalpy of the hydrolysis of the sulfate–phosphate anhydride ($\Delta G^{0\prime} = -71$ kJ mol^{-1}) is very much higher than that of the phosphate–phosphate anhydride bond in ATP ($\Delta G^{0\prime} = -31$ kJ mol^{-1}),

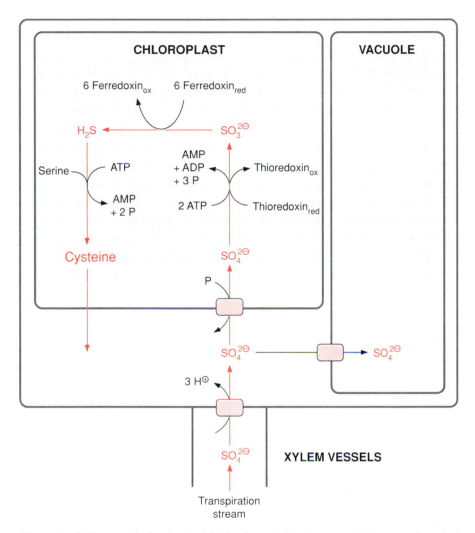

Figure 12.1 Sulfate metabolism in a leaf. Sulfate is carried by the transpiration stream into the leaves and is transported to the mesophyll cells. After being transported to the chloroplast via the phosphate translocator, sulfate is reduced there to H_2S and subsequently converted to cysteine. Sulfate can also be deposited in the vacuole.

the equilibrium of the reaction lies far towards ATP. This reaction can only proceed because pyrophosphate is withdrawn from the equilibrium by a high pyrophosphatase activity in the chloroplasts. But even then the concentration of the APS formed by ATP sulfurylase is only in the range of 10^{-6} mol l^{-1}.

The ribose of APS is esterified by *APS kinase* with a second phosphate residue. Since the free enthalpy of the hydrolysis of this phosphate ester is much lower than that of ATP, the equilibrium of this reaction lies far towards the product, 3-phospho-AMP sulfate (PAPS). APS kinase is already half saturated at an APS

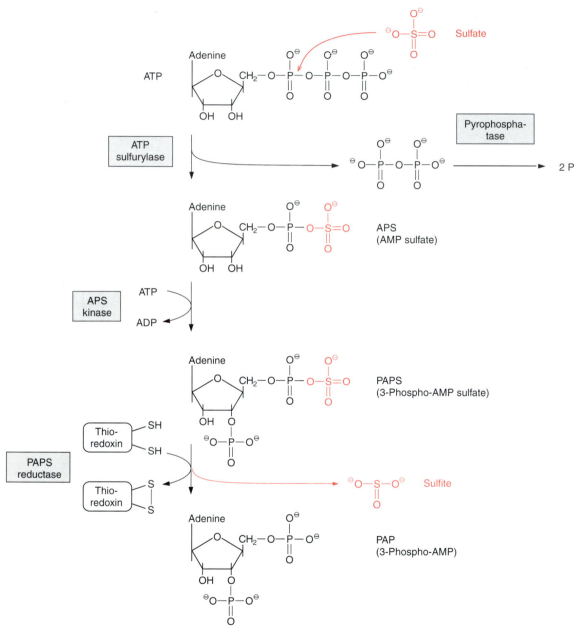

Figure 12.2 Reduction of sulfate to sulfite.

concentration of only 10^{-6} mol l^{-1} and the PAPS formed can reach a concentration in the range of 10^{-3} mol l^{-1}. In this way the second phosphorylation step results in a concentration of the activated sulfate.

Sulfate present in the form of PAPS is reduced by *thioredoxin* (section 6.6) to sulfite. The *PAPS reductase* involved in this reaction catalyses not only the reduction but also the following liberation of the sulfite. The redox potential difference from sulfate to sulfite is lowered, since the reduction of sulfate is driven by hydrolysis of the very energy-rich sulfite anhydride bond. The remaining 3-phospho-AMP is converted by hydrolysis to AMP.

Sulfite reductase is similar to nitrite reductase

As in nitrite reduction, six molecules of reduced ferredoxin are required as reductant for the reduction of sulfite in the chloroplasts (Fig. 12.3). The *sulfite reductase* is homologous to the nitrite reductase; it also contains a *sirohaem* (Fig. 10.5) and a *4Fe–4S cluster*. The enzyme is half saturated at a sulfite concentration of about 10^{-6} mol l^{-1} and thus is suitable to reduce efficiently the newly formed sulfite to hydrogen sulfide. (It may be noted that for a long time it has been debated whether there is an alternative pathway of sulfate and sulfite reduction in plants in which the sulfate residue is transferred from APS to glutathione or another carrier and in such a bound state is reduced to a bound hydrogen sulfide, which is then transferred from the carrier to *O*-acetyl serine. Although this so-called bound pathway is described in textbooks, conclusive evidence for its existence is still lacking.) The ferredoxin required by sulfite reductase, as in the case of nitrite reductase (Fig. 10.1), can be reduced by NADPH. This makes it possible for sulfite reduction to occur in heterotrophic tissues also.

H$_2$S is fixed in the form of cysteine

The fixation of the newly formed hydrogen sulfide requires the activation of serine and for this its hydroxyl group is acetylated by acetyl CoA (Fig. 12.4). The

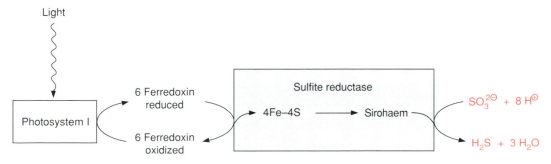

Figure 12.3 Reduction of sulfite to hydrogen sulfide by sulfite reductase in the chloroplasts. The reducing equivalents are delivered via ferredoxin from photosystem I.

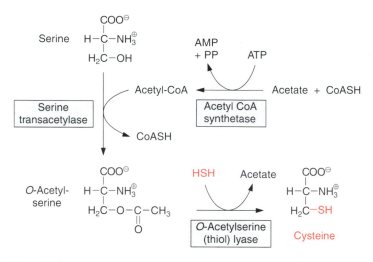

Figure 12.4 The hydrogen sulfide formed by sulfite reduction is incorporated into cysteine.

latter is formed from acetate and CoA with the consumption of ATP (which is converted to AMP) by the enzyme acetyl CoA synthetase. As pyrophosphate formed in this reaction is hydrolysed by the pyrophosphatase present in the chloroplasts, the activation of the serine costs the chloroplasts in total two energy-rich phosphates.

Fixation of hydrogen sulfide is catalysed by the enzyme *O-acetylserine (thiol) lyase*. The enzyme contains pyridoxal phosphate as a prosthetic group and has a high affinity for hydrogen sulfide and acetyl serine. The incorporation of the SH-group can be described as a cleavage of the ester linkage by hydrogen sulfide. In this way cysteine is formed as the end-product of sulfate assimilation.

Cysteine has an essential function in the structure and the activity of the catalytic site of enzymes, and cannot be replaced there by any other amino acid. Moreover, cysteine residues form iron–sulfur clusters (Fig. 3.26) and are constituents of thioredoxin (Fig. 6.25).

12.2 Glutathione serves the cell as an antioxidant and is an agent for the detoxification of pollutants

A relatively large proportion of the cysteine produced by the plant is used for synthesis of the tripeptide *glutathione* (Fig. 12.5). The synthesis of glutathione proceeds via two enzymatic steps: first an amide linkage between the γ-carboxyl group of glutamate with the amino group of the cysteine is formed by *γ-glutamyl-cysteine synthetase* accompanied by the hydrolysis of ATP, and second, a peptide bond between the carboxyl group of the cysteine and the amino group of the glycine is produced by *glutathione synthetase*, again with the consumption of ATP.

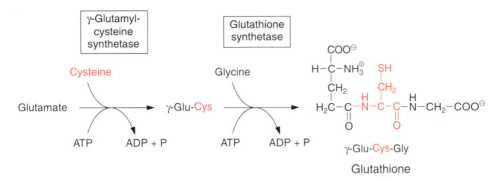

Figure 12.5 Biosynthesis of glutathione.

Glutathione, abbreviated GSH, is present in relatively high concentrations in all plant cells, where it has various functions. As an antioxidant it protects cell constituents against oxidation. Together with ascorbate it eliminates the oxygen radicals formed as by-products of photosynthesis (section 3.9). In addition, glutathione has a protective function for the plant in forming conjugates with xenobiotics and also as a precursor for the synthesis of phytochelatins, which are involved in the detoxification of heavy metals. Moreover, glutathione acts as a reserve of organic sulfur. If required, cysteine is released from glutathione by enzymatic degradation.

Xenobiotics are detoxified by conjugation

Toxic substances formed by the plant or which it has taken up (xenobiotics) are detoxified by reaction with glutathione. Catalysed by *glutathione-S transferases*, the reactive SH group of glutathione can form a thioether by reacting with carbon double bonds, carbonyl groups, and other reactive groups. *Glutathione conjugates* (Fig. 12.6) formed in this way are transported into the vacuole by a specific *glutathione translocator* against a concentration gradient. In contrast to the transport processes described so far, where metabolite transport against a gradient proceeds by secondary active transport, the uptake of glutathione conjugates into the vacuole proceeds by an ATP-driven primary active transport (Fig. 1.20). A single translocator is probably sufficient for the transport of the whole range of possible conjugates. The conjugates taken up are often modified, for instance by degradation to a cysteine conjugate, and are finally deposited in this form. In this way plants can also detoxify herbicides. Herbicide resistance, for instance resistance of maize to atrazine, can be due to the activity of a specific glutathione-S transferase. In an attempt to develop herbicides that selectively attack weeds and not crop plants, the plant protection industry has produced a variety of different substances that increase the tolerance of crop plants to certain herbicides. These protective substances are called *safeners*. Such safeners, as xenobiotics, stimulate the

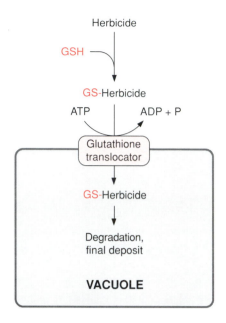

Figure 12.6 Detoxification of a herbicide. Glutathione (GSH) forms a conjugate with the herbicide, which is pumped into the vacuole by a specific glutathione translocator, to be finally deposited there after degradation.

increased expression of glutathione-S transferase and of the vacuolar glutathione translocator, and in this way the herbicides taken up into the plants are detoxified more rapidly. Formation of glutathione conjugates and their transport into the vacuole is also involved in the deposition of flower pigments (section 18.6).

Phytochelatins protect the plant against heavy metals

Glutathione is also a precursor for phytochelatins (Fig. 12.7). *Phytochelatin synthase*, a transpeptidase, transfers the amino group of glutamate to the carboxyl group of the cysteine of a second glutathione molecule, accompanied by liberation of one glycine molecule. The repetition of this process results in the formation of chains of up to 11 Glu–Cys residues. Phytochelatins have been found in all plants investigated so far, although sometimes in a modified form as *iso*-phytochelatins in which glycine is replaced by serine, glutamate, or β-alanine.

Phytochelatins protect plants against toxicity from heavy metals. Through the thiol groups of the cysteine residues, they form tight complexes with metal ions such as Cd^{2+}, Ag^+, Pb^{2+}, Cu^{2+}, Hg^{2+}, and Zn^{2+} (Fig. 12.8). The phytochelatin synthase present in the cytosol is activated by the ions of at least one of the heavy metals listed above. Thus upon the exposure of plants to heavy metals, in a very short time the phytochelatins required for detoxification are synthesized *de novo* from glutathione. Exposure to heavy metals can thus lead to a dramatic fall in the glutathione reserves in the cell. The phytochelatins loaded with heavy metals are pumped, in a similar manner to the glutathione conjugates, at the expense of ATP into the vacuoles. Because of the acidic environment in the vacuole, the

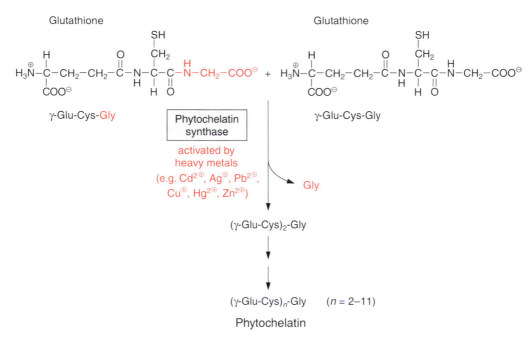

Figure 12.7 Phytochelatin synthesis. The phytochelatin synthase (a transpeptidase) cleaves the peptide bond between the cysteine and the glycine of a glutathione molecule and transfers the α-amino group of the glutamate residue of a second glutathione molecule to the liberated carboxyl group of the cysteine. Long-chain phytochelatins are formed by repetition of this reaction.

heavy metals are probably liberated from the phytochelatins and finally deposited there.

Phytochelatins are essential to protect plants against heavy metal poisoning. Recently mutants of *Arabidopsis* have been found with a defect in the phytochelatin synthase, which showed an extreme sensitivity towards Cd^{2+}.

12.3 Methionine is synthesized from cysteine

Cysteine is a precursor for *methionine*, another amino acid containing sulfur. *O*-Phosphohomoserine, which has already been mentioned as an intermediate of threonine synthesis (Fig. 10.14) reacts with cysteine, while a phosphate group is liberated to form cystathionine (Fig. 12.9). The thioether is cleaved by *cystathio-*

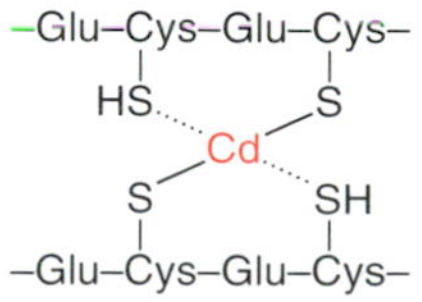

```
–Glu–Cys–Glu–Cys–
    HS     S
        Cd
     S      SH
–Glu–Cys–Glu–Cys–
```

Figure 12.8 Detoxification of heavy metals by phytochelatins: heavy metals are complexed by the thiol groups of the cysteine and thus rendered harmless.

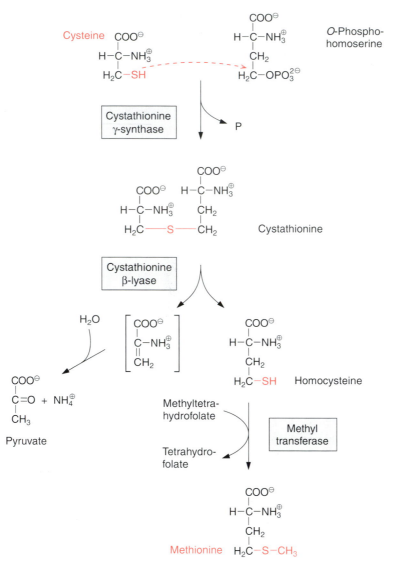

Figure 12.9 Biosynthesis of methionine from cysteine.

nine β-lyase to form homocysteine and an unstable enamine, which spontaneously degrades into pyruvate and NH_4^+. The sulfydryl group of homocysteine is
methylated by methyltetrahydrofolate (methyl-THF) (see Fig. 7.6) and thus the
end-product, methionine, is formed.

S-Adenosylmethionine is a universal methylation reagent

The origin of the methyl group provided by tetrahydrofolic acid (THF) is not
clear. It is possible that it is derived from formate molecules, reacting in an

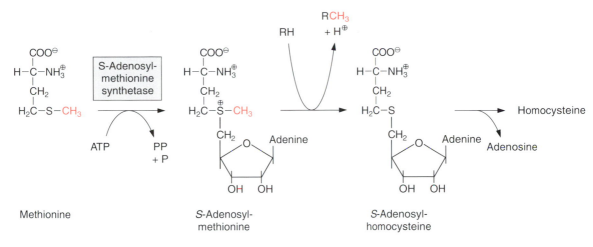

Figure 12.10 *S*-Adenosylmethionine, formed from methionine and ATP, is a methylating agent.

ATP-dependent reaction with THF to form formyl-THF, which is reduced by two molecules of NADPH to methyl-THF. Methyl-THF has only a low methyl transfer potential. *S-Adenosylmethionine*, however, has a more general role as methyl donor. It is involved in the methylation of nucleic acids, proteins, carbo-hydrates, membrane lipids, and many other substances and can therefore be regarded as a universal methylating agent of the cell.

S-Adenosylmethionine is formed by the transfer of an adenosyl residue from ATP to the sulfur atom of methionine with the release of phosphate and pyrophosphate (Fig. 12.10). The methyl group to which the positively charged S atom is linked, is activated and can thus be transferred by corresponding methyl transferases to other acceptors. The remaining *S*-adenosylhomocysteine is hydro-lysed to adenosine and homocysteine and from the latter methionine is recovered by reduction with methyltetrahydrofolate (Fig. 12.9).

S-Adenosylmethionine is a precursor for the phytohormone ethylene (section 19.5).

12.4 Excessive concentrations of sulfur dioxide in air are toxic for plants

SO_2, which is formed in particularly high amounts during the smelting of ores con-taining sulfur, but also during the combustion of fossil fuel, can cover the total nutritional sulfur requirement of a plant. However, in higher concentrations it leads to dramatic damage to plants. Gaseous SO_2 is taken up via the stomata into the leaves where it is converted to sulfite:

$$SO_2 + OH^- \rightarrow SO_3^{2-} + H^+$$

Plants possess protective mechanisms for removing the sulfite that has been formed. In one of these, sulfite is converted by the sulfite reductase, dealt with in section 12.1, to hydrogen sulfide and then further into cysteine. Cysteine formed in increasing amounts can be converted into glutathione. Thus one often finds an accumulation of glutathione in leaves of SO_2-polluted plants. Excessive hydrogen sulfide can leak out of the leaves through the stomata, although only in small amounts. Alternatively, sulfite can be oxidized, possibly by peroxidases in the leaf, to sulfate. Since this sulfate cannot be removed by transport from the leaves, it is finally deposited in the vacuoles of the leaf cells as potassium or magnesium sulfate. When the deposit site is full, the leaves are abscised. This explains in part the toxic effect of SO_2 on pine trees: the early loss of the pine needles of SO_2-polluted trees is to a large extent due to the fact that the capacity of the vacuoles for the final deposition of sulfate is exhausted. In cation-deficient soils the high cation demand for the final deposition of sulfate can lead to a serious K^+ or Mg^{2+} deficiency in leaves or pine needles. The bleaching of pine needles, often observed during SO_2 pollution, is partly attributed to a decreased availability of Mg^{2+} ions.

Further reading

Heber, U., Kaiser, W., Luwe, M., Kindermann, G., Veljovic-Iovanovic, S., Yin, Z.-H. *et al.* (1994). Air pollution, photosynthesis and forest decline. *Ecological Studies*, **100**, 279–96.

Howden, R., Goldsbrough, C. R., Anderson, C. R., and Cobbett, C. S. (1995). Cadmium-sensitive, *cad1* mutants of *Arabidopsis thaliana* are phytochelatin deficient. *Plant Physiology*, **107**, 1059–66.

Kreuz, K., Tommasini, R., and Martinoia, E. (1996). Old enzymes for a new job: herbicide detoxification in plants. *Plant Physiology*, **111**, 349–53.

Rauser, W. E. (1995). Phytochelatins and related peptides. Structure, biosynthesis and function. *Plant Physiology*, **109**, 1141–9.

Schmidt, A. and Jäger, K. (1992). Open questions about sulfur metabolism in plants. *Annual Review of Plant Physiology and Plant Molecular Biology*, **43**, 325–49.

Schwenn, J. D. (1994). Photosynthetic sulphate reduction. *Zeitschrift für Naturforschung*, **49c**, 531–9.

Zenk, M. H. (1996). Heavy metal detoxification in higher plants. *Gene*, **179**, 21–30.

chapter 13

Phloem transport

This chapter deals with the export of photoassimilates from the leaves to the other parts of the plant. Besides having the xylem as a long-distance translocation system for transport from the root to the leaves, plants possess a second long-distance transport system, the phloem, which exports the photoassimilates formed in the leaves to wherever they are required. The *xylem* and *phloem* together with the parenchyma cells form *vascular bundles* (Fig. 13.1). The xylem (*xylon*, Greek for wood) consists of lignified tubes which translocate water and dissolved mineral nutrients from the root via the transpiration stream (section 8.1) to the leaves. Several translocation vessels, arranged mostly on the outside of the vascular bundles, make up the phloem (*phloios*, Greek for bark) which transports photoassimilates from the site of formation (*source*), e.g. the mesophyll cell of a leaf, to the sites of consumption or storage (*sink*), e.g. roots, tubers, fruits, or areas of growth. The phloem system thus connects the sink and source tissues.

The phloem contains elongated cells joined by *sieve plates*, consisting of diagonal cell walls perforated by pores. The single cells are called *sieve elements* and a column of such cells is called a *sieve tube* (Fig. 13.2). The pores of the sieve plate are widened plasmodesmata lined with *callose* (section 9.6). The sieve elements can be regarded as living cells that have lost their nucleus, ribosomes, Golgi apparatus, and vacuoles, and contain only a few mitochondria, plastids, and some endoplasmic reticulum. The absence of many cell structures normally present in a cell, specializes the sieve tubes for the long-distance transport of carbon and nitrogen compounds synthesized in the leaves. In most plants sucrose is the main transport form of carbon, but some plants also transport oligosaccharides from the raffinose family or sugar alcohols (section 9.4). Nitrogen is transported in the sieve tubes

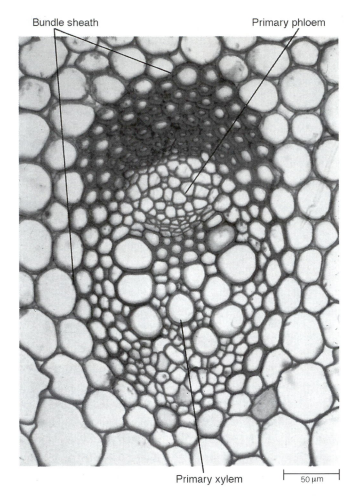

Bundle sheath   Primary phloem

Primary xylem   50 μm

Figure 13.1 Transverse section through a vascular bundle of *Ranunculus* (buttercup), a herbaceous dicot plant. The phloem and xylem are surrounded by bundle sheath cells. (From Raven, Evert, and Curtis (1985) *Biologie der Pflanzen*, De Gruiter Verlag, Berlin, by permission.)

almost exclusively in the organic form as amino acids. Organic acids, nucleotides, proteins, and phytohormones are present in the phloem sap in lower concentrations. Beside these organic substances, the sieve tubes transport inorganic ions, mainly K^+ ions.

The sieve elements of angiosperms are surrounded by *companion cells*, which contain all the constituents of a normal living plant cell, including the nucleus and, in particular, many mitochondria. Sieve elements and companion cells are connected to each other by many *plasmodesmata* (section 1.1) and are an important element of phloem loading.

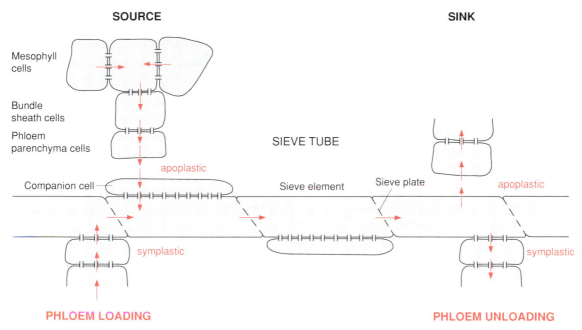

Figure 13.2 Scheme of the sieve tubes and their loading and unloading via the apoplastic and symplastic pathways. The plasmodesmata indicated by the double line allow unhindered diffusion of sugar and amino acids. The structures are not shown to scale.

13.1 There are two paths for phloem loading

Photoassimilates generated in the mesophyll cells, such as sucrose or other oligosaccharides and amino acids, diffuse via plasmodesmata to the *phloem parenchyma cells*. The further transport of photoassimilates from the phloem parenchyma cells to the sieve tubes can occur in two different ways:

1. Particularly in plants, such as squash plants where oligosaccharides from the raffinose family (section 9.4) are translocated in the sieve tubes, the phloem parenchyma cells are connected to the sieve tubes by a large number of plasmodesmata. Therefore, in these plants the transfer of the photoassimilates to the sieve tubes can proceed *symplastically* via plasmodesmata without involving translocators.

2. In contrast, in *apoplastic* phloem loading, found for instance in leaves of cereals, sugar beet, rapeseed, and potato, photoassimilates are first transported from the source cells via the phloem parenchyma cells to the extracellular compartment, the apoplast (Fig. 13.3). Since the concentration of sucrose and amino acids is very much higher in the source cells than in the apoplast, this export requires no energy expenditure. The translocators mediating this export have not yet been characterized.

PHLOEM
PARENCHYMA   APOPLAST   COMPANION CELL   SIEVE
CELL   ELEMENT

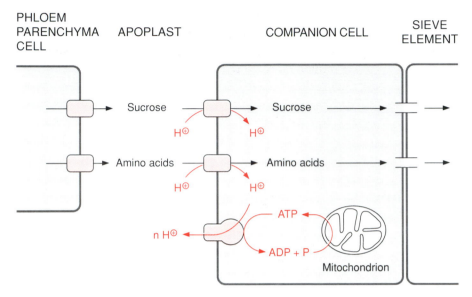

Figure 13.3 Apoplastic phloem loading. Transfer of the photoassimilates from the phloem parenchyma cells to the sieve tubes. Many observations indicate that active loading takes place in the plasma membrane of the companion cells and that the subsequent transfer to the sieve elements occurs by diffusion via plasmodesmata. However, recent results indicate that there can also be a direct transport to the sieve elements without passage through the companion cells or other accessory cells.

The transport of sucrose and amino acids from the apoplasts to the companion cells proceeds via proton symport (Fig. 13.3). This is driven by a proton gradient between the apoplast and the interior of the sieve tubes, which is generated by H^+-pumping ATPase (section 4. 4) present in the plasma membrane. Mitochondrial oxidation produces the necessary ATP. The H^+–sucrose translocator involved in phloem loading has recently been identified and characterized in several plants: in the vascular bundles of *Plantago major*, using specific antibodies, an H^+–sucrose translocator has been localized in the plasma membrane of companion cells. This and the large number of mitochondria in the companion cells indicate that phloem loading proceeds via the companion cells (Fig. 13.3). But this may not always be the case, as an H^+–sucrose translocator has been identified in the plasma membrane of the sieve elements of potato leaves. The substrates for mitochondrial respiration are probably provided by degradation of sucrose via sucrose synthase (see also Fig. 13.5), followed by glycolytic metabolism of the hexose phosphates then formed. Glutamate could also be a substrate for respiration (section 5.3). In many plants this amino acid is present in relatively high concentrations in the phloem sap.

Whereas in the plants with apoplastic phloem loading investigated so far, sucrose is the exclusive transport form for carbohydrates (hexoses are not transported), no special transport form exists for amino nitrogen. In principle, all protein amino acids are transported. Probably the transport of all the amino acids to the phloem is catalysed by just a few translocators with very broad substrate specificity. The relative amounts of the different amino acids are very similar in the phloem sap to that in the source cells. The amino acids most frequently found in the phloem sap are glutamate, glutamine, and aspartate, but in some plants also alanine. In contrast, the non-protein amino acid citrulline is often the predominant amino acid in the phloem sap of squash plants with symplastic phloem loading.

13.2 Phloem transport proceeds by mass flow

The proton co-transport results in very high concentrations of sucrose and amino acids in the sieve tubes. Depending on the plant and on growth conditions, the concentration of sucrose in the phloem sap amounts to 0.6–1.5 mol l^{-1}, and the sum of the amino acids to 0.05–0.5 mol l^{-1}. *Aphids* turned out to be useful helpers for obtaining phloem sap samples for such analyses. An aphid, after some attempts, can insert its stylet precisely into a sieve tube. As the phloem sap is under pressure, it flows through the tube of the stylet and is consumed by the aphid (Fig. 13.4). The aphid takes up more sucrose than it can metabolize and excretes the unused remainder as honey dew, which is the sticky sugary layer covering aphid-infested houseplants. When the stylet of a feeding aphid is severed by a laser beam, the phloem sap exudes from the sieve tube through the stump of the stylet. Although the amount of phloem sap obtained in this way is very low

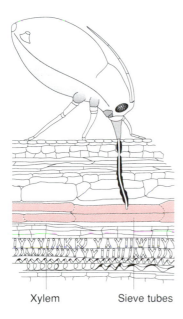

Xylem   Sieve tubes

Figure 13.4 Aphids know where to insert their stylet into the sieve tubes and feed themselves in this way from the exuding phloem sap. (Figure by A. F. G. Dixon (1975), *Encyclopaedia of plant physiology*, Vol. 1, Springer Verlag, Berlin, by permission.)

$(0.05–0.2 \times 10^{-6}\ 1\ h^{-1})$, modern techniques make a quantitative assay of the phloem sap in these samples possible.

In plants performing photosynthesis in the presence of radioactively labelled CO_2, phloem transport velocities of $30–150\ cm\ h^{-1}$ have been measured. This rapid transport proceeds by *mass flow*, driven on the one hand by very efficient pumping of sucrose and amino acids into the sieve tubes and, on the other hand, by their withdrawal at the sites of consumption. This mass flow is driven by many transverse osmotic gradients. The surge of this mass flow carries along substances present at low concentrations, such as phytohormones. The direction of mass flow is governed entirely by the consumption of the phloem contents. Depending on what is required, phloem transport can proceed in an upward direction, e.g. from the mature leaf to the growing shoot or flower, or downwards into the roots or storage tubers. Since the phloem sap is under high pressure and the phloem is highly branched, wounding the vascular tissue might result in the phloem sap 'bleeding'. Protective mechanisms prevent this. Due to the presence of nucleotides in the phloem sap and the enzymes sucrose synthase and callose synthase, which are probably membrane bound, the sieve pores of damaged sieve tubes are sealed by the formation of callose (section 9.6), and damaged sieve tubes are thus put out of action.

13.3 Sink tissues are supplied by phloem unloading

There are again two possibilities for phloem unloading (Fig. 13.2). In *symplastic unloading* the sucrose and amino acids reach the cells of the sink organs directly from the sieve elements via plasmodesmata. In *apoplastic unloading* the substances are first transported from the sieve tubes to the extracellular compartment and are then taken up into the cells of the sink organs. Electron microscopic investigations of the plasmodesmatal frequency indicate that in vegetative tissues, such as roots or growing shoots, phloem unloading proceeds primarily symplastically, whereas in storage tissues unloading is often, but not always, apoplastic.

In storage tissues the delivered carbohydrates are mostly converted to starch and stored as such. In apoplastic phloem unloading this may proceed by two alternative pathways. In the pathway coloured red in Fig. 13.5, the sucrose is taken up from the apoplast into the storage cells and converted there via *sucrose synthase* and *UDP-glucose pyrophosphorylase* to fructose and glucose 1-phosphate. In this reaction pyrophosphate is consumed and UTP is generated. It is still unresolved how the necessary pyrophosphate is formed. *Phosphoglucomutase* converts glucose 1-phosphate to glucose 6-phosphate. Alternatively the enzyme *invertase* first hydrolyses sucrose in the apoplast to glucose and fructose, and these two hexoses are then transported into the cell. A *fructokinase* and a *hexokinase* (the latter phosphorylating mannose as well as glucose) catalyse the formation of the corresponding hexose phosphates. Glucose 6-phosphate is transported via the glucose

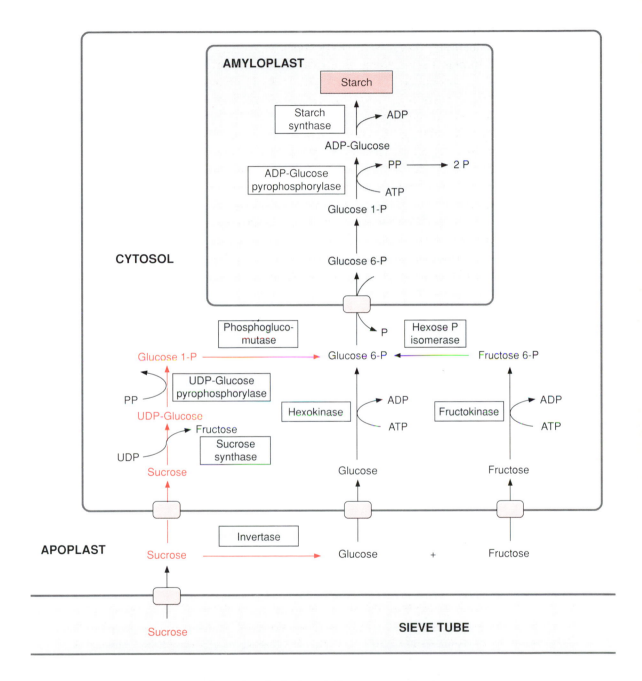

Figure 13.5 Apoplastic phloem unloading and synthesis of starch. Some storage cells take up sucrose, others glucose and fructose, which have been formed by the hydrolysis of sucrose catalysed by invertase. It is not yet known whether glucose and fructose are transported by the same or by different translocators (for details see section 9.1). Some amyloplasts transport glucose 1-phosphate in counterexchange for phosphate.

6-phosphate–phosphate translocator (section 8.2) in counter-exchange for phosphate to the amyloplast, where starch is formed via the synthesis of ADP-glucose (section 9.1). Some leucoplasts transport glucose 1-phosphate in counter-exchange for phosphate. The storage of starch in potato tubers probably proceeds mainly via sucrose synthase. However, the use of gene technology has made it possible to increase the invertase activity in the apoplast of potato tubers, which led to an increase in starch synthesis and thus also in the tuber yield. In the tap-roots of sugar beet the carbohydrates are stored as sucrose in the vacuoles. In some fruits, e.g. grapes, carbohydrates are stored in the vacuole as glucose.

Further reading

Baker, D. A. and Milburn, J. A. (ed.) (1989). *Transport of photoassimilates*. Longman Scientific and Technical, Harlow, England.

Eschrich, W. (1995). *Funktionelle Pflanzenanatomie*. Springer Verlag, Berlin–Heidelberg.

Eschrich, W. and Fromm, J. (1994). Evidence for two pathways of phloem loading. *Physiologia Plantarum*, **90**, 699–707.

Frommer, W. B. and Sonnewald, U. (1995). Molecular analysis of carbon partitioning in solanaceous species. *Journal of Experimental Botany*, **46**, 587–607.

Frommer, W. B., Kwart, M., Hirner, B., Fischer, W. N., Hummel, S., and Ninnemann, O. (1994). Transporters for nitrogenous compounds in plants. *Plant Molecular Biology*, **26**, 1651–70.

Lohaus, G., Winter, H., Riens, B., and Heldt, H. W. (1995). Further studies of the phloem loading process in leaves of barley and spinach. The comparison of metabolite concentrations in the apoplastic compartment with those in the cytosolic compartment and in the sieve tubes. *Botanica Acta*, **108**, 270–5.

Patrick, J. W. (1990). Sieve element unloading: cellular pathway, mechanism and control. *Physiologia Plantarum*, **78**, 298–308.

Stadler, R., Brandner, J., Schulz, A., Gahrtz, M., and Sauer, N. (1995). Phloem loading by the PmSUC2 sucrose carrier from *Plantago major* occurs in companion cells. *The Plant Cell*, **7**, 1545–54.

Stitt, M. (1994). Manipulation of carbohydrate partitioning. *Current Biology*, **5**, 137–43.

Turgeon, R. (1996). Phloem loading and plasmodesmata. *Trends in Plant Science*, **1**, 418–23.

van Bel, A. J. E. (1993). Strategies of phloem loading. *Annual Review of Plant Physiology and Plant Molecular Biology*, **44**, 253–81.

Zimmermann, M. H. and Milburn, J. A. (ed.) (1975). Phloem transport. In *Encyclopaedia of plant physiology*, Vol. 1, (ed. A. F. G. Dixon). Springer Verlag, Berlin–Heidelberg.

chapter 14

Plant storage proteins

Whereas the products of CO_2 assimilation are deposited in plants in the form of oligo- and polysaccharides, as dealt with in Chapter 9, the amino acids formed as products of nitrate assimilation are stored as proteins. These are mostly special *storage proteins* which have no enzymatic activity, and are often deposited in the cell within *protein bodies*. Protein bodies are surrounded by a single membrane and are derived from the endomembrane system of the endoplasmatic reticulum and the Golgi apparatus or the vacuoles. In potato tubers, storage proteins are also stored in the vacuole.

Storage proteins can be deposited in various plant organs, such as leaves, stems, and roots. They are stored in seeds and tubers and also in the cambium of tree trunks during winter, to enable rapid formation of leaves during seed germination and sprouting.

Storage proteins are located in the endosperm in cereal seeds and in the cotyledons in legume seeds. Whereas in cereals the protein content amounts to 10–15 per cent of the dry weight, in some legumes (e.g. soybean) it is as high as 40–50 per cent. About 85 per cent of these proteins are storage proteins.

Globally about 70 per cent of the human demand for protein is met by the consumption of seeds, either directly, or indirectly by feeding them to animals for meat production. Therefore plant storage proteins are the basis for human nutrition. However, in many plant storage proteins the content of certain amino acids essential for nutrition of man and animals is too low. In cereals, for example, the storage proteins are deficient in *threonine*, *tryptophan*, and particularly in *lysine*, whereas in legumes there is a deficiency of *methionine*. Since these amino acids cannot be synthesized by human metabolism, humans depend on being supplied with them in their food. In humans with an entirely vegetarian diet, such amino acid deficiencies can lead to irreparable physical and mental damage, especially in children. Such deficiencies can also be a serious problem in pig and poultry feed. A target of research in plant genetic engineering is to improve the amino acid composition of the storage proteins of harvest products.

Scientists have long been interested in plant proteins. As early as 1745 the Italian J. Beccari had isolated proteins from wheat. In 1924 at the Connecticut Agricultural Experimental Station T. B. Osborne classified plant proteins according to their solubility properties. He fractionated plant proteins into *albumins* (soluble in pure water), *globulins* (soluble in diluted salt solutions), *glutelins* (soluble in diluted solutions of alkali and acids), and *prolamins* (soluble in aqueous ethanol). When the structures of these proteins were determined later, it turned out that glutelins and prolamins were structurally closely related. Therefore in more recent literature glutelins are regarded as members of the group of prolamins. Table 14.1 shows some examples of various plant storage proteins.

14.1 Globulins are the most abundant storage proteins

Globulins occur in varying amounts in practically all plants. The most important globulins belong to the *legumin* and *vicilin* groups. Both these globulins are encoded by multigene families. Legumin is the main storage protein of leguminous seeds. In broad bean, for instance, 75 per cent of the total storage protein

Table 14.1 Some examples of plant storage proteins

Plant	Globulin	Prolamin (incl. glutelin)	2S-Protein
Rapeseed			Napin[a]
Pea, bean	Legumin, vicilin		
Wheat, rye		Gliadin, glutenin	
Maize		Zein	
Potato	Patatin		

[a] Structurally related to prolamins.

consists of legumin. Legumin is a hexamer with a molecular mass of 300–400 kDa. The monomers contain two different peptide chains (α, β), which are linked by a disulfide bridge. The large α-chain usually has a molecular mass of about 35–40 kDa, and the small β-chain about 20 kDa. Hexamers can be composed of different (α, β) monomers. Some contain methionine, others do not. In the hexamer the protein molecules are arranged in a very regular package and can be deposited in this form in the protein bodies. Protein molecules, in which some of the protein chains are not properly folded, do not fit into this package and are degraded by peptidases. Although it is relatively easy nowadays to exchange amino acids in a protein by genetic engineering, this has turned out to be difficult in storage proteins, as both the folding and the three-dimensional structure of the molecule may be altered by such exchanges. Successful improvement of the amino acid composition of storage proteins therefore will require detailed knowledge of the three-dimensional protein structure, which is not yet fully resolved.

Vicilin shows similarities in its amino acid sequence to legumin, but occurs mostly as a trimer, of which the monomers consist of one peptide chain only. Due to the lack of cysteine, the vicilin monomers are unable to form S–S bridges. In contrast to legumins, vicilins are often glycosylated; they contain carbohydrate residues such as mannose, glucose, and *N*-acetylglucosamine.

14.2 Prolamins are formed as storage proteins in grasses

Prolamins are only contained in grasses, such as cereals. They are present as a polymorphic mixture of many different subunits of 30–90 kDa each. Some of these subunits contain cysteine residues and are linked by S–S bridges. Also in *glutenin*, which occurs in the grains of wheat and rye, monomers are linked by S–S bridges. The glutenin molecules differ in size. The suitability of flour for bread-making depends on the content of high molecular weight glutenins, and therefore flour from barley, oat, or maize, lacking glutenin, is not suitable for baking bread. Since the glutenin content is a critical factor in determining the quality of bread grain, investigations are in progress to improve the glutenin content of bread grain by genetic engineering.

14.3 2S-Proteins are present in seeds of dicotyledonous plants

2S-Proteins are also widely distributed storage proteins. They represent a heterogeneous group of proteins, of which the sole definition is their sedimentation coefficient of about 2 svedbergs (S). Investigations of the structure have revealed that most 2S-proteins are structurally related to each other and are possibly derived, along with the prolamins, from a common ancestor protein. *Napin*, the predominant storage protein in rapeseed, is an example of a 2S-protein. This protein is of substantial economic importance since, after the oil has been extracted, the remainder of the rapeseed is used as fodder. Napin and other related 2S-proteins consist of two relatively small polypeptide chains of 9 kDa and 12 kDa, which are linked by S–S bridges. So far little is known about the packing of the prolamins and 2S-proteins in the protein bodies.

Recently, using gene transfer, methionine-rich 2S-proteins from brazil nuts have been successfully expressed in seeds of soybeans and other legumes, to compensate for a methionine deficiency when legumes are used as fodder.

14.4 Special proteins protect seeds from being eaten by animals

The protein bodies of some seeds contain other proteins, which, although acting as storage proteins, also protect the seeds from being eaten. For instance, the seeds of some legumes contain *lectins*, which bind to sugar residues, irrespective of whether these are free sugars or are constituents of glycolipids or glycoproteins. When these seeds are consumed by animals, the lectins bind to glycoproteins in the intestine and thus interfere with the absorption of food. The seeds of some legumes and other plants also contain *proteinase inhibitors*, which block the digestion of proteins by inhibiting proteinases in the animal digestive tract. Because of their content of lectins and proteinase inhibitors, many beans can only be eaten after denaturing by cooking. Castor beans contain the toxic protein ricin. Beans also contain *amylase inhibitors* which specifically inhibit the hydrolysis of starch by amylases in the digestive tract of certain insects. Recently, using genetic engineering, α-amylase inhibitors from beans have been successfully expressed in the seeds of pea. Whereas the larvae of the pea beetle normally cause large losses during storage of peas, the seeds of the genetically engineered plants were protected against these losses.

14.5 Synthesis of the storage proteins occurs on the rough endoplasmic reticulum

Storage proteins are formed by ribosomes on the rough endoplasmic reticulum (ER) (Fig. 14.1). The newly synthesized proteins occur in the lumen of the ER, and the storage proteins are finally deposited in *protein bodies*. In the case of 2S-proteins and prolamins, the protein bodies are formed by budding from the ER membrane. The globulins are moved from the ER by vesicle transfer via the Golgi apparatus (section 1.6) first to the vacuole, from which protein bodies are formed by fragmentation.

Figure 14.2 shows the formation of legumin in detail. The protein formed by

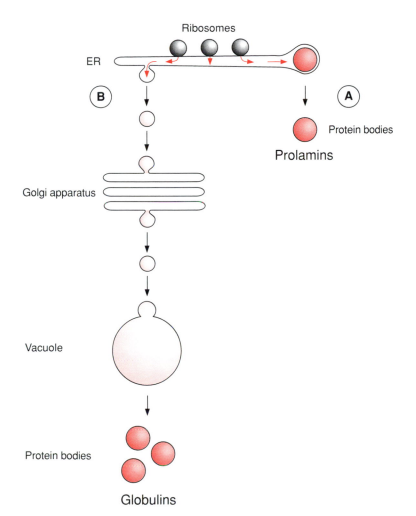

Figure 14.1 There are two ways of depositing storage proteins in protein bodies: (A) In the formation of prolamins in cereal grains the prolamin aggregates in the lumen of the ER and the protein bodies are formed by budding off from the ER membrane. (B) The proteins appearing in the lumen of the ER are transferred via the Golgi apparatus to the vacuole. The protein bodies are formed by fragmentation of the vacuole. This is probably the most common pathway.

the ribosome contains a hydrophobic section called a *signal sequence* at the N-terminus of the polypeptide chain. After the synthesis of this signal sequence translation comes to a halt, and the signal sequence forms a complex with three other components: a *signal recognition particle*, a *binding protein* located on the ER membrane, and a *pore protein* present in the ER membrane. The formation of this complex results in the opening of a pore in the ER membrane, translation continues and the newly formed protein chain (e.g. pre-pro-legumin) appears in the lumen of the ER and anchors the ribosome on the ER membrane for the

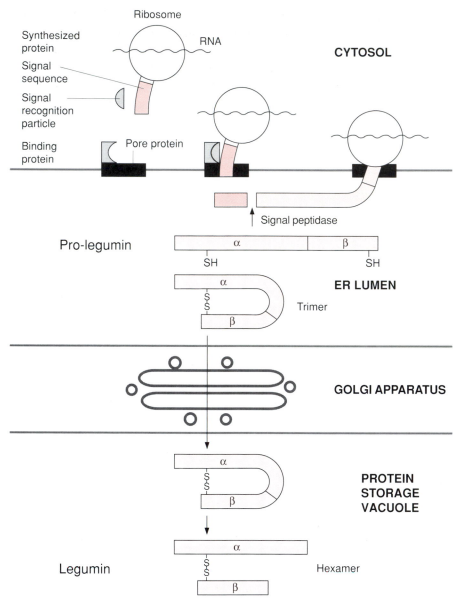

Figure 14.2 Legumin synthesis. The pre-form of the legumin (pre-pro-legumin) formed by the ribosome is processed first in the lumen of the ER and then further in the vacuole to give the end-product.

duration of protein synthesis. Immediately after the peptide chain enters the lumen, the signal sequence is removed by a *signal peptidase* located on the inside of the ER membrane. The remaining polypeptide, termed *pro-legumin*, contains the α- and β-chains of the legumin. An S–S linkage within the pro-legumin is formed in the ER lumen. Three pro-legumin molecules form a *trimer*. This is facilitated by specific proteins, probably chaperones (section 21.2). During this association a quality control process occurs: trimers without the correct conformation are degraded in the ER by proteinases. The trimers are then transferred by the Golgi apparatus to the vacuoles where the α- and β-chains are separated by a peptidase. The subunits of the legumins assemble to *hexamers* and are deposited in this form. The protein bodies, the final storage site of the legumins, are derived from fragmentation of the vacuole.

The pre-pro-forms of the newly synthesized 2S-proteins and prolamins, which occur in the lumen of the ER, also contain a signal sequence. The processing and aggregation of these proteins takes place in the lumen of the ER from which the protein bodies are formed by budding.

14.6 Proteinases mobilize the amino acids deposited in storage proteins

Our knowledge about the mobilization of the amino acids from the storage proteins derives primarily from investigations of processes during seed germination. In most cases germination is induced by the uptake of water, causing the protein bodies to fuse to form a vacuole. The hydrolysis of the storage proteins is catalysed by proteinases, which are in part deposited as inactive pro-forms together with the storage proteins in the protein bodies. Other proteinases are synthesized anew and transferred via the lumen of the ER and the Golgi apparatus to the vacuoles (Fig. 14.2). These enzymes are synthesized initially as inactive pro-forms. Activation of these pro-proteinases proceeds by limited proteolysis in which a section of the sequence is removed by a specific peptidase. The remainder of the polypeptide represents the active proteinase.

The degradation of the storage proteins is also initiated by limited proteolysis. A specific proteinase first removes small sections of the protein sequence, resulting in a change in the conformation of the storage protein. In cereal grains S–S bridges of storage proteins are cleaved by reduced thioredoxin (section 6.6). The unfolded protein is then susceptible to hydrolysis by various proteinases, e.g. exopeptidases, which split off amino acids one after the other from the end of the protein molecule, and endopeptidases, which cleave within the molecule. In this way storage proteins are completely degraded in the vacuole and the liberated amino acids are provided as building material to the germinating plant.

Further reading

Chrispeels, M. J. (1994). Sorting proteins in the secretory system. *Annual Review of Plant Physiology and Plant Molecular Biology*, **42**, 21–53.

Hoh, B., Hinz, G., Jeong, B.-K., and Robinson, D. (1995). Protein storage vacuoles form *de novo* during pea cotyledon development. *Journal of Cell Science*, **108**, 299–310.

Müntz, K., Jung, R., and Saalbach, G. (1993). Synthesis, processing and targeting of legume seed proteins. In *Seed storage compounds*, (ed. P. R. Shewry and K. Stobart), pp. 129–46. Clarendon Press, Oxford.

Okita, T. W. and Rogers, J. C. (1996). Compartmentation of proteins in the endomembrane system of plant cells. *Annual Review of Plant Physiology and Plant Molecular Biology*, **46**, 327–50.

Peumans, W. J. and van Damme, E. J. M. (1995). Lectins as plant defence proteins. *Plant Physiology*, **109**, 247–352.

Schroeder, H. E., Gollash, S., Moore, A., Tabe, L. M., Craig, S., Hardie, D. C., *et al.* (1995). Bean amylase inhibitor confers resistance to the pea weevil (*Bruchus pisorum*) in transgenic peas (*Pisum sativum L.*) *Plant Physiology*, **107**, 1233–9.

Shewry, P. R., Napier, J. A., and Tatham, A. S. (1995). Seed storage proteins: structures and biosynthesis. *The Plant Cell*, **7**, 945–56.

Staswick, P. E. (1994). Storage proteins of vegetative plant tissues. *Annual Review of Plant Physiology and Plant Molecular Biology*, **45**, 303–22.

chapter 15

Glycerolipids

Glycerolipids are fatty acid esters of glycerol (Fig. 15.1). Triacylglycerols (also named triglycerides) consist of a glycerol molecule that is esterified with three fatty acids. In contrast to animals, in plants triacylglycerols do not serve as an energy store but mainly as a carbon store in seeds. In polar glycerolipids the glycerol is esterified with only two fatty acids and a hydrophilic group is linked to the third -OH group. These polar lipids are the main constituent of membranes.

15.1 Polar glycerolipids are membrane constituents

The polar glycerolipids are *amphiphilic molecules*, consisting of a hydrophilic head and a hydrophobic tail. This quality enables them to form *lipid bilayers* in

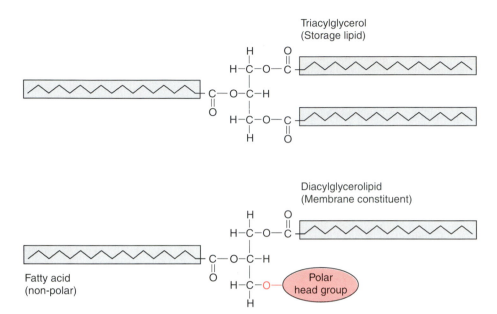

Triacylglycerol
(Storage lipid)

Diacylglycerolipid
(Membrane constituent)

Fatty acid
(non-polar)

Polar
head group

Figure 15.1 Triacylglycerols containing three fatty acids are of a non-polar nature. In contrast, polar lipids are amphiphilic substances since, besides the hydrophobic tail consisting of two fatty acids, they contain a hydrophilic head.

which the hydrocarbon tails are held together by hydrophobic interactions and the hydrophilic heads protrude into the aqueous phase, thus forming the basic structure of a membrane (Fig. 15.2). Since the middle C atom of the glycerol in a polar glycerolipid is asymmetrical, a distinction can be made between the two esterified groups of glycerol at the C-1 and C-3 positions.

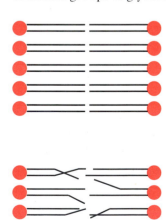

Figure 15.2 Membrane lipids with saturated fatty acids form a very regular lipid bilayer. The kinks caused in the hydrocarbon chain by *cis*-carbon–carbon double bonds in unsaturated fatty acids result in disturbances in the lipid bilayer and lead to an increase in its fluidity.

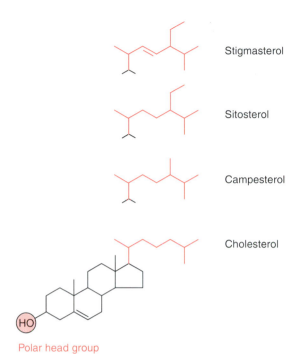

Stigmasterol

Sitosterol

Campesterol

Cholesterol

Polar head group

Figure 15.3 Cholesterol and related steroids (only side chains shown) are membrane constituents.

The steroids shown in Fig. 15.3 are also amphiphilic; the **-OH** groups form the hydrophilic head and the steran skeleton with the side chain serves as the hydrophobic tail. Besides the steroids shown here, plants contain a large variety of similar steroids as membrane constituents. They are contained primarily in the outer membrane of mitochondria, the ER membranes, and in the plasma membrane, and to a large extent determine the properties of these membranes (see below).

The polar glycerolipids mainly contain fatty acids with 16 or 18 carbon atoms (Fig. 15.4). The majority of these fatty acids are unsaturated and contain 1–3 carbon–carbon double bonds. These double bonds are present almost exclusively in the *cis*-configuration and only rarely in the *trans*-configuration. The double bonds are normally not conjugated. Figure 15.4 shows the number code for the structure of fatty acids in the following format: number of C atoms : number of double bonds, Δ position of the first C atom of the double bond, c = *cis*-configuration (t = *trans*).

Some plants also contain unusual fatty acids in their storage lipids, e.g. with conjugated double bonds, carbon–carbon triple bonds, or hydroxy-, keto-, or epoxy-groups. There are also glycerolipids in which the carbon chain is connected with the glycerol via an ether linkage. These substances are very rare and will not be dealt with here.

Fatty acids

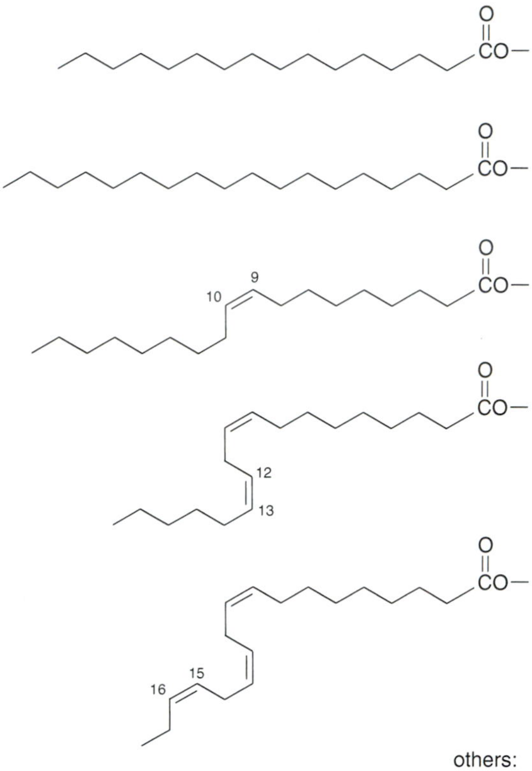

Palmitic acid
(16 : 0)

Stearic acid
(18 : 0)

Oleic acid
(18 : 1, Δ^{9c})

Linoleic acid
(18 : 2, $\Delta^{9,12c}$)

α-Linolenic acid
(18 : 3, $\Delta^{9,12,15c}$)

others:

γ-Linolenic acid
(18 : 3, $\Delta^{6,9,12c}$)

Arachidonic acid
(20 : 4, $\Delta^{5,8,11,14c}$)

Figure 15.4 Fatty acids as hydrophobic constituents of membrane lipids.

The fluidity of the membrane is governed by the proportion of unsaturated fatty acids and the content of steroids

The hydrocarbon chains in saturated fatty acids are packed in a regular layer (Fig. 15.2), whereas in unsaturated fatty acids the packing is disturbed due to kinks in the hydrocarbon chain caused by the *cis*-carbon–carbon double bonds, resulting in a more fluid layer. This is obvious when comparing the melting points of various fatty acids (Table 15.1). The melting point increases with increasing chain length as the packing becomes tighter, whereas with an increasing number of double bonds the melting point decreases. This also applies to the corresponding fats. Catalytically hydrogenated fat made from palm oil, for instance, consisting of saturated fatty acids only, is solid, whereas plant oils with a very high natural content of unsaturated fatty acids are liquid.

Table 15.1 Influence of chain length and the number of double bonds on the melting point of fatty acids

Fatty acid	Chain length : double bonds	Melting point
Lauric acid	12 : 0	40°C
Stearic acid	18 : 0	70°C
Oleic acid	18 : 1	13°C
Linoleic acid	18 : 2	−5°C
Linolenic acid	18 : 3	−11°C

Likewise, the fluidity of membranes is governed by the proportion of unsaturated fatty acids in the membrane lipids. This is why in some plants, during growth at a low temperature, more highly unsaturated fatty acids are incorporated into the membrane to compensate for the decrease in membrane fluidity caused by the low temperature. Recently the cold tolerance of tobacco has been enhanced by increasing the proportion of unsaturated fatty acids in the membrane lipids by genetic engineering. Steroids (Fig. 15.3) decrease the fluidity of membranes and probably also play a role in the adaptation of membranes to temperature.

Membrane lipids contain a variety of hydrophilic head groups

The head groups of the polar glycerolipids in plants are formed by a variety of compounds. However, these all fulfil the same function of providing the lipid molecule with a polar group (Fig. 15.5). In the *phospholipids* the head group consists of a phosphate residue that is esterified with a second alcoholic compound such as ethanolamine, choline, serine, glycerol, or inositol. Phosphatidic acid is not a significant membrane constituent.

The phospholipids mentioned already are found as membrane constituents in bacteria as well as in animals and plants. A feature of plants and cyanobacteria is that they contain, in addition to phospholipids, the *galactolipids* monogalactosyldiacylglycerol (MGDG), digalactosyldiacylglycerol (DGDG), and the sulfolipid sulfoquinovosyldiacylglycerol (SL) as membrane constituents. The latter contains a glucose moiety to which a sulfonic acid residue is linked at the C-6 position as the polar head group (Fig. 15.5B).

There are great differences between the lipid composition of the various membranes in a plant (Table 15.2). Chloroplasts contain galactolipids, which are the main constituents of the thylakoid membrane and also occur in the envelope membranes. The membranes of the mitochondria and the plasma contain no galactolipids, but have phospholipids as their main membrane constituents. Cardiolipin is a specific component of the inner mitochondrial membrane in animals and plants.

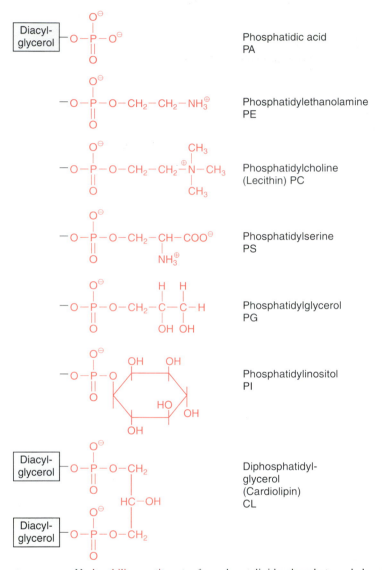

Figure 15.5A Hydrophilic constituents of membrane lipids: phosphate and phosphate esters.

In a green plant cell about 70–80 per cent of the total membrane lipids are constituents of the thylakoid membranes. Plants represent the largest part of the biosphere and this is why the galactolipids MGDG and DGDG are the most abundant membrane lipids on earth. Since plant growth is limited in many habitats by the phosphate content in the soil, it was probably advantageous for plants during evolution to be independent of the phosphate supply in the soil for the synthesis of the predominant part of their membrane lipids.

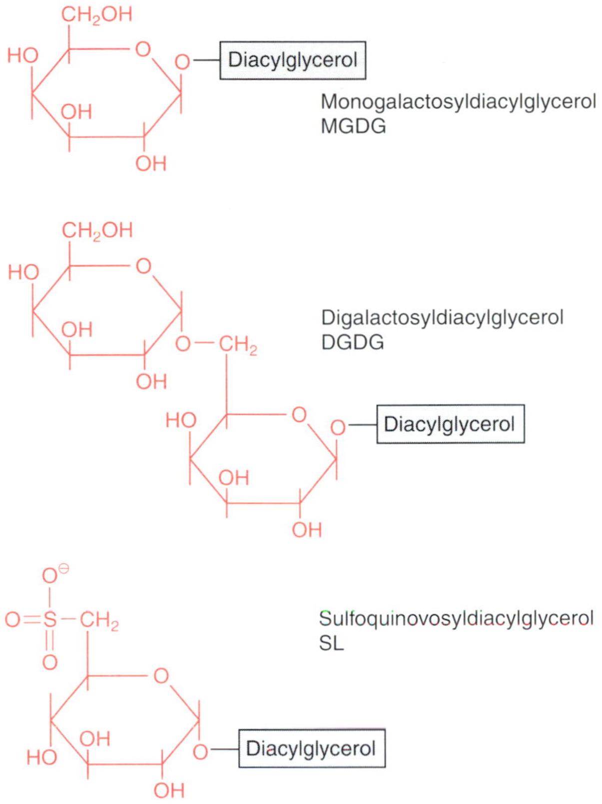

Figure 15.5B Hydrophilic constituents of membrane lipids: hexoses.

15.2 Triacylglycerols are storage substances

Triacylglycerols are contained primarily in seeds but also in some fruits, such as olives or avocados. The purpose of triacylglycerols in fruit is to attract animals to consume these fruits for distributing the seeds. The triacylglycerols in seeds function as a carbon store to supply the carbon required for biosynthetic processes during seed germination. Triacylglycerols have an advantage over carbohydrates as storage compounds, because their weight/carbon content ratio is much lower. A calculation illustrates this: in starch the glucose residue, containing six C atoms, has a molecular weight of 162 Da. The weight of one stored carbon atom thus amounts to 27 Da. In reality this value is higher since starch contains water molecules, as it is hydrated. A triacylglycerol with three palmitate residues contains 51 C atoms and has a molecular weight of 807 Da. The weight of one stored carbon atom thus amounts to only 16 Da. Since triacylglycerols, in contrast to starch, are not hydrated, carbon stored in the seed as fat requires less than half the weight as when stored as starch. Low weight is advantageous for seed dispersal.

Table 15.2 The composition of glycerolipids in various organelle membranes (after Harwood 1980)

Glycerolipids[a]	% of total acyl lipid content		
	Chloroplast thylakoid membrane	Mitochondrial inner membrane	Plasma membrane
MGDG	51	0	0
DGDG	26	0	0
SL	7	0	0
PC	3	27	32
PS	0	25	0
PE	0	29	46
PG	9	0	0
PI	1	0	19
CL	0	20	0

[a] For explanation of abbreviations see Fig. 15.5.

Triacylglycerols are deposited in *oil bodies* (Fig. 15.6). These consist of oil droplets, which are surrounded by a lipid monolayer. Proteins named *oleosins* are anchored to the lipid monolayer of the oil bodies and catalyse the mobilization of fatty acids from the triacylglycerol store during germination (section 15.6). These oleosins are only present in oil bodies contained in the endosperm and embryonic tissue of seeds. The oil bodies in the pericarp of olives or avocados, where the triacylglycerols are used not as a store, but only to lure animals, contain no oleosins and, with a diameter of 10–20 µm, are very much larger than the oil bodies of storage tissues (diameter 0.5–2 µm). As dealt with in section 15.5, synthesis of triacylglycerols occurs in the endoplasmic reticulum membrane. It is assumed that the newly synthesized triacylglycerol accumulates between the two-layered membrane of the ER, until the full size of the oil body is reached (Fig. 15.6). When the oil body buds from the ER membrane, it is surrounded by a single phospholipid layer. The oleosins are probably attached afterwards.

15.3 The *de novo* synthesis of fatty acids takes place in the plastids

The carbon fixed by CO_2 assimilation in the chloroplasts is the precursor not only for the synthesis of carbohydrates and amino acids, but also for the synthesis of fatty acids. Whereas the production of carbohydrates and amino acids by mesophyll cells is primarily destined for export to other parts of the plant, the synthesis of fatty acids, except in seed and fruit, is only for the cell's own requirements.

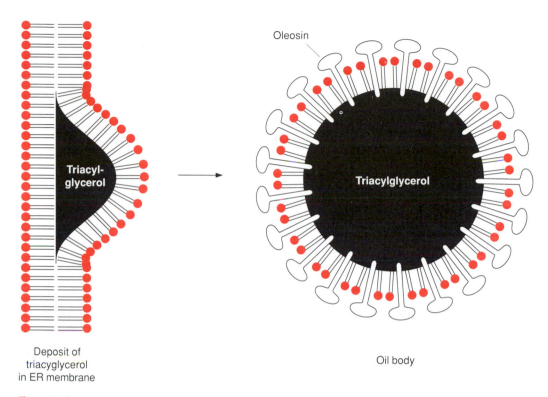

Deposit of
triacyglycerol
in ER membrane

Oil body

Figure 15.6 An oil body surrounded by a lipid monolayer is formed by the incorporation of triacylglycerols into an ER membrane. Oleosins are anchored to the lipid monolayer.

Plants are not capable of long-distance fatty acid transport. Since fatty acids are
present as constituents of membrane lipids in every cell, each cell must contain
the enzymes for the synthesis of membrane lipids and thus also for the synthesis
of fatty acids.

In plants the *de novo* synthesis of fatty acids always occurs in the plastids: in the
chloroplasts of green cells and the leucoplasts and chromoplasts of non-green
cells. Although in plant cells enzymes of fatty acid synthesis are also found in the
membrane of the endoplasmic reticulum, these enzymes, named *elongases*, only
catalyse a chain elongation of fatty acids which have been synthesized earlier in
the plastids.

Acetyl CoA is precursor for the synthesis of fatty acids

Acetyl CoA is the precursor for fatty acid synthesis. Like mitochondria (see
Fig. 5.4), plastids contain a *pyruvate dehydrogenase* by which pyruvate is oxidized
to acetyl CoA, accompanied by the reduction of NAD (Fig. 15.7).

  In chloroplasts photosynthesis provides the NADPH required for the synthesis
of fatty acids. But it is still unclear how acetyl CoA is formed from the products of

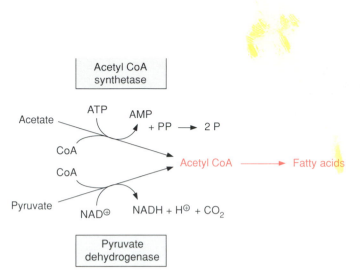

Figure 15.7 Acetyl CoA can be formed in two ways.

CO_2 fixation. In chloroplasts, depending on the developmental state of the cells, the activity of pyruvate dehydrogenase is often low. On the other hand, chloroplasts contain a high activity of *acetyl CoA synthetase* by which acetate, with consumption of ATP, can be converted to acetyl CoA. When radioactively labelled acetate is supplied to chloroplasts, the radioactivity is very rapidly incorporated into fatty acids. In many plants *acetate* is probably a major precursor for the formation of acetyl CoA in the chloroplasts and leucoplasts. The origin of acetate is still unclear. It could be formed in the mitochondria by hydrolysis of acetyl CoA, derived from the oxidation of pyruvate by mitochondrial pyruvate dehydrogenase.

In leucoplasts the reductive equivalents required for fatty acid synthesis are provided by the oxidation of glucose 6-phosphate. The latter is transported via a *glucose 6-phosphate–phosphate translocator* to the plastids (see Fig. 13.5). The oxidation of glucose 6-phosphate by the *oxidative pentose phosphate pathway* (Fig. 6.21) provides the NADPH required for fatty acid synthesis.

Fatty acid synthesis starts with the carboxylation of acetyl CoA by acetyl CoA carboxylase, with the consumption of ATP, to malonyl CoA (Fig. 15.8). In a subsequent reaction, CoA is exchanged for *acyl carrier protein* (ACP). ACP (Fig. 15.9) contains a serine residue to which a pantetheine residue is linked via a phosphate group. Since the pantetheine residue is also a functional constituent of CoA, ACP can be regarded as a CoA which is covalently bound to a protein. The enzyme *β-ketoacyl-ACP synthase III* catalyses the condensation of acetyl CoA with a malonyl ACP. The liberation of CO_2 makes this reaction irreversible. The acetoacetate thus formed remains bound as a thioester to ACP and is reduced by NADPH to β-D-hydroxyacyl-ACP. Following the release of water, the carbon–carbon double bond formed is reduced by NADPH to acyl-ACP. The product is a fatty acid which has been elongated by two carbon atoms.

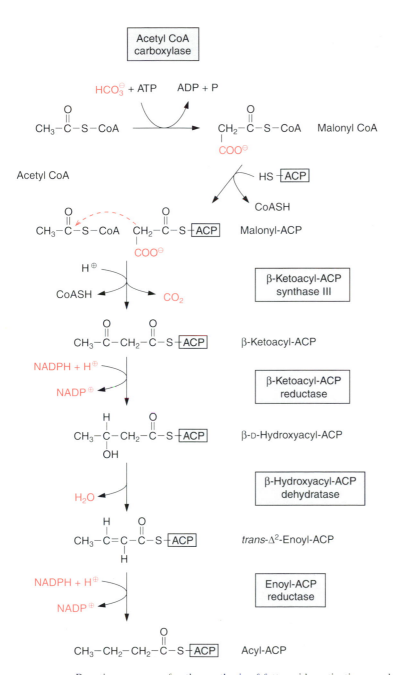

Figure 15.8 Reaction sequence for the synthesis of fatty acids: activation, condensation, reduction, release of water, and further reduction elongate a fatty acid by two carbon atoms.

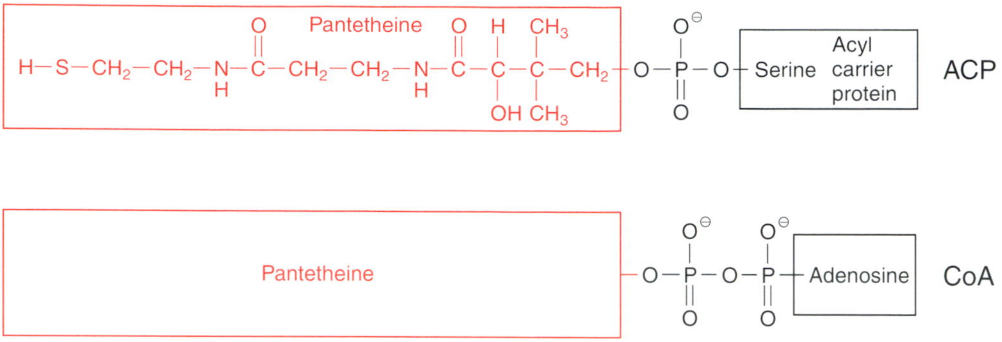

Figure 15.9 The acyl carrier protein (ACP) contains pantetheine, the same functional group as co-enzyme A.

Acetyl CoA carboxylase is the first enzyme of fatty acid synthesis

In the carboxylation of acetyl CoA *biotin* acts as a carrier for 'activated CO_2' (Fig. 15.10). Biotin is covalently linked by its carboxyl group to the ϵ-amino group of a lysine residue of the *biotin carboxyl carrier protein* and it contains an -NH- group which forms a carbamate with HCO_3^- (Fig. 15.11). This reaction is driven by the hydrolysis of ATP. Acetyl CoA carboxylation requires two steps:

(1) biotin is carboxylated at the expense of ATP by *biotin carboxylase*;

(2) bicarbonate is transferred to acetyl CoA by *carboxyl transferase*.

All three enzymes, biotin carboxyl carrier protein, biotin carboxylase, and carboxyl transferase form a single multienzyme complex. Since the biotin is attached to the carrier protein by a long, flexible hydrocarbon chain, it reacts alternately with the carboxylase and carboxyl transferase in the multienzyme complex (Fig. 15.11).

The acetyl CoA carboxylase multienzyme complex in the stroma of plastids

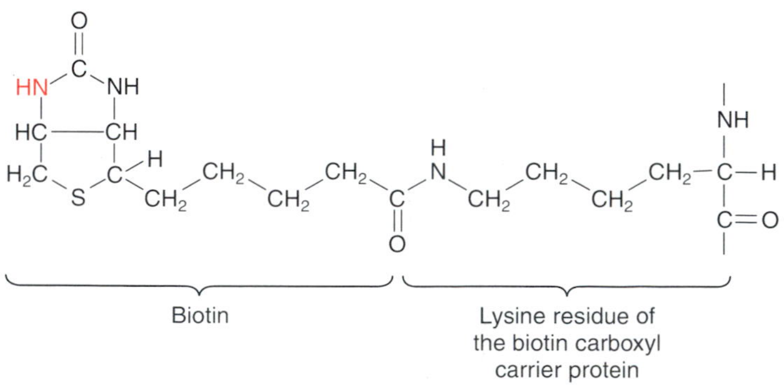

Figure 15.10 Biotin is linked via a lysine residue to the biotin carboxyl carrier protein.

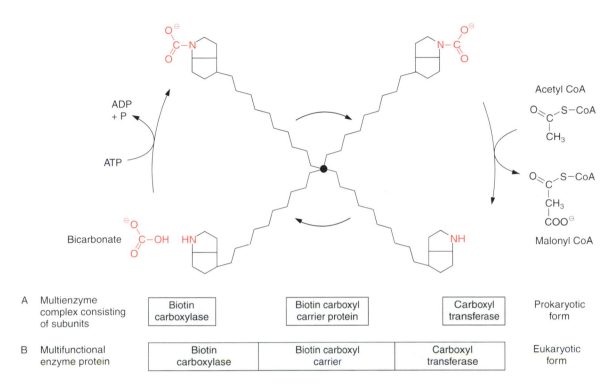

Figure 15.11 Acetyl CoA carboxylase: reaction scheme. The biotin linked to the biotin carboxyl carrier protein reacts in turn with biotin carboxylase and carboxyl transferase. The circular motion has been chosen for the sake of clarity; in reality there is probably a pendulum-like movement. The eukaryotic form of acetyl CoA carboxylase is present as a multifunctional protein.

consists of several subunits, resembling the acetyl CoA carboxylase in cyano-bacteria and other bacteria, and is referred to as the *prokaryotic form* of the acetyl CoA carboxylase. Acetyl CoA carboxylase is also present outside the plastids, probably in the cytosol. The malonyl CoA formed outside the plastids is used for chain elongation of fatty acids and is a precursor for the formation of flavonoids (see section 18.5). The extraplastidal acetyl CoA carboxylase, in contrast to the prokaryotic type, consists of a single large *multifunctional protein* in which the biotin carboxyl carrier, the biotin carboxylase, and the carboxyl transferase are located on different sections of the same polypeptide chain. Since this multi-functional protein also occurs in a very similar form in the cytosol of yeast and animals, it is referred to as the *eukaryotic form*. It should be emphasized, however, that both the eukaryotic and the prokaryotic forms of acetyl CoA carboxylase are encoded in the nucleus. Possibly only one protein of the prokaryotic enzyme is encoded in the plastid genome.

In Gramineae (grasses), including the various cereal species, the prokaryotic form is not present. In these plants the multifunctional eukaryotic acetyl CoA

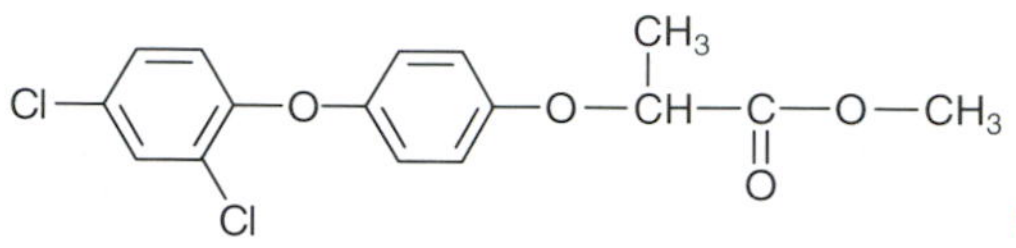

Diclofop methyl

Figure 15.12 Diclofop methyl, a herbicide (Hoe-Grass, AgrEvo), inhibits the eukaryotic multifunctional acetyl CoA carboxylase.

carboxylase is located in the cytosol as well as in the chloroplasts. The eukaryotic acetyl CoA carboxylase is inhibited by various arylphenoxypropionic acid derivatives, for example diclofop methyl (Fig. 15.12). Since eukaryotic acetyl CoA carboxylase in Gramineae is involved in the *de novo* fatty acid synthesis of the plastids, this inhibitor severely impairs lipid biosynthesis in this group of plants. Diclofop methyl, trade name Hoe-Grass (AgrEvo), and similar substances are therefore used as selective herbicides (section 3.6) to control grass weeds.

Acetyl CoA carboxylase, the first enzyme of fatty acid synthesis, is an important regulatory enzyme and its reaction is regarded as a rate-limiting step in fatty acid synthesis. In chloroplasts the enzyme is fully active only during illumination and is inhibited during darkness. In this way fatty acid synthesis proceeds mainly during the day when photosynthesis provides the necessary NADPH. The mechanism of light regulation of acetyl CoA carboxylase has not yet been resolved. Light-dependent changes in the Mg^{2+} concentration and of the pH in the stroma (see section 6.6) may be involved in the regulation.

Further steps of fatty acid synthesis are also catalysed by a multienzyme complex

β-Ketoacyl-ACP formed by the condensation of acetyl CoA and malonyl-ACP (Fig. 15.8) is reduced by NADPH to β-D-hydroxyacyl-ACP, and after the release of water the carbon–carbon double bond of the resulting enoyl-ACP is reduced again by NADPH to acyl-ACP. This reaction sequence resembles the reversal of the formation of oxaloacetate from succinate in the citrate cycle (Fig. 5.3). Fatty acid synthesis is catalysed by a multienzyme complex. Figure 15.13 shows a scheme of the interplay of the various reactions. The acyl residue bound as a thioester to the ACP located in the centre of the complex, is attached to a flexible chain and so is transferred from enzyme to enzyme during the reaction cycle.

The fatty acid is elongated by first transferring it to another ACP and then condensing it with malonyl-ACP. The enzyme *β-ketoacyl-ACP synthase I*, catalysing this reaction, enables the formation of fatty acids with a chain length of up to C-16. A further chain elongation to C-18 is catalysed by *β-ketoacyl-ACP synthase II* (Fig. 15.14).

It may be mentioned that in animals and fungi the enzymes of fatty acid synthesis shown in Fig. 15.13 are contained in only one or two multifunctional proteins which form a complex (eukaryotic fatty acids synthase complex). Since the

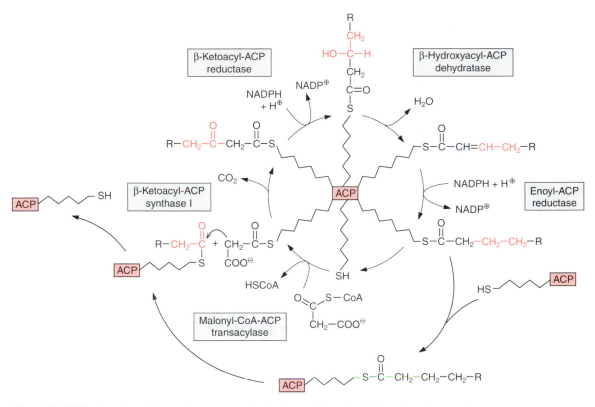

Figure 15.13 The interplay of the various enzymes during fatty acid synthesis. The acyl carrier protein (ACP), located in the centre, carries the fatty acid residue, bound as thioester, from enzyme to enzyme. The circular movement has been chosen for the sake of clarity but does not represent reality.

fatty acid synthase complex of the plastids, consisting of several proteins, is similar to those of many bacteria, it is called the *prokaryotic fatty acid synthase complex.*

The first double bond in a newly formed fatty acid is formed by a soluble desaturase

The stearoyl-ACP (18 : 0) formed in the plastid stroma is desaturated there to oleyl-ACP (18 : 1) (Fig. 15.15). This reaction can be regarded as a *mono-oxygenation*, in which one O atom from an O_2 molecule is reduced to water and the other is incorporated into the hydrocarbon chain of the fatty acid as hydroxyl group. A carbon–carbon double bond is formed by subsequent dehydration (analogous to the β-hydroxyacyl-ACP dehydratase reaction, Fig. 15.8) which, in contrast to fatty acid synthesis, has a *cis*-configuration. The mono-oxygenation requires two electrons which are provided by NADPH via reduced ferredoxin. Mono-oxygenases are widespread in bacteria, plants, and animals. In most cases O_2 is activated by a

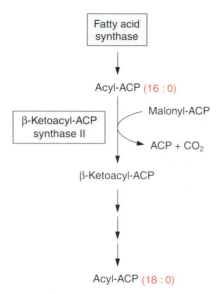

Acyl-ACP (16 : 0)

Malonyl-ACP

ACP + CO_2

β-Ketoacyl-ACP

Acyl-ACP (18 : 0)

Figure 15.14 Elongation and desaturation of fatty acids in plastids.

special cytochrome, cytochrome P_{450}. However, in the *stearoyl-ACP desaturase*, the O_2 molecule reacts with a *di-iron–oxo cluster* (Fig. 15.16). In previous sections we have dealt with iron–sulfur clusters as redox carriers, in which the Fe atoms are bound to the protein via cysteine and histidine residues (Fig. 3.26). In the di-iron–oxo cluster of the desaturase, two iron atoms are bound to the enzyme via the carboxyl groups of glutamate and aspartate. The two Fe atoms alternate between oxidation states +III and +II. An O_2 molecule is activated by the binding of the two Fe atoms.

Stearoyl-ACP desaturase in the plastids is so active that normally the newly

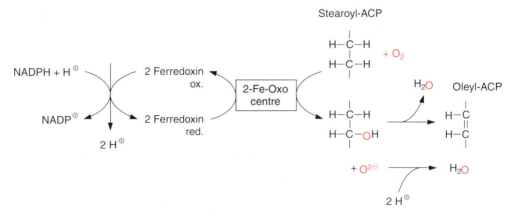

Figure 15.15 Stearoyl-ACP desaturase catalyses the desaturation of stearoyl-ACP to oleyl-ACP. The reaction can be regarded as a mono-oxygenation with the subsequent release of water.

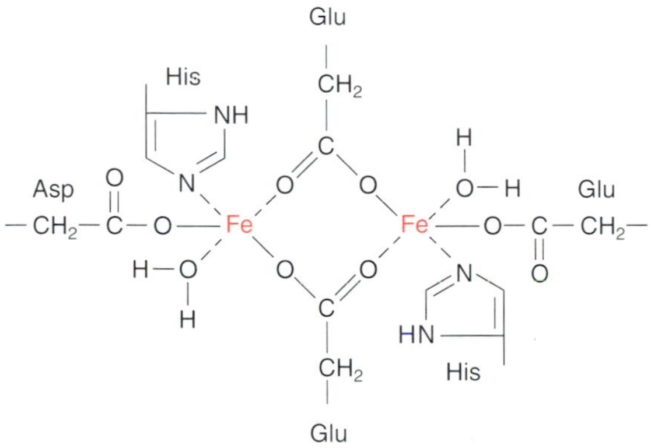

Figure 15.16 In a di-iron–oxo cluster two Fe atoms are bound to glutamate, aspartate, and histidine side chains of the protein. (After Karlin 1993.)

formed stearoyl-ACP is almost completely converted into oleyl-ACP (18 : 1) (Fig. 15.17). This soluble desaturase is only capable of introducing one double bond into fatty acids. The introduction of further double bonds is catalysed by other desaturases, which are integral membrane proteins and react only with fatty acids that are constituents of membrane lipids. These membrane-bound desaturases also require O_2 and reduced ferredoxin, but it is still not known how they function.

Acyl-ACP formed as product of fatty acid synthesis in the plastids serves two purposes

Acyl-ACP formed in the plastids has two important functions:

1. It acts as acyl-donor for the synthesis of plastid membrane lipids. The enzymes of glycerolipid synthesis are located in both the inner and outer envelope membranes, and in lipid biosynthesis there is a division of labour between these two membranes. To avoid going into too much detail, no distinction will be made between the lipid biosynthesis of the inner and outer envelope membranes in the following text.

2. For biosynthesis outside the plastids, acyl-ACP is hydrolysed by acyl-ACP thioesterases to free fatty acids, which then leave the plastids (Fig. 15.17). It is not known whether this export proceeds via non-specific diffusion or by specific transport. These free fatty acids are immediately captured outside the outer envelope membrane by conversion into acyl CoA, as catalysed by an acyl CoA synthetase with consumption of ATP. Since the thioesterases in the plastids hydrolyse primarily 16 : 0- and 18 : 1-acyl-ACP, and to a small extent 18 : 0-acyl-ACP, the plastids mainly provide CoA-esters with the acyl residues of 18 : 1 and 16 : 0 (also a low amount of 18 : 0) for lipid metabolism outside of the plastids.

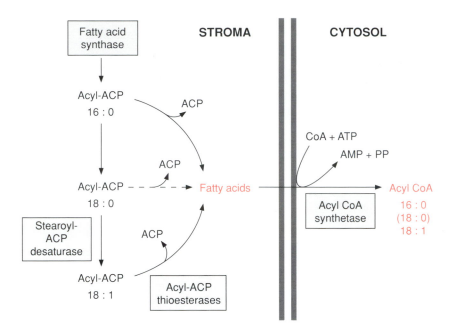

Figure 15.17 Acyl-ACP thioesterases release mainly 16 : 0 and 18 : 1 fatty acids, and only low amounts of 18 : 0 fatty acids. After the fatty acids leave the stroma and enter the cytosol they are immediately converted to acyl CoA.

15.4 Glycerol 3-phosphate is a precursor for the synthesis of glycerolipids

Glycerol 3-phosphate is a precursor for the synthesis of glycerolipids; it is formed by reduction of dihydroxyacetone phosphate with NADH as reductant (Fig. 15.18). Dihydroxyacetone phosphate reductases are present in the plastid stroma as well as in the cytosol. In plastid lipid biosynthesis the acyl residues are transferred directly from acyl-ACP to glycerol 3-phosphate. For the first acylation step mostly an 18 : 1-, less frequently a 16 : 0-, and more rarely an 18 : 0-acyl residue is esterified to position 1. The C-2 position, however, is always esterified with a 16 : 0-acyl residue. Since this specificity is also observed in cyanobacteria, the glycerolipid biosynthesis pathway of the plastids is called a *prokaryotic pathway*.

For glycerolipid synthesis in the ER membrane the acyl residues are transferred from acyl CoA. Here again the hydroxyl group in C-1 position is esterified with an 18 : 1-, 16 : 0- or 18 : 0-acyl residue, but in position C-2 always with a somewhat desaturated 18 : *n*-acyl residue. The glycerolipid pathway of the ER membrane is called a *eukaryotic pathway*.

Linkage of the polar head group to diacylglycerol proceeds mostly via an activation of the head group, but in some cases also by an activation of diacylglycerol.

Plastidal compartment: prokaryotic pathway

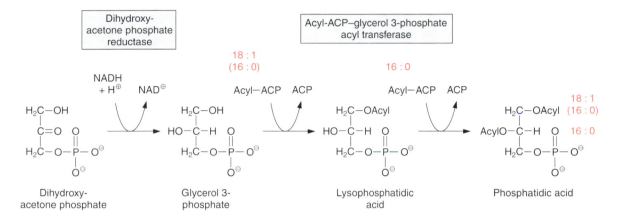

Endoplasmic reticulum: eukaryotic pathway

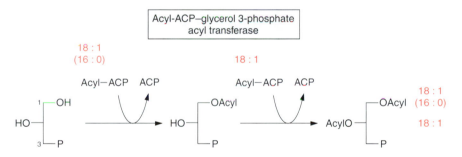

Figure 15.18 The membrane lipids synthesized in the plastids and at the endoplasmic reticulum have different fatty acid compositions.

Choline and ethanolamine are activated by phosphorylation via specific kinases and are then converted via cytidyl transferases by reaction with CTP into *CDP-choline* and *CDP-ethanolamine* (Fig. 15.19). A galactose head group is activated as *UDP-galactose* (Fig. 15.20). The latter is formed from glucose 1-phosphate and UTP via the UDP-glucose pyrophosphorylase (section 9.2) and UDP-glucose epimerase (Fig. 9.21). For the synthesis of DGDG from MGDG, a galactose residue is transferred from a second MGD molecule. Sulfoquinovose is also activated as a UDP derivative, but details of the synthesis of this moiety are still not known. The acceptor for the activated head group is diacylglycerol, which is formed from phosphatidic acid by the hydrolytic release of the phosphate residue.

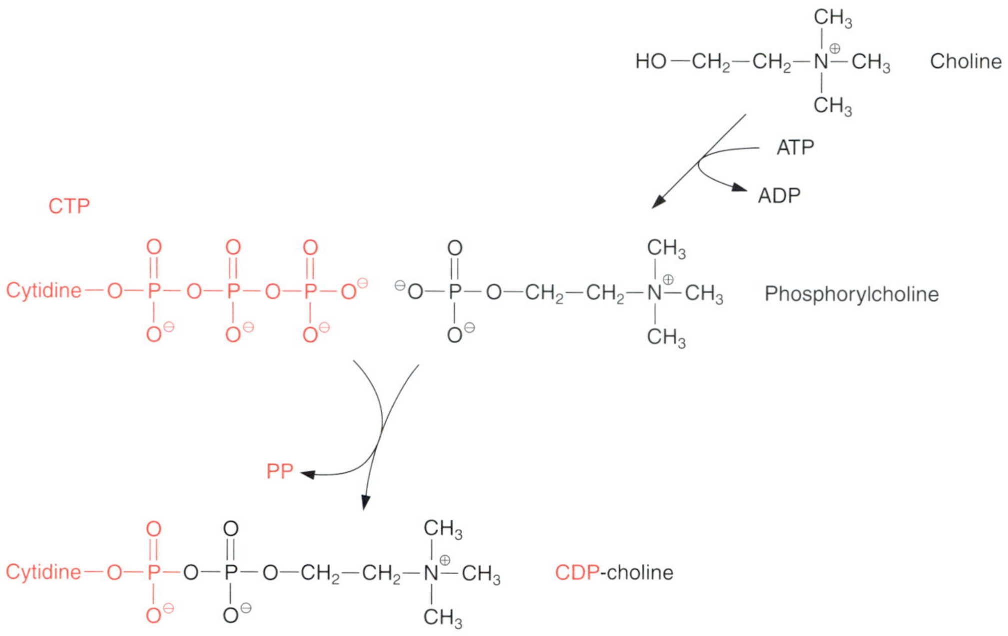

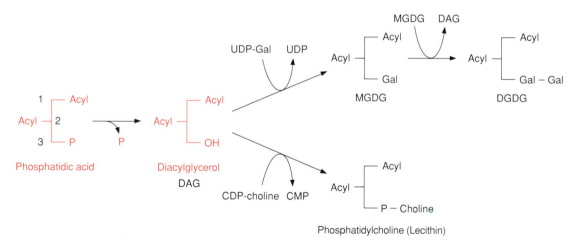

Figure 15.19 Formation of CDP-choline.

Figure 15.20 Overview of the synthesis of membrane lipids.

The ER membrane is the site of fatty acid elongation and desaturation

As shown in Fig. 15.17, 16 : 0-, 18 : 1- and to a lesser extent 18 : 0-acyl residues are produced by the plastids. However, some storage lipids contain fatty acids with a greater chain length. This also applies to waxes which are esters of a long chain

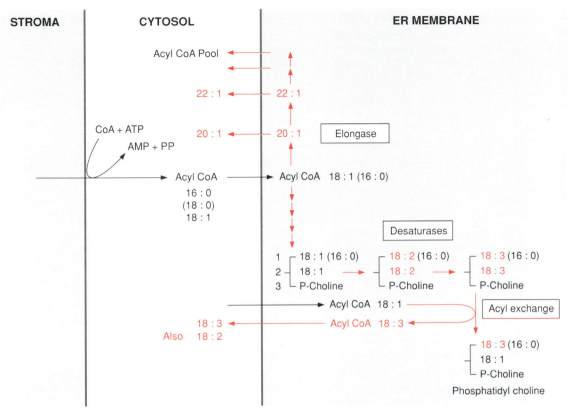

Figure 15.21 A pool of acyl CoA with various chain lengths and desaturation is present in the cytosol. Acyl residues, delivered from the plastids as acyl CoA, are elongated by elongases located at the ER. After incorporation into phosphatidyl choline, 18 : 1 acyl residues are desaturated by desaturases present in the ER membrane to 18 : 2 and 18 : 3. The more highly desaturated fatty acids in position 2 can be exchanged for 18 : 1-acyl CoA. In this way the cytosolic pool is provided with 18 : 2 and 18 : 3-acyl CoA.

fatty acid (C_{20}–C_{24}) with very long chain acyl alcohols (C_{24}–C_{32}). The elongation of fatty acids greater than C_{18} is catalysed by *elongases* which are located in the membranes of the endoplasmic reticulum (Fig. 15.21). Elongation proceeds in the same way as fatty acid synthesis, with the only differences being that different enzymes are involved and that the acyl and malonyl residues are activated as acyl CoA thioesters.

The ER membrane is also the site for further desaturations of the acyl residue. For desaturation the acyl groups are first incorporated into phospholipids such as phosphatidyl choline (Fig. 15.21). Desaturases bound to the membrane of the endoplasmic reticulum convert oleate (18 : 1) to linoleate (18 : 2) and then to linolenate (18 : 3). The 18 : 2- or 18 : 3-acyl residues in the C-2 position of glycerol can be exchanged for an 18 : 1-acyl residue and the latter can then be further desaturated.

The interplay of the desaturases in the plastids and the ER provides the cell with an *acyl CoA pool* to cover the various needs of the cytosol. The 16:0-, 18:0- and 18:1-acyl residues for this pool are delivered by the plastids, and the longer chain and more highly unsaturated acyl residues are provided by the ER membrane.

Some of the plastid membrane lipids are formed via the eukaryotic pathway

The synthesis of glycerolipids destined for the plastid membranes occurs in the envelope membranes of the plastids. Besides the prokaryotic pathway of glycerolipid synthesis, in which the acyl residues are directly transferred from acyl-ACP, glycerolipids are also synthesized via the eukaryotic pathway. The desaturases of the ER membrane can provide double unsaturated fatty acids for plastid membrane lipids via the synthesis of phosphatidyl choline with double unsaturated fatty acids as precursor (Fig. 15.21). This phosphatidyl choline is transferred to the envelope membrane of the plastids and hydrolysed to diacyl-glycerol (Fig. 15.22). In most cases the latter is provided with a head group consisting of one or two galactose residues. The acyl residues can be further desaturated to 18:3 by a desaturase present in the envelope membrane.

Some desaturases in the plastid envelope are able to desaturate lipid-bound 18:1- and 16:0-acyl residues. A comparison of the acyl residues in the 2-position (16:0 in the prokaryotic pathway and in 18:n the eukaryotic pathway) shows, however, that a large proportion of the highly unsaturated galactolipids in the plastids are formed via a detour through the eukaryotic pathway. The membrane lipids present in the envelope membrane are probably transferred by a special transfer protein to the thylakoid membrane.

15.5 Triacylglycerols are formed in the membranes of the endoplasmic reticulum

The lipid content of mature seeds can amount to 60 per cent of the dry weight. The precursor for the synthesis of triacylglycerol is again glycerol 3-phosphate. There are two pathways for its synthesis (Fig. 15.23):

1. Phosphatidic acid is formed by acylation of the hydroxyl groups in the C-1 and C-2 positions of glycerol, and after hydrolysis of the phosphate residue, that position is also acylated. The total cytosolic acyl CoA pool is available for these acylations, but because of the eukaryotic pathway position 2 is mostly esterified by a C_{18} acyl residue.

2. Phosphatidyl choline is formed first and its acyl residues are further desaturated. The choline phosphate residue is then liberated by hydrolysis and the corresponding diacylglycerol acylated. This pathway operates frequently in the synthesis of highly unsaturated triacylglycerols.

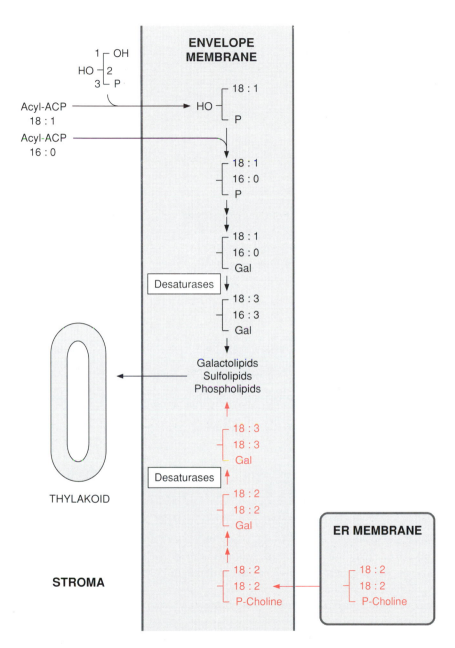

Figure 15.22 Part of the membrane lipids of the thylakoid membranes are synthesized via the eukaryotic pathway (red) and other parts via the prokaryotic pathway (black).

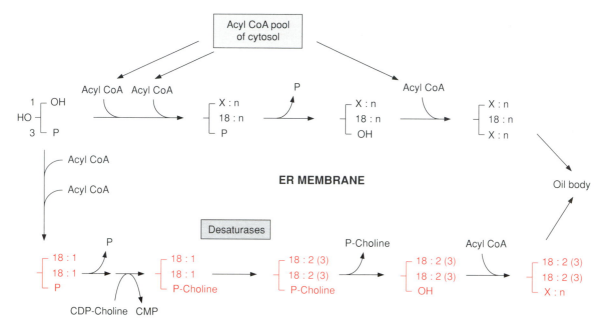

Figure 15.23 Overview of triacylglycerol synthesis in the ER membrane. The synthesis occurs either by acylation of glycerolphosphate (black) or via the intermediary synthesis of a phospholipid, of which the fatty acids are desaturated by desaturases in the ER membrane (red).

Cross-connections exist between the two pathways, but for the sake of simplicity these are not dealt with in Fig. 15.23.

Plant fat is used not only for nutrition but also as a raw material in industry

About 20 per cent of the human caloric nutritional uptake in industrialized countries is due to the consumption of plant fats. Plant fats have a higher content of unsaturated fatty acids than animal fats. Human metabolism requires unsaturated fatty acids with two or more carbon–carbon double bonds, but is not able to incorporate double bonds after the C-9 atom. This is why *linoleic* and *linolenic acid* (Fig. 15.4) are essential fatty acids that must form part of the human diet.

About 10 per cent of the world production of plant fats is utilized as a raw material for industrial purposes. Moreover, fatty acid methylesters synthesized from rapeseed oil are used in some countries as automobile fuel.

Table 15.3 shows that the fatty acid composition of various plant oils is variable. Triacylglycerols of some plants contain large quantities of rare fatty acids, which are used for industrial purposes (Table 15.4). Oil from palm kernels has a high content of the short-chained *lauric acid* (12 : 0), which is used as a raw material for the production of detergents and cosmetics. Large plantations of oil palms in

Table 15.3 Fatty acid composition of the most important fats

	% of total fatty acids				
	Soya oil	Palm oil	Rapeseed oil[a]	Sunflower oil	Peanut oil
16 : 0	11	42	4	5	10
18 : 0	3	5	1	1	3
18 : 1	22	41	60	15	50
18 : 2	55	10	20	79	30
18 : 3	8	0	9	0	0
20 : 1	0	0	2	0	3
Others	1	0	2	0	0
World production 1994 (10^6 metric tons year^{-1})[b]	18.6	14.7	10.3	7.9	4.2

[a]Erucic acid free varieties, after Baumann *et al.* (1988).
[b]Töpfer International, statistical information.

South-east Asia ensure its supply to the oleochemical industry. *Linolenic acid* from European linseed is used for the production of paints. *Ricinoleic acid*, a rare fatty acid, which contains a hydroxyl group in the C-12 position and makes up about 90 per cent of the fatty acids in castor oil, is used in industry as a lubricant and also as a means of surface protection. *Erucic acid* is used for similar purposes. Earlier varieties of rapeseed contained erucic acid in their triacylglycerols and for this reason rapeseed oil was then of inferior value in terms of nutrition. About 35 years ago successful crossings led to a breakthrough in the breeding of rapeseed free of erucic acid, making the rapeseed oil suitable for human consumption. The values in Table 15.3 are for this type, which is cultivated world-wide. However, rapeseed varieties containing a high percentage of erucic acid are again being cultivated to supply industrial demands.

Plant fats are customized by genetic engineering

The progress in gene technology now makes it possible to alter the quality of plant fats in a defined way by changing the enzymatic profile of the cell. The procedures for the introduction of a new enzyme into a cell, or for eliminating the activity of an enzyme present in the cell, will be described in detail in Chapter 22. Here three examples of the alteration of oil crops by genetic engineering will be illustrated.

The lauric acid present in palm kernel and coconut oil (12 : 0) is an important raw material for the production of soaps, detergents, and cosmetics. Recently, rapeseed plants which contained oil with a lauric acid content of 45 per cent were generated by genetic transformation. The synthesis of fatty acids is terminated by the hydrolysis of acyl-ACP (Fig. 15.17). An acyl-ACP thioesterase, which specifi-

Table 15.4 Industrial utilization of fatty acids from plant oils

	Main source	Utilization for
Lauric acid (12 : 0)	Palm kernel, coconut	Soap, detergents, cosmetics
Linolenic acid (18 : 3)	Linseed	Paints, lacquers
Ricinoleic acid (18 : 1, Δ^9, 12-OH)	Castor bean	Surface protectors, lubricant
Erucic acid (22 : 1, Δ^{13})	Rapeseed	Additive to lubricants, solvent softeners for textiles

cally hydrolyses lauroyl-ACP, was isolated and cloned from the seeds of the California bay tree (*Umbellularia californica*), which contains a very high proportion of lauric acid in its storage lipids. The introduction of the gene for this acyl-ACP thioesterase into rapeseed terminates its fatty acid synthesis at acyl (12 : 0), and the lauric acid released is incorporated into the seed oil. Field tests have shown that these plants grow normally and produce normal yields.

A relatively high content of stearic acid (18 : 0) improves the heat stability of fats for deep frying and for the production of margarine. The stearic acid content in rapeseed oil has been increased from 1–2 per cent to 40 per cent by decreasing the activity of stearoyl-ACP desaturase using the antisense technique (section 22.5).

Erucic acid (Fig. 15.24) could be an important industrial raw material, for instance for the synthesis of synthetic fibres. At the moment, however, its utilization is limited, since conventional breeding has not succeeded in increasing the erucic acid content to more than 50 per cent of the fatty acids in the seed oil. The cost of separating erucic acid and disposing of the other fatty acids is so high that for many purposes the industrial use of rapeseed oil as a source of erucic acid from the cultivars available at present is not economically viable. Attempts are being made to increase further the erucic acid content of rapeseed oil by overexpression of genes for elongases and by transferring the genes encoding enzymes which catalyse the specific incorporation of erucic acid to the C-2 position of triacylglycerols. If this were to be successful, present petrochemical-based industrial processes could be replaced with processes using rapeseed oil as a renewable raw material.

Erucic acid (22 : 1, Δ^{13}-*cis*)

Figure 15.24 Erucic acid, an industrial raw material.

15.6 During germination storage lipids are mobilized for the production of carbohydrates in the glyoxysomes

At the beginning of germination, storage proteins (Chapter 14) are degraded to amino acids from which the enzymes required for the mobilization of the storage lipids are synthesized. These enzymes include *lipases* which catalyse the hydrolysis of triacylglycerols to glycerol and fatty acids. Lipases bind to the oleosins of the oil bodies (section 15.2). The *glycerol* formed by hydrolysis of triacylglycerol can be fed into the gluconeogenesis pathway after phosphorylation to glycerol 3-phosphate and its subsequent oxidation to dihydroxyacetone phosphate (Chapter 9). The released *free fatty acids* are first activated as CoA-thioesters and then degraded to acetyl CoA by *β-oxidation* (Fig. 15.25). This process occurs in specialized peroxisomes called *glyoxysomes*.

Although in principle β-oxidation represents the reversal of the fatty acid synthesis shown in Fig. 15.8, there are distinct differences, enabling high metabolic fluxes through these two metabolic pathways which operate in opposite directions. The differences between β-oxidation and fatty acid synthesis are:

1. In acyl CoA dehydrogenation, hydrogen is transferred via an *FAD-dependent oxidase* to form H_2O_2. A catalase irreversibly eliminates the toxic H_2O_2 at the site of its production by conversion into water and oxygen (section 7.1).

2. β-L-Hydroxyacyl CoA is formed during the hydration of enoyl CoA, in contrast to the corresponding D-enantiomer during synthesis.

3. Hydrogen is transferred to NAD during the second dehydrogenation step. Normally the NAD system in the cell is highly oxidized, driving the reaction in the direction of hydroxyacyl CoA oxidation. It is not known which reactions utilize the NADH formed in the peroxisomes.

4. In an irreversible reaction CoASH-mediated thiolysis cleaves β-ketoacyl CoA to form one molecule of acetyl CoA and one of acyl CoA shortened by two C atoms.

During the degradation of unsaturated fatty acids, intermediate products are formed that cannot be metabolized by the reactions of β-oxidation. Δ^3-*cis*-enoyl CoA (Fig. 15.26), formed during the degradation of oleic acid, is converted by an *isomerase* shifting the double bond to Δ^2-*trans*-enoyl CoA, an intermediate of β-oxidation. In the β-oxidation of linoleic or of linolenic acid the second double bond in the corresponding intermediate is in the correct position, but in the *cis*-configuration, with the consequence that its hydration by enoyl CoA hydratase results in the formation of β-D-hydroxyacyl CoA. The latter is converted by an *epimerase* to the L-enantiomer, which is the intermediate of β-oxidation.

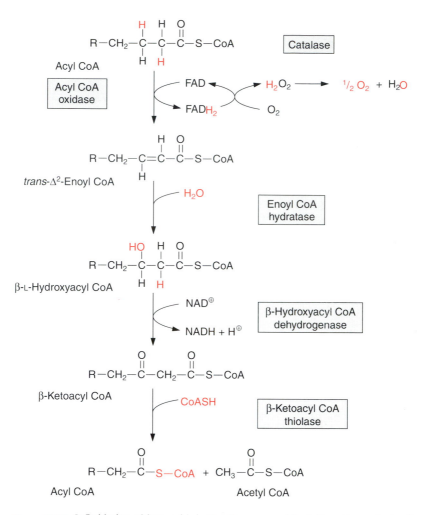

Figure 15.25 β-Oxidation of fatty acids in the glyoxysomes. The fatty acids are first activated as CoA-thioesters and then converted to acetyl CoA and a fatty acid shortened by two carbon atoms, involving dehydrogenation via an FAD-dependent oxidase, addition of water, a second dehydrogenation (by NAD), and thiolysis by CoASH.

The glyoxylate cycle enables plants to synthesize hexoses from acetyl CoA

In contrast to animals, which are unable to synthesize glucose from acetyl CoA, plants are capable of gluconeogenesis, as they possess the enzymes of the glyoxylate cycle (Fig. 15.27). Like β-oxidation this cycle is localized in the glyoxysomes, which are named after the cycle. The two starting reactions of the cycle are identical with those of the citrate cycle (see Fig. 5.3). Acetyl CoA condenses with oxaloacetate, catalysed by citrate synthase, to form citrate and the latter is

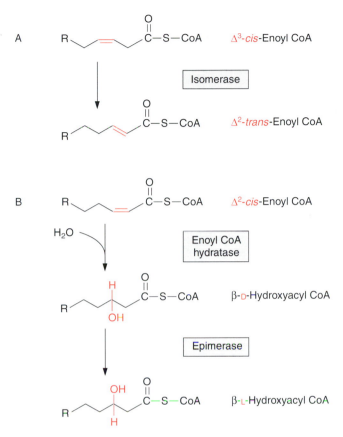

Figure 15.26 The intermediates formed during β-oxidation of unsaturated fatty acids are isomerized to enable subsequent degradation by β-oxidation.

converted by aconitase to isocitrate. The further reaction of isocitrate, however, is a speciality of the glyoxylate cycle: isocitrate is split by isocitrate lyase into succinate and glyoxylate (Fig. 15.28).

Malate synthase, the second special enzyme of the glyoxylate cycle, catalyses the instantaneous condensation of glyoxylate with acetyl CoA to form malate. The hydrolysis of the CoA-thioester makes this reaction irreversible. As in the citrate cycle, malate is oxidized by malate dehydrogenase to oxaloacetate, thus completing the glyoxylate cycle. In this way one molecule of succinate is generated from two molecules of acetyl CoA. The succinate is transferred to the mitochondria and converted there into oxaloacetate by a partial reaction of the citrate cycle. The oxaloacetate is released from the mitochondria by the oxaloacetate translocator and converted in the cytosol by phosphoenolpyruvate carboxykinase to phosphoenolpyruvate (Fig. 8.10). Phosphoenolpyruvate is a precursor for the synthesis of hexoses by the gluconeogenesis pathway and for other biosynthetic processes.

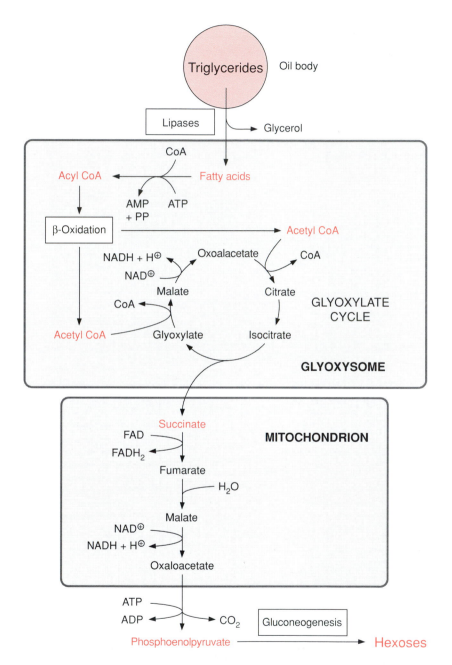

Figure 15.27 Mobilization of storage lipids for the synthesis of hexoses during germination. The hydrolysis of the triacylglycerols present in the oil bodies, as catalysed by oleosin-bound lipases, yields fatty acids which are activated in the glyoxysomes as CoA-thioesters and degraded by β-oxidation into acetyl CoA. From two molecules of acetyl CoA the glyoxylate cycle forms one molecule of succinate, which is converted by the citrate cycle in the mitochondria to oxaloacetate. Phosphoenolpyruvate formed from oxaloacetate in the cytosol is a precursor for the synthesis of hexoses via the gluconeogenesis pathway.

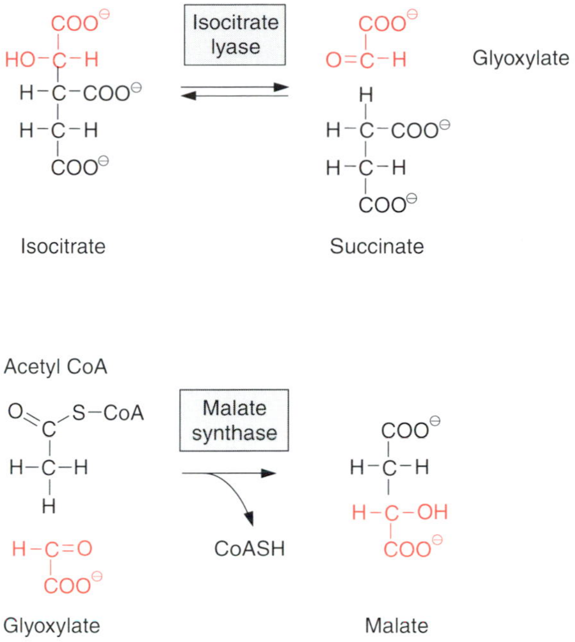

Isocitrate lyase

Glyoxylate

Isocitrate   Succinate

Acetyl CoA

Malate synthase

CoASH

Glyoxylate   Malate

Figure 15.28 Key reactions of the glyoxylate cycle.

Reactions with toxic intermediates take place in peroxisomes

There is a simple explanation for the fact that β-oxidation and the closely related glyoxylate cycle proceed in peroxisomes. As in photorespiration, in which peroxisomes participate (section 7.1), the toxic substances H_2O_2 and *glyoxylate* are also formed as intermediates in the conversion of fatty acids to succinate. Compartmentalization in the peroxisomes prevents these toxic substances from reaching the cytosol (section 7.4). β-Oxidation also takes place, although at a much lower rate, in leaf peroxisomes, where it serves the purpose of recycling the fatty acids that are no longer required or have been damaged. The hydrophobic amino acids leucine and valine are also degraded in peroxisomes.

In glyoxysomes, as in leaf peroxisomes (section 7.4), the transfer of metabolites across the boundary membrane is facilitated by porins (section 1.11).

15.7 Lipoxygenase is involved in the synthesis of aromatic defence and signal substances

In contrast to the mono-oxygenases dealt with in section 15.3, which catalyse the incorporation of one O atom of an O_2 molecule into hydrocarbons, *dioxygenases* mediate the incorporation of both oxygen atoms. *Lipoxygenase* is a dioxygenase widely distributed in higher plants. It catalyses the dioxygenation of multiple-unsaturated fatty acids, such as linoleic and linolenic acid, which contain a *cis, cis-1,4-pentadiene sequence* (coloured red in Fig. 15.29). At the end of the sequence a hydroperoxide group is introduced by reaction with O_2 and the neighbouring double bond is shifted by one C-position in the direction of the other double bond, thereby attaining a *trans*-configuration. The *hydroperoxide lyase* catalyses the cleavage of hydroperoxylinolenic acid into a 12-oxo-acid and a 3-*cis*-hexenal. More hexenals are formed by shifting of the double bond, and their reduction leads to the formation of hexenols (Fig. 15.29). In an analogous way hydroperoxy-linoleic acid yields hexanal and its reduction leads to hexanol.

Hexanals, hexenals, hexanols, and hexenols are volatile *aromatic substances*

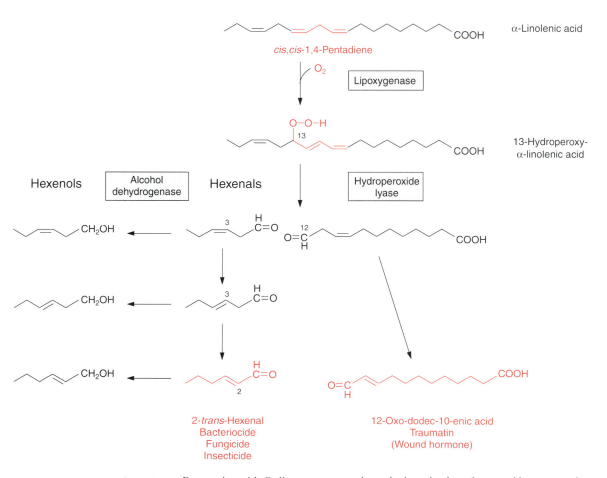

Figure 15.29 By reaction with O_2 lipoxygenase catalyses the introduction of a peroxide group at the end of a *cis,cis*-1,4-pentadiene sequence (red). Hydroperoxide lyase cleaves the C–C bond between C atoms 12 and 13. The hexenal thus formed can be isomerized by shifting of the double bond, probably due to enzymatic catalysis. The hexenals are reduced to the corresponding hexenols by an alcohol dehydrogenase. The 12-oxo-acid formed as a second product is isomerized to traumatin.

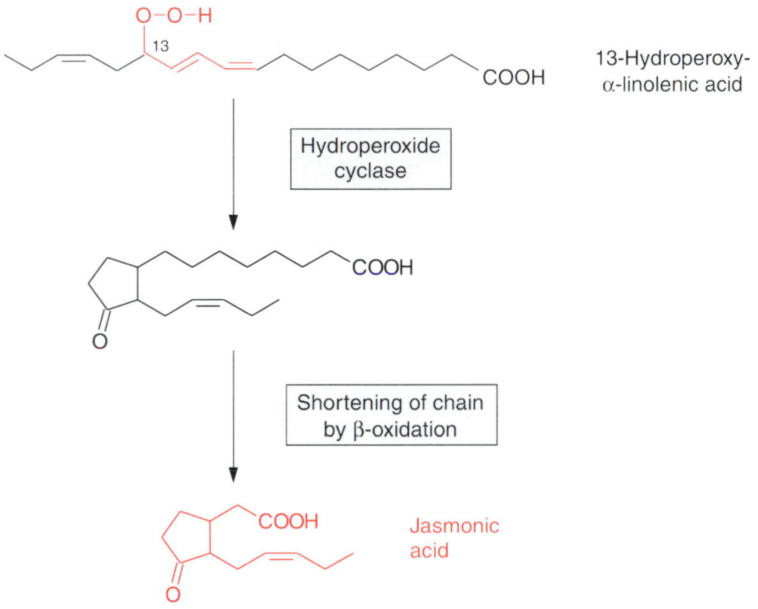

Figure 15.30 A hydroperoxide cyclase catalyses the cyclization of hydroperoxylinolenic acid accompanied by a shift of the oxygen. Shortening of the fatty acid chain by six carbon atoms via β-oxidation results in the formation of jasmonic acid, a phytohormone.

that are important components of the characteristic odour and taste of many fruits and vegetables. The wide range of aromas includes fruity, sweet, spicy, and grass-like. Thus hexanols contribute to the flavour of tomatoes. Work is in progress to improve the taste of tomatoes by increasing their hexanol content using genetic engineering. The quality of olive oil, for instance, depends on its content of hexenals and hexenols. Hexenals are responsible for the aroma of black tea. Green tea is processed to black tea by heat, whereby hexenals condense to aromatic compounds, which give black tea its typical taste. Large amounts of hexenals and hexenols are produced industrially as aromatic substances in the food industry or as ingredients in the preparation of perfumes.

The characteristic smell of freshly cut grass is caused primarily by the release of hexenals and hexenols, indicating that the activity of lipoxygenase and hydroperoxide ligase is greatly increased by cell wounding. This is part of a defence reaction: 2-*trans*-hexenal (coloured red in Fig. 15.29) is a strong bactericide, fungicide, and insecticide. 12-Oxo-dodec-10-enic acid, which is formed from the cleavage product of hydroperoxylinolenic acid by the shifting of a double bond, has the properties of a wound hormone and has therefore been named *traumatin*. Traumatin induces cell division in neighbouring cells. This results in the formation of calli and the wound is sealed. However, our knowledge of these defence processes is still rather fragmentary.

Hydroperoxide cyclase catalyses the cyclization of 13-hydroperoxy-α-linolenic acid (Fig. 15.30). Shortening of the hydrocarbon chain by β-oxidation of the product (Fig. 15.25) results in the formation of *jasmonic acid*. Jasmonic acid and its methylester are hormone-like signal substances, which activate defence genes for the synthesis of phytoalexins (section 16.1) and also initiate senescence (Chapter 19).

Further reading

Baumann, H., Bühler, M., Fochem, H., Hirsinger, F., Zoebelein, H., Falbe, J. (1988). Natürliche Fette und Öle – nachwachsende Rohstoffe für die chemische Industrie. *Angew. Chemie*, **100**, 41–62.

Browse, J. and Sommerville, C. (1991). Glycerolipid synthesis. Biochemistry and regulation. *Annual Review of Plant Physiology and Plant Molecular Biology*, **42**, 467–506.

Chasan, R. (1995). Engineering of fatty acids: The long and short of it. *The Plant Cell*, **7**, 235–7.

Cote, G. G. and Crain, R. C. (1993). Biochemistry of phosphoinosites. *Annual Review of Plant Physiology and Plant Molecular Biology*, **44**, 333–56.

Fox, B. G., Shanklin, J., Sommerville, C., and Münck, E. (1993). Stearoyl-acyl carrier protein Δ^9 desaturase from *Ricinus communis* is a diiron-oxo-protein. *Proceedings of the National Academy of Sciences, USA*, **90**, 2486–90.

Gerhardt, B. (1992). Fatty acid degradation in plants. *Progress in Lipid Research*, **31**, 417–46.

Harwood, J. L. (1980). Plant acyl lipids: Structure, distribution, and analysis. *The biochemistry of plants*. Vol. 4 (ed. P. K. Stumpf), pp. 2–56, Academic Press, New York.

Harwood, J. L. and Sanchez, J. (ed.) (1994). Biotechnological aspects of plant lipids. *Progress in Lipid Research*, **33**, 1–202.

Hatanaka, A. (1993). The biogeneration of green odour by green leaves. *Phytochemistry*, **34**, 1201–18.

Huang, A. (1996). Oleosins and oil bodies in seeds and other organs. *Plant Physiology*, **110**, 1055–62.

Kadar, J.-C. (1996). Lipid-transfer proteins in plants. *Annual Review of Plant Physiology and Plant Molecular Biology*, **47**, 627–54.

Karlin, K. (1993). Metalloenzymes, structural motifs and inorganic models. *Science*, **261**, 701–8.

Kates, M. (ed.) (1990). Glycolipids, phosphoglycolipids and sulfoglycolipids. *Handbook of lipid research*, Vol. 6, Plenum Press, New York.

Moore, T. S. (ed.) (1993). *Lipid metabolism in plants*. CRC Press, Boca Raton, Florida.

Murata, N., Ishizaki-Nishizawa, O., Higashi, S., Hayashi, H., Tasaka, Y., and Nishida, J. (1992). Genetically engineered alteration in the chilling sensitivity of plants. *Nature*, **356**, 710–13.

Murphy, D. J. (ed.) (1994). *Design of oil crops*. Verlag Chemie, Weinheim.

Ohlrogge, J. B. (1994). Design of new plant products. Engineering of fatty acid metabolism. *Plant Physiology*, **104**, 821–6.

Ohlrogge, J. B. and Browse, J. (1995). Lipid biosynthesis. *The Plant Cell*, **7**, 957–70.

Quinn, P. J. and Harwood, J. L. (ed.) (1990). *Plant lipids, biochemistry, structure and utilisation*. Portland Press Limited, London.

Ross, J. H. E., Sanchez, J., Millan, F., and Murphy, D. J. (1993). Differential presence of oleosins in oleogenic seed and mesocarp tissue in olive (*Olea europea*) and avocado (*Persea americana*). *Plant Science*, **93**, 203–10.

Siedow, J. N. (1991). Plant lipoxygenase: structure and function. *Annual Review of Plant Physiology and Plant Molecular Biology*, **42**, 145–88.

Stumpf, P. K. (ed.) (1987). Lipids: structure and function. *The Biochemistry of Plants*, Vol. 9. Academic Press, Orlando.

Töpfer, R., Martini, N., and Schell, J. (1995). Modification of plant lipid synthesis. *Science*, **268**, 681–5.

chapter 16

The function of secondary metabolites in plants

In addition to *primary metabolites*, such as carbohydrates, amino acids, fatty acids, cytochromes, chlorophylls, and metabolic intermediates of the anabolic and catabolic pathways, which occur in all plants and where they all have the same metabolic functions, plants also contain a large variety of substances, named *secondary metabolites*, with no apparent direct metabolic function. Certain secondary metabolites are restricted to a few plant species, where they fulfil specific ecological functions, such as attracting insects to transfer pollen, or animals to consume fruits and in this way to distribute seed, and last, but not least, to act as *natural pesticides*.

16.1 Secondary metabolites often protect plants from pathogenic micro-organisms and herbivores

Plants are an important source of food for many animals, such as insects, snails, and many vertebrates, because of their protein and carbohydrate content. Since plants cannot run away, they have had to evolve strategies, that make them indigestible or poisonous, to protect themselves from being eaten. A number of toxic plant products, such as amylase inhibitors and lectins, have already been dealt with in section 14.4. Groups of secondary metabolites will now be discussed comprising alkaloids (this chapter), isoprenoids (Chapter 17), and phenyl-propanoids (Chapter 18), all of which include natural pesticides that protect plants against herbivores and pathogenic micro-organisms. In some plants these natural pesticides amount to 10 per cent of the dry matter.

The plant forms phytoalexins in response to microbial infection

Defence substances against herbivores are usually part of the plant's permanent outfit; they are *constitutive*. (Section 18.7, however, describes a case where high quantities of a defence substance are only formed in response to browsing damage.) In contrast, defence substances against micro-organisms, especially fungi, are synthesized mostly in response to an infection. These inducible defence substances, formed within hours, are called *phytoalexins* (*alekein*, Greek, to defend). Phytoalexins comprise a large number of substances with very different structures, such as isoprenoids, flavonoids, and stilbenes, many of which act as antibiotics against a broad spectrum of pathogenic fungi and bacteria. In each case inducers for the synthesis of a phytoalexin are very distinctive substances, termed *elicitors*. Elicitors are often proteins excreted by the pathogens to attack plant cells, e.g. cell-degrading enzymes. In some cases polysaccharide segments of the cell's own wall, produced by degradative enzymes of the pathogen, or fragments from the cell wall of the pathogen released by defence enzymes of the plant, can also function as elicitors. These various elicitors bind to specific receptors on the outer surface of the plasma membrane of the plant cell. The binding of the elicitor results in the formation of a signal substance, a phosphory-lated protein for instance (section 19.7), which uses a signal transduction chain to induce the transcription of those genes which encode the enzymes of phytoalexin synthesis.

Elicitors may also cause an infected cell to die and the surrounding cells to die with it. In other words, the infected cells and those surrounding them commit suicide. This can be caused, for instance, by the infected cells producing phenols, with which they poison not only themselves but also their neighbours. This pro-grammed cell death, called a *hypersensitive response*, serves to protect the plant. The cell walls around the necrotic tissue are strengthened by increased biosynthesis of lignin and in this way the plant barricades itself against the infection spreading further.

Plant defence substances can also be a risk for man

Substances toxic for animals are, in many cases, also toxic for man. Plants culti-vated for human consumption usually have a relatively low toxic secondary metabolite content. This, however, has made cultivated plants more sensitive than wild plants to pests, necessitating the use of pest control, predominantly by chemical means. Attempts to breed more resistant culture plants by crossing them with wild plants can lead to problems: a newly introduced variety of insect-resistant potato had to be taken off the market as the highly toxic solanine content (an alkaloid, see following section) made these potatoes unsuitable for human consumption. In a new variety of insect-resistant celery cultivated in the USA, the tenfold increased content of psoralens (section 18.2) caused severe skin damage to people harvesting the plants. This illustrates that natural pest control is not without risk.

A number of plant constituents that are harmful to man, e.g. proteins such as lectins, amylase inhibitors, proteinase inhibitors, and also cyanogenic glycosides or glucosinolates, dealt with in this chapter, decompose when cooked. But most secondary metabolites are not destroyed in this way. In higher concentrations, a large proportion of plant secondary metabolites are carcinogenic. It has been estimated that in the industrialized countries more than 99 per cent of all carcinogenic substances, that man normally consumes with his diet, are plant secondary metabolites that are natural constituents of the food. However, experience has shown that human metabolism usually provides sufficient protection against these harmful natural substances.

16.2 Alkaloids comprise a variety of heterocyclic secondary metabolites

Alkaloids are basic secondary metabolites, that are synthesized from *amino acids* and contain one or several N atoms as constituents of *heterocycles*. Many of these alkaloids act as defence substances against animals and micro-organisms. They are stored in the protonated form, mostly in the vacuole, which is acidic. Since ancient times man has used alkaloids in the form of plant extracts as poisons, stimulants and narcotics, and, last but not least, as medicine. In 1806 the pharmacy assistant Friedrich Wilhelm Sertürner isolated morphine from poppy seeds. Another 146 years had to pass until the structure of morphine was finally resolved in 1952. More than 10 000 alkaloids of very different structures are now known. Their synthetic pathways are, to a large extent, still unknown and will not be dealt with here.

Figure 16.1 shows a small selection of important alkaloids. Alkaloids are classified according to their heterocycles. *Coniine*, a piperidine alkaloid, is a very potent poison in hemlock. Socrates died when he was forced to drink this poison. *Nicotine*, also very toxic, contains a pyridine and a pyrrolidine ring. It is formed in the roots of tobacco plants and is carried along with the xylem sap into the stems and leaves. Nicotine sulfate, a by-product of the tobacco industry, is used as a very potent insecticide, e.g. for fumigating greenhouses. No insect is known to be resistant to nicotine. *Cocaine*, the well-known narcotic, contains tropane as a hetero-

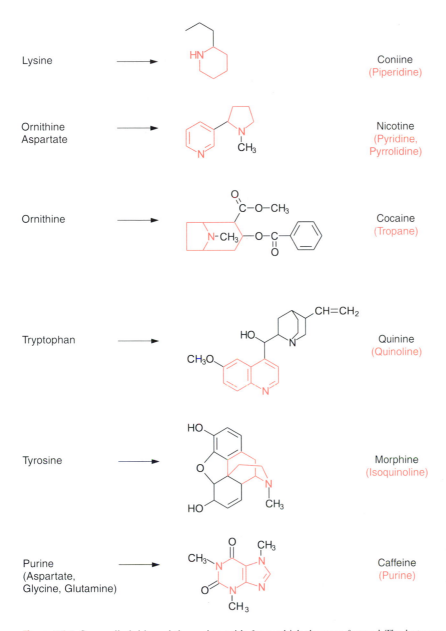

Lysine ⟶ Coniine (Piperidine)

Ornithine Aspartate ⟶ Nicotine (Pyridine, Pyrrolidine)

Ornithine ⟶ Cocaine (Tropane)

Tryptophan ⟶ Quinine (Quinoline)

Tyrosine ⟶ Morphine (Isoquinoline)

Purine (Aspartate, Glycine, Glutamine) ⟶ Caffeine (Purine)

Figure 16.1 Some alkaloids and the amino acids from which they are formed. The heterocycles after which the alkaloids are classified are coloured red with their names in brackets. For coniine a synthesis from acetyl CoA has also been described. Purine is synthesized from aspartate, glycine, and glutamine.

cycle, in which the N atom is a constituent of two rings. A further well-known tropane alkaloid is atropine (formula not shown), a poison contained in deadly nightshade (*Atropa belladonna*). In low doses it dilates the pupils of the eye and is therefore used in medicine for eye examination. Allegedly Cleopatra used extracts containing atropine to dilate her pupils in order to appear more attractive. *Quinine*, a quinoline alkaloid from the bark of *Chinchuna officinalis*, growing in South America, was known by the Spanish conquerors to be an antimalarial drug. The isoquinoline alkaloid *morphine* is an important pain killer and is also precursor for the synthesis of heroin. *Caffeine*, the stimulant contained in coffee, contains a purine as the heterocycle.

In the search for new medicines large numbers of plants are being analysed for their content of secondary metabolites. As a result of a systematic search, the alkaloid *taxol*, isolated from the yew tree *Taxus brevifolia*, is now used for cancer treatment. Derivatives of the alkaloid *camptothezine* from the Chinese 'happy tree', *Camptotheca acuminata*, are also being tested clinically as cancer therapeutics. The search for new medicines against malaria and viral infections still continues. Since large quantities of pharmacologically interesting substances often cannot be gained from plants, attempts are being made with the aid of genetic engineering either to increase production by the corresponding plants or to transfer the plant genes into micro-organisms in order to use the latter for production.

16.3 Some plants emit prussic acid when wounded by animals

Since *prussic acid* (HCN) inhibits cytochrome oxidase, which is the final step of the respiratory chain, it is a very potent poison (section 5.5). Ten per cent of all plants are estimated to use this poison as a defence against being eaten by animals. The consumption of peach kernels for instance, or of bitter almonds, can have fatal results for man. Plants possess a mitochondrial respiratory chain and, in order not to poison themselves, contain prussic acid in the bound form as *cyanogenic glycoside*. The amygdalin in the kernels and roots of peaches is an example of this (Fig. 16.2). The cyanogenic glycosides are stored as stable compounds in the vacuole. The *glycosidase* catalysing the hydrolysis of the glycoside is present in another compartment. If the cell is wounded by feeding animals, the compartmentalization is disrupted and the glycosidase comes into contact with cyanogenic glycoside. After the hydrolysis of the glucose residue, the remaining cyanhydrin is very unstable and decomposes spontaneously to prussic acid and an aldehyde. A *hydroxynitrile lyase* accelerates this reaction. The aldehydes formed from cyanogenic glycosides are often very toxic. For a feeding animal the detoxification of these aldehydes can be even more difficult than that of prussic acid. The formation of two different toxic substances makes the cyanogenic glycosides a very effective defence system.

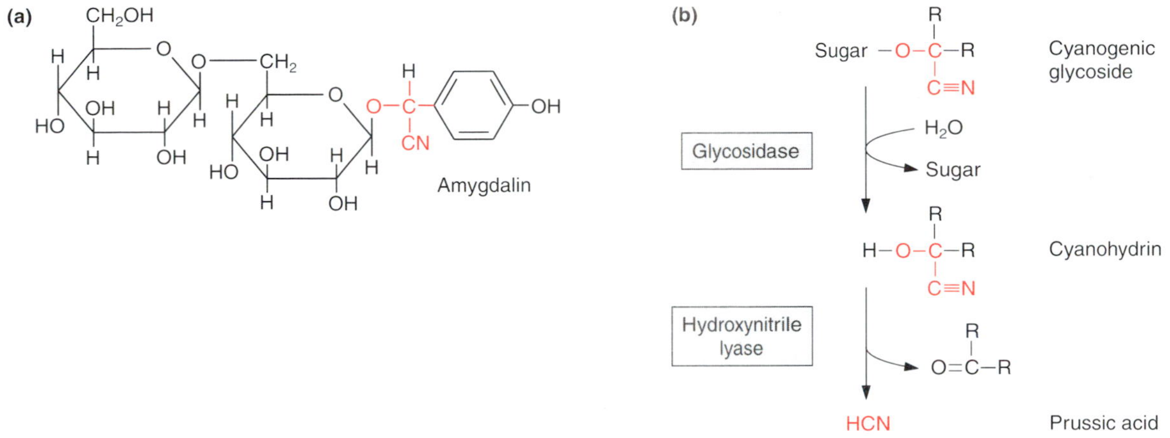

Figure 16.2 (a) Amygdalin, a cyanogenic glycoside, is contained in some stone fruit kernels. (b) After the sugar residue has been released by hydrolysis, cyanhydrin is formed from cyanogenic glycosides and decomposes spontaneously to prussic acid and a carbonyl compound.

16.4 When wounded, some plants emit volatile mustard oils

Glucosinolates, also named *mustard oil glycosides*, have a similar protective function against herbivores. Glucosinolates can be found, for instance, in radish, cabbage, and mustard plants. Cabbage contains the glycoside glucobrassicin (Fig. 16.3), which is formed from tryptophan. The hydrolysis of the glycoside by a *thioglycosidase* results in a very unstable product from which, after the liberation and rearrangement of the sulfate residue, an isothiocyanate, also termed *mustard oil*, is formed spontaneously. Mustard oils are toxic at higher concentrations. Like the cyanogenic glycosides (see above), glucosinolates and the hydrolysing enzyme thioglycosidase are located in separate compartments of the plant tissues. The enzyme comes into contact with its substrate only after wounding. When cells of these plants have been damaged, the pungent smell of mustard oil can easily be detected, e.g. in freshly cut radish. The high glucosinolate contents in early varieties of rapeseed made the pressed seed unsuitable for fodder. Successful breeding has produced rapeseed varieties without glucosinolate, which are cultivated nowadays.

16.5 Plants protect themselves by tricking herbivores with false amino acids

Many plants contain unusual amino acids with a structure very similar to that of protein amino acids, e.g. *canavanine* from Jack bean (*Canavalia ensiformis*), a structural analogue of *arginine* (Fig. 16.4). Herbivores take up canavanine with their food. During protein biosynthesis, the arginine-tRNAs of animals cannot

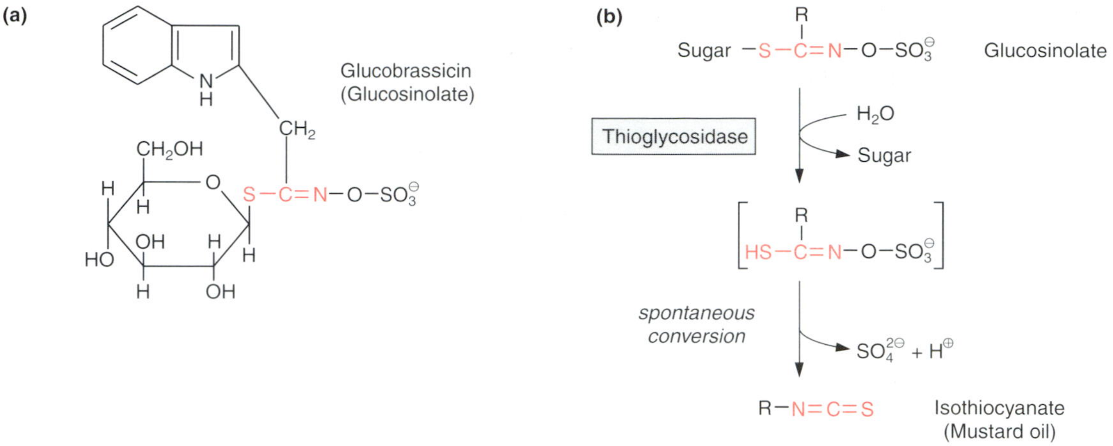

(a)

Glucobrassicin
(Glucosinolate)

CH_2

CH_2OH

$S-C=N-O-SO_3^\ominus$

HO   OH

(b)

$$Sugar-S-C=N-O-SO_3^\ominus \qquad Glucosinolate$$

with R above

Thioglycosidase

H_2O

Sugar

$$\left[HS-C=N-O-SO_3^\ominus \right]$$

with R above

spontaneous conversion

$SO_4^{2\ominus} + H^\oplus$

$R-N=C=S$ Isothiocyanate
(Mustard oil)

Figure 16.3 (a) Glucobrassicin, a glucosinolate from cabbage. (b) The hydrolysis of glycoside by thioglycosidase results in an unstable product which decomposes spontaneously into sulfate and isothiocyanate.

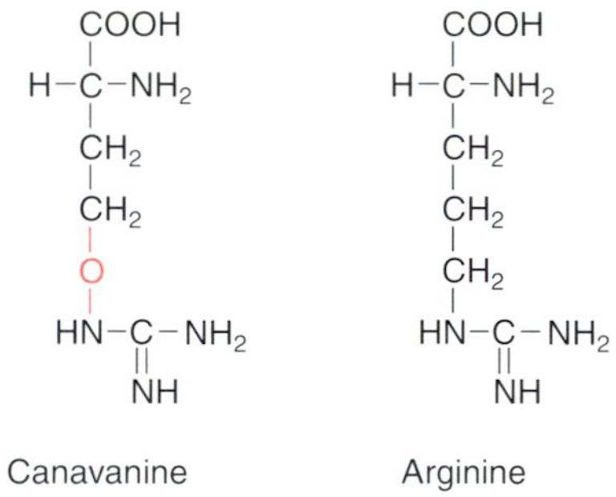

COOH
|
H-C-NH_2
|
CH_2
|
CH_2
|
O
|
HN-C-NH_2
‖
NH

Canavanine

COOH
|
H-C-NH_2
|
CH_2
|
CH_2
|
CH_2
|
HN-C-NH_2
‖
NH

Arginine

Figure 16.4 Canavanine is a structural analogue of arginine.

distinguish between arginine and canavanine and incorporate canavanine instead of arginine into proteins. This exchange can alter the three-dimensional structure of proteins which then lose their biological function partially or even completely. Therefore canavanine is toxic for herbivores. In those plants synthesizing canavanine, the arginine-tRNA does not react with canavanine, hence it is not toxic for these plants. This same protective mechanism is used by some insects which specialize in eating leaves containing canavanine. Plants contain a very large variety of false amino acids, which are toxic for herbivores in an analogous way to canavanine.

Further reading

Ames, B. N. and Gold, L. S. (1989). Pesticides, risk and applesauce. *Science*, **244**, 755–7.
Bell, E. A. and Charlwood, B. V. (eds.) (1980). Secondary plant products. *Encyclopedia of Plant Physiology*, Vol. 8, Springer Verlag, Berlin.

Hashimoto, T. and Yamada, Y. (1994). Alkaloid biogenesis: molecular aspects. *Annual Review of Plant Physiology and Plant Molecular Biology*, **45**, 257–85.

Kutchan, T. M. (1995). Alkaloid biosynthesis. The basis for metabolic engineering of medicinal plants. *The Plant Cell*, **7**, 1059–70.

Robinson, T. (1981). *The biochemistry of alkaloids*. Springer Verlag, Berlin.

Rosenthal, G. A. and Rhodes, D. (1984). L-Canavanin transport and utilisation in developing jack bean, *Canavalia ensiformis*. *Plant Physiology*, **76**, 541–4.

Scheel, D. and Parker, J. E. (1990). Elicitor recognition and signal transduction in plant defence gene activation. *Zeitschrift für Naturforschung*, **45c**, 569–75.

Southon, I. W. and Buckingham, J. (ed.) (1989). *Dictionary of alkaloids*. Chapman & Hall, London.

Tyler, V. E. (1994). *Herbs of choice*. Haworth Press, New York.

chapter 17

Isoprenoids

Isoprenoids are present in all living organisms, but with an unusual diversity in plants. By 1991, 22 000 different plant isoprenoids had been listed and new substances are constantly being identified. These isoprenoids have very many different functions (Table 17.1). In primary metabolism they function as membrane constituents, photosynthetic pigments, electron transport carriers, growth substances, and plant hormones. They act as glycosyl carriers in glycosylation reactions and are involved in the regulation of cell growth. In addition to this they have ecological functions, as secondary metabolites: the majority of the different plant isoprenoids are to be found in resins, latex, waxes, and oils and they make

Table 17.1 Isoprenoids of higher plants

Precursor	Class	Example	Function
C$_5$: Dimethylallyl-PP	Hemiterpene	Isoprene	Protection of the photosynthetic apparatus against heat
Isopentenyl-PP		Side chain of cytokinins	Growth regulator
C$_{10}$: Geranyl-PP	Monoterpene	Pinene	Defence substance
		Linalool	Attractant
C$_{15}$: Farnesyl-PP	Sesquiterpene	Capsidiol	Phytoalexin
C$_{20}$: Geranylgeranyl-PP	Diterpene	Gibberellin	Plant hormone
		Phorbol	Defence substance
		Casbene	Phytoalexin
C$_{30}$: 2 Farnesyl-PP	Triterpene	Cholesterol	Membrane constituents
		Sitosterol	
C$_{40}$: 2 Geranylgeranyl-PP	Tetraterpene	Carotenoids	Photosynthesis pigments
Geranylgeranyl-PP or farnesyl-PP	Polyprenols	Prenylated proteins	Regulation of cell growth
		Prenylation of plastoquinone, ubiquinone, chlorophyll, cyt-*a*	Membrane solubility of photosynthesis pigments and electron transport carriers
		Dolichols	Glycosyl carrier
		Rubber	

plants toxic or indigestible as a defence measure against herbivores. They act as antibiotics to protect the plant from pathogenic micro-organisms. Many isoprenoids are only formed in response to infection by bacteria or fungi. Some plants synthesize isoprenoids which inhibit the germination and development of competing plants. Other isoprenoids, in the form of pigments or scent in flowers or fruit, attract insects to distribute pollen or seed.

Plant isoprenoids are important commercially, e.g. as aroma substances for food, beverages and cosmetics, vitamins (A, D, E), natural insecticides (e.g. pyrethrin), solvents (e.g. turpentine), and as rubber and gutta. The plant isoprenoids are also important natural substances which are utilized as pharmaceuticals or their precursors. Genetic engineering investigations are in progress to increase plants' ability to synthesize isoprenoids.

Isoprenoids are also named *terpenoids*. Plant essential oils have long been of interest to chemists. A number of mainly cyclic compounds containing 10, 15, 20, or more C atoms, have been isolated from turpentine oil. Such substances have been found in many plants and have been given the collective name *terpenes*. Figure 17.1 shows some examples of terpenes. Limonene, an aromatic substance from lemon oil, is a terpene with 10 C atoms and is called a monoterpene. Carotene, with 40 C atoms, is accordingly a tetraterpene. Rubber is a polyterpene

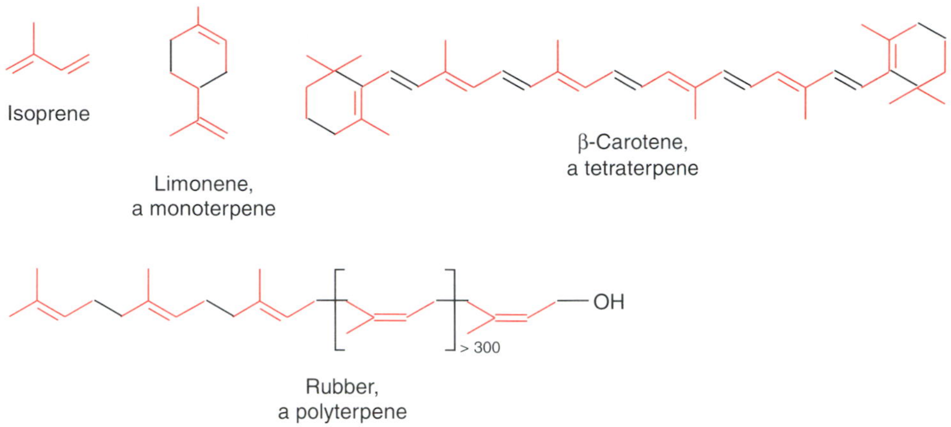

Figure 17.1 Various isoprenoids.

with about 1500 C atoms. It is obtained from the latex of the rubber tree, *Hevea brasiliensis*.

Otto Wallach (Bonn, Göttingen), who in 1910 was awarded the Nobel prize for chemistry for his basic studies on terpenes, recognized that isoprene is the basic constituent of terpenes (Fig. 17.1). Continuing these studies, Leopold Ruzicka (Zurich) found that isoprene is the universal basic element for the synthesis of very many natural substances, including steroids, and for this he was awarded the Nobel prize for chemistry in 1939. He postulated the biogenic isoprene rule, according to which all terpenoids (derivatives of terpenes) are synthesized via a hypothetical precursor which he named *active isoprene*. This speculation was verified by Feodor Lynen in Munich (Nobel prize for medicine, 1964), when he identified *isopentenyl pyrophosphate* as the much sought after 'active isoprene'.

17.1 Acetyl CoA is the precursor for the synthesis of isoprenoids

The basis for the elucidation of the biosynthetic pathway of isoprenoids was the discovery by Konrad Bloch (USA, a joint winner of the Nobel prize for medicine in 1964) that *acetyl CoA* is a precursor for the biosynthesis of steroids. Figure 17.2 shows the synthesis of the intermediary product isopentenyl pyrophosphate: two molecules of acetyl CoA react to produce *acetoacetyl CoA* and then with a further acetyl CoA to give *β-hydroxy-β-methylglutaryl CoA* (HMG-CoA). In yeast and animals these reactions are catalysed by two different enzymes, whereas in plants a single enzyme, *HMG-CoA synthase*, catalyses both reactions. The esterified carboxyl group of HMG-CoA is reduced by two molecules of NADPH to an alcohol group, accompanied by hydrolysis of the energy-rich thioester bond. Thus *mevalonate* is formed in an irreversible reaction. The formation of mevalonate from HMG-CoA is an important regulatory step of isoprenoid synthesis in

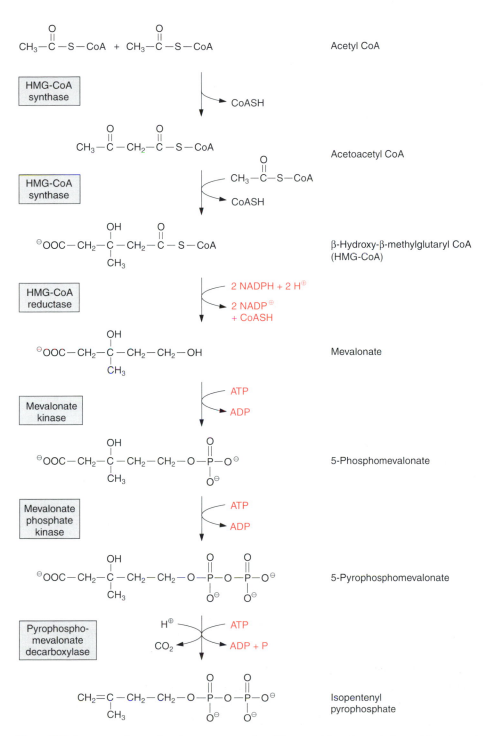

Figure 17.2 Isopentenyl pyrophosphate, precursor for all isoprenoid syntheses, is formed from acetyl-CoA.

animals. It has not yet been resolved whether this also applies to plants. A pyrophosphate ester is formed in two successive phosphorylation steps, catalysed by two different kinases. With consumption of a third molecule of ATP, involving the transitory formation of a phosphate ester, a carbon–carbon double bond is generated and the remaining carboxyl group is removed. Isopentenyl pyrophosphate thus formed is the basic element for the formation of an isoprenoid chain.

Very recently it was found that the mevalonate pathway described above is not the only pathway for isoprenoid biosynthesis in plants. In the non-mevalonate pathway, not shown in detail here, glyceraldehyde 3-phosphate reacts with pyruvate to form isopentenyl pyrophosphate, with the consumption of 3 molecules of NADPH and one of ATP and the release of CO_2. This novel non-mevalonate pathway was found to occur in plastids.

17.2 Prenyl transferases catalyse the association of isoprene units

Dimethylallyl pyrophosphate (dimethylallyl-PP), formed by isomerization of isopentenyl pyrophosphate, is the acceptor for successive transfers of isopentenyl residues (Fig. 17.3). With the liberation of the pyrophosphate residue, dimethylallyl-PP condenses with isopentenyl-PP to produce geranyl-PP. In an analogous way, chain elongation is attained by further head-to-tail condensations with isopentenyl-PP, and so farnesyl-PP and geranylgeranyl-PP are formed one after the other.

The transfer of the isopentenyl residue is catalysed by *prenyl transferases*. 'Prenyl residues' is a collective term for isoprene or polyisoprene residues. Recent results have shown that a special prenyl transferase is required for the production of each of the prenyl pyrophosphates mentioned. For example, the prenyl transferase termed *geranyl-PP synthase* catalyses only the synthesis of geranyl-PP. However, *farnesyl-PP synthase* synthesizes farnesyl-PP in two discrete steps: first geranyl-PP is formed, from dimethylallyl-PP and isopentenyl-PP, but this intermediate remains bound to the enzyme and reacts further with another isopentenyl-PP to give farnesyl-PP. Analogously *geranylgeranyl-PP synthase* catalyses all steps of the formation of geranylgeranyl-PP. Table 17.1 shows that each of these prenyl pyrophosphates is a precursor for the synthesis of certain isoprenoids. As these prenyl pyrophosphates are synthesized by separate enzymes, the synthesis of a certain prenyl pyrophosphate can be regulated by induction or repression of the corresponding enzyme.

The formation of a C–C linkage between two isoprenes proceeds by nucleophilic substitution (Fig. 17.4): an Mg^{2+} ion, bound to the prenyl transferase, facilitates the release of the negatively charged pyrophosphate residue from the acceptor molecule, whereby a positive charge remains at the terminal C atom (C-1), which is stabilized by the neighbouring double bond. The allyl cation thus formed reacts with the terminal C–C double bond of the donor molecule and a

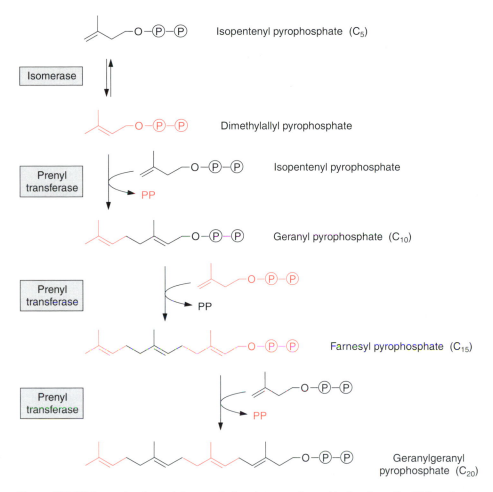

Figure 17.3 Higher molecular weight prenyl phosphates are formed by head to tail addition of active isoprene units.

new C–C bond is formed with the release of a proton. According to the same reaction mechanism, not only isoprene chains, but also rings are formed, producing
the unusual diversity of isoprenoids.

As shown in Table 17.1 the various prenyl phosphates are precursors for isoprenoids with very different structures and functions, which are classified as
hemiterpenes, monoterpenes, sesquiterpenes, etc.

17.3 Some plants emit isoprenes into the air

The *hemiterpene* isoprene is formed from dimethyallyl-PP upon the release of
pyrophosphate by an *isoprene synthase* which is present in many plants (Fig. 17.5).
Isoprene is volatile (boiling point 33 °C) and leaks from the plant in gaseous form.
Certain trees, such as oak, spruce, and poplar, emit isoprene during the day at tem-

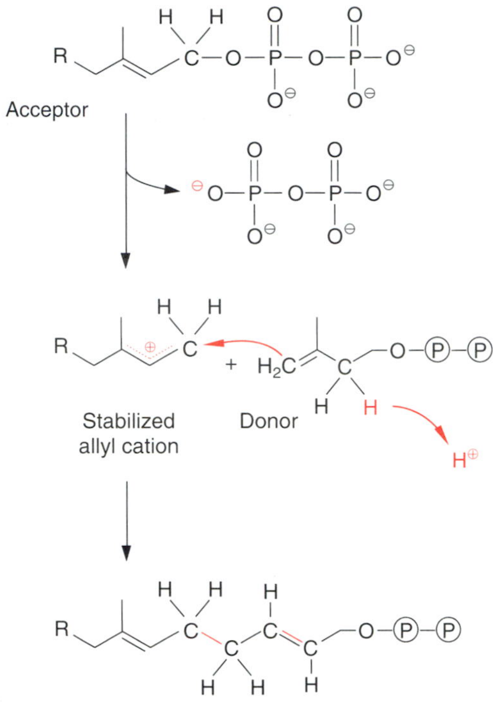

Figure 17.4 The head-to-tail addition of two prenyl phosphates by prenyl transferase is a nucleophilic substitution according to the S_N^1 mechanism. First pyrophosphate is released from the acceptor molecule. An allyl cation is formed which reacts with the double bond of the donor molecule and forms a new C–C bond. The double bond is restored by release of a proton, but shifted by one C atom. The reaction scheme is simplified.

peratures of 30 °C or above. At such high temperatures, as much as 15 per cent of the photosynthetically fixed carbon in oak leaves can be emitted as isoprenes. Isoprene emissions of up to 50 per cent of the total photoassimilate have been observed for the kudzu vine, a climbing plant which is grown in Asia as fodder. The isoprene emission of trees is responsible for the blue haze observed over forests at high temperatures.

Isoprene synthesis takes place in the chloroplasts. The precursor isopentenyl pyrophosphate was found to be synthesized by the non-mevalonate pathway (section 17.1). Isoprene synthase is induced when leaves are exposed to high temperatures. The physiological function of isoprene formation is not clear, but there are indications that low amounts of isoprenes stabilize photosynthetic membranes against high-temperature damage. The global isoprene emission by plants is considerable. It is several orders of magnitude above the anthropogenic hydrocarbon emission and is estimated to be about as high as the global methane emission.

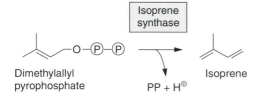

Dimethylallyl
pyrophosphate

Isoprene

PP + H$^{\oplus}$

Figure 17.5 By means of isoprene synthase some leaves form isoprene, which escapes as gas.

17.4 Many aromatic substances are derived from geranyl pyrophosphate

The monoterpenes comprise a large number of open chain and cyclic isoprenoids, many of which, due to their high volatility and their lipid character, are classed as essential oils. Mostly, they have a distinctive, often pleasant, scent and are, for example, responsible for the typical scent of pine needles, thyme, lavender, roses, and lily of the valley. As flower scents they attract insects for the distribution of pollen, but their main function is to repel insects and other animals and thus protect the plants from herbivores.

The hydrolysis of geranyl-PP results in the formation of the alcohol geraniol (Fig. 17.6), the main constituent of rose oil. Geraniol has the typical scent of freshly cut geraniums. Accompanied by a rearrangement, the hydrolysis of geranyl-PP can also yield linalool (Fig. 17.6), an aromatic substance in many flowers. In several members of the *Compositae* linalool is synthesized in the petals, where its biosynthesis commences simultaneously with the beginning of flowering.

The *cyclization* of geranyl-PP proceeds basically according to the same mechanism as the prenyl transferase reaction mentioned before, with the only difference that the reaction proceeds within the same molecule, often accompanied by an isomerization. Figure 17.7 shows as an example the synthesis of limonene by limonene synthase. Limonene is formed with a yield of 94 per cent by this enzyme-catalysed reaction, but also, as by-products, the very different compounds myrcene, α-pinene, and β-pinene are formed with yields of 2 per cent each. Thus a variety of substances can be synthesized by a single enzyme. These four monoterpenes are all major constituents of the resin of pine trees. They are toxic to many insects and protect against herbivores. Pine trees respond to an attack by bark beetles with a strong increase in cyclase activity which results in enhanced resin formation. Limonene is also found in the leaves and peel of lemons. Another example of a monoterpene is menthol (Fig. 17.6), the main constituent of peppermint oil. It serves the plant as an insect repellent. There are many other monoterpenes containing carbonyl and carboxyl groups, which are not dealt with here.

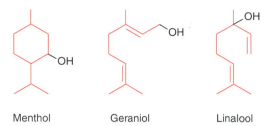

Menthol   Geraniol   Linalool

Figure 17.6 Menthol, a constituent of peppermint oil; geraniol, a constituent of rose oil, an aromatic substance in geraniums; and linalool, an aromatic substance of the Compositae.

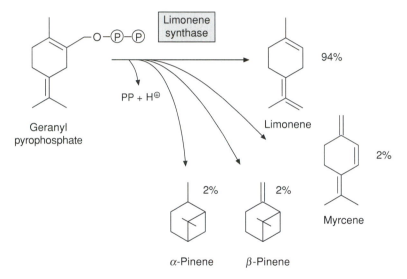

Figure 17.7 Examples of reactions catalysed by a monoterpene cyclase. The limonene synthase forms limonene with a yield of 94 per cent, and as by-products myrcene and α- and β-pinene, with yields of 2 per cent each.

17.5 Farnesyl pyrophosphate is the precursor for the formation of sesquiterpenes

The number of possible products is even greater for the cyclization of farnesyl-PP, proceeding according to the same mechanism as the cyclization of geranyl-PP described before. This is illustrated in Fig. 17.8. The reaction of the intermediary carbonium ion with the two double bonds of the molecule alone can lead to four different products. The number of possible products is multiplied by simultaneous rearrangements. Sesquiterpenes form the largest group of isoprenoids: they comprise more than 200 different ring structures. The sesquiterpenes include many aromatic substances such as eucalyptus oil and also several constituents of hops. Capsidiol (Fig. 17.9), a phytoalexin (section 16.1) formed in pepper and tobacco, is a sesquiterpene.

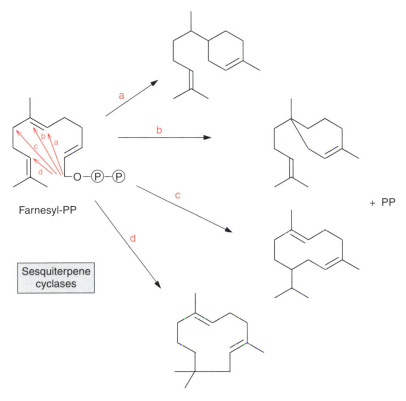

Figure 17.8 Without rearrangement of the double bonds there are four different possibilities for cyclization of farnesyl pyrophosphate.

Capsidiol

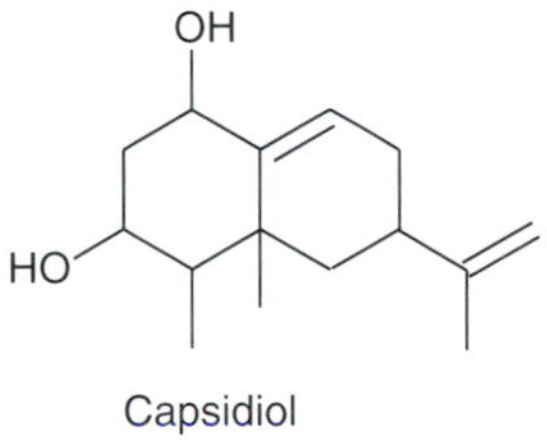

Figure 17.9 A phytoalexin from pepper and tobacco.

Steroids are synthesized from farnesyl pyrophosphate

The triterpene squalene is formed from two molecules of farnesyl-PP by an NADPH-dependent *reductive head-to-head condensation* (Fig. 17.10). Squalene is the precursor for membrane constituents such as cholesterol and sitosterol, the functions of which have been dealt with in section 15.1.

A class of glycosylated steroids, named saponins because of their soap-like properties (Fig. 17.11), function in plants as toxins against herbivores. The glycosyl moiety of the saponins consists of a branched oligosaccharide formed from

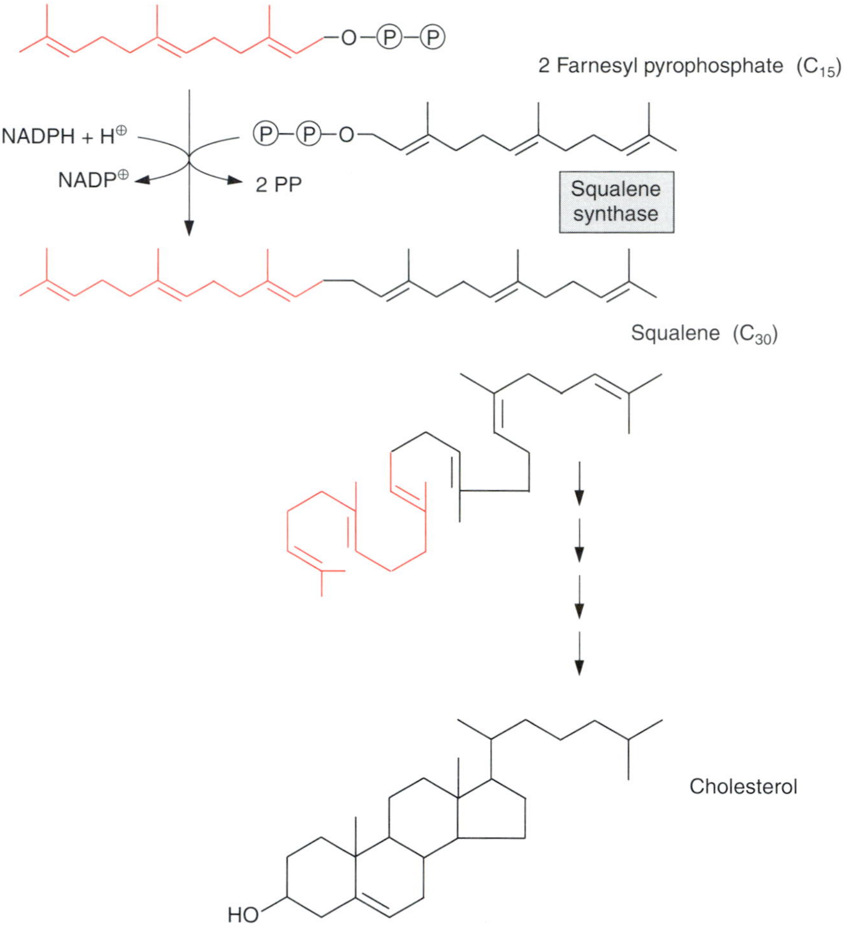

Figure 17.10 Squalene is formed from two farnesyl-PP molecules by head-to-head addition, accompanied by a reduction. After the introduction of an -OH group by a mono-oxygenase and a cyclization, cholesterol is formed in several reaction steps.

glucose, galactose, xylose, and other hexoses. The hydrophilic polysaccharide chain and the hydrophobic steroid give the saponins the properties of a *detergent*. The saponin toxicity damages the membranes. After ingestion, saponins can cause haemolysis of red blood cells. Saponins, as constituents of some grasses, are a hazard for grazing cattle. Yamonin, a saponin from the yam plant, is used in the pharmaceutical industry as a raw material for the synthesis of progesterones, components of contraceptive pills. A number of very toxic glycosylated steroids named *cardenolides*, which inhibit the Na^+–K^+ pump present in animals, also belong to the saponin family. A well-known member of this class of substances is digitoxigenin, a poison from foxgloves (formula not shown). Larvae of certain butterflies can ingest cardenolides without being harmed. They store these sub-

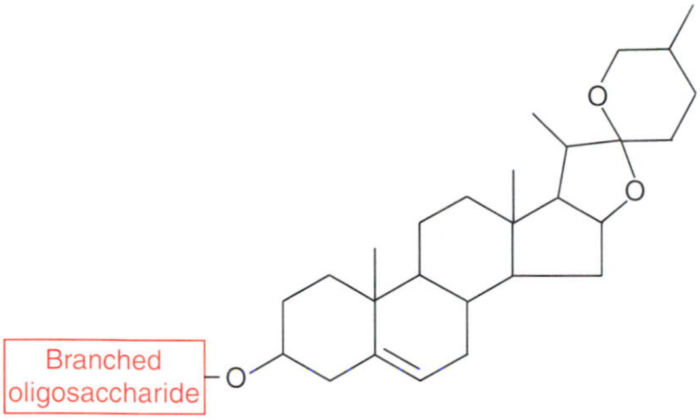

Yamonin, a saponin

Figure 17.11

stances, which then makes them poisonous for birds. In low doses cardenolides are used as a medicine against heart disease. Other defence substances are the *phytoecdysones*, a group of steroids with a structure similar to the insect hormone ecdysone. Ecdysone controls the moulting cycle of larvae. When insects eat plants containing phytoecdysone the moulting process is disturbed and the larvae die.

17.6 Geranylgeranyl pyrophosphate is the precursor for defence substances, phytohormones, and carotenoids

The cyclization of geranylgeranyl-PP leads to the formation of the diterpene *casbene* (Fig. 17.12). Casbene is formed as a phytoalexin (section 16.1) in castor bean. The diterpene *phorbol* is present as an ester in the latex of plants of the spurge family. Phorbol acts as a toxin against herbivores; even skin contact causes severe inflammation. Since phorbol esters induce the formation of tumours, they are widely used in medical research. Geranylgeranyl-PP is also a precursor for the synthesis of gibberellins, a group of phytohormones (section 19.2).

The function of *carotenoids* in photosynthesis has been dealt with in detail in Chapters 2 and 3. Additionally, carotenoids function as pigments, e.g. in flowers and paprika fruits. The synthesis of carotenoids requires two molecules of geranylgeranyl-PP, which, as in the formation of squalene, are linked by head-to-head condensation (Fig. 17.13). Upon release of the first pyrophosphate the intermediate pre-phytoene pyrophosphate is formed and release of the second pyrophosphate then results in the formation of *phytoene*. Here the two prenyl residues are linked to each other by a carbon–carbon double bond. The phytoene is converted to *lycopene*, probably catalysed by two different desaturases. According to recent results, these desaturations proceed via dehydrogenation reactions in which hydrogen is transferred to NADP or FAD. Cyclization of lycopene then results in the formation of *β-carotene*. Another cyclase generates α-

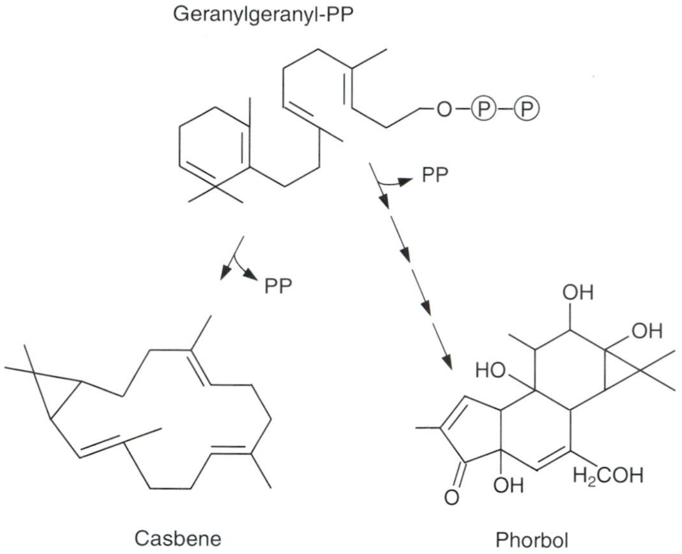

Figure 17.12 The phytoalexin casbene is formed in one step by cyclization from geranylgeranyl pyrophosphate. The synthesis of the defence substance phorbol requires several steps, including hydroxylations, which are in part catalysed by mono-oxygenases.

carotene. The hydroxylation of β-carotene leads to the xanthophyll *zeaxanthin*. The formation of the xanthophyll violaxanthin from zeaxanthin is described in Fig. 3.42.

17.7 A prenyl chain renders substances lipid soluble

Ubiquinone (Fig. 3.5), plastoquinone (Fig. 3.19), and cytochrome-*a* (Fig. 3.24) are anchored in membranes by isoprenoid chains of various sizes. In the biosynthesis of these electron carriers, the prenyl chains are introduced from prenyl phosphates by reactions similar to those catalysed by prenyl transferases. Chlorophyll (Fig. 2.4), tocopherols, and phylloquinone (Fig. 3.32), on the other hand, contain phytol side chains. These are formed from geranylgeranyl-PP by reduction with NADPH and similarly incorporated (Fig. 17.14).

Proteins can be anchored in a membrane by prenylation

Recently a large number of proteins have been found in yeast and animals which are linked at a cysteine residue near the C-terminus by a thioether bond to a farnesyl or geranyl residue (Fig. 17.15). The linking of these molecules is catalysed by a specific prenyl transferase. In many cases, the terminal amino acids linked to the cysteine residue are cut off after prenylation by a peptidase, and the carboxylic group of the cysteine is methylated. In this way the protein becomes

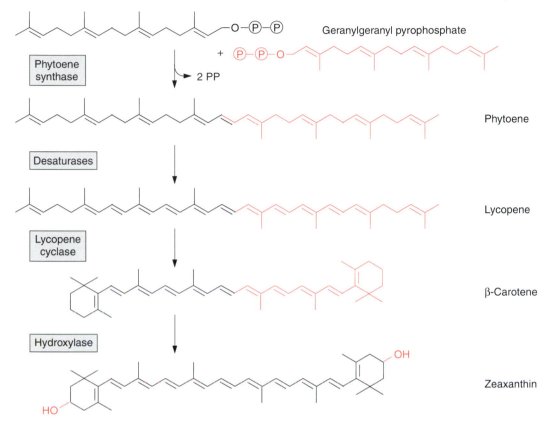

Figure 17.13 Carotenoid biosynthesis. The phytoene synthase catalyses the head-to-head addition of two geranylgeranyl-PPs to phytoene. The latter is converted by desaturases with neurosporene as intermediate (not shown) to lycopene. β-Carotene is formed by cyclization and zeaxanthin by hydroxylation.

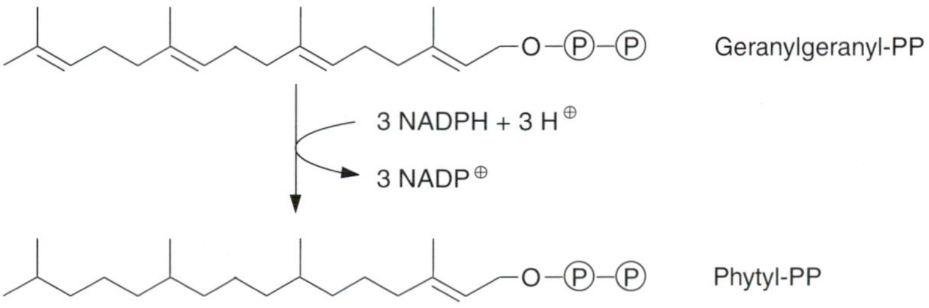

Figure 17.14 Synthesis of phytyl-PP from geranylgeranyl-PP.

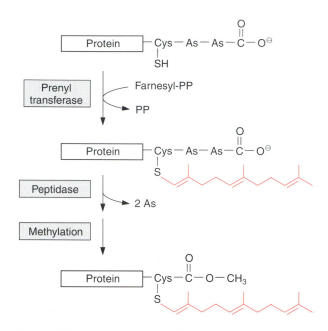

Figure 17.15 Prenylation of a protein. A farnesyl residue is transferred to the SH group of a cysteine residue of the protein by a prenyl transferase. After hydrolytic release of the terminal amino acids the carboxyl group of the cysteine is methylated. The prenyl residue provides the protein with a membrane anchor.

lipid soluble and can be anchored in a membrane. Recent results indicate that the prenylation of proteins also plays an important role in plants.

Dolichols mediate the glycosylation of proteins

Dolichols (Fig. 17.16) are isoprenoids with very long chains, occurring in the membranes of the endoplasmic reticulum and the Golgi network. They have an important function as *membrane-soluble carriers of oligosaccharides*. Many membrane proteins and secretory proteins are glycosylated by branched oligosaccharide chains. This glycosylation proceeds in the endoplasmic reticulum utilizing membrane-bound dolichol (Fig. 17.17). The oligosaccharide structure is first synthesized when attached to dolichol, and only after completion is transferred to an asparagine residue of the protein to be glycosylated. By subsequent modification in the Golgi network, in which certain carbohydrate residues are split off and others added, a large variety of oligosaccharide structures is generated.

17.8 The regulation of isoprenoid synthesis

The capacity for isoprenoid synthesis is widely distributed throughout the plant. Large amounts of hydrophobic isoprenoids are synthesized mostly in tissues specially adapted for this purpose, such as the glandular hairs of leaves (menthol)

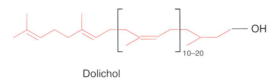

Dolichol

Figure 17.16 A polyprenol.

or the petals (linalool). The enzymes for synthesis of isoprenoids are usually present at three different locations in a cell: in the plastids, the mitochondria, and the cytosol. Therefore, each of these cellular compartments is essentially self-sufficient with respect to its isoprenoid content. Some isoprenoids, such as the phytohormone gibberellic acid, are synthesized in the plastids and then supplied to the cytosol of the cell. As mentioned in section 17.2, the various prenyl pyrophosphates, from which all the other isoprenoids are derived, are synthesized by different enzymes.

This spatial distribution of the synthetic pathways makes it possible that, in spite of their very large diversity, the different isoprenoids formed by basically similar

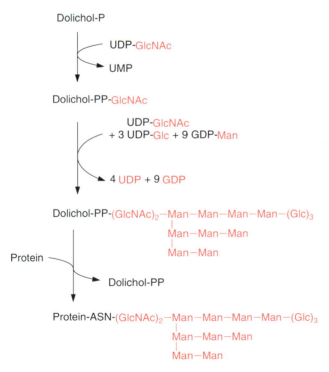

Figure 17.17 Dolichol as glycosyl carrier. For the synthesis of a branched oligosaccharide structure, N-acetylglucosamine (GlcNAc), mannose (Man), and glucose (Glc) are transferred from the corresponding UDP- and GDP-compounds to dolichol. The complete oligosaccharide is then transferred to an asparagine residue of the protein.

processes can be efficiently controlled in their rate of synthesis via regulation of the corresponding enzyme activities in the various compartments by specific control of gene expression. Results so far indicate that the synthesis of the different isoprenoids is mainly regulated by gene expression. This is especially obvious when, after infections or wounding, the isoprenoid metabolism is changed very rapidly by elicitor-controlled gene expression (section 16.1). This can lead to competition between the pathways. In tobacco cells, for instance, phytoalexin synthesis, induced by a fungal elicitor, blocks steroid synthesis. In such a case, the cell focuses its capacity for isoprenoid synthesis on defence.

17.9 Isoprenoids are very stable and persistent substances

Little is known about the catabolism of isoprenoids in plants. Biologically active derivatives, such as phytohormones, are converted by the introduction of further hydroxyl groups and by glycolysation into inactive substances, which are often finally deposited in the vacuole. It is questionable whether, after degradation, isoprenoids can be recycled in a plant. Some isoprenoids are remarkably stable. Large amounts of isoprenoids are found as relics of early life in practically all sedimentary rocks and also in crude oil. In archaebacteria the plasma membranes contain glycerol ethers with isoprenoid chains instead of fatty acid glycerol esters. Isoprenoids are probably constituents of very early forms of life.

Further reading

Bartley, G. E. and Scolnik, P. A. (1995). Plant carotenoids: pigments for photoprotection, visual attraction and human health. *The Plant Cell*, **7**, 1027–38.

Bell, E. A. and Charlwood, B. V. (ed.) (1980). Secondary plant products. *Encyclopedia of Plant Physiology*, Vol. 8. Springer-Verlag, Berlin.

Chappel, J. (1995). Biochemistry and molecular biology of the isoprenoid biosynthetic pathway in plants. *Annual Review of Plant Physiology and Plant Molecular Biology*, **45**, 521–47.

Connolly, J. D. and Hill, R. A. (1991). *Dictionary of terpenoids*. Chapman & Hall, London.

Gershenon, J. and Croteau, R. B. (1993). Terpenoid biosynthesis: the basic pathway and formation of monoterpenes, sesquiterpenes and diterpenes. In *Lipid metabolism in plants*, (ed. T. S. Moore), pp, 339–88. CRC Press, Boca Raton, Florida.

Kleinig, H. (1989). The role of plastids in isoprenoid biosynthesis. *Annual Review of Plant Physiology and Plant Molecular Biology*, **40**, 39–59.

Lichtenthaler, H. K., Schwender, J., Disch, A., and Rohmer, M. (1997). Biosynthesis of isoprenoids in higher plant chloroplasts proceeds via a mevalonate-independent pathway *FEBS·Letters*, **400**, 271–4.

McGarvey, D. J. and Croteau, R. (1995) Terpenoid metabolism. *The Plant Cell*, **7**, 1015–26.

Sharkey, T. D. (1996). Isoprene synthesis by plants and animals. *Endeavor*, **20**, 74–8.

chapter 18

Phenylpropanoids

Plants contain a large variety of phenolic derivatives. As well as simple phenols these comprise flavonoids, stilbenes, tannins, lignans, and lignin (Fig. 18.1). Together with long-chain carboxylic acids, phenolic compounds are also components of suberin and cutin. These rather varied substances have important functions as antibiotics, natural pesticides, signal substances for the establishment of symbiosis with rhizobia, attractants for pollinators, protective agents against ultraviolet (UV) light, insulating materials to make cell walls impermeable to gas and water, and as structural material to give plants stability (Table 18.1). All these substances are derived from phenylalanine, and in some plants, also from tyrosine. Phenylalanine and tyrosine are formed by the shikimate pathway, described in section 10.4. Since the phenolic compounds derived from the two amino acids

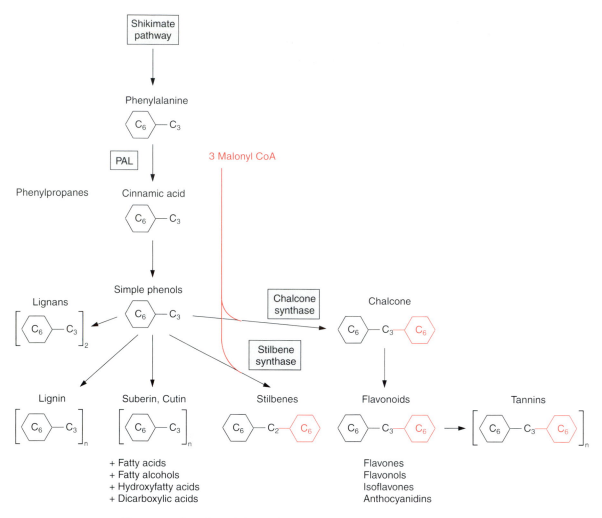

Figure 18.1 Overview of the products of phenylpropanoid metabolism. Cinnamic acid, formed from phenylalanine by phenylalanine ammonia lyase (PAL) is precursor for the various phenylpropanoids. In some plants 4-hydroxycinnamic acid is formed from tyrosine in an analogous way (not shown in the figure). An additional aromatic ring is formed either by chalcone synthase or stilbene synthase from three molecules of malonyl CoA.

contain a phenyl ring with a C_3 side chain they are collectively termed *phenyl-propanoids*. The flavonoids, including flavones, isoflavones, and anthocyanidins, contain, as well as the phenylpropane structure, a second aromatic ring which is formed from three molecules of malonyl CoA (Fig. 18.1). This also applies to the stilbenes, but here, after the introduction of the second aromatic ring, one C atom of the phenylpropane is split off.

Table 18.1 Some functions of phenylpropanoids

Coumarins	Antibiotics, toxins against browsing animals
Lignans	Antibiotics, toxins against browsing animals
Lignin	Cell wall constituent
Suberin and cutin	Formation of impermeable layers
Stilbenes	Antibiotics, especially fungicides
Flavonoids	Antibiotics, signal for interaction with symbionts, flower pigments, light protection substances
Tannin	Tanning agents, fungicides, protection against herbivores

18.1 Phenylalanine ammonia lyase catalyses the initial reaction of phenylpropanoid metabolism

Phenylalanine ammonia lyase, abbreviated to PAL, catalyses a deamination of phenylalanine (Fig. 18.2): a carbon–carbon double bond is formed with the release of NH_3, yielding *trans*-cinnamic acid. In some grasses tyrosine is converted into 4-hydroxycinnamic acid in an analogous way by *tyrosine ammonia lyase*. The released NH_3 is probably refixed by the glutamine synthetase reaction (Fig. 7.9).

   PAL is one of the most intensively studied enzymes of plant secondary metabolism. The enzyme consists of a tetramer with subunits of 77–83 kDa. The formation of phenylpropanoid phytolalexins after fungal infection involves a very rapid induction of PAL. PAL is inhibited by its product *trans*-cinnamic acid. The phenylalanine analogue aminoxyphenylpropionic acid (Fig. 18.3) is also a very potent inhibitor of PAL.

18.2 Mono-oxygenases are involved in the synthesis of phenols

The introduction of the hydroxyl group into the phenyl ring of cinnamic acid

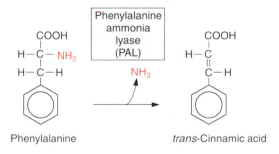

Phenylalanine   *trans*-Cinnamic acid

Figure 18.2 Formation of *trans*-cinnamic acid.

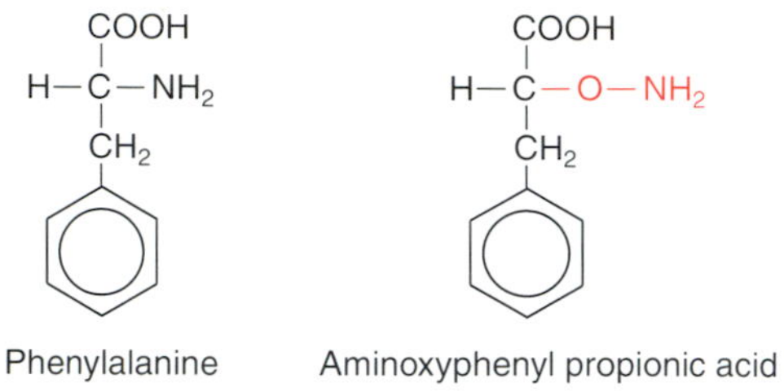

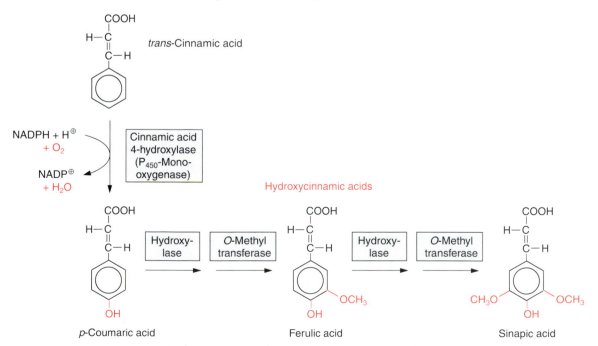

Figure 18.3 Aminoxyphenylpropionic acid, a structural analogue of phenylalanine, inhibits PAL.

(Fig. 18.4) proceeds via a *mono-oxygenase-catalysed reaction* utilizing cytochrome P_{450} as the O_2 binding site according to:

$$NADPH + H^+ + R\text{-}CH_3 + O_2 \rightarrow NADP^+ + RCH_2\text{-}OH + H_2O$$

Like other P_{450} mono-oxygenases this enzyme is bound at the membranes of the endoplasmic reticulum. With consumption of NADPH one O atom of the O_2 molecule is reduced to water and the other is transferred to the phenyl ring. *p*-Coumaric acid formed in this way can be hydroxylated further in positions 3 and 5 by hydroxylases, for which the reaction mechanisms have not yet been resolved. The -OH groups thus generated are mostly methylated via *O*-methyl transferases with *S-adenosylmethionine* as methyl donor (Fig. 12.10). In this way ferulic and sinapic acids are formed, which, together with *p*-coumaric acid, are precursors for the synthesis of lignin (section 18.3).

Figure 18.4 Synthesis of various hydroxycinnamic acids from *trans*-cinnamic acid.

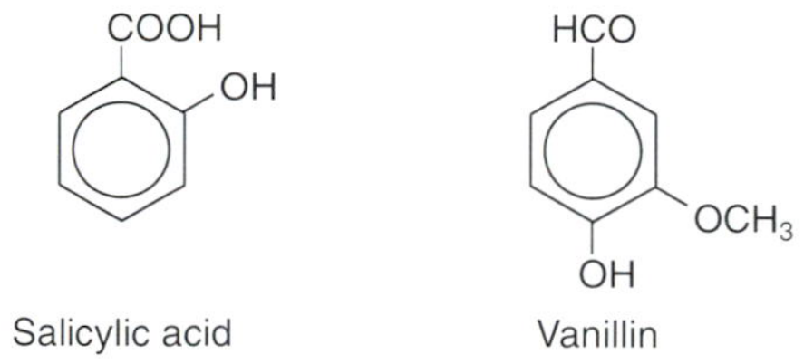

Figure 18.5 Salicylic acid and vanillin are formed from phenylpropanoids.

Benzoic acid derivatives are formed by cleavage of a C_2 fragment from the phenylpropanes. *Salicylic acid* (Fig. 18.5) acts as a signal substance. After infection or UV irradiation, many plants increase their salicylic acid content and this presumably induces the biosynthesis of defence substances. Salicylic acid induces heat production in the spadix of the foul-smelling voodoo lily by stimulating mitochondrial alternative oxidase (section 5.7). There are also reports that in some plants salicylic acid may be involved in flower induction.

The acetyl ester of salicylic acid is widely used as aspirin, as a remedy against pain and fever. The name salicylic acid is derived from *salix*, the Latin name for willow, since it was first isolated from the bark of the willow tree, where it occurs in particularly high amounts. Since ancient times the salicylic acid content of willow bark has led to its being used as medicine in the Old and New Worlds. In the fourth century BC Hippocrates gave women willow bark to chew to relieve pain during childbirth. The north American Indians also used extracts of willow bark as painkillers. The aroma substance vanillin is a benzaldehyde (Fig. 18.5).

7-Hydroxycoumarin, also called *umbelliferone*, is synthesized from *p*-coumaric acid by hydroxylation and formation of an intermolecular ester, a lactone (Fig. 18.6). The introduction of a C_2 group into umbelliferone yields psoralen, a

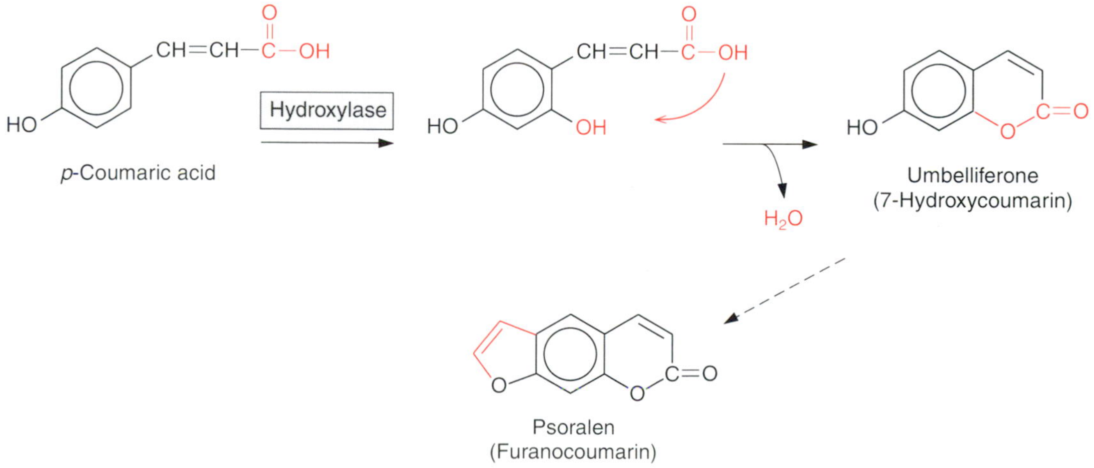

Figure 18.6 Umbelliferone, which is a precursor for the synthesis of the defensive substance psoralen, is formed by hydroxylation of *p*-coumaric acid and the formation of a ring.

furanocoumarin. Illumination with UV light turns psoralen into a toxic substance. The illuminated psoralen reacts with the pyrimidine bases of DNA, causing blockage of transcription and DNA repair mechanisms and finally resulting in the death of the cell. As mentioned in section 16.1, some celery varieties contain very high concentrations of psoralen and have given workers involved in harvesting severe skin inflammation. Many furanocoumarins have antibiotic properties. In some cases they are constitutive components of the plants and in other cases they are formed only after infection or wounding as phytoalexins.

18.3 Phenylpropanoid compounds polymerize to form macromolecules

It has already been mentioned in section 1.1 that, after cellulose, *lignin* is the second most abundant natural substance on earth. The basic components for lignin synthesis are *p*-coumaryl, sinapyl, and coniferyl alcohols, collectively termed *monolignols* (Fig. 18.7). Synthesis of the monolignols requires reduction of the carboxylic group of the corresponding acids to an alcohol. When discussing the glyceraldehyde phosphate dehydrogenase reaction in section 6.3 it was shown

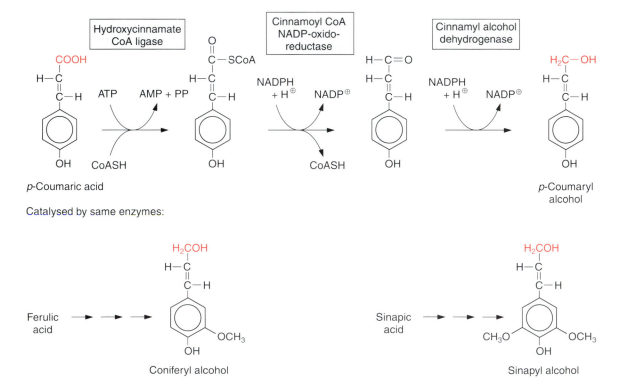

Figure 18.7 Reduction of the hydroxycinnamic acids to the corresponding alcohols (monolignols).

that a carboxyl group can only be reduced by NADPH to an aldehyde if it is activated beforehand via the formation of a thioester. For the reduction of *p*-coumaric acid by NADPH (Fig. 18.7) a similar activation occurs. The required thioester is formed with CoA at the expense of ATP in a reaction analogous to fatty acid activation described in section 15.6. The cleavage of the energy-rich thioester bond drives the reduction of the carboxylate to the aldehyde. In the subsequent reduction to an alcohol, NADPH is again reductant. The same three enzymes that catalyse the conversion of *p*-coumaric acid to *p*-coumaryl alcohol (Fig. 18.7) may also catalyse the formation of sinapyl and coniferyl alcohols from sinapic and ferulic acids.

Lignans act as defence substances

The dimerization of monolignols leads to the formation of *lignans* (Fig. 18.8). This takes place mostly by a reductive linkage of the side chains at the 8,8-position, but

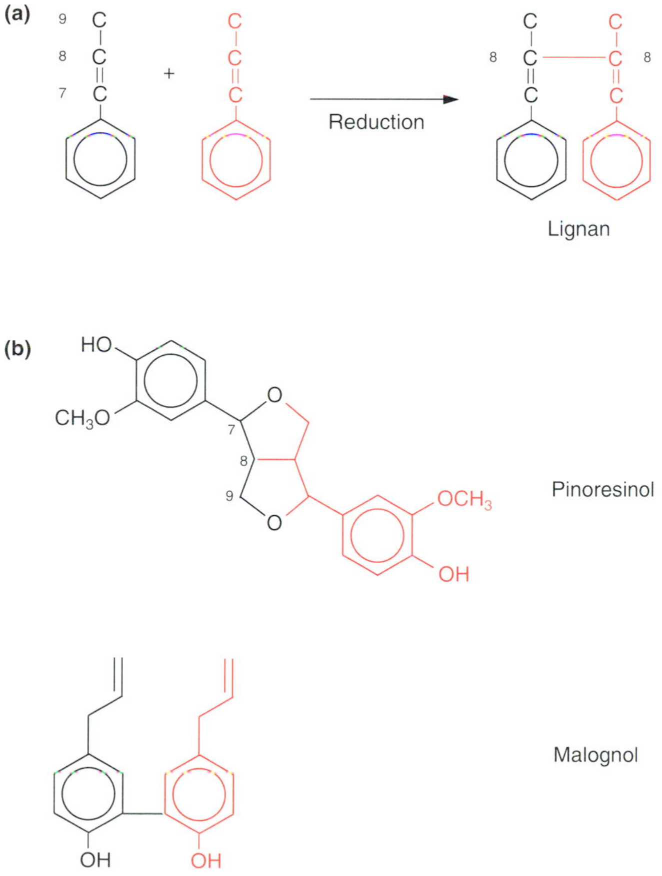

Figure 18.8 (a) Lignans are formed by dimerization of monolignols; (b) examples of two lignans.

sometimes also by a condensation of the two phenol rings. The mechanism of lignan formation is still unclear. Free radicals are probably involved (see next section). Plant lignans also occur as higher oligomers. In the plant world lignans are widely distributed as defensive substances. The lignan pinoresinol is a constituent of the resin of forsythia and is formed when the plant is wounded. Its toxicity to micro-organisms is caused by an inhibition of cAMP phosphodiesterase. Pinoresinol thus disrupts the action of cAMP which, according to our present state of knowledge, acts as a messenger substance in cells of animals and fungi, but not of plants. Malognol inhibits the growth of bacteria and fungi.

Some lignans have interesting pharmacological effects. Two examples may be used to illustrate this: podophyllotoxin, from *Podophyllum*, a member of the family *Berberidacea* occurring in America, is a mitosis toxin. Derivatives of podophyllotoxin are used to cure cancer. Arctigenin and tracheologin (from tropical climbing plants) have antiviral properties. Investigations are under way to try to utilize this property to cure AIDS.

Lignin is formed by radical polymerization of phenylpropanoid derivatives

Lignin is formed by polymerization of a mixture of the three monolignols *p*-coumaryl, sinapyl, and coniferyl alcohol. The synthesis of lignin takes place outside the cell, but the mechanism by which the monolignols are secreted from the cell for lignin synthesis is still unknown. There are indications that the monolignols are exported as glucosides which are hydrolysed outside the cell by glucosidases, but this is still a matter of controversy. The mechanism of lignin formation also remains unclear. Several results indicate that *peroxidases* are involved in linking the monolignols, although the origin of the required H_2O_2 is uncertain. As shown in Fig. 18.9a the oxidation of a phenol by H_2O_2 presumably results in the formation of a resonance-stabilized phenol radical. These phenol radicals can dimerize and finally polymerize (Fig. 18.9b). Due to the various resonance structures, many combinations are possible in the polymer. Usually monolignols react to form several C–C or C–O–C linkages, forming an unordered, highly branched phenylpropanoid polymer. Free hydroxyl groups are present in only a few side chains of lignin and are sometimes oxidized to aldehyde and carboxyl groups.

The composition of lignin varies greatly in different plants. Lignin of conifers, for instance, has a high coniferyl alcohol content, whereas in the straw of cereals the coumaryl moiety prevails. Lignin is covalently bound to cellulose in the cell walls. Lignified cell walls have been compared to reinforced concrete in which the cellulose fibres are the steel and lignin the concrete. As well as to give mechanical strength to plant parts such as stems or twigs, or to provide stability for the vascular tissues of the xylem, lignin also has a function in defence. Its mechanical strength and chemical composition make plant tissues difficult for herbivores to

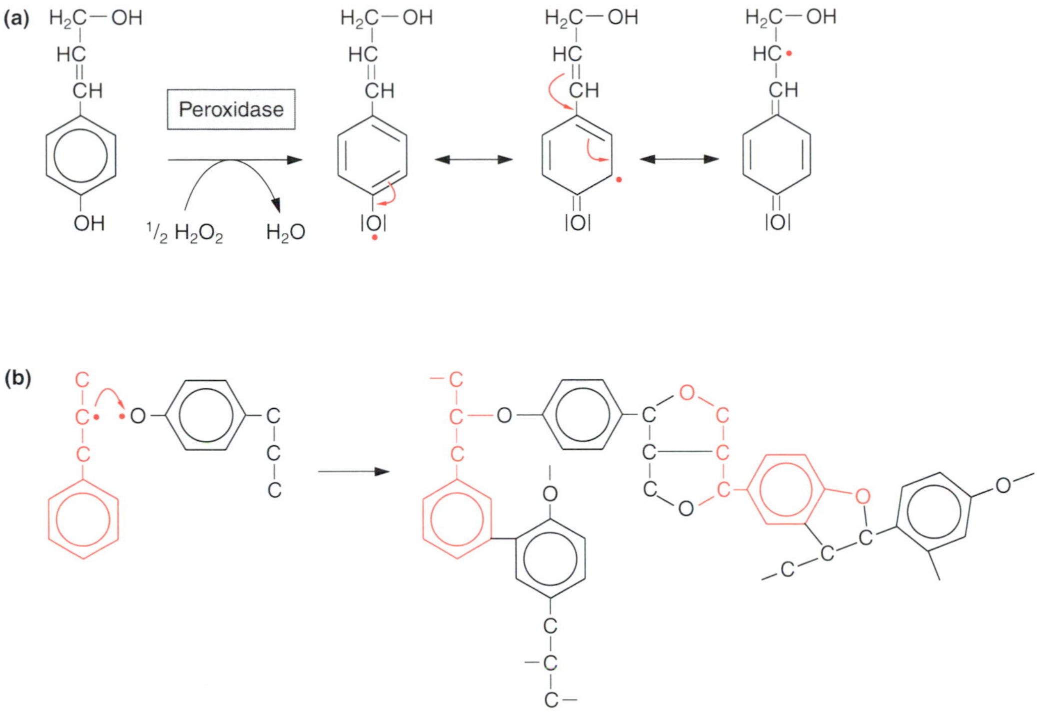

Figure 18.9 (a) Oxidation of a monolignol by H_2O_2 results in the formation of a phenol radical. The unpaired electron is delocalized and can react with various resonance structures of the monolignol. (b) Two monolignols can form a dimer and polymerize further. Finally, highly branched lignan polymers are formed.

digest. In addition, lignin inhibits the growth of pathogenic micro-organisms. Lignin is formed in many plants in response to wounding. Only a few bacteria and fungi are able to cleave lignin. A special role in the degradation of lignin is played by woodrot fungi which are involved in the rotting of tree trunks.

Suberins form gas- and water-impermeable layers between cells

Suberin is a polymeric compound of phenylpropanoids, long-chain fatty acids and fatty alcohols (C_{18}–C_{30}), and also of hydroxyfatty acids and dicarboxylic acids (C_{14}–C_{20}) (Fig. 18.10). In suberin the phenylpropanoids are partly linked with each other as in lignin. However, the 9'-OH groups are mostly not involved in these linkages and instead form esters with the fatty acids mentioned above. Often two phenylpropanoids are connected by dicarboxylic acids via ester linkages, and fatty acids and hydroxyfatty acids can also form esters with each other. The mechanism of suberin synthesis is to a large extent still unknown.

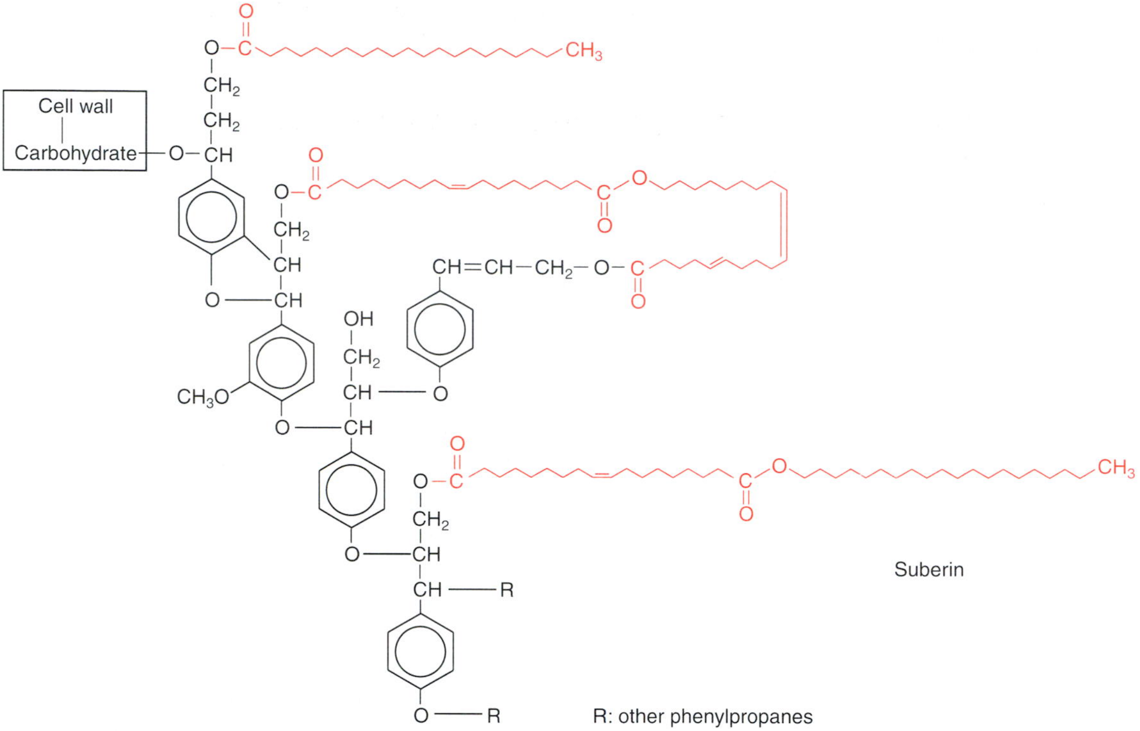

Figure 18.10 In suberin the monolignols are connected similarly to those in lignin, but the 9′-OH groups do not participate. Instead they form esters with long-chain fatty acids and hydroxyfatty acids. Carboxylic acid esters provide a link between two monolignols.

Suberin is a cell wall constituent that forms gas- and water-tight layers. It is probably contained in the *Casparian strip* of the root endodermis, where it may represent the barrier between the apoplast of the root cortex and the central cylinder. Suberin is present in many C_4 plants as an impermeable layer between the bundle sheath and mesophyll cells. Cork tissue, consisting of dead cells surrounded by alternating layers of suberin and wax, has a particularly high suberin content. *Cork cells* are found in a secondary protective layer called the periderm and also in the bark of trees. Cork layers containing suberin protect plants against loss of water, infection by micro-organisms, and heat exposure. Due to this, some plants even survive short fires and afterwards are able to continue growing.

Cutin is a gas- and water-impermeable constituent of the cuticle

The epidermis of leaves and other shoot organs is surrounded by a gas- and water-impermeable cuticle (Chapter 8). This consists of a cell wall which is impregnated with cutin and in addition covered by a wax layer. Cutin is a polymer similar to suberin, but with a relatively small proportion of phenylpropanoids and dicarboxylic acids, consisting mainly of esterified hydroxyfatty acids (C_{16}–C_{18}).

18.4 For the synthesis of flavonoids and stilbenes a second aromatic ring is formed from acetate residues

Probably the largest group of phenylpropanoids is that of the flavonoids, in which a second aromatic ring is linked to the 9′-C atom of the phenylpropanoid moiety. A precursor for the synthesis of flavonoids is chalcone (Fig. 18.11), synthesized by *chalcone synthase* (CHS) from *p*-coumaryl CoA and three molecules of malonyl CoA. This reaction is also named the *malonate pathway*. The release of three CO_2 and four CoA molecules makes chalcone synthesis an irreversible process. In the overall reaction the new aromatic ring is formed from three acetate residues. Since CHS represents the first step of flavonoid biosynthesis, this enzyme has been thoroughly investigated. In some plants one or two different isoforms of the enzyme have been found, while in others there are up to nine. CHS is the most abundant enzyme of phenylpropanoid metabolism in plant cells, probably due to the fact that this enzyme has only a low catalytic activity. As in the case of phenylalanine ammonia lyase (section 18.1), the *de novo* synthesis of CHS is subject to multiple controls of gene expression by internal and external factors, including elicitors.

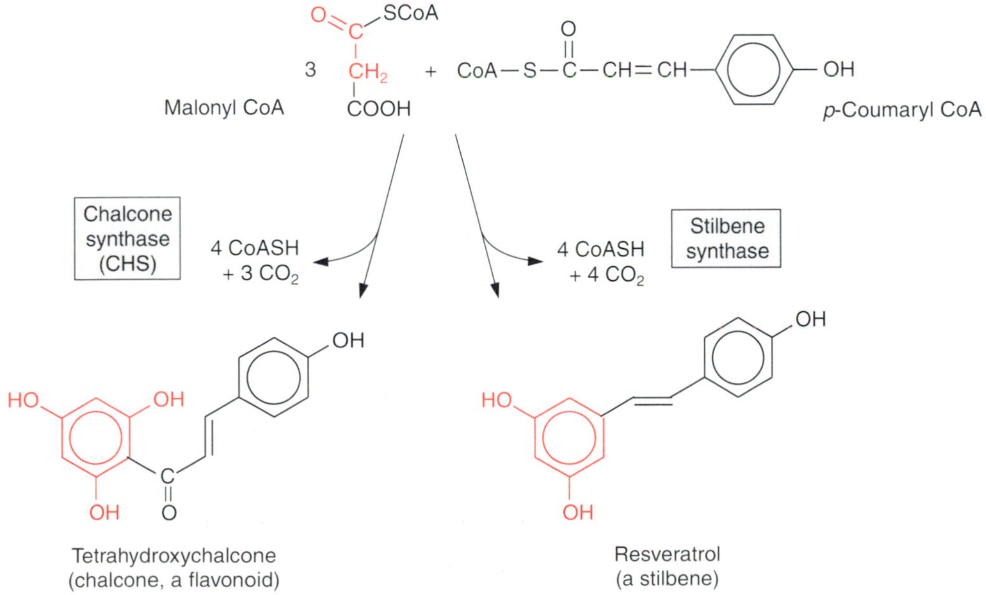

Figure 18.11 An additional aromatic ring is formed by chalcone synthase and stilbene synthase.

The stilbenes include very potent natural fungicides

Some plants, including pine, grapevine, and peanuts, possess a *stilbene synthase* activity, by which *p*-coumaryl CoA reacts with three molecules malonyl CoA. In contrast to CHS, the 9'-C atom of the phenylpropane is released as CO_2 (Fig. 18.11). Resveratrol, synthesized by this process, is a phytoalexin belonging to the stilbene group. A number of very potent plant fungicides derive from the stilbenes, including *viniferin* (Fig. 18.12), which is formed in grapevine. The elucidation of stilbene synthesis has opened new possibilities to combat fungal infections. Recently, a gene from grapevine for the formation of resveratrol has been expressed by genetic engineering in tobacco, and the resultant transgenic tobacco plants were resistant to the pathogenic fungus *Botrytis cinerea*.

18.5 Flavonoids have multiple functions in plants

Chalcone is converted to flavanone by *chalcone isomerase* (Fig. 18.13). The ring structure is formed by the addition of a phenolic hydroxyl group to the double bond of the carbon chain connecting the two phenolic rings. Flavanone is precursor for a variety of flavonoids, the synthesis of which will not be described here. As a key enzyme of flavonoid synthesis, the synthesis of the enzyme protein of chalcone isomerase is subject to a strict control. It is induced together with PAL and CHS by elicitors.

The flavonoids include many phytoalexins. An example of this is the isoflavone medicarpin from alfalfa (Fig. 18.14). Flavonoids also serve as signals for interactions of the plant with symbionts. Flavones and flavonols are emitted as signal substances from leguminous roots in order to induce in rhizobia the expression of the genes required for nodulation (section 11.1).

Flavones and flavonols have an absorption maximum in the ultraviolet region. As *protective pigments* they shield plants from the damaging effect of ultraviolet (UV) light. The irradiation of leaves with ultraviolet light induces a strong

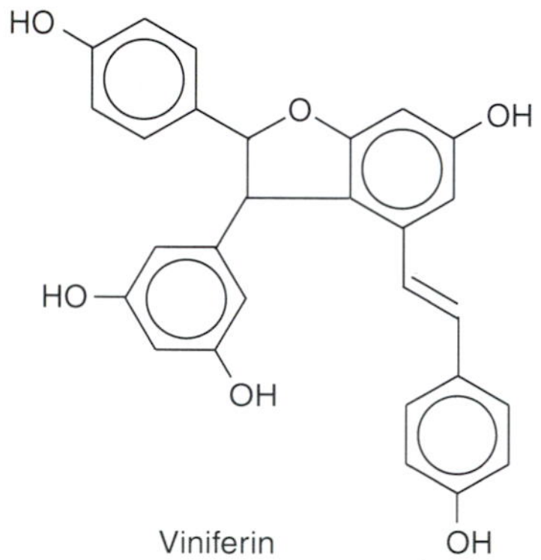

Viniferin

Figure 18.12 A natural fungicide from grapevine.

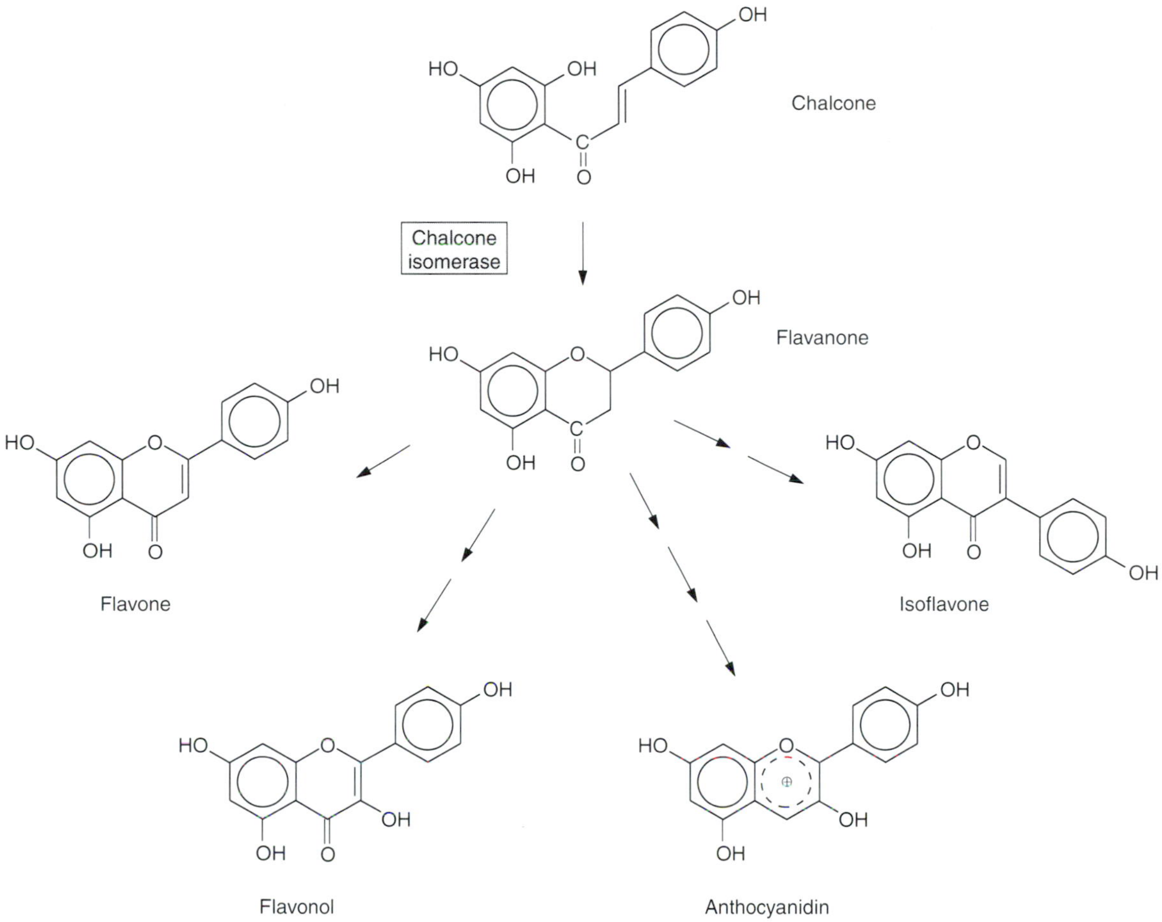

Figure 18.13 Chalcone is the precursor for the synthesis of various flavonoids.

increase in flavonoid biosynthesis. Mutants of *Arabidopsis thaliana* (section 20.1), which, because of a defect in either chalcone synthase or chalcone isomerase, are not able to synthesize flavones, are extremely sensitive to the damaging effects of UV light. In some plants fatty acid esters of sinapic acid (section 18.2) can also act as protective pigments against UV light.

18.6 Anthocyanins are flower pigments and protect plants against excessive light

We have already discussed carotenoids as yellow- and orange-coloured flower pigments (section 17.6). Other widely distributed flower pigments are the yellow *chalcones*, light yellow *flavones*, and red and blue *anthocyanins*. Anthocyanins are glucosides of anthocyanidins (Fig. 18.15) in which the sugar component, consisting

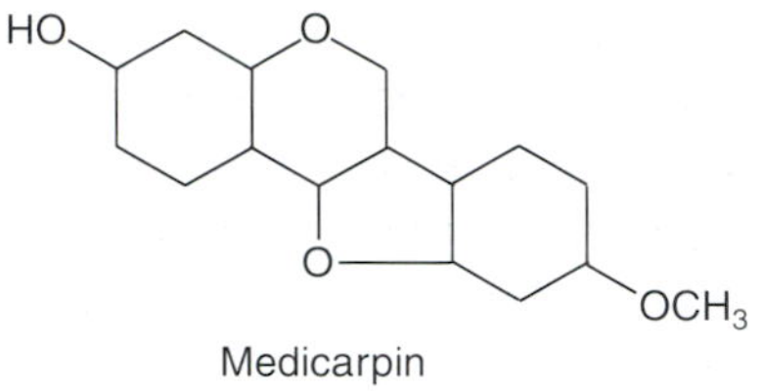

Medicarpin

Figure 18.14 A phytoalexin from alfalfa.

of one or more hexoses, is linked to the -OH group of the pyrrylium ring. Anthocyanins are deposited in the vacuole. It has been shown recently that anthocyanins are transported via the glutathione translocator to the vacuole as glutathione conjugates (section 12.2). The anthocyanin pelargonin, shown in Fig. 18.15, contains *pelargonidin* as chromophore. The introduction of two -OH groups in 3′ and 5′ positions of the phenyl residue and their successive methylation yields five additional flower pigments, each with a different colour. Substitutions at other positions result in even more pigments. Moreover, the colour of the pigment is altered by the formation of complexes with metal ions. Thus upon complex formation with Al^{3+} or Fe^{3+} the colour of pelargonin changes from orange-red to blue. These various pigments and their mixtures lead to the many colour nuances of flowers. With the exception of pelargonidin, all the pigments listed in Fig. 18.15 are found in the the flowers of petunia. To date, 35 genes have been isolated from petunia which are involved in the colouring of flowers.

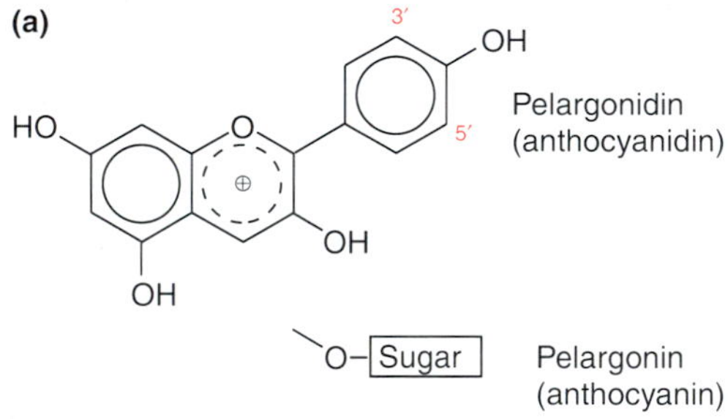

(a)

Pelargonidin
(anthocyanidin)

Pelargonin
(anthocyanin)

(b)

Anthocyanidin	Substituent	Colour
Pelargonidin	—	Orange-red
Cyanidin	3′-OH	Red
Peonidin	3′-OCH$_3$	Pink
Delphinidin	3′-OH, 5′-OH	Bluish-purple
Petunidin	3′-OCH$_3$, 5′-OH	Purple
Malvidin	3′-OCH$_3$, 5′-OCH$_3$	Reddish-purple

Figure 18.15 (a) Pelargonidin, an anthocyanidin, is a flower pigment. It is present in the petals as a glucoside, named pelargonin. (b) More plant pigments are formed by additional -OH groups in 3′ and 5′ positions and their methylation.

Anthocyanins not only act as flower pigments to attract pollen-transferring insects, but they also function as protective pigments for shading leaf mesophyll cells. Plants in which the growth is limited by environmental stress factors, for instance phosphate deficiency, cold, or a high salt content of the soil, often have red-coloured leaves, due mainly to the accumulation of anthocyanins. Stress conditions, in general, reduce the utilization of NADPH and ATP which are provided by the light reactions of photosynthesis. Shading the mesophyll cells by anthocyanins decreases the light reactions and thus prevents overenergization and over-reduction of the photosynthetic electron transport chain (section 3.10).

18.7 Tannins bind tightly to proteins and therefore have defence functions

'Tannins' is a collective term for a variety of plant polyphenols used in the tanning of raw hides to produce leather. Tannins are widely distributed in plants and occur in especially high amounts in the bark of certain trees (e.g. oak) and in galls. Tannins are classified as condensed or hydrolysable tannins (Fig. 18.16). The *condensed tannins* (Fig. 18.16a) are flavonoid polymers and are thus products of phenylpropanoid metabolism. Radical reactions are probably involved in their synthesis, but little is known about details of the biosynthesis pathways. The *hydrolysable tannins* consist of glycosylated *gallic acids* (Fig. 18.16b). Many of these gallic acids are linked to hexose molecules. Gallic acid in plants is formed mainly from shikimate (Fig. 10.19).

The phenolic groups of the tannins bind so tightly to proteins, by forming hydrogen bonds with the -NH groups of peptides, that these bonds cannot be cleaved by digestive enzymes. In the tanning process tannin binds to the collagen of the animal hides and thus produces leather able to withstand the attack of degrading micro-organisms. Tannins have a sharp unpleasant taste; binding of tannins to the proteins of the mucous membranes and saliva draws the mouth together. In this way animals are discouraged from eating plants containing tannin. In addition, when an animal eats plant material, destruction of plant cells results in binding of tannins to plant proteins, making them less digestible and thus unsuitable for fodder. Tannins also react with enzymes of the herbivore digestive tract. For these reasons tannins are very effective in protecting leaves from being eaten by animals. As an example of this, in the South African savannah the leaves of the acacia are the main source of food for the kudu antelope. These leaves contain tannin, but in such low amounts that it does not affect the nutritional quality. Trees injured by feeding animals emit gaseous ethylene (section 19.5), and within 30 minutes the synthesis of tannin is induced in the leaves of neighbouring acacias. If too many acacia leaves are eaten, the tannin content can increase to such a high level that the kudu die when feeding from these leaves. Thus the acacias protect themselves from complete defoliation by a collective warning system.

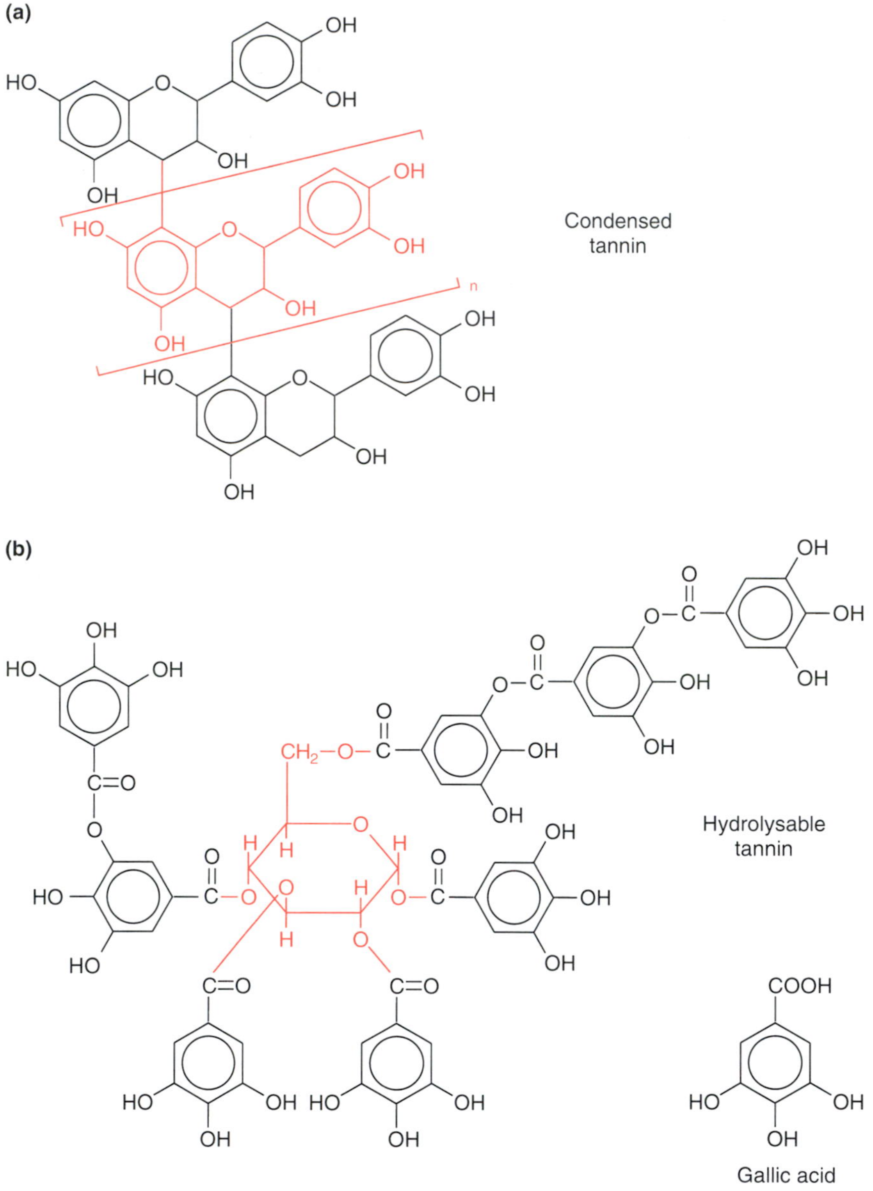

Figure 18.16 (a) General composition of a condensed tannin (n = 1–10). The terminal phenyl residue can also contain three hydroxyl groups. (b) Example of a hydrolysable tannin. The hydroxyl groups of a hexose are esterified with gallic acids (from *Anarcadia* plants).

Tannins also protect plants against attack by micro-organisms. Infection of plant cells by micro-organisms is often initiated by secretion of enzymes for lytic digestion of plant cell walls. The formation of tannins causes these aggressive enzymes to be inactivated.

Further reading

Davin, L. B. and Lewis, N. G. (1992). Phenylpropanoid metabolism: biosynthesis of mono-lignols, lignans and neolignans, lignins and subarins. In *Phenolic metabolism in plants*, (ed. H. A. Stafford and R. K. Ibrahim), pp. 325–75. Plenum Press, New York.

Dixon, R. A. and Paiva, N. L. (1995). Stress-induced phenylpropanoid metabolism. *The Plant Cell*, **7**, 1085–97.

Douglas, C. J. (1996). Phenylpropanoid metabolism and lignin biosynthesis: from weeds to trees. *Trends in Plant Science*, **1**, 171–8.

Hahlbrock, K. and Scheel, D. (1989). Physiology and molecular biology of phenylpropanoid metabolism. *Annual Review of Plant Physiology and Plant Molecular Biology*, **40**, 347–69.

Holton, T. A. and Cornish, E. C. (1995). Genetics and biochemistry of anthocyanin biosyn-thesis. *The Plant Cell*, **7**, 1071–83.

Ibrahim, R. K. and Varin, L. (1993). Flavonoid enzymology. In *Methods in plant biochem-istry*. Volume 9: *Enzymes of secondary metabolism*, (ed. P. J. Lea), pp. 99–135. Academic Press, London.

Lewis, N. G. and Yamamoto, E. (1990). Lignin: occurrence, biogenesis and biodegradation. *Annual Review of Plant Physiology and Plant Molecular Biology*, **41**, 455–96.

Lois, R. and Buchanan, B. B. (1994). Severe sensitivity to ultraviolet radiation in *Arabidopsis* mutants deficient in flavonoid accumulation. *Planta*, **194**, 504–9.

Lynn, D. G. and Chang, M. (1990). Phenolic signals in cohabitation: implications for plant development. *Annual Review of Plant Physiology and Plant Molecular Biology*, **41**, 497–526.

Marrs, K. A., Altonito, M. R., Lloyd, A. M., and Walbot, V. (1995). Glutathione-S-transferase involved in vacuolar transfer encoded by the maize gene Bronze-2. *Nature*, **375**, 397–400.

Raskin, J. (1992). Role of salicylic acid in plants. *Annual Review of Plant Physiology and Plant Molecular Biology*, **43**, 439–63.

Shirley, B. W. (1996). Flavonoid biosynthesis: new functions for an old pathway. *Trends in Plant Science*, **1**, 377–82.

Stafford, H. (1990). *Flavonoid metabolism*. CRC Press, Boca Raton, Florida.

Strid, A., Chow, W. S., and Anderson, J. M. (1994). UV-B damage and protection at the mol-ecular level in plants. *Photosynthesis Research*, **39**, 475–89.

chapter 19

Signals regulating the growth and development of plant organs

In complex multicellular organisms such as higher plants and animals, metabolism, growth, and development of the various organs are co-ordinated by the emission of signal substances. In animals such signals can be *hormones* which are secreted by specialized glandular cells and are distributed, for instance via the blood circulation, to the cells in the different parts of the body. In plants signal substances are also formed in certain cells and distributed to various organs, for instance via the xylem, for controlling processes of metabolism and gene expression. However, these plant signalling substances, called *phytohormones*, are very different in their structure and function from the animal hormones. Besides the phytohormones, the earth's gravity and the quality of the light, in particular the intensity of the red, blue, and ultraviolet light, also control the growth and the differentiation of plants. Major light sensors are *phytochromes* which recognize red and far red light. The corresponding receptors for blue and ultraviolet light are still largely unknown.

The signal chain between the binding of a certain hormone to the corresponding receptor and its effect on specific parameters, such as the transcription of genes or the activity of enzymes, is now known for many animal hormones. In contrast, the signal transduction chain has not been fully resolved for any of the phyto-hormones or the light sensors, although partial results indicate that components of the signal transduction chain in plants may be similar to those in animals. Therefore, signal transduction chains of animal metabolism will also be dealt with in this chapter.

The phytohormones (Fig. 19.1) encompass substances of very diverse structure

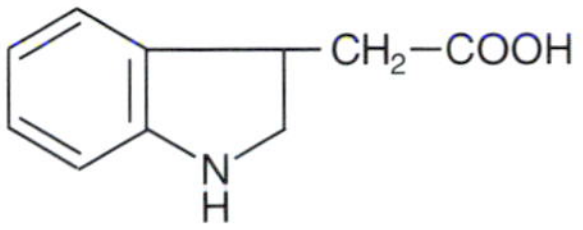

Indole-3′-acetic acid, an auxin

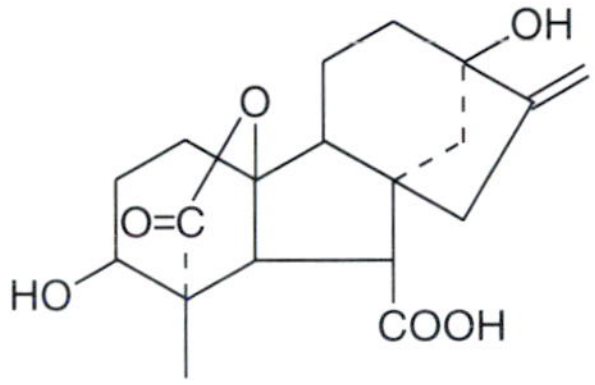

Gibberellin GA_1

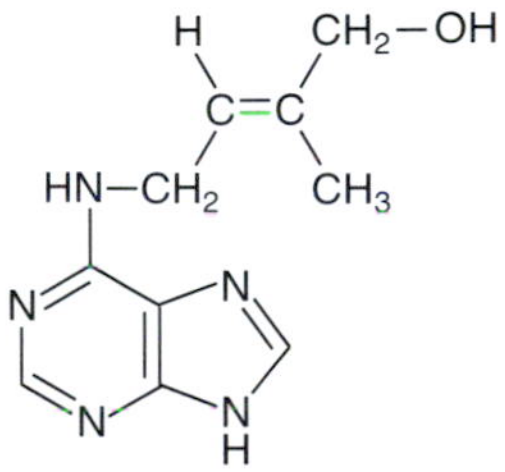

Zeatin, a cytokinin

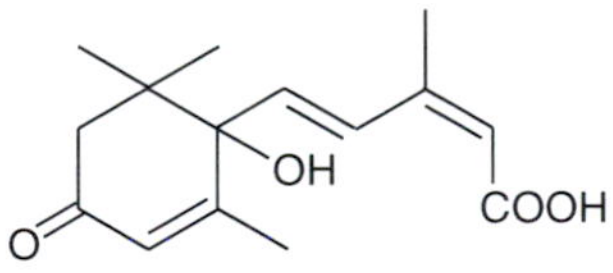

Abscisic acid

Ethylene

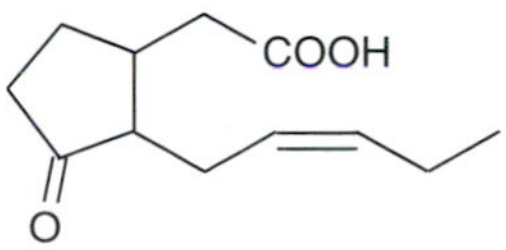

Jasmonic acid

Figure 19.1 Important phytohormones.

and function. Indoleacetic acid, an *auxin*, derived from indole, stimulates cell elongation. *Gibberellins*, derivatives of gibberellane, stimulate elongation of internodes. Zeatin, a *cytokinin*, is a prenylated adenine and stimulates cell division. *Abscisic acid*, formed from carotenoids, regulates water balance. *Ethylene* and *jasmonic acid* (a derivative of fatty acids, section 15.7) enhance senescence. Let us now look at the phytohormones in detail.

19.1 Auxin stimulates shoot elongation

Charles Darwin and his son Francis had already noted in 1880 that growing plant seedlings bend towards sunlight. They found that illumination of the tip initiated the bending of seedlings of canary grass (*Phalaris canariensis*). Since the growth zone is a few millimetres distant from the tip, they assumed that a signal is transmitted from the tip to the growth zone. In 1926 the Dutch researcher Frits Went isolated from the tip of oat seedlings a growth-stimulating substance, which he named *auxin* and which was later identified as *indoleacetic acid* (IAA). The synthesis of IAA proceeds in different ways in different plants. Figure 19.2 shows two synthetic pathways starting from tryptophan. But precursors of tryptophan synthesis (Fig. 10.19) may also act as precursors of IAA synthesis. Besides IAA, some other substances with auxin properties are known, e.g. phenylacetic acid (Fig. 19.3). The synthetic auxin *2,4-dichlorophenoxyacetic acid* (2,4-D, Rohm and Haas) is used as a *herbicide*. This substance is not directly toxic. It kills plants by acting as an especially powerful auxin, resulting in disordered morphogenesis. 2,4-D is a selective herbicide that destroys dicot plants. Monocots are insensitive to it, since they eliminate the herbicide by oxidation. For this reason, 2,4-D is used for combating weeds in cereal crops.

One effect of IAA is to enhance the elongation of cells (see below). Therefore the highest IAA concentrations are found in the main growth zones of the shoot. However, IAA is formed primarily at the tip of the shoot. From there it is transported from cell to cell by an energy-dependent *polar transport*. During the curvature of the coleoptile, IAA is transported laterally to one side. The resulting differential stimulation of cell elongation at one side of the shoot only, leads to the bending. IAA is also transported via the phloem from the leaves to other parts of the plant. The action of IAA on cell growth in the shoot can be traced back to two effects:

1. A few minutes after IAA has been added to a cut tissue, a loosening of the normally rigid cell wall can be observed. How this happens has not been resolved unequivocally. One hypothesis is that acidification of the cell wall region activates enzymes which loosen the rigid cell wall and thus enable a turgor-driven cell enlargement. However, this hypothesis is a matter of controversy.

2. About 1 hour after the addition of IAA, proteins required for growth are synthesized.

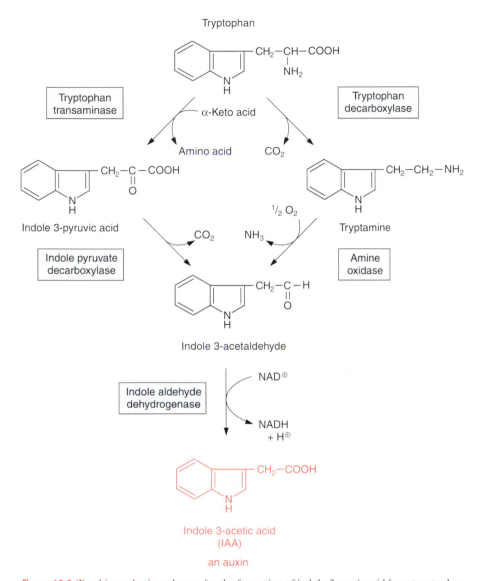

Figure 19.2 Two biosynthetic pathways for the formation of indole-3-acetic acid from tryptophan.

The signal transduction chain for the response to IAA is still unresolved. Recently IAA binding proteins have been identified which may act as receptors.

IAA shows rather diverse effects in different tissues. IAA stimulates cell division in the cambium, enhances *apical dominance* by suppression of lateral bud growth, and controls embryo development. Moreover, IAA prevents the formation of an abscission layer for leaves and fruits and is thus an antagonist to ethylene (section 19.5). On the other hand, increased IAA concentrations can induce the synthesis of ethylene.

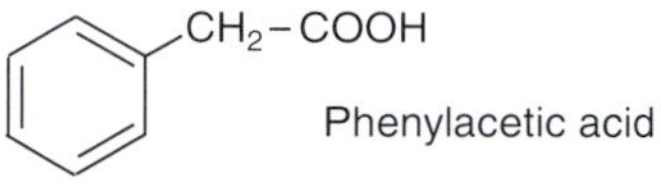

CH₂–COOH

Phenylacetic acid

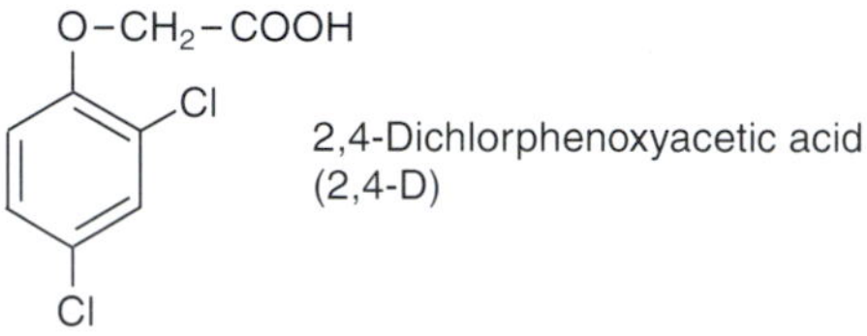

O–CH₂–COOH

Cl

2,4-Dichlorphenoxyacetic acid
(2,4-D)

Cl

Figure 19.3 Phenylacetic acid, another auxin, and 2,4-D, a structural analogue acting as a herbicide.

19.2 Gibberellins regulate stem elongation

The discovery of gibberellins goes back to a plant disease. The infection of rice by the fungus *Gibberella fujikuroi* results in the formation of extremely tall plants which fall over and bear no seed. In Japan this disease was called 'foolish seedling disease'. In 1926 Eiichi Kurozawa and collaborators in Japan isolated a substance from this fungus which induces unnatural growth. It was named *gibberellin*. These results were only known in the West after the Second World War. Structural analysis revealed that gibberellin is a mixture of various substances with similar structures, which also occur in plants and act there as phytohormones.

Gibberellins are derived from the hydrocarbon *ent*-gibberellane (Fig. 19.4). About 100 gibberellins are now known, which are numbered in the order of their identification. Therefore the numbering gives no information about structural relationships or functions. Many of these gibberellins are intermediates or by-products of the biosynthetic pathway. Only a few of them have been shown to act as phytohormones. Whether other gibberellins have a physiological function is not known. The most important gibberellin is GA_1 (Fig. 19.5a). The synthesis of GA_1 occurs from geranylgeranyl pyrophosphate in 13 steps, with *ent*-kaurene as an intermediate.

Similarly to IAA, gibberellins *stimulate elongation*, especially in the internodes of the stems. A pronounced gibberellin effect is the induction of the bolting of rosette plants (e.g. spinach or lettuce) for the formation of flowers. Additionally,

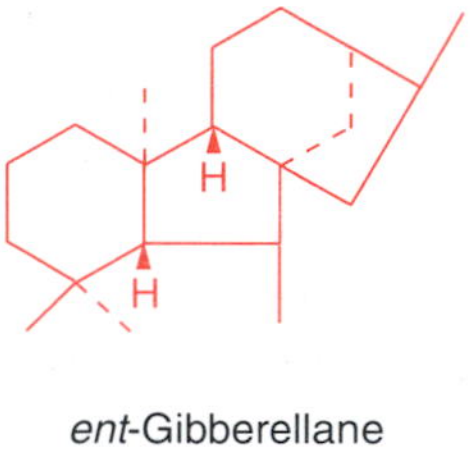

ent-Gibberellane

Figure 19.4 Hydrocarbon from which gibberellins are derived.

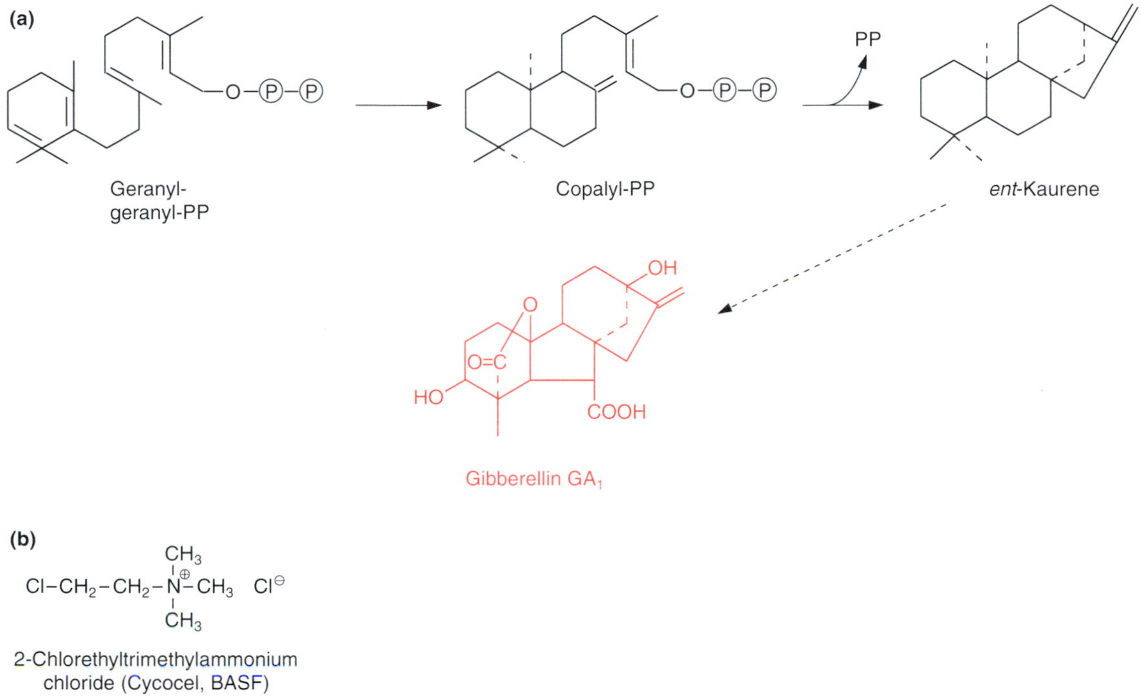

(a)

Geranyl-
geranyl-PP

Copalyl-PP

PP

ent-Kaurene

Gibberellin GA$_1$

(b)

$$Cl-CH_2-CH_2-\overset{\overset{\displaystyle CH_3}{|}}{\underset{\underset{\displaystyle CH_3}{|}}{N^{\oplus}}}-CH_3 \quad Cl^{\ominus}$$

2-Chlorethyltrimethylammonium
chloride (Cycocel, BASF)

Figure 19.5 (a) Synthesis of gibberellin GA$_1$. (b) Cycocel (BASF), a growth retardant which decreases the formation of stems in wheat and other cereals, inhibits kaurene synthesis and thus also the synthesis of gibberellins.

gibberellins have a number of other functions such as the preformation of fruits and the stimulation of their growth. Gibberellins also initiate seed germination by the expression of genes for the necessary enzymes, e.g. amylases.

The use of gibberellins is of economic importance for the production of long, seedless grapes. In these grapes GA$_1$ causes not only extension of the cells, but also parthenocarpy (the generation of fruit as a result of parthenogenesis). Moreover, in the malting of barley for beer brewing, gibberellin is added in order to induce the formation of α-amylase in the barley grains. The gibberellin GA$_3$, produced by the fungus *Gibberella fujikuroi* mentioned above, is generally used for these purposes. Inhibitors of gibberellin biosynthesis are commercially used as *retardants* (growth inhibitors). A number of substances that inhibit the synthesis of the gibberellin precursor *ent*-kaurene, such as chloroethyltrimethylammonium chloride (trade name: Cycocel, BASF) (Fig. 19.5b) are sprayed on cereal fields in order to decrease the growth of the stems. This enhances the strength of the cereal stems and, at the same time, also increases the proportion of total dry matter in seeds (harvest index). Slowly degradable gibberellin synthesis inhibitors are used in horticulture to keep houseplants small.

19.3 Cytokinins stimulate cell division

Cytokinins are prenylated derivatives of adenine. In *zeatin*, which is the most common cytokinin, the amino group of adenine is linked with the hydroxylated isoprene residue in the *trans*-position (Fig. 19.6). Cytokinins enhance plant growth by stimulating cell division and increase the sprouting of lateral buds. As cytokinins override apical dominance, they are antagonists of the auxin IAA. Cytokinins retard senescence and thus counteract the phytohormone ethylene which will be described later. Recently, cytokinin binding proteins have been found which are possibly involved in the otherwise unknown signal transduction chain of the response to cytokinin.

Mature, i.e. differentiated, plant cells normally stop dividing. By adding cytokinin and auxin, differentiated cells can be induced to start dividing again. When a leaf piece is placed on a solid culture medium containing auxin and cytokinin, leaf cells start unrestricted growth, resulting in the formation of a callus which can be propagated in tissue culture. Upon the application of a certain cytokinin/auxin ratio, complete plants can be regenerated from single cells of this callus. The use of tissue culture for the generation of transgenic plants will be described in section 22.3.

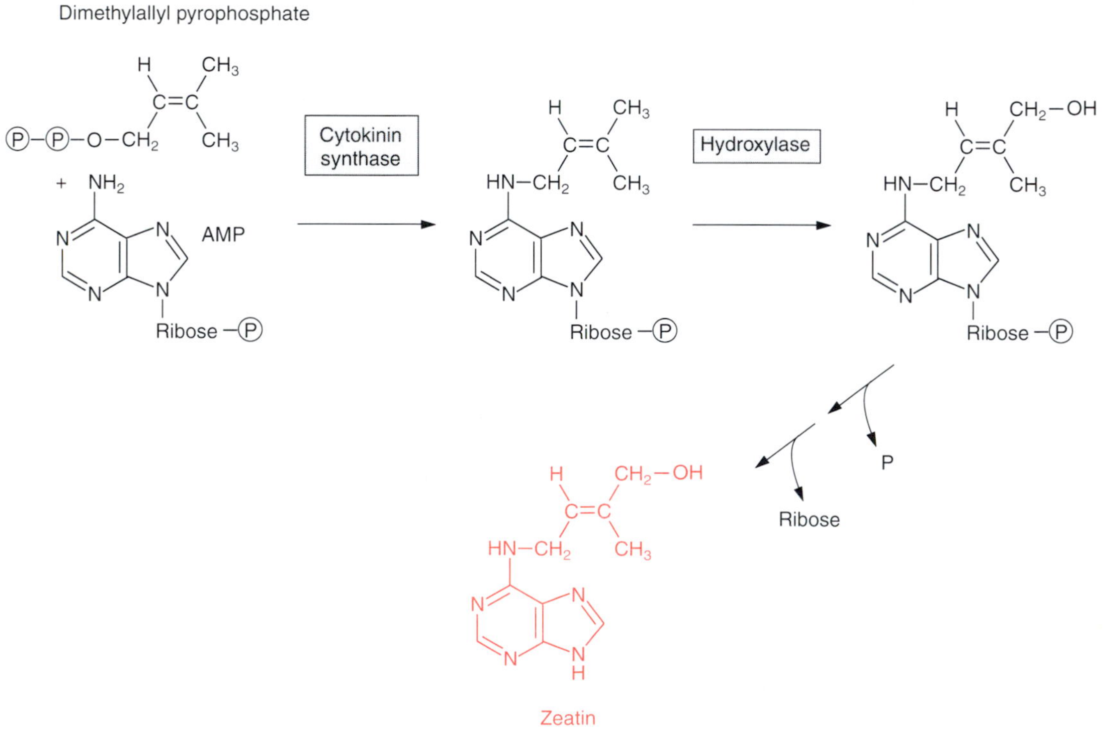

Figure 19.6 Synthesis of zeatin, a cytokinin.

Some bacteria and fungi produce auxin and cytokinin in order to induce unrestricted cell division which results in tumour growth in a plant. The formation of the crown gall induced by *Agrobacterium tumefaciens* (section 22.2) is caused by stimulation of the production of cytokinin and auxin. For this purpose genes for the biosynthesis of cytokinin and auxin are transferred from the Ti plasmid of the bacterium to the plant genome.

Zeatin is formed from AMP and dimethyallyl pyrophosphate (Fig. 19.6). The isoprene residue is transferred by *cytokinin synthase*, a prenyl transferase (section 17.2), to the amino group of the AMP and is then hydroxylated. Cytokinin synthesis takes place primarily in the meristematic tissues. Recently transgenic tobacco plants have been generated in which the activity of cytokinin synthase in the leaves is increased. The leaves of these tobacco plants have a much longer life span than normal, since their senescence is suppressed by the enhanced production of cytokinin.

19.4 Abscisic acid controls the water balance of the plant

When searching for substances that cause the abscission of leaves and fruits, abscisic acid (ABA) (Fig. 19.7) was found to be an inducing factor and was named accordingly. Later it turned out that the formation of the abscission layer for leaves and fruit is induced primarily by ethylene (described in the following section). An important function of the phytohormone ABA, however, is the induction of *dormancy* (endogenic rest) of seeds and buds by affecting gene expression. Moreover, ABA has a major function in maintaining the *water balance of plants*, since it induces the closure of the stomata during water shortage (section 8.2). In addition, ABA prevents seed embryos from germinating before the seeds

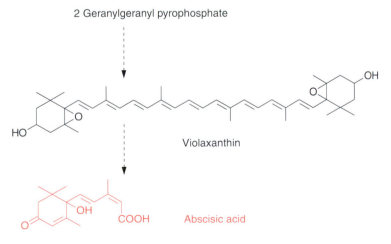

Figure 19.7 Abscisic acid is formed in several steps by oxidative degradation of violaxanthin.

are mature (vivipary). Mutants deficient in ABA have been found in tomatoes, which, due to the resulting disturbance of the water balance, bear wilty leaves and fruits. In these wilty mutants the immature seeds germinate in the tomato fruits while they are still attached to the mother plant.

Abscisic acid is a product of isoprenoid metabolism. The synthesis of ABA proceeds via oxidation of violaxanthin (Fig. 19.7, see also Figs 3.42 and 17.13). ABA synthesis occurs in leaves and also in roots where a water shortage is directly monitored. ABA can be transported by the transpiration stream via the xylem vessels from the roots to the leaves where it induces closure of the stomata (section 8.2). There are indications that in the signal transduction chain in response to ABA, Ca^{2+} ions participate as a second messenger and that protein kinases are involved.

19.5 Ethylene makes fruit ripen

It has already been mentioned in the previous section that ethylene is involved in the induction of senescence. During senescence the degradation of leaf material is initiated. Proteins are degraded to amino acids which, together with certain ions, e.g. Mg^{2+}, are withdrawn from the senescing leaves via the phloem system for re-utilization. In perennial plants, these substances are stored in the stem or in the roots and in annual plants they are utilized to enhance the formation of seeds. The effect of ethylene in the induction of the synthesis of tannins has already been dealt with in section 18.7.

Beside stimulating the abscission of fruit, ethylene has a general function in fruit ripening. This effect of gaseous ethylene can be demonstrated by placing a ripe apple and a green tomato together in a plastic bag; ethylene produced by the apple accelerates the ripening of the tomato. Bananas are harvested green and transported halfway around the globe under conditions which suppress ethylene synthesis (low temperature, CO_2 atmosphere). Before being sold, these bananas are ripened by gassing them with ethylene. Also, tomatoes are often only ripened prior to sale by exposure to ethylene.

S-Adenosylmethionine (Fig. 12.10) is a precursor for the synthesis of ethylene (Fig. 19.8). The positive charge of the sulfur atom in *S*-adenosylmethionine enables its cleavage to form a cyclopropane, in a reaction catalysed by amino-cyclopropane carboxylate synthase, abbreviated to *ACC synthase*. Subsequently, *ACC oxidase* catalyses the oxidation of the cyclopropane to ethylene, CO_2, prussic acid, and water. Prussic acid is immediately detoxified by conversion to β-cyanoalanine (reaction not shown).

Recently, genetic engineering has been employed to suppress ethylene synthesis in tomato fruits in two different ways:

(1) the activity of ACC synthase and ACC oxidase was decreased by an antisense technique (section 22.5);

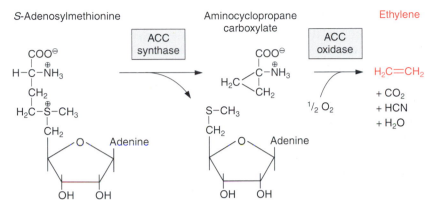

Figure 19.8 Synthesis of ethylene.

(2) by introducing a bacterial gene into the plants, an enzyme was expressed which degraded the ACC formed in the tomato fruits so rapidly that it could no longer be converted to ethylene by the ACC oxidase.

When ethylene biosynthesis is decreased in these ways, the ripening process continues in the unpicked tomatoes, but in harvested fruits it is retarded to such an extent that ripe tomatoes can be transported for several days without decaying. The aim of this genetic engineering is to produce tomatoes that have developed their natural taste by ripening on the plant instead of being picked green and later ripened artificially by ethylene gassing. It may be noted that transgenic tomato plants have also been generated in which the durability of the harvested fruits is prolonged by an antisense repression of the polygalacturonidase enzyme, which lyses the cell wall. In the USA these tomatoes are already being distributed under the trade name Flavr Savr (Calgene).

The signal transduction chain for the action of ethylene has not yet been fully resolved. Recent investigations of ethylene-insensitive mutants from *Arabidopsis thaliana* (section 20.1) indicate that several protein kinases are involved in the signal transduction chain. The receptor for ethylene is probably a histidine protein kinase.

Jasmonic acid (Fig. 15.30) and its methylester are found in very many plants, where they probably function as phytohormones. The effects of jasmonic acid show some resemblance to the action of abscisic acid and of ethylene. In a bioassay jasmonic acid inhibits auxin-induced elongation growth, cytokinin- and auxin-induced callus growth, and also seed germination. Like ethylene, it enhances senescence. Moreover, jasmonic acid may function as a signal substance in the induction of the synthesis of certain phytoalexins.

19.6 Phytochromes function as light sensors

Light controls plant development from germination to the formation of flowers in many different ways. Important light sensors in this control are *phytochromes*,

which sense red light. Phytochromes are involved when light initiates the germination and greening of the seedling, and in the adaptation of the photosynthetic
apparatus of the leaves to full sunlight or shade. For their adaptation to the full
spectrum of sunlight, plants also possess photoreceptors for blue and ultraviolet
light, but the chromophores of these photoreceptors have not yet been unequivocally identified. Flavins and pteridins probably participate in the perception of
blue light.

Since the structure and function of phytochromes have been studied intensively
in the past, they offer a good example for a more detailed discussion of the problems of signal transduction in plants. Phytochromes are soluble *dimeric proteins*
which probably function without being bound to a membrane. The monomer consists of an apoprotein (molecular weight 120–130 kDa), in which the sulfydryl
group of a cysteine residue is linked to an open-chain tetrapyrrole, functioning as
a chromophore (Fig. 19.9). The autocatalytic binding of the tetrapyrrole to the
apoprotein results in the formation of a phytochrome, P_r (r = red) with an absorption maximum at about 660 nm (red light) (Fig. 19.10). The absorption of this light
results in a change in the chromophore: a double bond between the two pyrrole
rings changes from the *trans-* to the *cis-*configuration (coloured red in Fig. 19.9),
resulting in a change in the conformation of the protein. The phytochrome in this

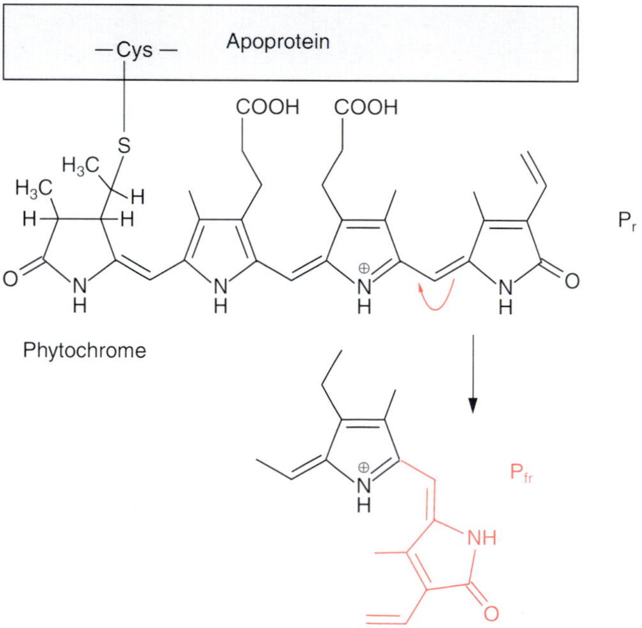

Figure 19.9 The chromophore of phytochrome consists of an open-chain tetrapyrrole which is linked
via a thioether bond to the apoprotein. The absorption of red light results in a *cis–trans*-isomerization
of a double bond, causing a change in the position of one pyrrole ring (coloured red).

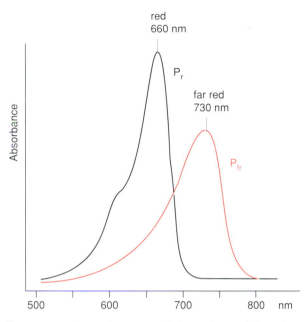

Figure 19.10 Absorption spectra of the two forms of phytochrome, P_r and P_{fr}.

new conformation has an absorption maximum at about 730 nm (far red light) and in this state is named P_{fr}. P_{fr} represents the active form of the phytochrome. It signals the state of illumination. P_{fr} is reconverted to P_r by the absorption of far red light. Since the light absorption of P_r and P_{fr} overlaps (Fig. 19.10), depending on the colour of the irradiated light, a reversible equilibrium between P_r and P_{fr} is attained. Thus with light of 660 nm, 88 per cent of the total phytochrome is present as P_{fr} and at 720 nm, only 3 per cent. In bright sunlight, where the red component is stronger than the far red one, the phytochrome is present primarily as P_{fr} and signals the state of illumination to the plant.

Altogether five different phytochromes (A–E) have been identified in the well-characterized model plant *Arabidopsis thaliana* (section 20.1). *Phytochrome A* is present primarily in etiolated tissues where it is responsible for the induction of photomorphogenic processes resulting in the formation of green leaves. The active form, P_{fr}, has a half-life of 30–60 minutes (Fig. 19.11). It is irreversibly degraded by proteolysis after reacting with *ubiquitin*, a highly conserved small protein (molecular weight 8.5 kDa), present in all eukaryotic cells. Ubiquitin binds in an ATP-dependent mode to proteins, thereby marking them for pro-teolytic degradation. Binding to ubiquitin terminates the P_{fr} signal function. In comparison with phytochrome A, phytochromes B–E have a longer lifetime, and are probably mainly involved in the light adaptation of those cells that are already green. So far little is known about the degradation of these phytochromes .

The active phytochrome P_{fr} affects protein biosynthesis. According to present

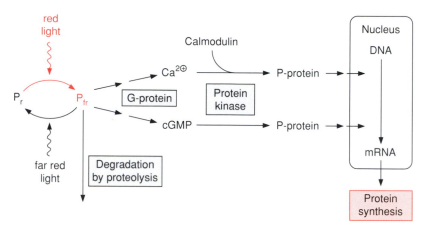

Figure 19.11 Phytochrome is converted by irradiation with red light to the active form, P_{fr}, and is reconverted by far red light to the inactive P_r. P_{fr} has an indirect effect on gene expression in the nuclear genome via a signal transduction chain in which a G-protein, cGMP or Ca^{2+} plus calmodulin, and also protein kinases are presumably involved. P_{fr} is degraded irreversibly by proteolysis.

knowledge it activates factors that influence the transcription of DNA in the nucleus. Such *transcription factors* are often proteins which are either activated or inactivated by phosphorylation via protein kinases. Probably this also applies to the indirect activation of transcription factors by P_{fr}.

At present not much is known about the signal transduction chain from P_{fr} to the activation of proteins. Important objects for the study of signal transduction chains have been cells of tomato mutants, which, due to the lack of phytochrome, are unable to green. The micro-injection of Ca^{2+} ions and calmodulin into single cells of these mutants simulated the action of phytochromes in inducing the synthesis of light harvesting complexes. These results indicate that Ca^{2+} is involved as a messenger in signal transduction from phytochromes (Fig. 19.11). Phytochromes also induce the enzyme chalcone synthase (Fig. 18.11). Such an induction of chalcone synthase can also be achieved by injection of cGMP (Fig. 19.12) in the mutant cells already mentioned. This led to the conclusion that cGMP participates as an alternative messenger to Ca^{2+} in P_{fr}-induced signal transduction. Further investigations have indicated that a *G-protein* (see next section) is involved in both transduction chains.

19.7 Signal transduction chains known from animal metabolism could be models for signal transduction in plants

G-proteins act as molecular switches

A family of proteins which can alternate between two conformational states by binding of GTP or GDP is widely distributed in the animal and plant kingdoms.

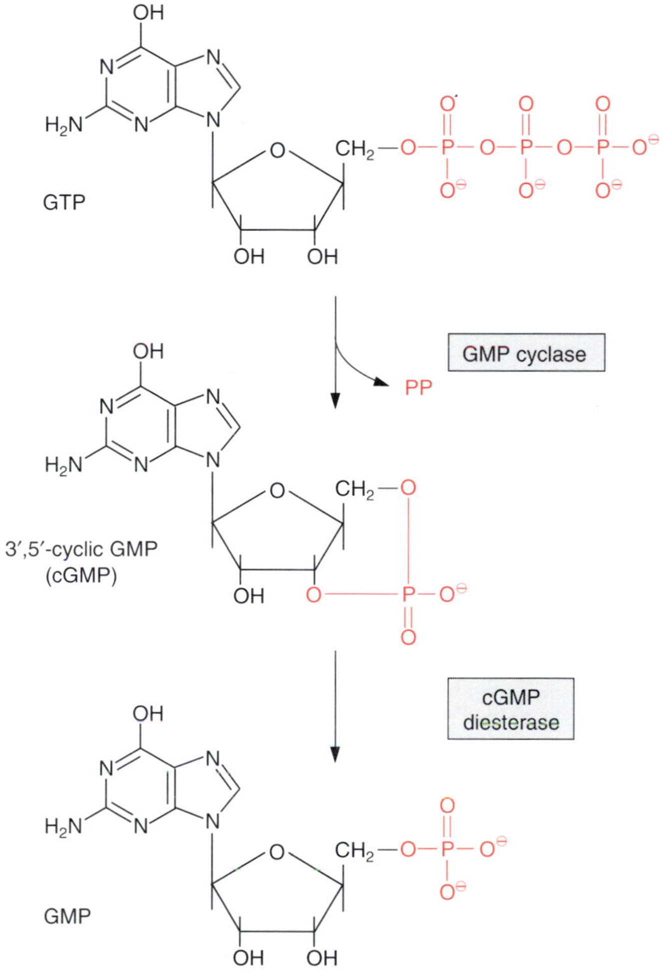

Figure 19.12 cGMP is formed by GMP-cyclase from GTP and degraded by a diesterase to GMP. The enzyme is known in animal metabolism.

These proteins are called *GTP binding proteins*, or G-proteins for short. Our knowledge about G-proteins derives primarily from investigations in animals. In the case of plants it is known that G-proteins definitely exist, but so far little is known about their function. *Heterotrimeric G-proteins* are composed of three different subunits: G_α (molecular weight 45–55 kDa), G_β (molecular weight 35 kDa), and G_γ (molecular weight 8 kDa) (Fig. 19.13). Subunit G_α contains a binding site which can be occupied by either GDP or GTP. In animals, binding of the heterotrimer to a receptor, e.g. an adrenaline receptor occupied by adrenaline, enables the exchange of bound GDP for GTP. It is feasible that such an exchange also occurs when a G-protein is bound to P_{fr}.

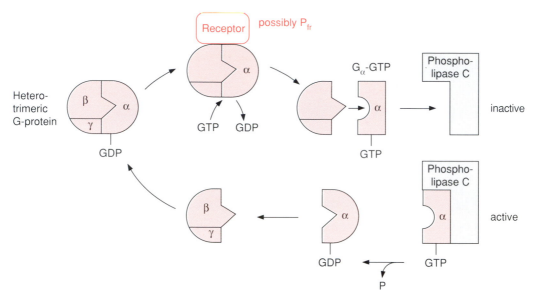

Figure 19.13 Scheme of the function of a G-protein in analogy to animal metabolism.

The binding of GTP results in a *conformational change* of the G_α subunit and thereby in its dissociation from the trimer. The liberated G_α unit, loaded with GTP, functions as an activator of enzymes forming messengers for signal transduction. Thus in animal cells the G_α-GTP stimulates adenylate cyclase, catalysing the synthesis of the messenger cyclic AMP (cAMP) from ATP. However, present evidence suggests that cAMP is not involved in signal transduction in plants. None the less it is feasible that the formation of cGMP in plants is stimulated in an analogous way by G_α-GTP (Fig. 19.12). This would explain the results mentioned above about the participation of cGMP in the signal transduction chain from P_{fr}. In animal cells G_α-GTP also activates phospholipase C. The function of this re-action in the liberation of Ca^{2+} as a messenger will be dealt with in the following section.

G_α-GTP has a life of only a few minutes. Bound GTP is hydrolysed by an intrinsic GTPase activity to GDP and the resulting conformational change causes G_α to lose its activator function. It binds once more to the trimer and a new cycle can begin. The short life of the signal enables signal transduction to be very efficient.

It may be noted that all eukaryotes also contain *small G-proteins*, which consist of only one subunit and act as molecular switches analogously to large G-proteins. The small G-proteins include the proteins of the Ras superfamily, which in animals, fungi, algae, and higher plants have various functions in cell differentiation and vesicle transport. For the sake of brevity these will not be dealt with further.

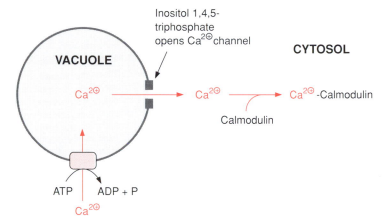

Figure 19.14 Plant vacuoles contain an ATP-dependent Ca^{2+} translocator which pumps Ca^{2+} against a concentration gradient into the vacuole. Ca^{2+} can be released by an IP_3-dependent Ca^{2+} channel from the vacuole to the cytosol.

Ca^{2+} acts as a messenger in signal transduction

In animal as well as in plant cells, the cytosolic concentration of free Ca^{2+} is normally lower than 10^{-6} mol l^{-1}. These very low Ca^{2+} concentrations are maintained by ATP-dependent pumps, which accumulate Ca^{2+} in the lumen of the endoplasmic reticulum (in animals) and in the vacuole (in plants) or transport Ca^{2+} via the plasma membrane to the extracellular compartment (Fig. 19.14). Signals induce the Ca^{2+} channels to open, resulting in a rapid increase in the cytosolic concentration of free Ca^{2+}, which stimulates, in almost all cells, the regulatory enzymes, including protein kinases.

The phosphoinositol pathway controls the opening of Ca^{2+} channels

Ca^{2+} channels can be controlled by the phosphoinositol signal transduction cascade which has been resolved in animal metabolism (Fig. 19.15). *Phosphatidyl inositol* is present, although in relatively low amounts, as a constituent of cell membranes. In animal cells the two fatty acids of phosphatidyl inositol are usually stearic and arachidonic acids. The inositol residue is phosphorylated at the hydroxyl groups in the 4′ and 5′ positions by a kinase. *Phospholipase C*, stimulated by a G-protein, cleaves the lipid to inositol 1,4,5-triphosphate (IP_3) and diacylglycerol (DAG). The G-protein frequently interacts with a hormone receptor. IP_3, as well as diacylglycerol, participates in signal transduction: IP_3 causes a rise in Ca^{2+} concentration, whereas diacylglycerol stimulates a Ca^{2+}-dependent protein kinase. In animal cells the binding of three IP_3 molecules to a Ca^{2+} channel of the ER membrane causes it to open. The rapid influx of Ca^{2+} into

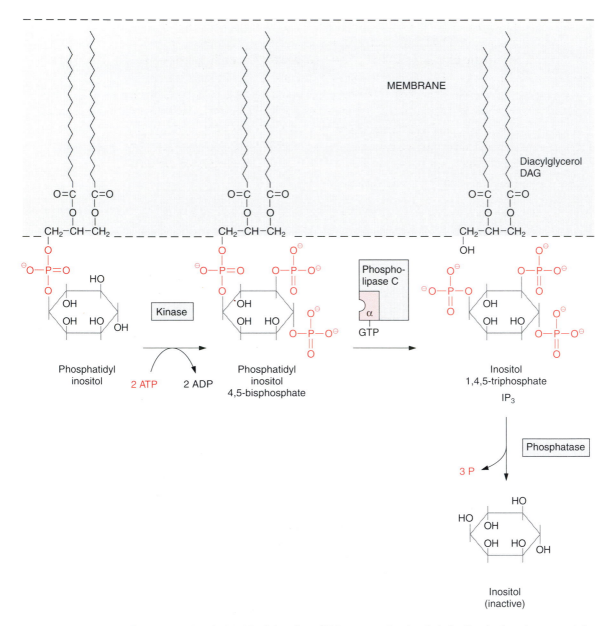

Figure 19.15 Inositol 1,4,5-triphosphate (IP₃) as part of a signal chain. Two hydroxyl groups of the inositol residue of a membrane phospholipid are phosphorylated by a kinase and the resultant IP₃ is liberated by a G-protein-dependent phospholipase C. The messenger IP₃ formed in this way is degraded by phosphatases. This metabolic chain has been resolved in animal metabolism.

the cytosol is limited by the very short life of IP_3 (in animals often less than 1 s). The rapid elimination of IP_3 proceeds either via further phosphorylation of inositol or a hydrolytic liberation of the phosphate groups by a phosphatase. The short life of IP_3 enables very efficient signal transduction.

Components of the phosphoinositol cascade have also been identified in plant cells. Results obtained so far indicate that the phosphoinositol cascade plays an important role in signal transduction in plants. Opening of Ca^{2+} channels by IP_3 has been demonstrated by patch clamp studies (section 1.10) with plant vacuoles. It is feasible that P_{fr} induces an increase in Ca^{2+} concentration in plants via a G-protein dependent phosphoinositol cascade (Fig. 19.11).

Calmodulin mediates the messenger function of Ca^{2+} ions

Often Ca^{2+} acts as a messenger not directly but by binding to calmodulin. Calmodulin is a soluble protein (17 kDa) occurring in animals as well as in plants. It is a highly conserved protein: the homology between the calmodulin from wheat and cattle is 91 per cent. Calmodulin is present mainly in the cytosol. It consists of two domains connected by a flexible α-helix. Each domain contains two binding sites for Ca^{2+} (Fig. 19.16). The occupation of all four Ca^{2+} binding sites results in a conformational change of calmodulin by which the hydrophobic domain is exposed, and forms a complex with certain protein kinases, which in turn are activated. Although the function of calmodulin in plants has not yet been thoroughly investigated, it is most likely that its function in plants and animals is alike.

Phosphorylated proteins form elements of signal transduction

Protein kinases and phosphatases, of which several examples have already been dealt with, are important elements in the regulation of intracellular processes. Since proteins change by phosphorylation and dephosphorylation between two states of different effectiveness, and many protein kinases are also switched on or off by phosphorylation, protein kinases represent a network of on–off switches in

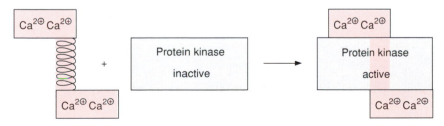

Figure 19.16 The protein calmodulin contains two Ca^{2+} binding domains which are connected by a flexible α-helix. Ca^{2+}-calmodulin activates certain protein kinases.

Table 19.1 Some members of the eukaryotic protein kinase superfamily

Protein kinase	Regulated by	Modulator
Protein kinase A	Messenger	cAMP
Protein kinase G		cGMP
Protein kinase C		Ca^{2+}
Protein kinase CaM		$(Ca^{2+})_4$–Calmodulin
Cyclin-dependent kinase (CDK)	Proteins or protein kinases	Cyclin
Mitogen-activated protein kinase (MAPK)		Mitogen
MAPK-activated protein kinase (MAPK-PK)		MAPK

the cell, comparable to those of computer chips, and they control differentiation, metabolism, defence against pests, and many other cell processes. It is estimated that in a eukaryotic cell 1–3 per cent of the functional genes encode protein kinases. So far protein kinases have been investigated mainly in yeast and animals. Until the end of 1994 the genes for only about 70 protein kinases had been identified in plants, and the physiological function of only a few of them is known. Thus we are still at the initial stage of an undoubtedly very important research area.

Most protein kinases in eukaryotes contain 12 conserved regions. These protein kinases phosphorylate mainly the -OH group of serine and/or threonine, and in some cases also of tyrosine. Since all these protein kinases are homologous, they are placed in a *superfamily of eukaryotic protein kinases*. Table 19.1 shows some members of this family. Up to now only a few Ca^{2+}-dependent and one Ca^{2+}–calmodulin-dependent protein kinase have been identified in plants, but their function is not yet known. Protein kinases involved in signal transduction from phytochrome (Fig. 19.11) have not yet been identified. Several protein kinases have been found which are regulated in plants by proteins, including protein kinases. Some examples of such regulatory proteins are cyclins and mitogens, an extensive group of proteins which, by activating protein kinases, participate in the formation of the mitotic spindle and other processes of cell division. Protein kinase cascades are involved in these reactions. For example the mitogen-activated protein kinase (MAPK) is activated by a further protein kinase (MAPK-PK).

Recently, protein kinases have been identified which phosphorylate histidine residues of proteins and which do not belong to the above-mentioned superfamily. A histidine protein kinase may be involved in the function of the ethylene receptor.

Further reading

Blatt, M. R. and Thiel, G. (1993). Hormonal control of ion channel gating. *Annual Review of Plant Physiology and Plant Molecular Biology*, **44**, 543–67.

Bleecker, A. B. (1996). The mechanism of ethylene perception. *Plant Physiology*, **111**, 653–60.

Bowler, C. and Chua, N. H. (1994). Emerging themes of plant signal transduction. *The Plant Cell*, **6**, 1529–41.

Bush, D. S. (1993). Regulation of cytosolic calcium in plants. *Plant Physiology*, **103**, 7–13.

Cote, G. C. and Crain, R. C. (1993). Biochemistry of phosphoinositides. *Annual Review of Plant Physiology and Plant Molecular Biology*, **44**, 333–56.

Davies, W. J., Tardieu, F., and Trejo, C. (1995). How do chemical signals work in plants that grow in drying soil? *Plant Physiology*, **104**, 309–14.

Ecker, J. E. (1995). The ethylene signal transduction chain in plants. *Science*, **268**, 667–75.

Evans, D. E. (1994). Calmodulin-stimulated calcium pumping ATPases located at higher plants interacellular membranes: a significant divergence from other eukaryotes? *Physiologia Plantarum*, **90**, 420–6.

Fabry, S. (1995). Kleine G-Proteine: Universelle Schalter und Regler im Zellgeschehen. *Biologie in unserer Zeit*, **25**, 44–50.

Furuya, M. (1993). Phytochromes: their molecular species, gene families and functions. *Annual Review of Plant Physiology and Plant Molecular Biology*, **44**, 617–45.

Graebe, J. E. (1987). Gibberellin biosynthesis and control. *Annual Review of Plant Physiology and Plant Molecular Biology*, **38**, 419–65.

Kaufmann, L. S. (1993). Transduction of blue-light signals. *Plant Physiology*, **102**, 333–7.

Kende, H. (1993). Ethylene biosynthesis. *Annual Review of Plant Physiology and Plant Molecular Biology*, **44**, 283–307.

Normanly, J., Slovin, J. P., and Cohen, J. D. (1995). Rethinking auxin biosynthesis and metabolism. *Plant Physiology*, **107**, 323–9.

Palme, K. (1992). Molecular analysis of plant signalling elements: relevance of eukaryotic signal transduction models. *International Reviews of Cytology*, **132**, 223–83.

Quail, P. H., Boylan, M. T., Parks, B. M., Short, T. W., Xn, Y., and Wagner, D. (1995). Phytochromes: photosensory perception and signal transduction. *Science*, **268**, 675–80.

Roberts, D. M. and Marmon, A. C. (1995). Calcium-modulated proteins: targets of intracellular calcium signals in higher plants. *Annual Review of Plant Physiology and Plant Molecular Biology*, **43**, 375–414.

Schäfer, E. and Furuya, M. (1996). Photoperception and signalling of induction reactions by different phytochromes. *Trends in Plant Science*, **1**, 301–7.

Short, T. W. and Briggs, W. (1994). The transduction of blue light signals in higher plants. *Annual Review of Plant Physiology and Plant Molecular Biology*, **45**, 143–71.

Stone, J. M. and Walker, J. C. (1995). Plant protein kinase families and signal transduction. *Plant Physiology*, **108**, 451–7.

Taiz, L. and Zeiger, E. (1991). *Plant physiology*. Benjamin/Cummings, Redwood City, California.

Takahashi, Y., Ishida, S., and Nagata, T. (1995). Auxin-regulated genes. *Plant Cell Physiology*, **36**, 383–90.

TIBS (1994). Protein phosphorylation (a special issue). *Trends in Biochemical Sciences*, **19**, 439–518.

Vierstra, R. D. (1993). Illuminating phytochrome functions: there is light at the end of the tunnel. *Plant Physiology*, **103**, 679–84.

chapter 20

The genomes of plant cells

A plant cell contains three genomes: in the *nucleus*, the *mitochondria*, and the *plastids*. Table 20.1 lists the size of the three genomes in three plant species and, in comparison, the two human genomes. The size of the genomes is given in base pairs (bp). As dealt with in Chapter 1, the genetic information of the mitochondria and plastids is located on one (sometimes on several) circular DNA double strands, with many copies present in each organelle. During the multiplication of the organelles by division these copies are distributed randomly between the daughter organelles. Each cell contains a large number of plastids and mitochondria, which are also randomly distributed during cell division between both daughter cells. The structures and functions of the plastid and mitochondrial genomes are dealt with in sections 20.6 and 20.7 respectively.

Table 20.1 Size of the genome in plants and in man

	Number of base pairs in a single genome			
	Arabidopsis thaliana	*Zea mays* (maize)	*Vicia faba* (broad bean)	*Homo sapiens* (human)
Nucleus (haploid chromosome set)	7×10^7	390×10^7	1450×10^7	280×10^7
Plastid	156×10^3	136×10^3	120×10^3	
Mitochondrion	370×10^3	570×10^3	290×10^3	17×10^3

20.1 In the nucleus the genetic information is divided among several chromosomes

For almost the whole of their developmental cycle, eukaryotic cells, as they are *diploid*, normally contain two chromosome sets, one set from the mother and the other from the father. Only the generative cells (egg, pollen) are *haploid*, that is they possess just one set of chromosomes. The DNA of the genome is replicated during the interphase of mitosis, and thus generates a quadruple chromosome set. During anaphase this set is distributed by the spindle apparatus to two opposite poles of the cell. Thus, after cell division each daughter cell contains a double chromosome set and is diploid once more. *Colchicine*, an alkaloid from the autumn crocus, inhibits the spindle apparatus and thus interrupts the distribution of the chromosomes during anaphase. In such a case all the chromosomes of the mother cell can end up in only one of the daughter cells, which then possesses four sets of chromosomes, making it *tetraploid*. Occasionally this tetraploidy occurs spontaneously due to mitosis malfunction.

Tetraploidy is often stable and is then inherited by the following generations via somatic cells. Hexaploid plants can be generated by crossing tetraploid plants with diploid plants, although this is seldom successful. Tetraploid and hexaploid (polyploid) plants often show a higher growth rate and this is the reason why many crop plants are polyploid. Polyploid plants can also be generated by proto-plast fusion, a method which can produce hybrids between two different breeding lines. When very different species are crossed, the resulting diploid hybrids are often sterile due to incompatibilities in their chromosomes. In contrast to this, polyploid hybrids are generally fertile.

The chromosome content of various plants is listed in Table 20.2. The crucifer *Arabidopsis thaliana* (Fig. 20.1), an inconspicuous weed growing at the roadside in central Europe, has only 2×5 chromosomes with altogether just 7×10^7 base pairs. *Arabidopsis* is a typical dicot plant and, when grown in a growth chamber, has a life cycle of only 6 weeks. The breeding of a defined line of *Arabidopsis*

Table 20.2 Number of chromosomes

		$n^a \times m^b$
A	**Dicot plants**	
	Arabidopsis thaliana	2×5
	Vicia faba (broad bean)	2×6
	Glycine max (soybean)	2×20
	Brassica napus (rapeseed)	2×19
	Beta vulgaris (sugar beet)	6×19
	Solanum tuberosum (potato)	4×12
	Nicotiana tabacum (tobacco)	4×12
B	**Monocot plants**	
	Zea mays (maize)	2×10
	Hordeum vulgare (barley)	2×7
	Triticum aestivum (wheat)	6×7
	Oryza sativa (rice)	2×12

[a] *n*: Ploidy

[b] *m*: Number of chromosomes in the haploid genome

thaliana by the botanist Friedrich Laibach (Frankfurt) in 1943 marked the beginning of the world-wide use of *Arabidopsis* as a model plant for the investigation of plant functions. Currently, laboratories in many parts of the world are co-operating in the task of obtaining a complete sequence analysis of the genome of *Arabidopsis thaliana* and, hopefully, in a few years the total genetic information of this plant will be known.

The much higher number of chromosomes sometimes found in other plants is in part the result of combining genomes. Hence rapeseed (*Brassica napus*) with 2×19 chromosomes is the result of crossing *Brassica campestris* (2×10 chromosomes) and *Brassica oleracea* (2×9 chromosomes) (Table 20.2). In this case a diploid genome is the result of the crossing. In wheat, on the other hand, the successive crossing of three wild forms resulted in hexaploidy: crossing of einkorn wheat and goat grass (2×7 chromosomes each) resulted in wild emmer wheat (4×7 chromosomes) and this crossed with another wild wheat, resulted in the wheat (*Triticum aestivum*) with 6×7 chromosomes cultivated nowadays. Tobacco (*Nicotiana tabacum*) is also a cross between two species (*Nicotiana tomentosiformis* and *N. sylvestris*), each with 2×12 chromosomes.

The nuclear genome of broad bean with 14.5×10^9 base pairs has a 200-fold higher DNA content than *Arabidopsis*. However, this does not mean that the number of protein-encoding genes (structural genes) in the broad bean genome is

Figure 20.1 *Arabidopsis thaliana*, an inconspicuous small weed, growing in central Europe at the roadside, from the family Brassicaceae (crucifers), has become the most important model plant world-wide, because of its small genome. In the growth chamber with continuous illumination, the time from germination until the formation of mature seed is only 6 weeks. The plant reaches a height of 30–40 cm. Each fruit contains about 20 seeds. (By permission of M. A. Estelle and C. R. Somerville.)

200-fold higher than in *Arabidopsis*. Presumably the number of structural genes in both plants does not differ by more than a factor of 2–3. The difference in the size of the genome is mainly due to a different number of identical DNA sequences of various sizes arranged in sequence, termed *repetitive DNA*, of which a very large part may have no coding function at all. In broad bean for example, 85 per cent of the DNA is repetitive sequence. This includes the *tandem repeats*, a large number (sometimes thousands) of identical repeated DNA sequences (of a unit size of 170–180 bp, sometimes also 350 bp). The tandem repeats are spread over the entire chromosome, often arranged as blocks, especially at the beginning and the end of the chromosome, and sometimes also in the interior. This highly repetitive DNA is called *satellite DNA*. Its sequence is genus- or even species-specific. In some plants more than 15 per cent of the total nuclear genome consists of satellite DNA. So far its function is not known. Perhaps it plays a role in the segregation of species. The sequence of the satellite DNA can be used as a species-specific marker in generating hybrids by protoplast fusion in order to check the outcome of the fusion in cell culture.

The genes for ribosomal RNAs also occur as repetitive sequences and, together with the genes for some transfer RNAs, are present in the nuclear genome in several thousand copies.

In contrast, structural genes are present in only a few copies, sometimes just one (*single-copy gene*). Structural genes encoding for structurally and functionally related proteins often form a *gene family*. Such a gene family, for instance, is formed by the genes for the small subunit of the ribulose bisphosphate carboxylase, which exist several times in a slightly modified form in the nuclear genome (e.g. five times in tomato). So far 14 members of the light harvesting complex (LHC) gene family (section 2.4) have been identified in tomato. Zein, present in maize kernels as a storage protein (Chapter 14), is encoded by a gene family of about 100 genes.

20.2 The DNA of the nuclear genome is transcribed by three specialized RNA polymerases

Of the two DNA strands only the *template strand* is transcribed (Fig. 20.2). The DNA of the nuclear genome is transcribed by three specialized RNA polymerases (I, II, III) (Table 20.3). The division of labour between the three RNA polymerases, along with many details of gene structure and principles of gene regulation, apply to all eukaryotic cells. RNA polymerase II catalyses the transcription of the structural genes and is strongly inhibited by α-amanitin at a concentration as low as 10^{-8} mol l^{-1}. α-Amanitin is the deadly poison found in the toadstool *Amanita phalloides* (also called death cap). People frequently die from eating this toadstool.

The transcription of structural genes is regulated

It is estimated that a plant contains about 10 000–30 000 structural genes. However, most of these genes are switched off and are activated only in certain organs, and then often only in certain cells. Moreover, many genes are only switched on at specific times, e.g. the genes for the synthesis of phytoalexins after pathogenic infection (section 16.1). Therefore the transcription of most structural genes is subject to very complex and specific regulation. The genes for enzymes of metabolism

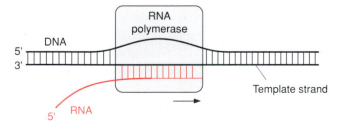

Figure 20.2 The template strand of DNA is transcribed.

Table 20.3 RNA polymerases

RNA polymerase	Transcript	Inhibition by α-amanitin
Type I	Ribosomal RNA (5.8S, 18S, 25S rRNA)	None
Type II	Messenger RNA precursors, small nuclear RNA (snRNA)	In concentrations of 10^{-8} mol l^{-1}
Type III	Transfer RNA, ribosomal RNA (5S rRNA)	Only at higher concentrations (10^{-6} mol l^{-1})

or protein biosynthesis, which proceed in all cells, are transcribed more often. The genes which every cell needs for such basic functions, independent of its specialization, are called *housekeeping genes*.

Promoter and regulatory sequences regulate the transcription of genes

Figure 20.3 shows the basic design of a structural gene. The section of the DNA on the left of the transcription starting point is termed 5′ or upstream and that to the right, 3′ or downstream. The encoding region of the gene is mostly distributed among several exons which are interrupted by introns.

About 25 bp upstream from the transcription start site is situated a promoter element, which is the position where RNA polymerase II binds. The sequence of this promoter element can vary greatly between genes and between species, but can be depicted as a consensus sequence. (A consensus sequence is an idealized sequence where each base is found in the majority of the promoters. Most promoter elements differ in their DNA sequences only by one or two bases from this consensus sequence.) This consensus sequence is named the *TATA box* (Fig. 20.4). About 80–110 bp upstream another consensus sequence is often found, the *CAAT box* (Fig. 20.4), which influences the rate of transcription. The housekeeping genes mentioned earlier often contain a C-rich region instead of the

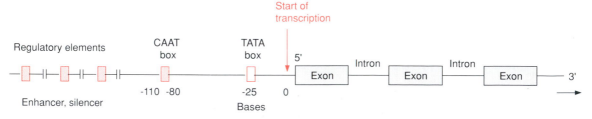

Figure 20.3 Sequence elements of a eukaryotic gene. The numbers mark the distance of the bases from the transcription start.

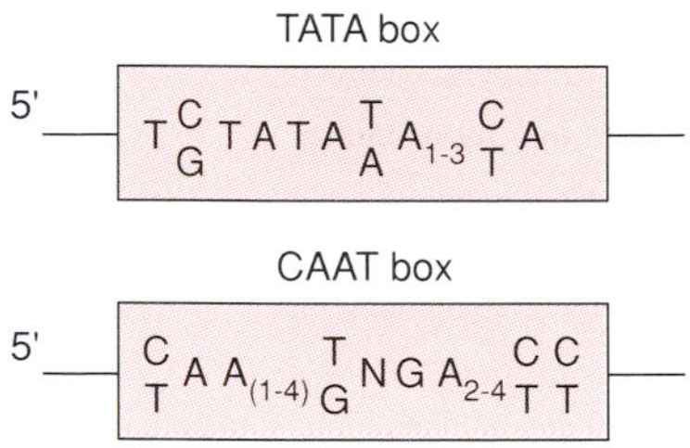

Figure 20.4 Consensus sequences for two promoter elements of the eukaryotic gene (see Fig. 20.3).

CAAT box. Additionally, sometimes more than 1000 bp upstream, several sequences can be present which function as *enhancer* or as *silencer* regulatory elements.

Transcription factors regulate the transcription of a gene

The regulatory elements contain binding sites for *transcription factors*, which are proteins modifying the rate of transcription. Different transcription factors often have certain structures in common. One type of transcription factor consists of a peptide chain containing two cysteine residues and, separated from these by 12 amino acids, two histidine residues (Fig. 20.5). The two cysteine residues bind covalently to a zinc atom, which is also coordinately bound to the imidazole rings of two histidine residues, thus forming a so-called *zinc finger*. Such a finger binds to a base triplet of a DNA sequence. Zinc finger transcription factors usually contain several (up to 9) fingers and so are able to cling tightly to certain DNA sections.

Another type of transcription factor consists of a dimer of DNA binding pro-

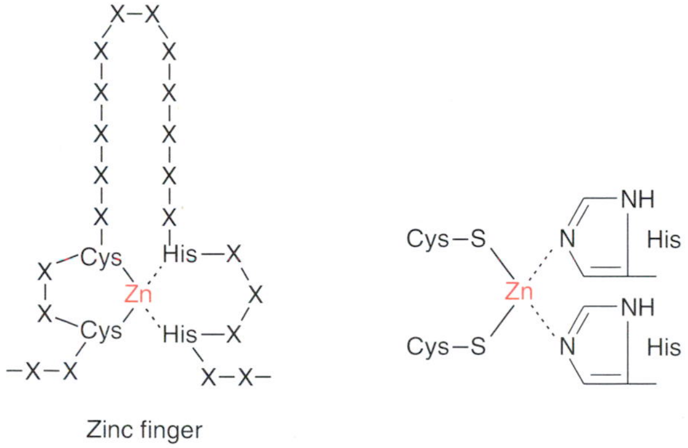

Zinc finger

Figure 20.5 Many transcription factors have the structure of a zinc finger. An amino acid sequence (X = any amino acid) contains two cysteine residues separated by 2–4 other amino acids and, after a further 12 amino acids, two histidine residues which are separated from each other by 3–4 amino acids. A zinc ion is bound between the cysteine and the histidine residues. A transcription factor contains 3–9 such zinc fingers, with each finger binding to a sequence of three bases on the DNA.

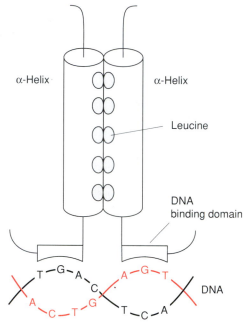

Figure 20.6 A frequent structural motif of transcription factors is the leucine zipper. The factor consists of two polypeptide chains, each with an α-helix, which contains a leucine residue at about every seventh position. The leucine residues, which are all located on one side of the α-helix, hold the two α-helices together by hydrophobic interactions in the manner of a zipper. The two DNA binding domains contain basic amino acids which enable binding to the DNA.

teins where each monomer contains a DNA binding domain and an α-helix with 4–5 leucine residues (Fig. 20.6). The hydrophobic leucine residues of the two α-helices are arranged so that they are exactly opposite each other and are held together by hydrophobic interaction like a zipper. This typical structure of a transcription factor has been named the *leucine zipper*.

The transcription of structural genes requires a complex transcription apparatus

RNA polymerase II consists of 8–14 subunits but, on its own, it is unable to start transcription. Transcription factors are required to direct the enzyme to the start position of the gene. The *TATA binding protein*, which recognizes the TATA box and binds to it, has a central function in transcription (Fig. 20.7). The interaction of the TATA binding protein with RNA polymerase requires a number of additional transcription factors (designated as A, B, F, E, H in the figure). They are all essential for transcription and are termed *basal factors*.

The transcription apparatus is a complex of many protein components, around which the DNA is wrapped in a loop. In this way, regulatory elements positioned far upstream or downstream from the encoding gene are able to influence the activity of RNA polymerase. The rate of transcription is determined by transcription factors, either *activators* or *repressors*, which bind to the upstream regulatory elements (Fig. 20.3). These transcription factors interact through a number of *co-activators* with the TATA binding protein and modulate function on RNA TATA

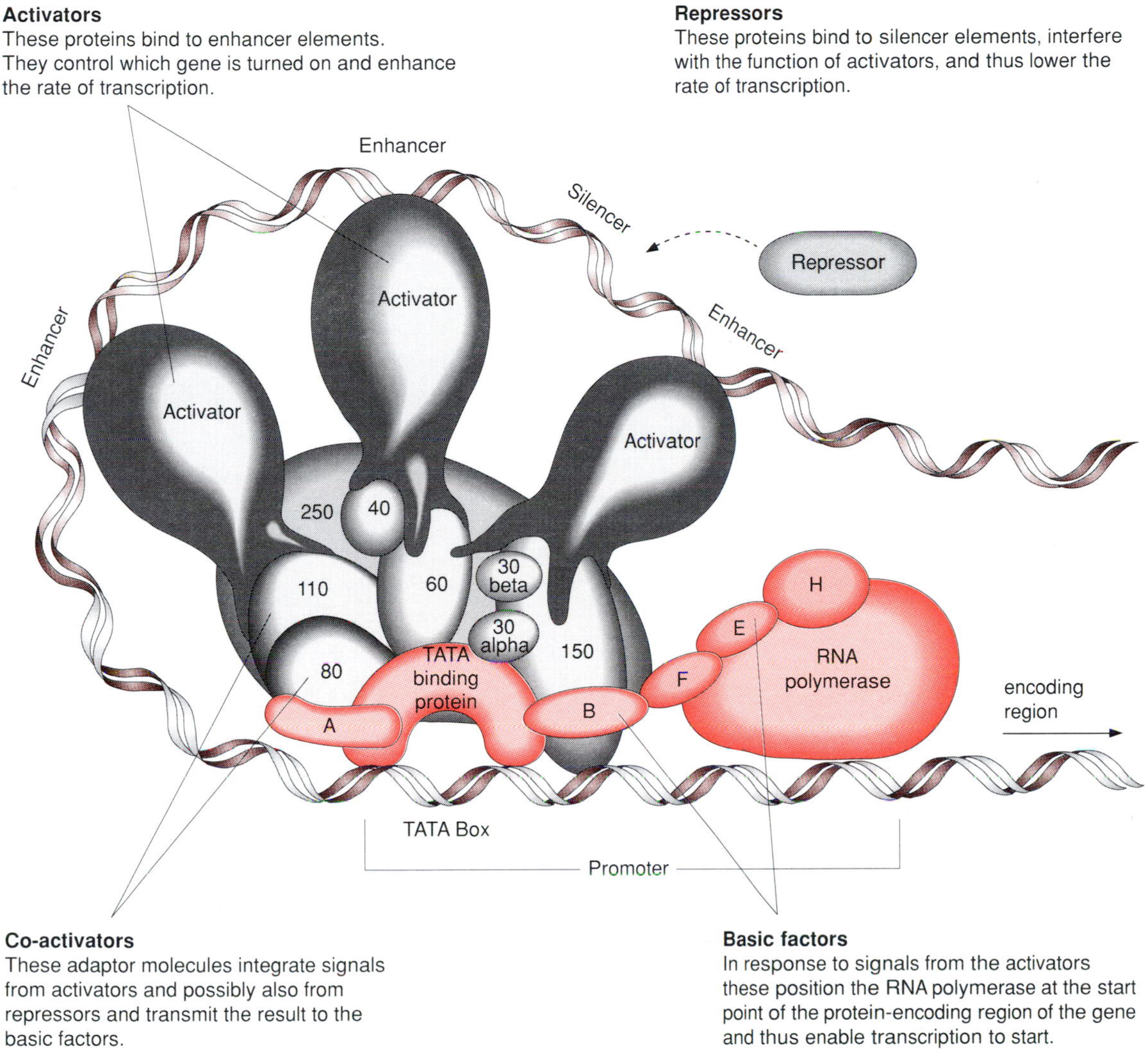

Activators
These proteins bind to enhancer elements.
They control which gene is turned on and enhance
the rate of transcription.

Repressors
These proteins bind to silencer elements, interfere
with the function of activators, and thus lower the
rate of transcription.

Co-activators
These adaptor molecules integrate signals
from activators and possibly also from
repressors and transmit the result to the
basic factors.

Basic factors
In response to signals from the activators
these position the RNA polymerase at the start
point of the protein-encoding region of the gene
and thus enable transcription to start.

Figure 20.7 Eukaryotic transcription apparatus of mammals, which is thought to be similar to the transcription apparatus of plants, which has not yet been resolved in detail. The basal factors (TATA binding protein, peptides A–H, coloured red) are indispensable for transcription, but they can neither enhance nor slow down the process. This is brought about by regulatory molecules, activators and repressors, of which the combination is different for each gene. They bind to DNA regulatory sequences, termed enhancers or silencers, which are located far upstream from the transcription start. Activators (and possibly repressors) communicate with the basal factors via co-activators which form tight complexes with the TATA binding protein. This complex docks first to the nuclear promoter, a control region close to the protein gene. The co-activators are designated according to their molecular weight (given in kDa). (From R. Tijan (1995). *Spektrum der Wissenschaft*, **4**, with permission.)

binding protein's polymerase. Various combinations of activators and repressors thus lead to activation or inactivation of gene transcription. The scheme of the transcription apparatus shown in Fig. 20.7 was derived from investigations with animals. Results obtained so far in plants indicate that the regulation of transcription occurs there in essentially the same way.

Knowledge of the promoter and enhancer/silencer sequences is very important for genetic engineering of plants (Chapter 22). It can be sufficient for practical applications if the DNA region that is positioned upstream of the structural gene and influences its transcription in a specific way, is identified and isolated. Promoter sequences have been identified which determine where a gene is to be transcribed, e.g. only in the leaf mesophyll cells. In such cases the specificity of gene expression is explained by the effect of cell-specific transcription factors on the corresponding promoters.

The formation of messenger RNA requires processing

The transcription of DNA in the nucleus by RNA polymerase II yields a *primary transcript* (pre-mRNA, Fig. 20.8) which is processed in the nucleus to mature

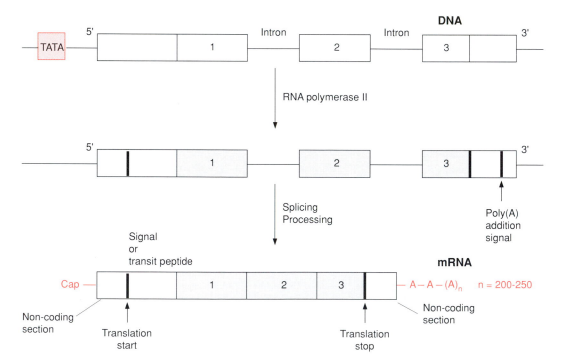

Figure 20.8 Transcription and post-transcriptional processing of a eukaryotic structural gene. The introns are removed from the primary transcript formed by RNA polymerase II by splicing (Figs 20.10, 20.11). The mature mRNA is formed by the addition of a cap sequence (Fig. 20.9) to the 5′ end and a poly(A) sequence to the 3′ end of a cleavage site marked by a poly(A) addition signal.

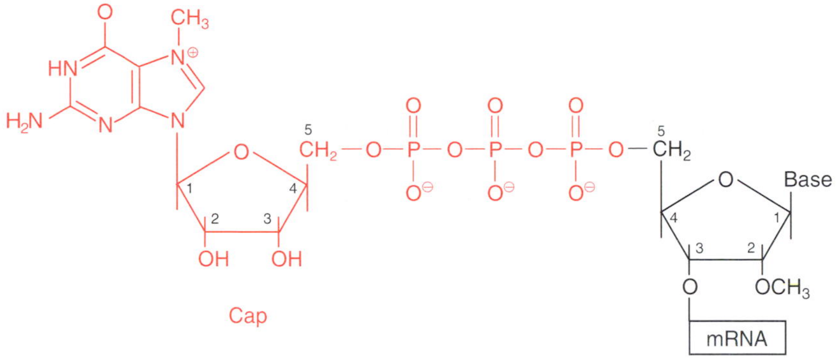

Figure 20.9 The cap sequence consists of a 7'-methylguanosine triphosphate which is linked to the 5' terminus of the mRNA. The ribose residues of the last two nucleotides of the mRNA are often methylated at the 2' position.

mRNA. During transcription, a special GTP molecule is linked by a 5'–5' pyrophosphate bridge to the RNA terminus (Fig. 20.9). Moreover, guanine and the second ribose (sometimes also the third, not shown in the figure) are methylated using *S*-adenosylmethionine as a donor (Fig. 12.10). This modified GTP at the beginning of RNA is called the *cap*, and is only contained in mRNA. It functions as a binding site in the formation of the initiation complex during the start of protein biosynthesis by the ribosomes (section 21.1) and probably also provides protection against degradation by exoribonucleases.

The introns have to be removed for further processing of the pre-mRNA. Their size can vary from 50 to over 10 000 bases. The beginning and end of an intron are defined by certain base sequences. The last two bases on the exon are mostly AG and the first bases of the intron are GU (Fig. 20.10). The intron ends with AG. About 20–50 bases upstream from the 3' end of the intron an adenyl residue, known as the *branching site*, is in the sequence.

The excision of introns (*splicing*) is catalysed by *riboprotein complexes*, composed of RNA and proteins. Five different RNAs of 100–190 bases, called snRNAs (sn = small nuclear), are involved in the splicing procedure. Together with proteins and the RNA to be spliced, these snRNAs form the *spliceosome*

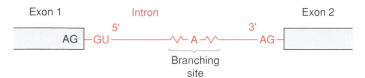

Figure 20.10 The beginning of an intron is often marked by the sequence AG/GU and the end by the sequence AG. At about 50 bases before the end of the intron, there is a consensus sequence of about seven nucleotides containing an A, which forms the branching site during splicing (see Fig. 20.11).

particle (Fig. 20.11). The first step is that the 2′-OH group of the ribose at the adenyl residue of the branching site forms a phosphate ester with the phosphate residue linking the end of exon 1 with the start of the intron, cleaving the ester bond between exon 1 and the intron. It is followed by a second ester formed between the 3′-OH group of exon 1 and the phosphate residue at the 5′-OH of exon 2, accompanied by a cleavage of the phosphate ester with the intron, thus

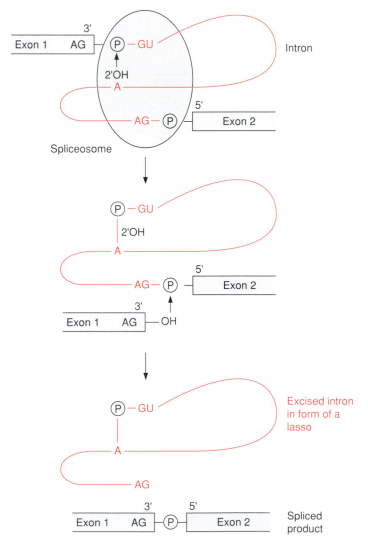

Figure 20.11 In the splicing procedure, several RNAs and proteins assemble at the splicing site of the RNA to form a spliceosome. The terminal phosphate at the 5′ end of the introns forms an ester linkage with the 2′-OH group of the A in the branching site, and the RNA chain after the exon 1/intron junction is cleaved. The 3′ end of exon 1 forms a new ester linkage with the phosphate residue at the 5′ end of exon 2, releasing the intron in the form of a lasso.

Poly(A) addition signal

Figure 20.12 Consensus sequence for the poly(A) addition signal.

completing the splicing. The intron remains in the form of a lasso and later is degraded by ribonucleases.

As a further step in RNA processing the 3′ end of the pre-mRNA is cleaved behind a poly(A) addition signal (Fig. 20.12) by an endonuclease, and a *poly(A) sequence* of up to 250 bp is added at the cleaving site. The resulting mature mRNA is bound to special proteins and leaves the nucleus as a DNA–protein complex.

rRNA and tRNA are synthesized by RNA polymerases I and III

Eukaryotic ribosomes of plants contain four different rRNA molecules, named 5S, 5.8S, 18S, and 25S rRNA according to their sedimentation coefficients. The genes for 5S rRNA are present in very many copies, arranged in tandem on certain sections of the chromosomes. The transcription of these genes and the tRNA genes is catalysed by RNA polymerase III. The three remaining ribosomal RNAs are encoded by a continuous gene sequence, again in tandem and in very many copies. These genes are transcribed by RNA polymerase I. The primary transcript is subsequently processed by methylation, especially of -OH groups of ribose residues, and the cleavage of RNA, to produce mature 18S, 5.8S, and 25S rRNA (Fig. 20.13). The excised RNA spacers between these rRNAs are degraded. Because of their rapid evolution, comparative sequence analyses of these spacer regions can be used to establish a phylogenetic classification of various plant species within a genus.

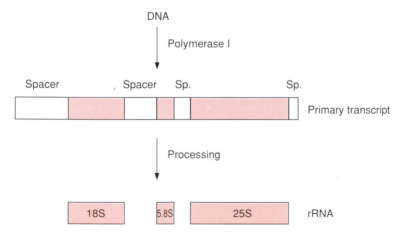

Figure 20.13 Three of the four rRNAs are transcribed polycistronically. The spacers are removed during processing

20.3 DNA polymorphism yields genetic markers for plant breeding

An organism is defined by the DNA base sequences of its genome. Differences between the DNA sequences (DNA polymorphisms) exist not only between different species but also between individuals of the same species. Between two varieties of a cultivated plant species, often 0.1–1 per cent of the bases in a structural gene are changed, mainly in the introns.

Plants are usually selected for breeding purposes by external features, e.g. for their yield or resistance to certain pests. In the end these characteristics are all due to differences in the base sequence of structural genes. Selecting plants for breeding would be much easier if it were not necessary to wait until the phenotypes of the next generation were evident, but if instead the corresponding DNA sequences could be analysed. A complete comparative analysis of these genes is not practical since most of the genes involved in the expression of the phenotypes are not known and DNA sequencing is expensive and slow. Other techniques are available which are easier and cheaper to perform.

Individuals of the same species can be differentiated by restriction fragment length polymorphism

It is possible to detect differences in the genes of individuals within a species even without a detailed sequence comparison and to relate these differences empirically to analysed properties. One of the methods for this is the analysis of *r*estriction *f*ragment *l*ength *p*olymorphism (RFLP). This is based on the use of bacterial *restriction endonucleases*, which cleave a DNA at a palindromic recognition sequence known as a *restriction site* (Fig. 20.14). The various restriction endonucleases have specific restriction sites of 4–8 bp. Since these restriction sites appear at random in DNA (those recognition sequences with 4 bp appearing more frequently than those with 8 bp), it is possible to cleave the genomic DNA of a plant into thousands of defined DNA fragments by using a particular restriction endonuclease (usually enzymes with 6 bp restriction sites). A restriction site can be eliminated or newly formed by exchanging only one base in the DNA, so that a very slight change in the base sequence of the DNA can result in a polymorphism of the restriction fragment length.

To select fragments of certain sections of the genome, labelled *DNA probes* are required. Probes are prepared by first finding a certain DNA section of a chromosome of about 10–20 kb, located as near as possible to the gene responsible for the trait of interest, or which is even part of that gene. This DNA section is introduced into bacteria (usually *Escherichia coli*) using plasmids or bacteriophages as a vector and is propagated there (Chapter 22). The plasmids or bacteriophages are isolated from the bacterial suspension, the multiplied DNA sections are cut out

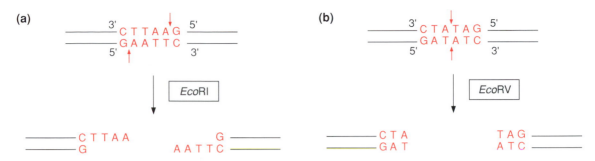

Figure 20.14 Restriction endonucleases of type II (only these are dealt with here) cleave the DNA at restriction sites which consist of a palindromic recognition sequence. As an example the restriction sites for two enzymes from an *Escherichia coli* strain are shown. (a) The restriction endonuclease *Eco*RI causes staggered cuts of the two DNA strands, leaving four nucleotides of one strand unpaired at each resulting end. These unpaired ends are called sticky ends because they can form a base pair with each other, or with complementary sticky ends of other DNA fragments. (b) In contrast, the restriction endonuclease *Eco*RV produces blunt ends.

again, isolated, and afterwards radioactively labelled or provided with a fluorescent label. Such probes used for this purpose are called *RFLP markers*.

The analysis of DNA restriction fragments by the labelled probes described above is carried out by the *Southern blot* method developed by Edward Southern in 1975. The restriction fragments are first separated according to their length by electrophoresis in an agarose gel (the shortest fragment moves the furthest). The separated DNA fragments in the gel are transferred to a nitrocellulose or nylon membrane by placing the membrane on the gel. A buffer solution is drawn through the gel and the membrane by covering them with a stack of tissue paper and the displaced DNA fragments are bound to the membrane and dissociated by the buffer into single strands (Fig. 20.15). When a labelled DNA probe is added, it hybridizes to complementary DNA sequences on the membrane. Only those DNA fragments that are complementary to the probe are labelled and, after removal of the non-bound DNA probe molecules by washing, subsequently identified by autoradiography (in the case of a radioactive probe) or by fluorescence measurement. The position of the band on the blot is then related to its migration and hence its size.

Figure 20.16 explains the principles of RFLP. Figure 20.16a shows in (i) a section of a gene with three restriction sites (R_1, R_2, R_3). Since the probe binds only to the DNA region between R_1 and R_3, just two noticeable restriction fragments result from the treatment with the restriction endonuclease. Due to their different lengths, they are separated by gel electrophoresis and detected by hybridization with a probe (Fig. 20.16b(i)). Upon the exchange of one base (point mutation) (ii) the restriction site R_2 is eliminated and therefore only one labelled restriction fragment is detected, which, because of its larger size, migrates in gel electrophoresis for a lesser distance than the fragments of (i). When a DNA section is

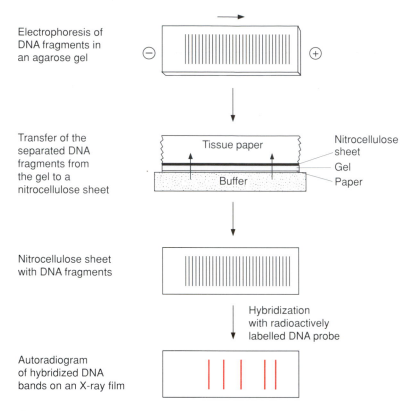

Electrophoresis of
DNA fragments in
an agarose gel

Transfer of the
separated DNA
fragments from
the gel to a
nitrocellulose sheet

Tissue paper

Nitrocellulose
sheet
Gel
Paper

Buffer

Nitrocellulose sheet
with DNA fragments

Hybridization
with radioactively
labelled DNA probe

Autoradiogram
of hybridized DNA
bands on an X-ray film

Figure 20.15 The Southern blot procedure.

inserted between restriction sites R_2 and R_3 (iii), the corresponding fragment is longer.

The RFLPs represent *genetic markers* which are inherited according to Mendelian laws and can be employed to characterize a certain variety. Normally several probes are used in parallel measurements. RFLP is also used in plant systematics to establish phylogenetic trees. Moreover, defined restriction fragments can be used as labelled probes in order to localize certain genes on the chromosomes. In this way chromosome maps have been established for several plants (e.g. *Arabidopsis*, potato, tomato, and maize).

The RAPD technique is an especially simple method for investigating DNA polymorphism

An alternative method for analysing the differences between DNA sequences of individuals or varieties of a species is the amplification of randomly obtained DNA fragments (*random amplified polymorphic DNA*, RAPD). This method, which has only been in use since 1990, is much easier to work with than the RFLP

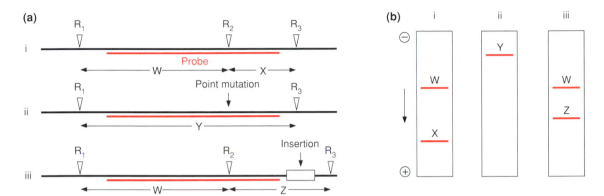

Figure 20.16 The formation of restriction fragments in different genotypes. (a) The restriction sites for the restriction endonuclease in genotypes i, ii, and iii are numbered R_1, R_2, and R_3. The probe by which the fragments are identified, is marked red. (b) The electrophoretic movement of the fragments (W, X), that are labelled by the probe. (ii) By point mutation one restriction site is eliminated and only one fragment is formed (Y) which, because of its larger size, migrates less during electrophoresis. (iii) After insertion, fragment X is enlarged (Z).

technique mentioned above, and its application has become widespread in a very short time.

The basis for the RAPD technique is the *polymerase chain reaction* (PCR). The method enables selected DNA fragments of a length of up to 2–3 kb to be amplified by DNA polymerase. This necessitates an *oligonucleotide primer*, which binds to a complementary sequence of the DNA and forms the starting point for the synthesis of a DNA daughter strand at the template of the DNA mother strand. In the polymerase chain reaction, two primers (A, B) are needed to define the beginning and end of the DNA strand which is to be amplified. Figure 20.17 shows the principle of the reaction. In the first step, by heating to about 95°C, the DNA double strands are separated into single strands. During a subsequent cooling period, the primers hybridize with the DNA single strands and thus enable, in a third step at a medium temperature, the synthesis of DNA from deoxynucleoside triphosphates. A DNA polymerase originally isolated from the thermophilic bacterium *Thermus aquaticus* living in hot wells (*Taq polymerase*) is usually used, since this enzyme is not affected by the preceding heat treatment. Subsequently, the DNA double strands thus formed are separated again by being heated, the primer binds during an ensuing cooling period and this is followed by another cycle of DNA synthesis by *Taq* polymerase. By continuous alternating heating and cooling this reaction can be continued for 30–40 cycles, with the amount of DNA being doubled in each cycle. In the first cycle the length of the newly formed DNA is not yet restricted at one end. By the primer binding to the complementary base sequence of the newly formed DNA strand, in the next DNA synthesis cycle a product is formed which is restricted in its length by both primers. With the

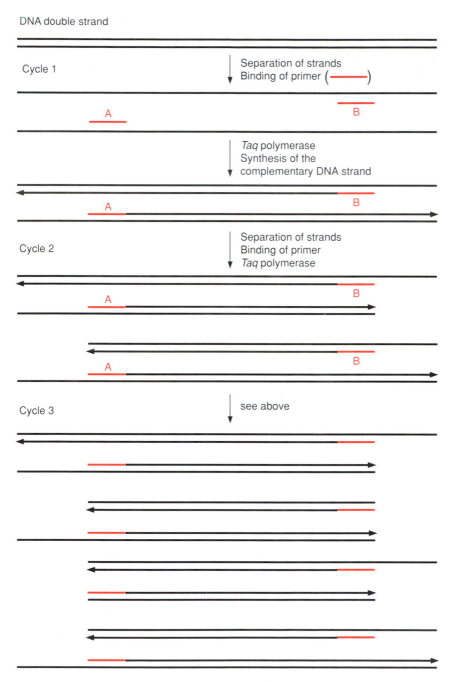

Figure 20.17 Principle of the polymerase chain reaction.

increasing number of cycles, DNA fragments of uniform length are amplified. Since in the polymerase chain reaction the number of the DNA molecules formed is multiplied exponentially by the number of cycles (e.g. after 25 cycles by the factor 8×10^6), very small DNA samples (in the extreme case, a single molecule) can be multiplied *ad libitum*.

In the RAPD technique mentioned above, genomic DNA and only one oligonucleotide primer, typically consisting of 10 bases, are required for the polymerase chain reaction. Since the probability of the exact match of 10 complementary bases on the genomic DNA is low, the primer binds only at a few sites of the genomic DNA. The characteristics of the DNA polymerases used for the amplification require the distance between the two bound primers to be no larger than 2–3 kbp. Therefore, only a few sections of the genome are amplified by the polymerase chain reaction and a further selection of the fragments by a probe is not necessary. The amplification produces such large amounts of single DNA fragments that, after being separated by gel electrophoresis and stained with ethidium bromide, the fragments can be detected as fluorescent bands under UV light. Point mutations that eliminate primer binding sites or form new ones, and deletions or insertions, which shift bands, can all change the band pattern of the DNA fragments in an analogous way to that of the RFLP technique (Fig. 20.16). Changing the primer sequence can generate different DNA fragments. Defined primers of 10 bases are commercially available in very many variations. In the RAPD technique different primers are tried until, by chance, bands of DNA fragments are obtained which cosegregate with a certain trait. The RAPD technique is less work than the RFLP technique since it requires neither the preparation of probes nor the time-consuming procedure of a Southern blot. It has the additional advantage that only very small amounts of DNA, e.g. the amount that can be isolated from the embryo of a plant, are required for analysis. However, it is not possible to say from which gene these fragments derive. Since the RAPD technique allows differentiation between varieties of a species, it has become an important tool in breeding.

20.4 Transposable DNA elements roam through the genome

In certain maize varieties, a cob may contain some kernels with different pigmentation from the others, indicating that a mutation has changed the pigment formation. Snapdragons normally have red flowers, but occasionally have mutated progeny in which parts of the flower no longer form red pigment, resulting in white stripes in the flowers. Sometimes the descendants of these defective cells regain the ability to synthesize the red pigment, forming flowers with not only white stripes but also red dots.

Barbara McClintock (USA) studied these phenomena for many years in maize using the methods of classical genetics. In the genome of maize she found mobile DNA elements, which jump into a structural gene and thus inactivate it. Generally this mobile element does not stay there permanently, but sooner or later jumps into another gene, whereby, in most cases, the function of the first structural gene is restored. For these important findings Barbara McClintock was awarded the Nobel prize for medicine in 1983. Later it became apparent that these transposable elements, which were to be named *transposons*, are not unique to plants, but also occur in bacteria, fungi, and animals.

Figure 20.18 shows the structure of the transposon Ac (activator) from maize, consisting of double-stranded DNA with 4600 base pairs. Both ends contain a 15 bp long inverted repeat sequence (IR$_A$, IR$_B$). Inside the transposon is a structural gene, which encodes *transposase*, an enzyme catalysing the transposition of the gene. This enzyme binds to the flanking inverted repeats and catalyses the transfer of the transposon to another location. It can happen that the excision of the transposon is sometimes imprecise, so that after its exit the remaining gene may have a slightly modified sequence, which could result in a lasting mutation.

In maize, besides the transposon Ac, a very similar transposon has been found, named Ds. However, in this element the structural gene for the transposase is eliminated. Therefore, the transposon Ds is only mobile in the presence of the transposon Ac, which encodes the transposase required for the transposition of the Ds.

A transposon can thus be regarded as an autonomous unit encoding the proteins required for jumping. There are controversial opinions about the origin and

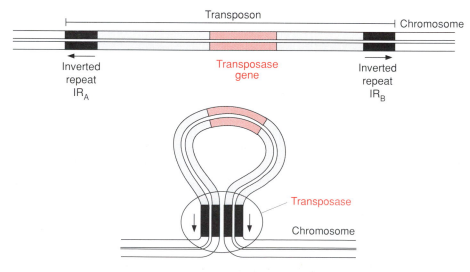

Figure 20.18 A transposon is defined by inverted repeats at both ends. The structural gene for transposase is included in the transposon. When a transposon leaves a chromosome, the two inverted repeats bind to each other and the remaining gap in the chromosome is closed. In an analogous way the transposon enters the chromosome at another site.

function of the transposons. Many see the transposons as a kind of parasitic DNA, with features comparable to the viruses, which exploit the cell to multiply themselves. But it is also possible that the transposons offer the cells a selection advantage by increasing the mutation rate in order to enhance adaptation to changed environmental conditions.

As the transposons can be used for *tagging genes*, they have become an interesting tool in biotechnology. It has already been discussed that the insertion of a transposon in a structural gene results in the loss of the encoding function. For example, when a transposon jumps into a gene for anthocyanin synthesis in snapdragons, the red flower pigment can no longer be formed. The transposon inserted in this inactivated gene can be used as a DNA probe (marker) in order to isolate and characterize a gene of the anthocyanin biosynthesis pathway. The relevant procedures will be dealt with in section 22.1.

20.5 Most plant cells contain viruses

With the exception of meristematic cells, all other plant cells are generally infected by viruses. In many cases viruses do not kill their host since they depend on the host's metabolism to reproduce themselves. The viruses encode only a few special proteins and use the energy metabolism and the biosynthetic capacity of the host cell to multiply. This often weakens the host plant and lowers the yield of virus-infected cultivars. Infection by some viruses can lead to the destruction of the entire crop. Zucchini (courgettes) and melons are extremely susceptible to the cucumber mosaic virus. In some provinces of Brazil, 75 per cent of the orange trees were destroyed by the tristeza virus within 12 years.

The virus genome consists of RNA or DNA surrounded by a protein coat. In the majority of plant viruses the genome consists of a single-stranded RNA, called the plus RNA strand. In some viruses, for instance brome mosaic virus which infects certain cereals, the plus RNA strand shows the characteristics of an mRNA and is translated by the host. In other viruses, e.g. the *tobacco mosaic virus* (TMV), the plus RNA strand is first transcribed to a complementary minus RNA strand and the latter then serves as a template for the formation of mRNAs (Fig. 20.19). The translation products of these mRNAs encompass *replicases* which catalyse the replication of the plus and minus RNAs, *movement proteins* which enable the spreading of the viruses from cell to cell (section 1.1), and *coat proteins* for packing the viruses. Normally a virus reaches a cell by wounds which, for instance, have been caused by insects, such as aphids feeding on the plant (section 13.2). Once viruses have entered the cell, their movement proteins widen the plasmodesmata between the single cells enough to let them pass through and spread over the whole symplast.

In the retroviruses, the genome consists of a single-stranded RNA, but in this case, when the cell has been infected, the RNA is transcribed by a *reverse tran-*

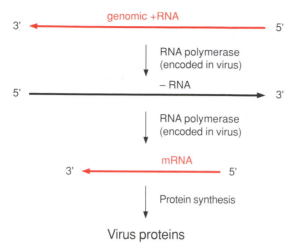

genomic +RNA

3'   ◄────────────────────   5'

RNA polymerase
(encoded in virus)

─ RNA

5'   ────────────────────►   3'

RNA polymerase
(encoded in virus)

mRNA

3'   ◄────────────────   5'

Protein synthesis

Virus proteins

Figure 20.19 The single-stranded genomic RNA of many viruses is first transcribed to a minus strand RNA, and the latter then to mRNAs for the synthesis of proteins.

scriptase into DNA, which is in part integrated in the nuclear genome. So far, infections by retroviruses are only known in animals.

The *cauliflower mosaic virus* (CaMV), which causes pathogenic changes in the leaves of cauliflower and related plants, is somewhat similar to a retrovirus. The genome of the cauliflower mosaic virus consists of a double-stranded DNA of about 8 kb, with gaps in it (Fig. 20.20). When a plant cell is infected, the virus loses its protein coat and the gapped DNA strands are filled by repair enzymes of the host. The viral genome acquires a double helical structure and forms a chromatin-like aggregate with the histones in the nucleus. This permits the viral genome to stay in the nucleus as a *mini-chromosome*. The viral genome contains promoter sequences which are similar to those of nuclear genes. They possess a TATA and a CAAT box and also enhancer elements. The virus promoter is recognized by the RNA polymerase II of the host cell and is transcribed at a high rate. The transcript is subsequently processed into individual mRNAs, which encode the synthesis of six virus proteins, including the coat protein and the reverse transcriptase. The strong CaMV 35S promoter is often used as a promoter for the expression of foreign genes in transgenic plants (Chapter 22).

The transcript formed by RNA polymerase II is also transcribed in reverse by the virus-encoded reverse transcriptase into DNA and, after synthesis of the complementary strand, packed as double-stranded genome into a protein coat. This completed virus is now ready to infect other cells.

Retrotransposons are degenerate retroviruses

Besides the transposons described in the previous section there is another class of

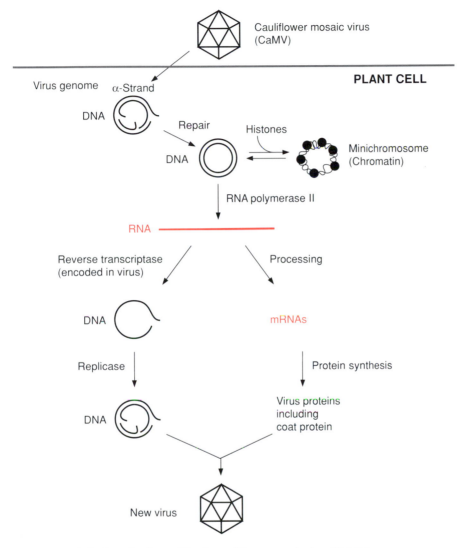

Figure 20.20 Infection of a plant cell by the cauliflower mosaic virus (CaMV).

mobile elements which are derived from the retroviruses. They do not jump out of a gene like the transposons but just multiply. Retrotransposons contain at both ends sequences that carry the signals for the transcription of the retrotransposon DNA by the RNA polymerase of the host cell. The retrotransposon RNA encodes some proteins, but no coat protein. It encodes a reverse transcriptase, which is homologous to the retroviral enzyme and transcribes the retrotransposon RNA into DNA (Fig. 20.21). This DNA is then inserted into another site of the genome. It is assumed that these retrotransposons are retroviruses which have lost the

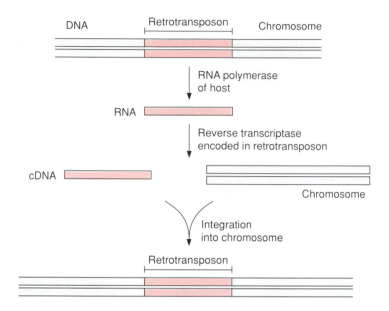

Figure 20.21 Retrotransposons consist of a DNA sequence integrated into a chromosome. Flanking sequences contain the recognition signal for transcription by an RNA polymerase of the host cell. The RNA is transcribed into cDNA by a reverse transcriptase encoded in the retrotransposon, and integrated into another section of the genome.

ability to form a protein coat. Three different retrotransposons containing all the constituents for their multiplication have been found in *Arabidopsis*. But so far an insertion of a retrotransposon into a gene of *Arabidopsis* has never been monitored. About 0.1 per cent of the genome of *Arabidopsis* consists of these retrotransposons, suggesting that these do multiply, albeit slowly.

20.6 Plastids possess a circular genome

Many arguments support the hypothesis that plastids have evolved from prokaryotic endosymbionts (section 1.3). The circular genome of the plastids is similar to the genome of the prokaryotic cyanobacteria, although much smaller. The DNA of the plastid (chloroplast) genome is named ctDNA. In the majority of the plants investigated so far, the size of the circular plastid genome is 120–160 kb. Depending on the plant, this is only 0.001–0.1 per cent of the size of the nuclear genome (Table 20.1). But the cell contains very many copies of the plastid genome, for two reasons. First, each plastid contains many genome copies. In young leaves the number of ctDNA molecules per chloroplast is about 100, whereas in older leaves, it is between 15–20. Secondly, a cell contains a large number of plastids, for instance 20–50 in a mesophyll cell. Thus, in spite of the small size of the plastid genome, plastid DNA can amount to 5–10 per cent of the total cell DNA.

The first complete analysis of the base sequence of a plastid genome was carried out in 1986 by the group of Katzuo Shinozaki in Nagoya with chloroplasts from tobacco and also by Kanji Ohyama in Kyoto with chloroplasts from the liverwort *Marchantia polymorpha*. Although the two investigated plants are very distantly related, their plastid genomes are rather similar in gene composition and arrangement. Obviously the plastid genome has changed little during recent evolution. Analysis of the DNA sequence of plastid genes from many plants supports this notion.

Figure 20.22a shows a complete gene map of the chloroplast genome of tobacco and Figs 20.22b and c show schematic representations of the plastid genomes of other plants. The plastid genome of most plants contains so-called *inverted repeats* (IRs), which divide the remaining genome into large or small single copy regions. The repeats IR_A and IR_B each contain the genes for the four ribosomal RNAs and also the genes for some transfer RNAs, and can vary in size from 20 to 50 kbp. These inverted repeats are not found in the plastid genomes of pea, broad bean, and other legumes (Fig. 20.22c), where they have probably been lost during the course of evolution.

Analysis of the ctDNA sequence of tobacco has revealed that the genome contains 122 genes (146 if the genes of each of the two inverted repeats are counted) (Table 20.4). The gene for the large subunit of ribulose bisphosphate carboxylase/

Table 20.4 Identified genes in the genome of maize chloroplasts (Shinozaki *et al.* 1986)

Name of the gene	Gene product (protein or RNA)
Photosynthetic apparatus	
rbcL	RubisCO: large subunit
atpA, -B, -E, -F, -H, -I	F-ATP synthase: subunits α, β, ϵ, I, III, IV
psaA, -B, -C	Photosystem I: subunit A1, A2, 9 kDa protein
psbA, -B, -C, -D, -E, -F, -G, -H, -I	Photosystem II: subunit D1, 51 kDa, 44 kDa, D2
	Cyt-b_{559}−9 kDa, −4 kDa, G, 10Pi, I-protein
petA, -B, -D	Cyt-b_6/f complex: cyt-f, cyt-b_6, subunit IV
ndhA, -B, -C, -D, -E, -F	NADH dehydrogenase (ND) subunits 1, 2, 3, 4, ND4L, 5
Protein synthesis	
rDNA	Ribosomal RNAs (16S, 23S, 4.5S, 5S)
trn	Transfer RNAs (30 species)
rps2, -3, -4, -7, -8, -11	30S ribosomal proteins (CS) 2, 3, 4, 7, 8, 11
-12, -14, -15, -16, -18, -19	12, 14, 16, 18, 19
rpl2, -14, -16, -20, -22	50S ribosomal proteins (CL) 2, 14, 16, 20, 22
-23, -33, -36	CL 23, 33, 36
infA	Initiation factor 1
Gene transcription	
rpoA, -B, -C	RNA polymerase-α, -β, -β'
ssb	ssDNA binding protein

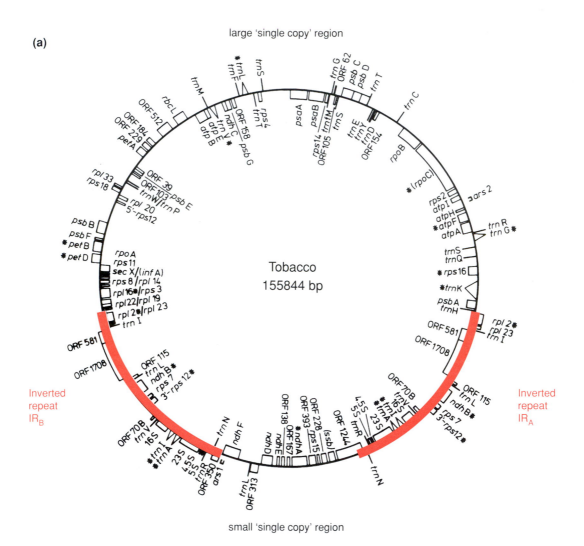

(a)

large 'single copy' region

Tobacco
155844 bp

Inverted
repeat
IR_B

Inverted
repeat
IR_A

small 'single copy' region

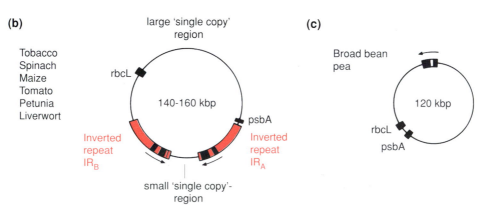

(b)

Tobacco
Spinach
Maize
Tomato
Petunia
Liverwort

large 'single copy'
region

rbcL

140-160 kbp

psbA

Inverted
repeat
IR_B

Inverted
repeat
IR_A

small 'single copy'-
region

(c)

Broad bean
pea

120 kbp

rbcL

psbA

oxygenase (RubisCO, section 6.2) is located in the large single copy region, whereas the gene for the small subunit is present in the nuclear genome. The single copy region of the plastid genome also encodes six subunits of F-ATP synthase, whereas the remaining genes of F-ATP synthase are encoded in the nucleus. Also encoded in the plastid genome are part of the subunits of photosystems I and II, of the cytochrome-b_6/f complex, and of an NADH dehydrogenase (which also occurs in mitochondria, see sections 3.8 and 5.5) and, furthermore, proteins of plastid protein synthesis and gene transcription. Some of these plastid structural genes contain introns. In addition, there are putative genes on the genome with so-called *open reading frames* (ORF), where the encoded proteins are not yet known. The plastid genome encodes only a fraction of the plastid proteins, as the majority are encoded in the nucleus. It is assumed that many genes of the original endosymbiont have been transferred during evolution to the nucleus, but there are also indications for gene transfer between the plastids and the mitochondria (section 20.7).

All four rRNAs, that are constituents of the plastid ribosome (4.5S, 5S, 16S, and 23S rRNA) are encoded in the plastid genome. The plastid ribosomes (sedimentation coefficient 70S) are smaller than the eukaryotic ribosomes (80S) contained in the cytosol, but are similar in size to the ribosomes of bacteria. As in bacteria, these four rRNAs are contained in the plastid genome in one transcription unit (Fig. 20.23). Between the 16S and 23S rRNA is situated a large spacer containing the sequence for one or two tRNAs. In total about 30 tRNAs are encoded in the plastid genome. It is assumed that additional tRNAs needed in the plastids are encoded in the nucleus.

The transcription apparatus of the plastids resembles that of bacteria

In the plastids probably two RNA polymerases are active, of which only one is encoded in the plastid genome. This plastid RNA polymerase is a multienzyme

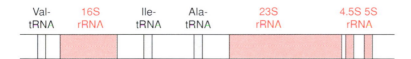

Figure 20.23 In the plastids of maize all four ribosomal rRNAs and two tRNAs are transcribed as one transcription unit.

complex resembling that of bacteria. But in contrast to the RNA polymerase of bacteria, the plastid enzyme is insensitive towards *rifampicin,* a synthetic derivative of an antibiotic from *Streptomyces.* As in bacteria, the plastid genes contain a box 10 bp upstream from the transcription start with the consensus sequence TATAAT and at −35 bp a further promoter site with the consensus sequence TTGACA. Some structural genes are polycistronic, which means that several are contained in one transcription unit, and are transcribed together in a large primary transcript, as also often happens with bacterial genes. In some cases, the primary transcript is subsequently processed by ribonucleases of which many details are still unknown.

20.7 The mitochondrial genome of plants varies largely in size

In contrast to animals, plants possess a very large mitochondrial (mt) genome. In *Arabidopsis* it is 20 times, and in melon 140 times, larger than in humans (Table 20.5). The plant mitochondrial genome contains also more genetic information: the number of encoding genes in a plant mt-genome is about seven times higher than in humans.

The size of the mt-genome varies largely in higher plants, even within a family. *Citrullus lanatus* (330 kbp), *Cucurbita pepo* (850 kbp), and *Cucumis melo* (2400 kbp), listed in Table 20.5, are all members of the Cucurbitaceae (squash plants). However, in most plants the size of the plastid genome is relatively constant at 120–160 kbp.

In plants the mitochondrial genome often consists of one large and several smaller DNA molecules. In some mitochondrial genomes this partitioning may be permanent, but in many cases the fragmentation of the mt-genome seems to be

Table 20.5 Size of the mitochondrial DNA (mtDNA) in plants compared with human mtDNA

Organism	Size of the mtDNA (kbp)
Arabidopsis thaliana	370
Vicia faba (broad bean)	290
Zea mays (maize)	570
Citrullus lanatus (water melon)	330
Cucurbita pepo (pumpkin)	850
Cucumis melo (honey melon)	2400
Marchantia polymorpha (liverwort)	170
Homo sapiens (human)	17

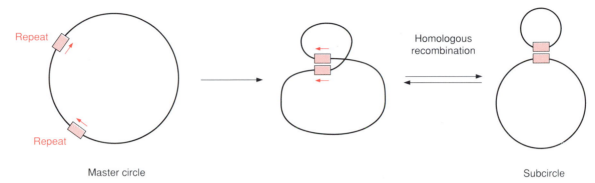

Figure 20.24 From a mitochondrial genome (master circle) two subcircles can be formed by recombination of two repetitive sequences by reversible homologous recombination.

derived from homologous recombination of repetitive elements (e.g. maize contains six such repeats). Figure 20.24 shows how an interaction of two repeats can lead by *homologous recombination* to fragmentation of a DNA molecule into two parts. The 570 kbp genome of maize is present in the form of a *master circle* as well as up to four *subcircles* (Fig. 20.25). The recombination of DNA molecules can also form larger units. This may explain the large variability in the size of the mitochondrial genome in plants.

The number of mitochondria in a plant cell can be between 50 and 2000, with each mitochondrion containing 1–100 genomes. Therefore at each cell division the mitochondrial genome is inherited in very many copies.

In animals and in yeast the mtDNA is normally circular, as is the DNA in bacteria. Plant mtDNA is also thought to be circular. This is undisputed for the small mtDNA molecules (subcircles, Fig. 20.25), but it is still unclear whether this

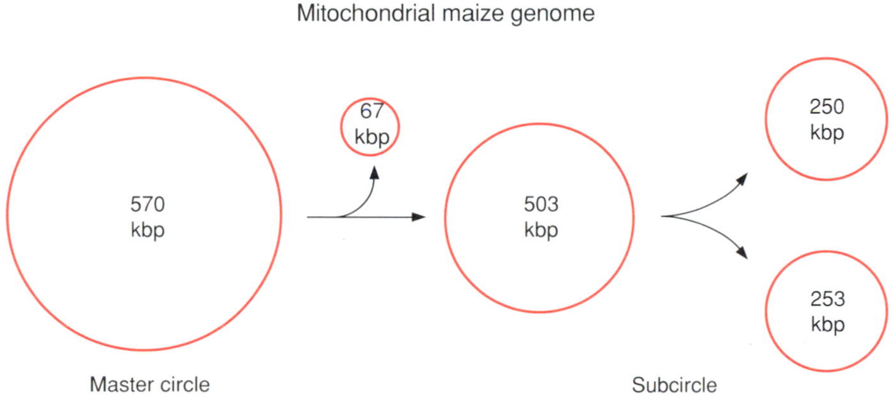

Mitochondrial maize genome

570 kbp — Master circle

67 kbp

503 kbp — Subcircle

250 kbp

253 kbp

Figure 20.25 Due to homologous recombination, the mitochondrial genome from maize can be present as a continuous large genome as well as in the form of several subcircles.

circular structure also generally applies to the master mtDNA. There are indications that the master genome can also occur as open strands.

The base sequence of the mtDNA of humans and of brewer's yeast (*Saccharomyces cerevisiae*) is fully known. For plants only the base sequence of the mt-genome of the liverwort *Marchantia polymorpha* has been analysed so far. Work on the analysis of the mt-genome of *Arabidopsis thaliana* is in progress and may be completed soon. Without a complete analysis of the base sequence, single genes can still be identified on the mt-genome by using DNA probes (section 20.3).

Table 20.6 shows the average content of genes in the mt-genome of a higher plant. A comparison with the number of genes in the plastid genome (Table 20.4) shows that the mt-genome of the plant, although mostly very much larger than the plastid genome, contains much less genetic information. The relatively low information content of the mt-genome in relation to its size is due to a high content of repetitive sequences, which are probably derived from gene duplication. The mt-genome contains much DNA with no recognizable function. Some researchers regard this as *junk DNA*, which has accumulated in the mt-genome during evolution. Part of this junk DNA has its origin in plastid DNA and another part in nuclear DNA. The mitochondrial genome, like the nuclear genome, apparently tolerates a large portion of senseless sequences and inherits these. It is interesting that a large part of the mtDNA is transcribed. About 30 per cent of the mt-genome of *Brassica campestris* (218 kbp) is transcribed, as is that of the six times larger genome of *Cucumis melo* (2400 kbp). Why so many transcripts are formed is obscure, considering that in the two above-mentioned mt-genomes the total number of the encoded proteins, tRNAs, and rRNAs amounts only to about 60.

The mitochondrial genome encodes parts of the translation machinery, including three rRNAs, 16 tRNAs, and about 10 ribosomal proteins. Furthermore, various hydrophobic membrane proteins are encoded in the mt-genome, e.g. some subunits of the respiratory chain (section 5.5), of F-ATP synthase (section 4.3), and also at least four enzymes of cytochrome-*c* synthesis (section 10.5). About 95

Table 20.6 Identified genes in the genome of plant mitochondria (after Schuster and Brennicke 1994)

Translation apparatus	5S, 18S, 26S rRNA, 10 ribosomal proteins 16 tRNAs
NADH dehydrogenase	9 subunits
Cytochrome-*b*/*c*$_1$ complex	1 subunit
Cytochrome-*a*/*a*$_3$ complex	3 subunits
F-ATP synthase	3 subunits
Cytochrome-*c* biogenesis	>3 genes
Conserved open reading frame of unknown coding	>10 genes

per cent of the mitochondrial proteins, including most subunits of the respiratory chain and F-ATP synthase, as well as several tRNAs, are encoded in the nucleus. Considering that mitochondria have derived from endosymbionts, it has to be assumed that the greatest part of the genetic information of the endosymbiont genome has been transferred to the nucleus. But a gene transfer has apparently also occurred from the plastids to the mitochondria. From their base sequence, several tRNA genes of the mt-genome seem to originate in the plastid genome.

The promoters of plant mitochondrial genes are heterogeneous. The sequences signalling the start and end of transcription are quite variable, even for the genes of the same mitochondrion. In plant mitochondria the mt-RNA polymerase and the corresponding transcription factors have not yet been unequivocally characterized. Most of the mitochondrial genes are transcribed monocistronically.

Mitochondrial DNA contains incorrect information which is corrected after transcription

A comparison of the amino acid sequences of proteins encoded in mitochondria with the corresponding nucleotide sequences of the encoding genes revealed strange discrepancies: the amino acid sequences did not correspond to the DNA sequences of the genome according to the rules of the genetic code. At sites of the DNA sequence where, according to the protein sequence, a T was to be expected, a C was found, and sometimes vice versa. More detailed studies showed that the transcription of mtDNA yielded an mRNA which did not contain the correct information for the protein to be synthesized. It was all the more astonishing to discover that subsequently this incorrect mRNA is processed in the mitochondria by several replacements of C by U, but sometimes also of U by C, until the correct mRNA is reconstructed as a template for synthesizing the proper protein. This process is named *RNA editing*.

Subsequent editing of the initially incorrect mRNA to the correct, translatable mRNA is not an exception only taking place in some exotic genes, but is the rule for the mitochondrial genes of higher plants. In some mRNAs produced in the mitochondria, 40 per cent of the Cs are replaced by U in the editing process. Mitochondrial tRNAs are also edited in this way. Such RNA editing has been observed in the mitochondria of all higher plants investigated so far. The question arises whether, by differences in the editing, a structural gene can be translated into different proteins. Now and again mitochondrial proteins have been found which had been translated from only partially edited mRNA. Since these proteins normally are non-functional they are probably rapidly degraded.

RNA editing was first shown in the mitochondria of trypanosomes, the unicellular pathogen of sleeping sickness. It occurs also in mitochondria of animals and, in a few cases, has also been found in plastids.

Only since 1989 has RNA editing been known to exist in plant mitochondria, leaving many questions still open. What mechanism is used for RNA editing? How are the bases exchanged? Also the question of the physiological meaning of RNA editing is still unanswered. Is a higher mutation rate of the maternally inherited mt-genome corrected by the editing? Where does the information for the proper base sequence of the mRNA come from? Is this information provided by the nucleus or is it also contained in the mitochondrial genome? It is feasible that the very large mitochondrial genome contains, in addition to the structural genes, single fragments of these genes, the transcripts of which are utilized for the correction of the mRNA, as has actually been observed in the mitochondria from trypanosomes.

Male sterility of plants caused by the mitochondria is an important tool in hybrid breeding

When two selected inbreeding lines are crossed, the resulting *F₁ hybrids* are normally larger, more robust, and produce higher yields of harvest products than the parent plants. This effect, called *hybrid vigour*, was observed before the rediscovery of the Mendelian rules of inheritance and was first utilized in 1906 for breeding hybrid maize by George Schull in the laboratory at Cold Spring Harbor, USA. The success of these studies brought about a revolution in agriculture. Based on the results of Schull's research, private seed companies bred maize F_1 hybrids which provided much higher yields than the usual varieties. In 1965, 95 per cent of the maize grown in the corn belt of the USA was hybrid. The use of F_1 hybrids was, to large extent, responsible for the increase in maize yields per acre by a factor of 3.5 between 1940 and 1980 in the USA.

F_1 hybrids cannot be further propagated, since according to the Mendelian laws the offspring of the F_2 generation is heterogeneous. Most of the second-generation (F_2) plants have some homozygosity, resulting in yield depression. Each year, therefore, farmers have to buy new hybrid seed from the seed companies. Hybrid breeding has been put to use for the production of many varieties of cultivated species. Taking maize as an example, the following describes the principles and problems of hybrid breeding.

For the production of F_1 hybrid seed (Fig. 20.26) the pollen of a paternal line A is transferred to the pistil of a maternal line B, and only the cobs of B are harvested for seed. These crossings are mostly carried out in the field. Plants of lines A and B are planted in separate, but neighbouring rows, so that the pollen is transferred from A to B by the wind. To prevent the pistils of line B from being fertilized by the pollen of the same line, the plants of line B are emasculated. Since in maize the pollen producing male flowers are separated from the female flowers in a panicle, it is possible to remove only the male flowers by cutting them off. To produce hybrid seed in this way on a commercial scale, however, requires a great expenditure in manual labour. This method is totally unworkable on any practical

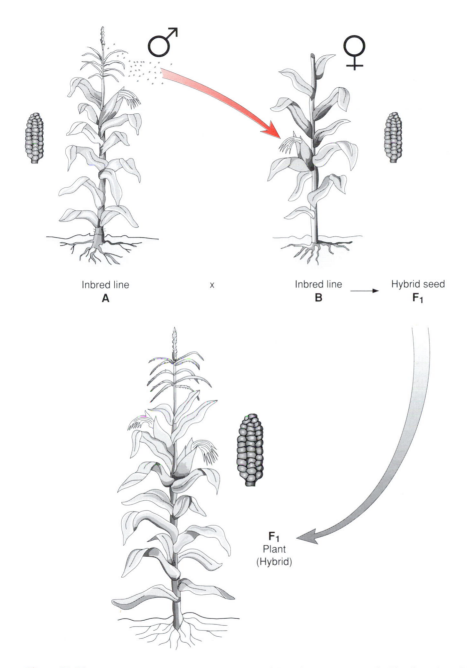

Figure 20.26 Principle of hybrid breeding shown with maize as an example. The lines A and B are inbred lines, which produce a relatively low yield of harvest products. When these lines are crossed, the resulting F_1 progeny is much more robust and produces high yields. The seeds are obtained from line B, which is pollinated by line A. To prevent B being fertilized by its own pollen, the male flowers of B are removed by cutting them off. (From Patricia Nevers, Pflanzenzüchtung aus der Nähe gesehen, Max Planck Institut für Züchtungsforschung, Köln, by permission.)

scale in plants such as rye, where the male and the female parts of the flower are combined.

It was a great step forward in the production of hybrids when maize mutants were found which produced sterile pollen. In these male-sterile plants the fertility of the pistil was not affected as long as it was fertilized by pollen of other lines. This male sterility is inherited maternally by the genome of the mitochondria. Several male-sterile mutants of maize and other plants are known which are the result of the mutation of mitochondrial genes.

The relationship between the mutation of a mitochondrial gene and the male sterility of a plant has been thoroughly investigated in the maize mutant T (Texas). The mitochondria of this mutant contain a gene designated as T-*urf*13, which encodes a *13 kDa protein*. This gene is probably the product of a mutation. The 13 kDa protein has no apparent effect on the metabolism of the mitochondria under normal growth conditions, and the mutants grow normally. Only the formation of pollen is disturbed by this protein, for reasons not known so far. A possible explanation might be that the *tapetum cells* of the pollen sac, which are involved in pollen production, have an unusual abundance of mitochondria and apparently depend very much on mitochondrial metabolism. Therefore, in these cells a mitochondrial defect, which normally did not affect metabolism, might interfere with pollen production.

The use of these male-sterile mutants for breeding is based on a second discovery: maize lines were found which contain so-called *restorer genes* in their nucleus. These encode proteins that repress the expression of the T-*urf*13 gene in the mitochondria (Fig. 20.27). The crossing of a paternal plant A containing these

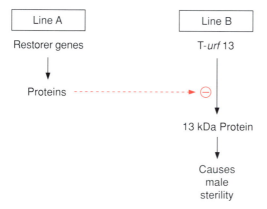

Figure 20.27 A maize mutant T (here designated as line B) contains in the mitochondrial genome a gene named T-*urf*13. The product of this gene, a 13 kDa protein, prevents the formation of sterile pollen and thus causes male sterility in this mutant. Another maize line (A) contains in its nucleus one or several so-called restorer genes, encoding proteins which suppress the expression of the T-*urf*13 gene in the mitochondria. Thus after a crossing of A with B the pollen is fertile again and the male sterility is abolished.

restorer genes, with a male-sterile maternal plant B results in an F_1 generation in which the fertility of the pollen is restored and corn cobs are produced normally.

The crossing of male-sterile T maize lines with lines containing restorer genes enables F_1 hybrids to be produced very efficiently. Unfortunately, the 13 kDa protein encoded by the T-*urf*13 gene makes a maize plant more sensitive to the toxin of the fungus *Bipolaris maydis T*, the pathogen of the much-dreaded fungal disease 'southern corn blight', which in 1971 destroyed a large part of the American maize crop. The 13 kDa protein reacts with the fungal toxin to form a pore in the inner mitochondrial membrane and thus eliminates mitochondrial ATP production. In order to continue hybrid seed production it was then necessary to return to the manual removal of the male flowers. Male-sterile lines are now known not only in maize, but also in many other plants, in which the sterility is caused by proteins encoded in the mt-genome, and also other lines which suppress the formation of the inhibiting protein by nuclearly encoded proteins. Nowadays these lines are used for the production of fertile F_1 hybrid seed. Presumably the effect of nuclearly encoded proteins on the expression of mitochondrial genes such as T-*urf*13 is a normal feature of mitochondrial metabolism in plants and therefore, after the production of corresponding mutants, can be used in many ways to generate male sterility.

Today intensive research is being carried out all over the world to find ways of generating male sterility in plants by genetic engineering. Some success can be noted. Using a specific promoter, it is possible to express a ribonuclease from the bacterium *Bacillus amyloliquefaciens* exclusively in the tapetum cells of the pollen sac in tobacco and rapeseed. This ribonuclease degrades the mRNA formed in tapetum cells, thus preventing the development of pollen. Other parts of the plants are not affected and the plants grow normally. For the generation of a restorer line the gene of a ribonuclease inhibitor (from the same bacterium) was transferred to the tapetum cells. The great advantage of such a synthetic system is its potential for general application. In this way male sterility can be introduced into species in which this cannot be achieved by manual removal of the stamen, and where male sterility due to mutants is not available. Genetically engineered rapeseed hybrids have been licensed in the European Union for commercial cultivation. It is to be expected that the generation of male-sterile plants by genetic engineering will become very important for the production of hybrid seeds.

Further reading

Bachmann, K. (1994). Molecular markers in plant ecology. *New Phytologist*, **126**, 403–18.
Brennicke, A. and Kück, U. (ed.) (1993). *Plant mitochondria with emphasis on RNA editing and cytoplasmic male sterility*. Verlag Chemie, Weinheim.
Brown, T. A. (1990). *Gene cloning*. Chapman & Hall, London.

Brown, T. A. (1992). *Genetics: a molecular approach*. Chapman & Hall, London.

Chrispeels, M. J. and Sadava, D. E. (1994). *Plants, genes and agriculture*. Jones and Bartlett, Boston, London.

Dean, C. and Schmidt, R. (1995). Plant genomes: a current molecular description. *Annual Review of Plant Physiology and Plant Molecular Biology*, **46**, 395–418.

Estelle, M. A. and Somerville, C. R. (1986). The mutants of *Arabidopsis*. *Trends in Genetics*, **2**, 89–93.

Fosket, D. E. (1994). *Plant growth and development. A molecular approach*. Academic Press, San Diego.

Gray, M. W., Hanic-Joyce, P. J., and Covello, P. S. (1992). Transcription, processing and editing in plant mitochondria. *Annual Review of Plant Physiology and Plant Molecular Biology*, **43**, 145–75.

Hanson, M. R., Sutton, C. A., and Lu, B. (1996). Plant organelle gene expression altered by RNA editing. *Trends in Plant Sciences*, **1**, 57–64.

Leewings III, C. S. and Vasil, I. K. (ed.) (1995). *The molecular biology of plant mitochondria*. Kluver Academic, Dordrecht.

Maréchal-Drouard, L., Weil, J. H., and Dietrich, A. (1993). Transfer RNAs and transfer RNA genes in plants. *Annual Review of Plant Physiology and Plant Molecular Biology*, **44**, 13–32.

Mayfield, S. P., Yohn, C. B., Cohen, A., and Danon, A. (1995). Regulation of chloroplast gene expression. *Annual Review of Plant Physiology and Plant Molecular Biology*, **46**, 147–66.

Newton, K. J. (1988). Plant mitochondrial genomes: organisation, expression and variation. *Annual Review of Plant Physiology and Plant Molecular Biology*, **39**, 503–32.

Ohyama, K., Fukuzuwa, H., Kochi, T., *et al.* (1986). Chloroplast gene organisation deduced from the complete sequence of liverwort *Marchantia polymorpha* chloroplast DNA. *Nature*, **322**, 572–4.

Saedler, H. and Gierl, A. (1996). Transposable elements. *Current Topics in Microbiology and Immunology*, **204**. Springer Verlag, Heidelberg.

Schuster, W. and Brennicke, A. (1994). The plant mitochondrial genome: physical structure, information content, RNA editing and gene migration to the nucleus. *Annual Review of Plant Physiology and Plant Molecular Biology*, **45**, 61–78.

Shinozaki, K., Ohme, M., Wakasugi, T., *et al.* (1986). The complete nucleotide sequence of the chloroplast genome: its gene organisation and expression. *EMBO Journal*, **9**, 2043–9.

Smart, C. J., Monéger, F., and Leaver, C. J. (1994). Cell-specific regulation of gene expression in mitochondria during anther development in sunflower. *The Plant Cell*, **6**, 811–25.

Sundaresan, V. (1996). Horizontal spread of transposon mutagenesis: new uses of old elements. *Trends in Plant Science*, **1**, 184–90.

Watson, J. D., Gilman, M., Witkowski, J., and Zoller, M. (1992). *Recombinant DNA*. Freeman, New York.

chapter 21

Protein biosynthesis

During protein biosynthesis the base sequence of mRNA is translated into an amino acid sequence. The 'interpreters' are *transfer ribonucleic acids* (tRNAs), small RNAs of 75–85 ribonucleotides which have a defined structure with three hairpin loops. The middle loop contains the *anticodon* which is complementary to the mRNA *codon*. For each amino acid there is at least one and sometimes even several tRNAs. The covalent binding of the amino acid to the corresponding tRNA is catalysed by its specific *aminoacyl-tRNA synthetase* with the consumption of ATP, and the mixed anhydride aminoacyl-AMP is formed as an intermediate (Fig. 21.1).

In a plant cell protein biosynthesis takes place at three different sites. The translation of the nuclearly encoded mRNA proceeds in the cytosol and that of the mRNAs encoded in the plastidal or mitochondrial genome takes place in the plastid stroma and mitochondrial matrix, respectively.

21.1 Protein synthesis is catalysed by ribosomes

Ribosomes are large riboprotein complexes which contain three or four different rRNA molecules and a large number of proteins. In the interval between the end of the translation of one mRNA and the start of the translation of another mRNA, the ribosomes dissociate into two subunits. The ribosomes of the cytosol, plastids, and mitochondria are different in size and composition (Table 21.1). The cytosolic ribosomes, with a sedimentation coefficient of 80S, are larger than those

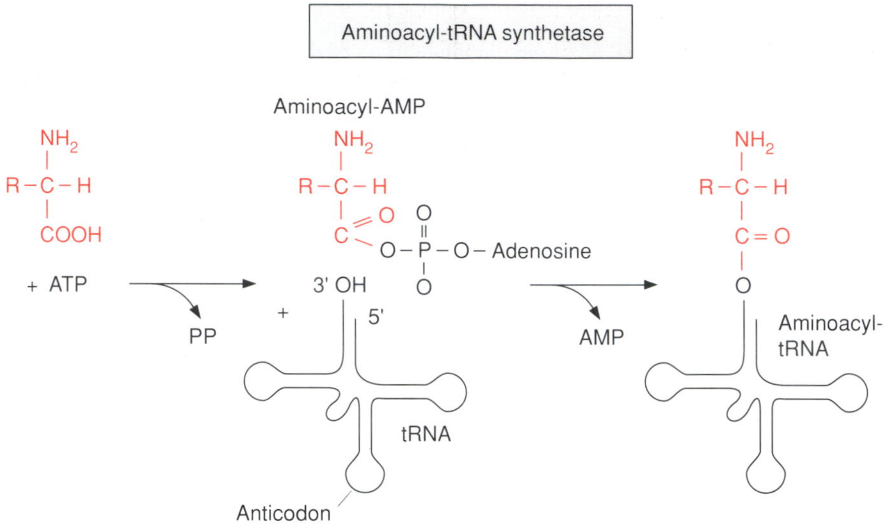

Aminoacyl-tRNA synthetase

Aminoacyl-AMP

NH_2
|
$R - C - H$
|
COOH

+ ATP

NH_2
|
$R - C - H$
|
$C \diagdown^O$
$O - \overset{\overset{\displaystyle O}{\|}}{P} - O -$ Adenosine
$\overset{\displaystyle |}{O}$

3' OH

5'

+

PP

tRNA

Anticodon

AMP

NH_2
|
$R - C - H$
|
$C = O$
|
O

Aminoacyl-tRNA

Figure 21.1 A tRNA is loaded with its corresponding amino acid by the amionoacyl-tRNA synthetase during which ATP is consumed. In this reaction aminoacyl-AMP is formed as an intermediate.

of the mitochondria (78S, variable between species) and the plastids (70S). The ribosomes contained in the cytosol are called *eukaryotic ribosomes*, and those in the plastids and in the mitochondria, because of their relationship to bacterial ribosomes, are regarded as *prokaryotic ribosomes*. Ribosomes of the bacterium *Escherichia coli* have a sedimentation coefficient of 70S. Let us first look at the translation occurring in the cytosol.

Table 21.1 Composition of the ribosomes in the cytosol, chloroplast stroma, and mitochondrial matrix in plants

	Complete ribosome	Ribosomal subunits	r-RNA components	Proteins
Cytosol	80S	1. 40S	18S rRNA	c. 30
(eukaryotic ribosome in plants)		2. 60S	5S rRNA	c. 50
			5.8S rRNA	
			25S rRNA	
Chloroplast	70S	1. 30S	16S rRNA	c. 24
(prokaryotic ribosome)		2. 50S	4.5S rRNA	c. 35
			5S rRNA	
			23S rRNA	
Mitochondrion (prokaryotic ribosome)	78S	1. ≈ 30S	18S rRNA	c. 33
		2. ≈ 50S	5S rRNA	c. 35
			26S rRNA	

At the beginning of translation mRNA forms an *initiation complex* with a ribosome. A number of *initiation factors* participate in this process. The cap sequence present at the 5′ end of the mRNA (see Fig. 20.9) binds to the *eukaryotic initiation factor 4* (eIF4). (This factor consists of several protein components which will not be described in detail here). Subsequently the initiation factor eIF3 is bound (Fig. 21.2). The adduct thus formed is transferred to the small 40S subunit of the ribosome. The 18S RNA present in the 40S subunit is involved in binding the cap

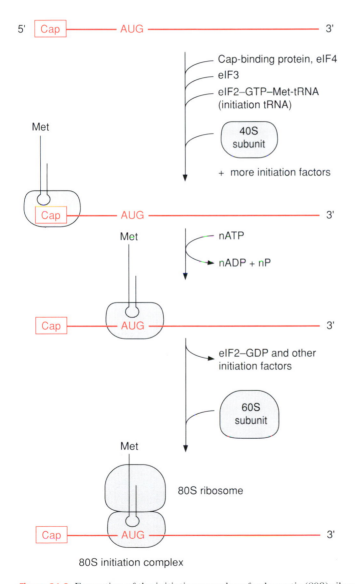

Figure 21.2 Formation of the initiation complex of eukaryotic (80S) ribosomes.

sequence. Another initiation factor (eIF2) binds GTP and the *initiation tRNA*, which recognizes the start codon, AUG. The initiation tRNA carries methionine. This adduct is also bound to the 40S subunit. Driven by hydrolysis of ATP, the 40S subunit migrates downstream $(5' \rightarrow 3')$ until it finds an AUG start codon. Normally, but not always, the first AUG triplet on the mRNA is the one recognized. In some mRNAs the translation starts at a later AUG triplet. Probably the neighbouring sequences on the mRNA decide which AUG triplet is recognized as the start codon. The large 60S subunit is then bound to the 40S subunit, accompanied by the dissociation of several initiation factors and of GDP. The formation of the initiation complex is now completed and the resulting ribosome is able to translate.

The mitochondrial and plastid mRNA have no cap sequence. Plastid mRNA has a special ribosome binding site for the initial binding to the small subunit of the ribosome, consisting of a purine-rich sequence of about 10 bases. This sequence, called the *Shine–Dalgarno sequence*, binds to the rRNA of the small ribosome subunit. A Shine–Dalgarno sequence is also found in bacterial mRNA, but it is not known whether it also plays this role in the mitochondria. In mitochondria, plastids, and bacteria the initiation tRNA is loaded with *N*-formylmethionine (instead of methionine as in the cytosol). After peptide formation the formyl residue is cleaved from the methionine.

A peptide chain is synthesized

A ribosome, completed by the initiation process, contains two sites where the tRNAs can bind to the mRNA. The peptidyl (P) site allows the binding of the initiation tRNA to the AUG start codon (Fig. 21.3). The aminoacyl (A) site covers the second codon of the gene and at first is unoccupied. On the other side of the P site is the exit (E) site where the empty tRNA is released. The elongation begins after the corresponding aminoacyl-tRNA occupies the A site by forming base pairs with the second codon. Two *elongation factors* participate in this. The *e*ukaryotic *e*longation factor 1α (eEF1α) binds GTP and guides the corresponding aminoacyl-tRNA to the A site, during which the GTP is hydrolysed to GDP and P. The cleavage of the energy-rich anhydride bond in GTP enables the aminoacyl-tRNA to bind to the codon at the A site. Afterwards the GDP, still bound to eEF1α, is exchanged for GTP, as mediated by the elongation factor eEF1βγ. The eEF1α–GTP is now ready for the next cycle.

Subsequently a peptide linkage is formed between the carboxyl group of methionine and the amino group of the amino acid of the tRNA bound to the A site. The *peptidyl transferase* catalysing this reaction is a complex enzyme consisting of several ribosomal proteins. The 25S rRNA has a decisive function in the catalysis. The enzyme facilitates the *N*-nucleophilic attack on the carboxyl group, whereby the peptide bond is formed with the release of water. This results in the formation of a dipeptide bound to the tRNA at the A site (Fig. 21.4a).

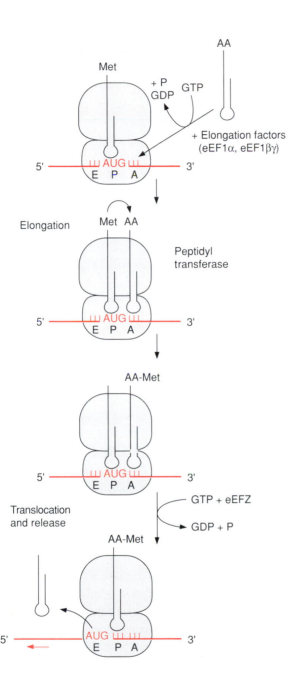

Figure 21.3 Elongation cycle of protein biosynthesis. After binding of the corresponding aminoacyl-tRNA to the A site, a peptide bond is formed by peptidyl transferase. By subsequent translocation the remaining empty tRNA is moved to the E site and released, while the tRNA loaded with the peptide chain now occupies the P site. The binding of a new tRNA to site A starts another elongation cycle. AA = amino acid.

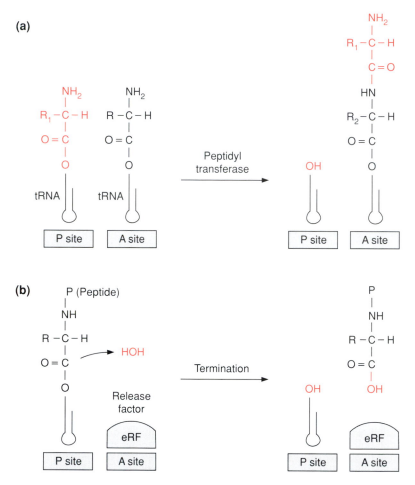

Figure 21.4 (a) Formation of a peptide bond by peptidyl transferase. (b) Termination of peptide synthesis by the binding of a release factor (eRF) to the stop codon at the A site. The peptide is released from the tRNA by hydrolysis.

Accompanied by the hydrolysis of one molecule of GTP to form GDP and P, the elongation factor eEF2 facilitates the *translocation* of the ribosome along the mRNA to three bases downstream (Fig. 21.3). In this way the free tRNA arrives at site E, is released, and the tRNA loaded with the peptide now occupies the P site. The third aminoacyl-tRNA binds to the now vacant the A site and a further elongation cycle can begin.

After several elongation cycles, the 5′ end of the mRNA is no longer bound to the ribosome and can start a new initiation complex. An mRNA that is translated simultaneously by several ribosomes is called a *polysome* (Fig. 21.5).

Translation is terminated when the A site finally binds to a *stop codon* (UGA,

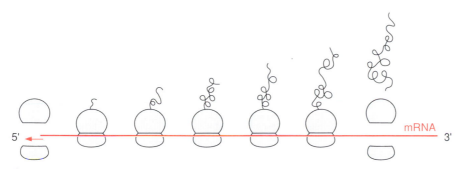

Figure 21.5 Several ribosomes that simultaneously translate the same mRNA are called a polysome.

UAG, or UAA) (Fig. 21.4b). These stop codons bind the *release factor* (eRF) accompanied by hydrolysis of GTP to form GDP and P. Binding of eRF to the stop codon alters the specificity of the peptidyl transferase: water instead of an amino acid is now the acceptor for the peptide chain. In this way the formed protein is released from the tRNA.

Specific inhibitors of translation can be used to decide whether a protein is encoded in the nucleus or in the genome of plastids or mitochondria

Elongation, translocation, and termination in prokaryotic ribosomes occur in an analogous way to that described above with only minor differences, but in this case the termination requires no GTP. However, eukaryotic and prokaryotic translation can react differently to certain antibiotics (Fig. 21.6, Table 21.2). *Puromycin*, an analogue of tRNA, is a general inhibitor of protein synthesis, whereas *cycloheximide* only inhibits protein synthesis by eukaryotic ribosomes. *Chloramphenicol*, *tetracycline*, and *streptomycin* primarily inhibit protein synthesis by prokaryotic ribosomes. These inhibitors can be used to define whether a certain protein is encoded in the nucleus or in the genome of plastids or mito-chondria. A relatively simple method for monitoring protein synthesis is to measure the incorporation of a radioactively labelled amino acid, e.g. ^{35}S-labelled methionine, into proteins. If the incorporation of this amino acid into a particular protein is inhibited by cycloheximide, this indicates that the protein is encoded in the nucleus. Likewise, inhibition by chloramphenicol shows that the corre-sponding protein is encoded in the genome of plastids or mitochondria.

Translation is regulated

The synthesis of many proteins is specifically regulated at the level of translation. This regulation may involve protein kinases by which proteins participating in translation are phosphorylated. Since the rate-limiting step of translation is usually initiation, this step is especially suited for regulation of translation.

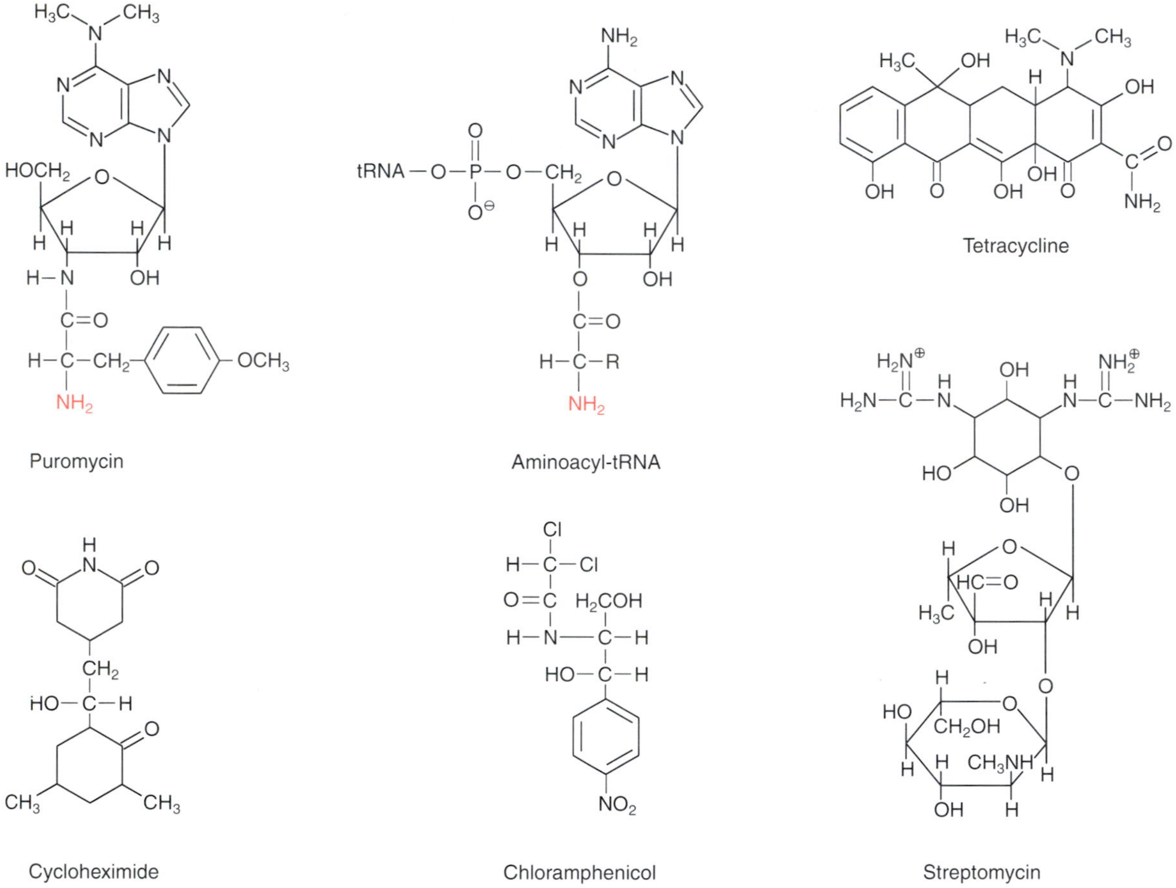

Figure 21.6 Antibiotics as inhibitors of protein synthesis. Their mode of action is described in Table 21.2.

In animals the initiation factor eIF2 is inactivated by phosphorylation and initiation is therefore inhibited. Little is known about the regulation of translation in plants.

21.2 Proteins attain their three-dimensional structure by controlled folding

Protein biosynthesis by ribosomes first yields an unfinished product which, in order to attain its biological function, has to be folded into its correct three-dimensional (native) structure. The resultant functional protein is called a *native protein*. The three-dimensional protein structure of the native protein normally represents the lowest energy state of the molecule and is determined to a large extent by the amino acid sequence of the molecule.

Table 21.2 Antibiotics as inhibitors of protein synthesis. The listed antibiotics are all derived from streptomycins

Antibiotic	Inhibitor action
Puromycin	Binds as an analogue of an aminoacyl-tRNA to the A site and participates in all elongation steps, but prevents the formation of a peptide bond, thus terminating protein synthesis in prokaryotic and eukaryotic ribosomes
Cycloheximide	Inhibits peptidyl transferase in eukaryotic ribosomes
Chloramphenicol	Inhibits peptidyl transferase in prokaryotic ribosomes
Tetracycline	Binds to the 30S subunit and inhibits the binding of aminoacyl-tRNA to prokaryotic ribosomes much more than to eukaryotic ones
Streptomycin	The interaction with 70S ribosomes results in an incorrect recognition of mRNA sequences and thus inhibits initiation in prokaryotic ribosomes

The folding of a protein is a multistep process

Theoretically there are about 10^{100} possible conformations for a peptide with 100 amino acids, which is a rather small molecule of about 11 kDa. Since the reorientation of a single bond takes about 10^{-13} s, it would take the absurdly long time of 10^{87} s to try out all possible folding states one after the other. By comparison, the age of the earth is about 1.6×10^{17} s. In reality a protein attains its native form within seconds or minutes. The folding of the molecule apparently proceeds in a multistep process. It begins by forming secondary structures such as *α-helices* or *β-sheets*. These consist of 8–15 amino acid residues, and are formed or dissolved again within milliseconds (Fig. 21.7). The secondary structures then associate stepwise with increasingly larger associates and in this way also stabilize the regions of the molecule that do not form secondary structures. The hydrophobic interaction between the secondary structures is the driving force in these folding processes. After further conformational changes, the correct three-dimensional structure of the molecule is attained rapidly by this co-operative folding procedure. In proteins with several subunits, the subunits associate to form a quaternary structure.

Proteins are protected during the folding process

The folding process can be severely disturbed when the secondary structures in the molecule associate incorrectly, or particularly when secondary structures of different molecules associate resulting in an undesirable aggregation of proteins (hydrophobic collapse). This danger is especially high during protein synthesis, when the incomplete protein is still attached to the ribosome (Fig. 21.5), or during the transport of an unfolded protein through a membrane, when only part of the peptide chain has reached the other side. Moreover, incorrect intermolecular

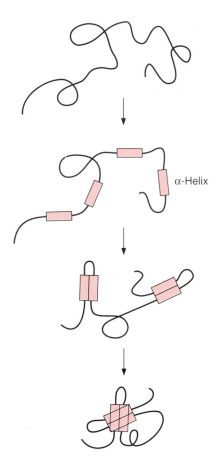

α-Helix

Figure 21.7 Protein folding is a stepwise hierarchic process. First, secondary structures are formed, which then aggregate successively until finally, after slight corrections to the folding, the tertiary conformation of the native protein is attained.

associations are likely to occur when the concentration of a newly synthesized protein is very high, as can be the case in the lumen of the rough endoplasmic reticulum (Chapter 14).

To prevent such incorrect folding, a family of proteins present in the various cell compartments helps newly formed protein molecules to attain their correct conformation by avoiding incorrect associations. These proteins have been named *chaperones*.

Heat shock proteins protect against high temperatures

Chaperones not only have a function in protein biosynthesis but also protect cell proteins that have been denatured by exposure to high temperatures, against aggregation, thus assisting their reconversion to the native conformation. Bacteria, animal, and plant cells react to a temperature increase of about 10 per cent above the temperature optimum with a very rapid synthesis of so-called *heat shock proteins*, most of which are chaperones. Many plants can survive otherwise lethal high temperatures if they have been previously exposed to a smaller temperature increase, which induces the synthesis of heat shock proteins. This phenomenon is called *acquired thermal tolerance*. Investigations with soybean seedlings showed that such tolerance coincides with an increase in the content of heat shock proteins. However, most of these heat shock proteins are constitutive, which means that they are also present in a normal cell, and under all conditions have important functions in the folding of proteins.

Chaperones bind to unfolded proteins

Since chaperones were initially characterized as heat shock proteins, they, and also protein factors modulating the chaperone function, are commonly designated by the abbreviation Hsp followed by the molecular weight in kDa.

Chaperones of the *Hsp70 family* have been found in bacteria, mitochondria, chloroplasts, the cytosol of eukaryotes, and also in the endoplasmic reticulum. These are highly conserved proteins. Hsp70 has a binding site for adenine nucleotides which can be occupied either by ATP or ADP. When occupied with ADP, Hsp70 forms with the chaperone Hsp40 a tight complex with unfolded segments of a protein, but not with native proteins (Fig. 21.8). The ADP bound to Hsp70 is replaced by ATP. The resultant ATP–Hsp70 complex has only a low binding affinity and therefore dissociates from the protein segment. Due to the

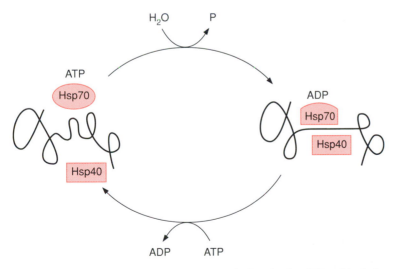

Figure 21.8 The Hsp70 chaperone contains a binding site for ATP and hydrolyses ATP to ADP. The Hsp70 paired with ADP binds tightly with Hsp40 to an unfolded segment of a protein. The ADP bound to Hsp70 is exchanged for ATP. The Hsp70 paired with ATP has only a low binding affinity for the protein, which is therefore released. A protein can be bound again only after ATP hydrolysis. This simplified scheme does not deal with intermediates involved and also does not represent the real structure of the binding complex.

subsequent hydrolysis of the bound ATP to ADP, Hsp70 is ready to bind once more to an unfolded peptide segment. In this way Hsp70 only binds to a protein for a short time, dissociates from it, and, if necessary, binds to the protein again. This stabilizes an unfolded protein without restricting its folding capacity. The mechanism of the ADP-dependent binding of Hsp70 to unfolded peptides as mediated by Hsp40 has been conserved during evolution. Fifty per cent of the amino acids in the sequences of the Hsp70 protein in *E. coli* and in humans are identical.

The proteins of the *Hsp60 family*, contained in bacteria, plastids, and mitochondria, also bind to unfolded proteins. They were first identified as the GroEL factor in *E. coli* and as RubisCO binding protein in chloroplasts, until it was realized that the two proteins were homologous and act as chaperones. A GroES factor, called *Hsp10* in mitochondria and in chloroplasts, is involved in binding the Hsp60 chaperones. In bacteria, 14 GroEL and 7 GroES molecules are assembled to a *super-chaperone complex*, forming a large cavity into which an unfolded protein fits (Fig. 21.9, 21.10). The protein is temporarily bound to Hsp60 molecules of the cavity analogously to the binding to Hsp70 in Fig. 21.8. Correct folding to the native protein is aided by several ATP hydrolysis cycles, involving dissociation and rebinding of the protein segments. In this way the unfolded protein can reach the native conformation by avoiding association with other proteins.

Presumably Hsp70 as well as Hsp60 and Hsp10 participate in the protein folding in plastids and mitochondria (Fig. 21.10). Hsp70 protects single segments of the growing peptide chain during protein synthesis, and the superchaperone complex from Hsp60 and Hsp10 enables undisturbed folding of the total protein.

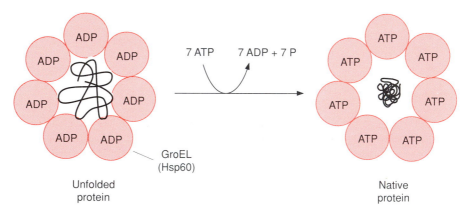

Figure 21.9 Section through the superchaperone complex of prokaryotes consisting of 14 molecules of GroEL (Hsp60) and seven molecules of GroES (Hsp10). The chaperone molecules, paired with ADP, bind unfolded segments of the newly formed protein. Repeated release of the protein after exchange with ATP, and consecutive binding after ATP hydrolysis (see previous figure), enables the protein to fold. The native protein is released because it no longer binds to chaperones.

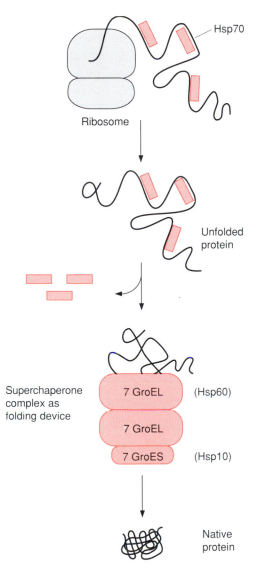

Figure 21.10 The folding of proteins in prokaryotes, plastids, and mitochondria. The unfolded protein is protected by being bound to an Hsp70 chaperone paired with ADP and is then folded to the native protein in the cavity of the superchaperone complex which consists of GroEL (Hsp60) and GroES (Hsp10). ATP is consumed in this reaction.

The chaperone *Hsp90*, regarded as the most abundant cytosolic protein, is found in very high concentrations in the cytosol of eukaryotes. Hsp90 is believed to play a central role in the folding and assembling of cytosolic proteins. Moreover, chaperones named CCT (cytosolic complex T), somewhat resembling the prokaryotic Hsp60, have recently been identified in the cytosol of eukaryotes.

They act as a folding device by forming oligomeric chaperone complexes, which are probably similar to the Hsp60–Hsp10 superchaperone complex (Fig. 21.10).

Furthermore there are proteins facilitating other processes which limit protein folding. Such limiting processes include the formation of disulfide bridges and the cis–trans isomerization of the normally non-rotatable prolyl peptide bonds.

Since thorough investigation of chaperones began only a few years ago, many questions about their structure and function are still unsolved.

21.3 Nuclearly encoded proteins are distributed throughout various cell compartments

The ribosomes contained in the cytosol also synthesize proteins destined for cell organelles, such as plastids, mitochondria, peroxisomes, and vacuoles, and also proteins to be secreted from the cell. To reach their location these proteins have to be specifically transported across various membranes.

Proteins destined for the vacuole are transferred during their synthesis to the lumen of the ER (section 14.5). This is aided by a signal sequence at the terminus of the synthesized protein, which binds with a signal recognition particle to a pore protein present in the ER membrane and thus directs the protein to the ER lumen. In such a case the ribosome is attached to the ER membrane during protein synthesis and the synthesized protein appears immediately in the ER lumen (Fig. 14.2). This process is called *co-translational protein transport*. These proteins are then transferred from the ER lumen by vesicle transfer across the Golgi apparatus to the vacuole or are exported by secretory vesicles from the cell.

In contrast, protein uptake into plastids, mitochondria, and peroxisomes occurs mainly, if not exclusively, by *post-translational transport*, which means the proteins are transported across the membrane only after completion of protein synthesis and their release from the ribosomes. So far, transport into mitochondria has been investigated most thoroughly and therefore will be described first.

Most of the proteins imported into the mitochondria have to cross two membranes

More than 95 per cent of the mitochondrial proteins in a plant are encoded in the nucleus and translated in the cytosol. Our present knowledge about the import of proteins from the cytosol into the mitochondria derives primarily from studies with yeast. In order to direct proteins from the cytosol to the mitochondria they have to be provided with a *targeting signal*. Some proteins destined for the mitochondrial inner membrane or the inter membrane compartment, as well as all the proteins for the mitochondrial outer membrane, contain internal targeting signals which have not yet been identified. Other proteins of the mitochondrial inner membrane and the proteins of the mitochondrial matrix are synthesized in the cytosol as *precursor proteins*, which contain as a targeting signal a *signal sequence* of 12–70 amino acids at their amino terminus. These targeting presequences have a high content of positively charged amino acids and are able to form α-helices in which one side is positively charged and the other side is hydrophobic. The three-dimensional structure of the amphipatic α-helices, rather than a certain amino acid sequence, forms the targeting signal. The directing function of this pre-sequence can be demonstrated in an experiment: when a foreign protein, such as dihydrofolate reductase from mouse, is provided with a targeting signal sequence for the mitochondrial matrix, this protein is taken up into the mitochondrial matrix.

For the import of proteins into the mitochondrial matrix, both the outer and inner membrane have to be traversed (Fig. 1.12). This protein import occurs primarily at so-called *translocation sites* where the inner and outer membrane are closely attached to each other (Fig. 21.11). Each membrane contains its own translocation apparatus which transfers the proteins in the unfolded state through the membranes.

The precursor proteins formed by the ribosomes associate in the cytosol with chaperones (e.g. ctHsp70, ct = cytosol) in order to prevent premature folding or aggregation of the often hydrophobic precursor proteins. The association with ctHsp70 is accompanied by the hydrolysis of ATP (Fig. 21.8). The transport across the outer membrane is catalysed by a so-called TOM (*t*ranslocase *o*uter mito-chondrial *m*embrane) complex consisting of at least eight different subunits. The TOM20 and TOM22 subunits function as receptors for the targeting signal sequence. An electrostatic interaction between the positively charged side of the α-helix of the presequence and the negative charge on the surface of TOM22 is probably involved in the specific recognition of the targeting signal. TOM22 and TOM20 then mediate the threading of the polypeptide chain into the transloca-tion pore. Another receptor for the transport of proteins is TOM70. This receptor, together with TOM37, mediates the uptake of the ATP–ADP translocator protein and other translocators of the inner membrane, which contain an internal targeting signal instead of a presequence. Probably TOM40 as well as the small subunits TOM5, 6, 7 (not shown in Fig. 21.11) participate in the formation of the translocation pore.

The subsequent transport across the inner membrane is catalysed by the TIM (*t*ranslocase *i*nner mitochondrial *m*embrane) complex, consisting of the proteins TIM17, 23, 44, and several others not yet identified. A precondition for protein transport across the inner membrane is the presence of a *membrane potential*, $\Delta\Psi$ (section 5.6). Presumably the positively charged presequence is driven through the translocation pore by the negative charge at the matrix side of the inner mem-brane. The protein chain appearing in the matrix is first bound to TIM44 and is then bound with hydrolysis of ATP (Fig. 21.8) to an mtHsp70 chaperone and also to other chaperones not dealt with here. It is assumed that Brownian movement

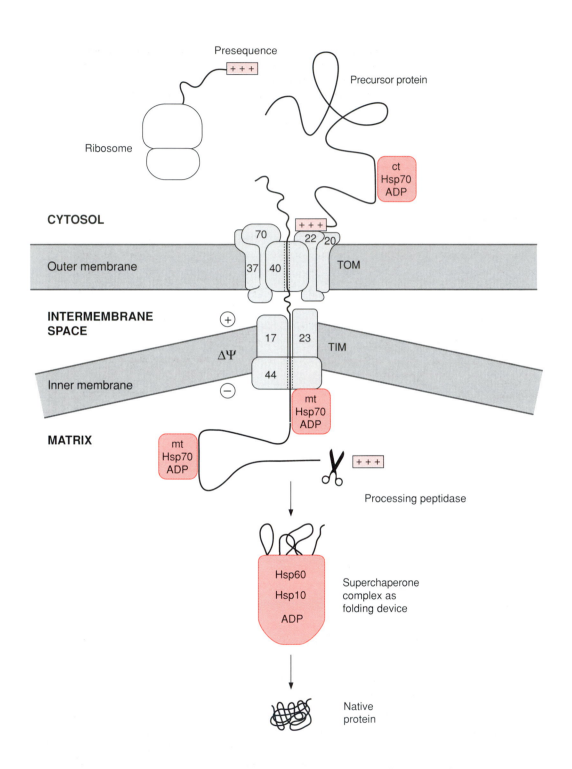

Figure 21.11 Protein import into mitochondria (after Lill and Neupert). The precursor protein formed at the ribosomes present in the cytosol is stabilized in its unfolded conformation by the cytosolic Hsp70 chaperone. A positively charged presequence binds to the receptors TOM20 and TOM22. The presequence threads the precursor protein into the translocation pore of the outer and the inner membrane. Mitochondrial Hsp70 chaperones bind to the peptide chain appearing in the matrix, and thus enable the chain to slide through the translocation pore. The presequence is cut off by a matrix processing peptidase and afterwards the protein attains its native conformation in a superchaperone complex.

causes a section of the protein chain to slip through the pore, which is then immediately bound to the mtHsp70 inside, preventing the protein from slipping back. It is postulated that repeated binding of Hsp70 converts a random movement of the protein chain in the translocation channel into a unidirectional motion. According to this model of a *molecular ratchet*, the ATP required for the reversible binding of mtHsp70 is probably required not for pulling the polypeptide chain through the pore but, instead, to change its free diffusion across the two translocation pores into unidirectional transport. But an alternative hypothesis is also under discussion, according to which the protein entering the pore is pulled into the matrix by ATP-dependent conformational changes of the mtHsp70 bound to the protein.

When the peptide chain arrives in the matrix, the presequence is immediately cleaved from the protein by a *processing peptidase* (Fig. 21.11). The folding of the matrix protein probably occurs via a superchaperone folding apparatus consisting of the chaperones Hsp60 and Hsp10 (see Figs 21.9, 21.10). Proteins destined for the mitochondrial outer membrane can, after being bound to the receptors of the TOM complex, be inserted directly into the membrane.

In most cases proteins destined for the mitochondrial inner membrane are, after transport through the outer membrane, guided directly to their location by an *internal targeting sequence*. In some cases, proteins destined for the inner mitochondrial membrane contain a presequence that guides them first into the mitochondrial matrix. After removal of the presequence by the processing peptidase, the proteins are directed by a second targeting signal sequence into the inner membrane.

The import of proteins into chloroplasts requires several translocators

Proteins of the chloroplasts are also mainly encoded in the nucleus. Although much less is known about protein transport into chloroplasts than into mitochondria, parallels as well as differences between the two transports processes are apparent. Transport into the chloroplasts also proceeds *post-translationally*. The precursor proteins formed in the cytosol contain a targeting signal sequence with 30–100 amino acid residues at the N terminus. As in the mitochondria, the targeting signal does not consist of a specific amino acid sequence, but its function is due to the secondary structure of the signal presequence. During its passage through the cytosol the precursor protein is stabilized by Hsp70 chaperones.

In order to be imported into the stroma the protein must cross two membranes (Fig. 21.12). The translocation apparatus of the outer chloroplast envelope membrane contains four functional proteins which, according to their molecular weight (in kDa), are named OEP (*outer envelope protein*) 86, 75, 70, and 34, and together represent about 30 per cent of the total membrane proteins of the outer envelope membrane. OEP70 is homologous to the Hsp70 chaperone. OEP75 seems to form the translocation pore for the passage of the unfolded peptide chain. OEP86 acts as receptor for the precursor protein. So far little is known about the subsequent transport across the inner envelope membrane in which the proteins IEP (*inner envelope protein*) 36, 44, and 97 are involved.

In contrast to mitochondrial protein transport, a membrane potential is not required for protein transport into the chloroplast stroma. Instead, ATP is consumed for phosphorylation of a protein, probably the receptor OEP86. The protein transport is regulated by the binding of GTP to OEP86 and OEP34. OEP34 has GTPase activity. It is assumed that transport of a protein through the translocation pore is accompanied by the repeated binding of GTP with subsequent hydrolysis to GDP and a GTP/GDP exchange at OEP34. This illustrates marked differences from the translocation machinery of the mitochondria. But also in the chloroplasts a unidirectional motion of the unfolded peptide chain through the translocation pore is probably caused by repeated binding of Hsp70 chaperones according to the model of a molecular ratchet, accompanied by the hydrolysis of ATP.

After it has delivered the protein chain to the stroma, the presequence is removed by a processing peptidase. The resulting protein is folded to the native conformation, probably with the aid of an Hsp60–Hsp10 superchaperone complex, and is then released. In this way also the small subunit of RubisCO (section 6.2) is delivered to the stroma, where it is assembled with the large subunit encoded in the chloroplasts.

Figure 21.12 Protein import into chloroplasts (after Soll 1995). A protein formed in the cytosol and destined for the thylakoid lumen contains two presequences as a targeting signal. The first presequence (coloured red) binds to the receptor OEP86 of the translocation apparatus of the outer envelope membrane. The translocation probably requires the receptor protein to be phosphorylated by ATP. Additionally, it is presumed that transport is regulated by the exchange of GDP for GTP and the subsequent hydrolysis of GTP. The membrane protein OEP75 is probably the translocation pore. The exact composition of the translocation apparatus of the inner envelope membrane, consisting of at least three proteins (IEP36, 44, 97), is not yet known. The peptide chain appearing in the stroma is bound to several chloroplastic Hsp70 chaperones and in this way makes it easier for the unfolded chain to slide through the translocation pore. After cleavage of the first presequence (red), the second presequence (black) serves as targeting signal for transport across the thylakoid membrane. The second sequence is removed by a membrane-bound thylakoid processing peptidase.

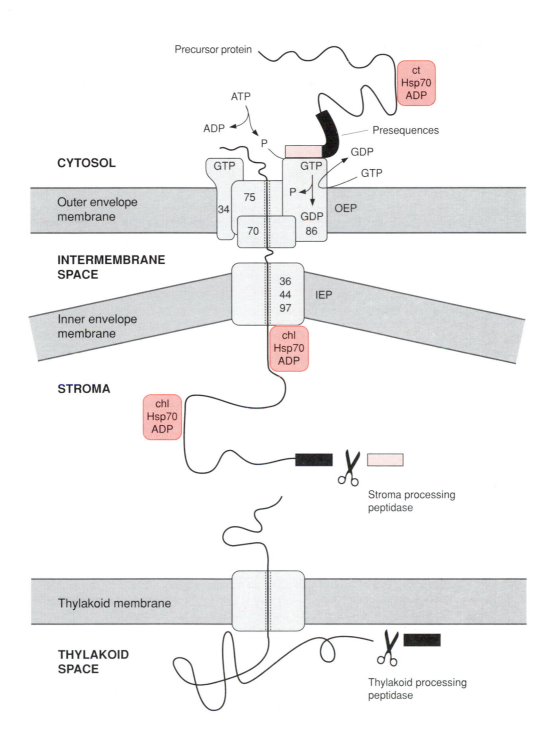

Those proteins destined for the thylakoid membrane are first delivered to the stroma and then directed by internal targeting signals into the thylakoid membrane. Some proteins of the thylakoid membrane, including proteins of the water-splitting apparatus, are first transported into the thylakoid space before being incorporated into the membrane. Three different transport systems are now known to carry this out. The precursor protein of plastocyanin, destined for the thylakoid space, contains two targeting signal sequences. The first directs the protein into the stroma and is cut off by the stroma processing endopeptidase, and the second targeting signal sequence is for transport across the thylakoid membrane, which is afterwards cut off by a thylakoid membrane-bound processing peptidase. The transport of plastocyanin and also of some proteins of the water-splitting apparatus is driven by the hydrolysis of ATP. Since this mode of transport occurs also in cyanobacteria, it is called a prokaryotic pathway. Other proteins of the water-splitting apparatus, comprising those which are not present in the cyanobacteria, are transported by a so-called eukaryotic pathway, driven by a pH gradient across a thylakoid membrane.

Proteins are imported in the folded state into peroxisomes

It is widely assumed that peroxisomes multiply by division, similar to plastids and mitochondria. Since peroxisomes do not contain their own genome (section 1.5), all peroxisomal proteins have to be delivered from the cytosol. The transport of protein into peroxisomes is accompanied by ATP hydrolysis. In plants, animals, and yeast the targeting sequence *serine–lysine–leucine* (SKL) has been found at the C terminus in a number of proteins destined for the peroxisomal matrix. This SKL sequence is not removed after the uptake of the proteins into the peroxisomes. When this SKL sequence is added by gene manipulation at the C terminus to a cytosolic protein, this protein is transported into the peroxisomes.

So far little is known about the mechanism of ATP-driven protein transport into peroxisomes, although more than 10 different components that appear to be involved in this transport have already been identified by genetic studies. In protein transport into mitochondria, plastids, and the endoplasmic reticulum described so far, the proteins are always transferred across a membrane in the *unfolded state*. In contrast, according to present knowledge, proteins are imported into the peroxisomes in the folded state. Chaperones are probably not involved in this import. It seems that protein import into the peroxisomes is entirely different from protein transport into the ER, mitochondria, and plastids.

Further reading

Bruce, B. D. and Keegstra, K. (1994). Translocation of proteins across chloroplast membranes. *Advances in Molecular and Cell Biology*, **10**, 389–430.

Gatenby, A. A. and Viitanen, P. V. (1994). Structural and functional aspects of chaperonin-mediated protein folding. *Annual Review of Plant Physiology and Plant Molecular Biology*, **45**, 469–91.

Hartl, F.-U., Hlodan, R., and Langer, T. (1994). Molecular chaperones in protein folding: the art of avoiding sticky situations. *Trends in Biochemical Sciences*, **19**, 20–6.

Jacob, U. and Buchner, J. (1994). Assisting spontaneity: the role of Hsp90 and small Hsps as molecular chaperones. *Trends in Biochemical Sciences*, **19**, 185–229.

Lill, R. and Neupert, W. (1995). Biogenese von Mitochondrien: Import, Sortierung und Faltung von Proteinen. *Biospektrum* **1**, 28–32.

Lodish, H., Baltimore, D., Berk, A., Zipurski, S. L., Matsudaira, P., and Darnell, J. (ed.) (1995). *Molecular cell biology*. Scientific American Books, Freeman and Co., New York.

McNew, J. A. and Goodmann, J. M. (1996). The targeting and assembly of peroxisomal proteins: some old rules do not apply. *Trends in Biochemical Sciences*, **21**, 41–80.

Moore, A. L., Wood, C. K., and Watts, F. Z. (1994). Protein import into plant mitochondria. *Annual Review of Plant Physiology and Plant Molecular Biology*, **45**, 545–75.

Morimoto, R. J., Tissières, A., and Georgopoulos, C. (ed.) (1994). *The biology of heat shock proteins and molecular chaperones*. Cold Spring Harbor Laboratory Press, Cold Spring Harbor, NY.

Nover, L. (1991). Induced thermotolerance. In *Heat shock response*, (ed. L. Nover), pp. 409–52. CRC Press, Boca Raton, FL.

Pfanner, N., Douglas, M. G., Endo, T., Hogenraad, N. J., Jensen, R. E., Meijer, M., *et al.* (1996). Uniform nomenclature for the protein transport machinery of the mitochondrial membranes. *Trends in Biochemical Sciences*, **21**, 51–2.

Schatz, G. and Dobberstein, B. (1996). Common principles of protein translocation across membranes. *Science*, **271**, 1519–26.

Schneider, H., Berthold, J., Bauer, M. F., Dietmeier, K., Guiard, B., Brunner, M., *et al.* (1994). Mitochondrial Hsp70–MIM44 complex facilitates protein import. *Nature*, **371**, 768–74.

Soll, J. (1995). New insights into the protein import machinery of the chloroplast's outer envelope. *Botanica Acta*, **108**, 277–82.

Vierling, E. (1991). The roles of heat shock proteins in plants. *Annual Review of Plant Physiology and Plant Molecular Biology*, **42**, 579–620.

chapter 22

Gene technology in plants

Recent years have witnessed spectacular developments in plant gene technology. In 1984 the group of Jeff Schell and Marc van Montagu in Cologne and Gent, and the group of Robert Horsch and collaborators of the Monsanto Company in St. Louis (USA) simultaneously published procedures for the transfer of foreign DNA into the genome of plants utilizing the Ti plasmids of *Agrobacterium tume-faciens*. This method has made it possible to alter the protein complement of a plant specifically to meet special requirements: for example, to render plants resistant to pests or herbicides, to achieve a qualitative or quantitative improvement in the productivity of crop plants, and to adapt plants so that they can produce defined sustainable raw materials for the chemical industry.

In the decade after the first publication of this transformation procedure, an immense number of transformed plants have been generated. By 1995 14 transgenic cultivars had been licensed to be grown commercially, of which examples will be described later in this chapter. It is to be expected that the application of plant genetic engineering will become of great economic importance and will change agriculture fundamentally.

The following sections will describe how a plant can be altered by genetic engineering. From the abundance of established procedures only the principles of some major methods can be outlined here. For the sake of brevity, details or complications in methods will be omitted. Some practical examples will show how genetic engineering can be used to alter crop plants.

22.1 A gene is isolated

Let us consider the case where a transgenic plant (A) is to be generated which synthesizes a foreign protein, e.g. a protein from another plant (B). For this, the gene encoding the corresponding protein first has to be isolated from plant B. Since a plant probably contains between 10 000 and 30 000 structural genes, it will be difficult to isolate a single gene from this very large number.

A gene library is required for the isolation of a gene

To isolate a particular gene from the great number of genes existing in the plant genome, it is advantageous to make these genes available in the form of a gene DNA library. Two different kinds of gene libraries can be prepared.

To prepare a *genomic DNA library*, the total genome of the organism is cleaved by restriction endonucleases (section 20.3) into fragments of about 15 to several 100 kbp. Digestion of the genome in this way results in a very large number of DNA sequences which frequently contain only parts of genes. These fragments are inserted into a vector (e.g. a plasmid or a bacteriophage) and then each fragment is amplified by cloning, usually in bacteria.

To prepare a *cDNA library*, the mRNA molecules present in a specific tissue are first isolated and then transcribed into corresponding cDNAs by *reverse transcriptase* (section 20.5). The cDNAs are inserted into a vector and amplified by cloning. The mRNA is isolated from a tissue in which the corresponding gene is thought to be expressed to a high extent. In contrast to the fragments of the genomic library, the resulting cDNAs contain no introns and can therefore, after transformation, be expressed in prokaryotes to synthesize proteins. Since a cDNA contains no promoter regions, such an expression requires a prokaryotic promoter to be added to the cDNA.

To prepare a cDNA library from leaf tissue for example, all the RNA is isolated of which the mRNA may amount to only 2 per cent. To separate the mRNA from the other RNA species one makes use of the fact that eukaryotic mRNA contains a *poly(A) tail* at the 3' terminus (see Fig. 20.8). This allows mRNA to be separated from the other RNAs by affinity chromatography. The column material consists of solid particles of cellulose or other material to which a polydeoxythymidine oligonucleotide (poly(dT)) is linked. When an RNA mixture extracted from leaves is applied to the column, the mRNA molecules bind to the column by hybridization of their poly(A) tail to the poly(dT) of the column material,

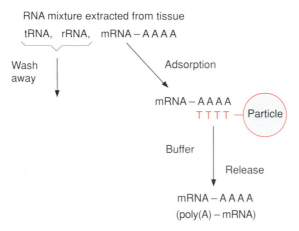

Figure 22.1 Separation of mRNA from an RNA mixture by binding to poly(dT) sequences which are linked to particles.

whereas the other RNAs run through (Fig. 22.1). With a suitable buffer, the bound mRNA is eluted from the column.

To synthesize by reverse transcriptase a cDNA strand complementary to the mRNA, a poly(dT) is used as primer (Fig. 22.2). Subsequently the mRNA is hydrolysed by a ribonuclease either completely or, as shown in the figure, only partly. The latter has the advantage that the mRNA fragments can serve as primers for the synthesis of the second cDNA strand by DNA polymerase. By using DNA polymerase I these mRNA fragments are successively replaced by DNA fragments and these are linked to each other by DNA ligase. A short RNA section remains which is not replaced at the end of the second cDNA strand, but this is of minor importance, since in most cases the mRNA at the 5′ terminus contains a non-coding region (see Fig. 20.8).

The double-stranded cDNA molecules thus formed from the mRNA molecules are amplified by cloning. *Plasmids* or *bacteriophages* can be used as *cloning vectors*. Nowadays there is a large variety of made-to-measure phages and plasmids commercially available for many special purposes. A distinction is made between vectors which only multiply DNA, and *expression vectors* by which the proteins encoded by the multiplied genes can also be synthesized.

A gene library can be kept in phages

Figure 22.3 shows the insertion of cDNA into the DNA of a λ *phage*. In the example shown here the phage DNA possesses a cleavage site for the restriction endonuclease *Eco*RI (section 20.3). To protect the restriction sites within the cDNA, the cDNA double strand is first methylated by an *Eco*RI methylase at the *Eco*RI restriction sites. DNA ligase is then used to link chemically synthesized

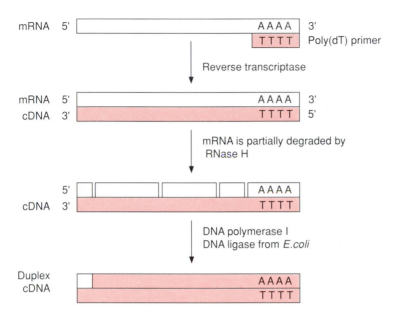

Figure 22.2 Transcription of mRNA to double-stranded cDNA.

double-stranded oligonucleotides with an inbuilt restriction site (in this case for *Eco*RI) to both ends of the double-stranded cDNA. These oligonucleotides are called *linkers*. The restriction endonuclease *Eco*RI cleaves this linker as well as the λ phage DNA and thus generates *sticky ends* at which the complementary bases of the cDNA and of the phage DNA can anneal by base pairing. The DNA strands are then linked by DNA ligase, and in this way the cDNA is inserted into the vector.

The phage DNA with the inserted cDNA is packed *in vitro* into a phage protein coat (Fig. 22.4), using a packing extract from phage-infected bacteria. After the bacteria have been infected, these phages can be multiplied to an unlimited extent, whereby each cDNA in the library is cloned.

The bacteria are infected by mixing them with the phages and they are then plated on agar plates containing cultivation medium. At first the infected bacteria grow on the agar plates to produce a bacterial lawn, but they are then lysed by the phages which have multiplied within the bacteria. The lysed bacterial colonies appear on the agar plate as clear spots called *plaques*. These plaques contain newly formed phages which can be multiplied further. It is customary to plate a typical cDNA gene library on about 10–20 agar plates. The concentration of the phages (titre) is adjusted, so that about 20000–30000 plaques are formed on each agar plate. Ideally each of these plaques contains only one clone. From these plaques, the clone containing the cDNA of the desired gene is selected, using specific probes, as described later.

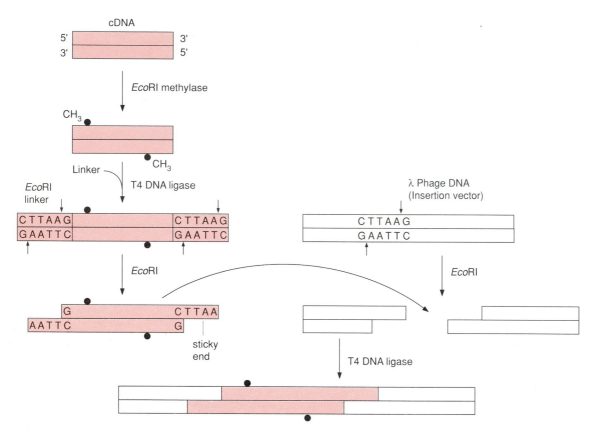

Figure 22.3 Insertion of cDNA in a λ phage insertion vector.

A gene library can also be kept in plasmids

To clone a gene library in plasmids, cDNA is inserted into plasmids via a restriction cleavage site in more or less the same way as in the insertion into phage DNA (Fig. 22.5). The plasmids are then transferred to *E. coli* cells. The transfer is brought about by treating the cells with $CaCl_2$ to make their membrane more permeable to the plasmid. The cells are then mixed with plasmid DNA and exposed to a short heat shock. In order to select from the large majority of untransformed cells those bacterial cells which have been transformed by a plasmid, the plasmid vector contains an *antibiotic resistance gene* which makes the bacteria resistant to an antibiotic such as ampicillin or tetracycline. When the corresponding antibiotic is added to the culture medium, cells containing the plasmid survive and grow, whereas the other non-transformed cells are killed. After plating on an agar culture medium, bacterial colonies develop which can be recognized as spots.

In order to check whether a plasmid actually contains an inserted DNA sequence (insert), plasmid vectors have been constructed in which the restriction

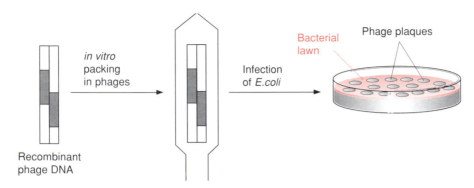

Figure 22.4 A recombinant phage DNA is packed into a virus particle. *E. coli* cells are infected with the formed phage and plated on agar plates. The cells of the infected colonies are lysed by the multiplying phages and show as transparent spots (plaques) in the bacterial lawn.

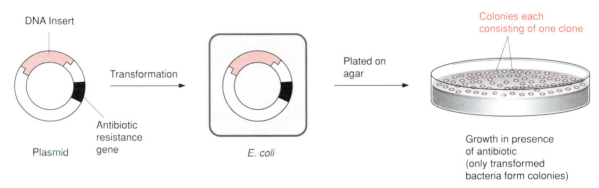

Figure 22.5 cDNA can be propagated via a plasmid vector in *E. coli*. An antibiotic resistance gene on the plasmid enables the selection of the transformed cells.

cleavage site for insertion of the foreign DNA is located inside a gene which encodes the enzyme β-galactosidase (Fig. 22.6). This enzyme hydrolyses the colourless compound X-Gal into an insoluble blue product. When X-Gal is added to the agar plate culture medium, all the cells which do not contain a DNA insert, and therefore contain an intact β-galactosidase gene, form blue colonies. If a DNA segment is inserted into the cleavage site of the β-galactosidase gene, this gene is interrupted and no longer able to encode a functional β-galactosidase. Therefore the corresponding colonies are not stained blue but remain white (blue/white selection).

A gene library is screened for a gene

Specific probes are employed to screen the bacterial colonies or phage plaques for the desired gene. A blot is made of the various agar plates by placing a nylon or nitrocellulose membrane on top of them (Fig. 22.7). Some of the phages

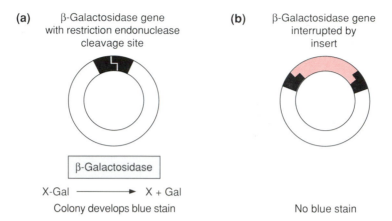

(a) β-Galactosidase gene with restriction endonuclease cleavage site

β-Galactosidase

X-Gal ⟶ X + Gal

Colony develops blue stain

(b) β-Galactosidase gene interrupted by insert

No blue stain

Figure 22.6 To check whether the plasmid of a bacterial colony contains a DNA insert, the cleavage site of the plasmid vector is contained within a β-galactosidase gene. In colonies with the intact gene (i.e. no insert), the colourless chemical X-Gal (5-bromo-4-chloro-3-indolyl-β-D-galactopyranoside) is hydrolysed by the corresponding gene product into galactose and an indoxyl derivative, which oxidizes to form a blue-coloured dimer. This results in blue staining of the colony. When the function of the β-galactosidase gene is disrupted by the DNA insert, the gene product can no longer be formed and the corresponding colonies are not stained blue.

contained in the plaques, or the bacteria contained in the colonies, bind to the blotting membrane, although most of the contents of the plaques and the colonies remain on the agar plate.

Two kinds of probes can be used to screen the phage or bacterial clones bound to the blotting membrane:

(1) specific antibodies to identify the protein formed as gene product of the desired clone;

(2) specific DNA probes to label the cDNA of the desired clone by hybridization.

A clone is identified by antibodies against the gene product

Antibodies against a certain protein are often used to identify the corresponding gene. This method requires sufficient amounts of the corresponding protein to be purified beforehand in order to obtain *polyclonal antibodies* by immunization of animals. If antibodies are to be used to identify a gene product, the cDNA library must be inserted in an appropriate expression vector. This is a cloning vector which facilitates transcription of the inserted DNA molecule, and the resultant mRNA is then translated into the corresponding protein which might be recognized by the antibody. The vector contains a promoter sequence which controls initiation of transcription of the inserted gene, and often also a sequence for termination of the transcription at its end.

In practice, bacteria bound to the blotting membrane are disrupted and the

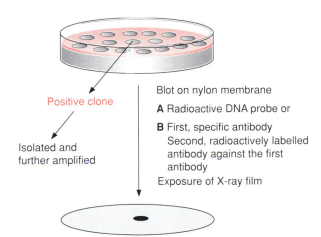

Positive clone

Isolated and
further amplified

Blot on nylon membrane

A Radioactive DNA probe or

B First, specific antibody
Second, radioactively labelled
antibody against the first
antibody

Exposure of X-ray film

Figure 22.7 For screening the gene library, a blot is prepared by placing a nylon or nitrocellulose membrane on the agar plate. By means of a radioactively labelled probe (DNA probe or antibody) and subsequent autoradiography, the desired DNA or the corresponding gene product is detected on the blot as a dark spot on the X-ray film. The corresponding clone is identified on the agar plate by comparing the positions and can be picked up with a toothpick for further propagation. Antibodies can also be detected by colour or fluorescence label.

released bacterial proteins are fixed to the membrane. When phages are used as vectors, cell disruption is not required since the phages themselves lyse the cell and thus liberate the cell proteins which are then fixed to the blotting membrane. Afterwards antibodies are added which bind specifically to the corresponding protein but are washed off from all other parts of the membrane. Usually a second antibody, which recognizes the first antibody, is used to detect the bound antibody (Fig. 22.8). The second antibody could be, for instance, an antibody from chicken, raised against the first antibody, e.g. from rabbit. The second antibody may be labelled, e.g. by radioactive iodine, or can be conjugated with an enzyme or a coloured product or a fluorescent label, which will not be described in detail here. After incubation with the radioactive second antibody the blotting membrane is

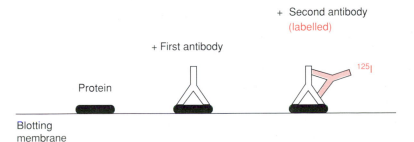

Figure 22.8 An antibody bound to a protein is recognized by a second antibody which is, for instance, radioactively labelled. The desired gene product can then be detected by autoradiography.

washed, dried and, in the case of a radioactive label, exposed to an X-ray film. A positive colony can be recognized as a dark spot on the autoradiograph (Fig. 22.7). The position of the positive clone on the agar plate can be identified from the position of the spot on the blotting membrane. After the first screening this apparently positive clone may actually contain several clones, due to a high density of bacteria. For this reason the colony, picked up from the positive region with a toothpick, is diluted and plated on to another agar plate. By repeating the screening procedure described above (*rescreening*), single pure clones are finally obtained. Positive phage plaques are also regrown in bacteria and plated again in order to obtain pure clones.

A clone can also be identified by DNA probes

In this procedure the phages or bacteria present on the blotting membrane are first lysed and the proteins removed. The remaining DNA is then dissolved into single strands which are tightly bound to the membrane. Complementary DNA sequences, which are radioactively labelled by ^{32}P-labelled deoxynucleotides, are used as probes. These probes bind by hybridization to the desired cDNA present on the blotting membrane. The identification of the positive clones proceeds via autoradiography, similar to the detection procedures described above.

Chemically synthesized oligonucleotides of about 20 bases are also employed as DNA probes. These probes are radioactively labelled at the 5' end by ^{32}P-phosphate and are used particularly when only low quantities of the purified proteins are available, which are not sufficient for the generation of antibodies. However, with very low amounts of protein it is possible to determine by micro-sequencing part of the N-terminal amino acid sequence of the protein. From such a partial amino acid sequence the corresponding DNA sequence can be deduced, according to the genetic code, in order to produce oligonucleotide probes by chemical synthesis using automatic synthesizers. However, the degeneracy of the genetic code means that amino acids are often encoded by more than one nucleotide triplet (Fig. 22.9). This is taken into account during the design of oligonucleotides. When, for example, the third base of the triplet encoding lysine can be either an A or G, a mixture of precursors for both is added to the synthesizer during the linking of the third base of this triplet to the oligomer. In order to introduce the third base of the alanine triplet, a mixture of all four nucleotide precursors is added to the synthesizer. The 'degenerate' oligonucleotide shown in Fig. 22.9 is thus, in fact, a mixture of 48 different oligonucleotides, of which only one contains the correct sequence of the desired gene.

Figure 22.9 A degenerate oligonucleotide contains all the possible sequences for encoding the given amino acid sequence.

When the protein encoded by the desired gene has not been purified, a corresponding gene section from related organisms can sometimes be used as probe. Domains with specific amino acid sequences are often conserved in enzymes or other proteins (e.g. translocators), even from distantly related plants, and are encoded by correspondingly similar gene sequences.

After a cDNA has been successfully isolated, usually the next step is to determine its base sequence. To an increasing extent DNA sequencing is carried out by automatic analysers which will not be described here. It can be recognized from the DNA sequence whether the isolated cDNA represents a full-length clone encoding a complete mRNA, including the cap and poly(A) sequence (Fig. 20.8). When this is not the case, the partial sequence obtained is used as a DNA probe in order to search for full-length clones in the cDNA library.

Genes encoding unknown proteins can be isolated by complementation

In some cases it has been possible to isolate the cDNA encoding an unknown protein that is only defined by its function. One method for identifying the gene encoding such an unknown protein is by the complementation of *deficiency mutants* of bacteria or yeast after transformation with plasmids from a cDNA library (Fig. 22.10). Plasmids that can be amplified in *E. coli* as well as in yeast are available as cloning vectors and can be used to express the encoded protein. Several plant translocators, including the sucrose translocator involved in phloem loading (section 13.1), have been identified by complementation. To identify the gene of the sucrose translocator, a yeast deficiency mutant was employed, which had lost the ability to take up sucrose and therefore could no longer use sucrose as a nutritional source. This mutant was transformed with plasmids from a plant cDNA library, which were expressed under the control of a yeast promoter. After plating the transformed yeast cells on a culture medium with sucrose as the only carbon source, a yeast clone was found which grew on the sucrose cultivation medium. This indicated that transformation with the corresponding plasmid from the plant cDNA library had generated yeast cells that produced the plant sucrose translocator and incorporated it into their cell membrane. After this positive yeast clone had been amplified, plasmid cDNA was isolated and sequenced. The cDNA sequence yielded the amino acid sequence of the previously unknown sucrose translocator.

Genes can be tracked down with the help of transposons

Another possible way to identify genes aside from the function of their gene products is to use transposons. Transposons are DNA sequences that can 'jump' within the genome of a plant, which sometimes results in the recognizable elimination of a gene function (section 20.4). This could, for example, be the loss of the ability to

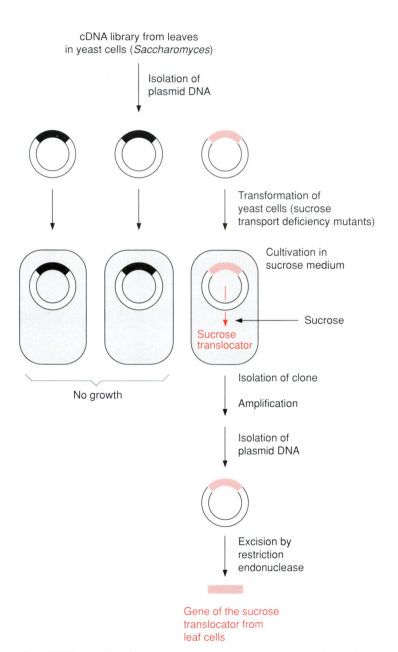

cDNA library from leaves
in yeast cells (*Saccharomyces*)

Isolation of
plasmid DNA

Transformation of
yeast cells (sucrose
transport deficiency mutants)

Cultivation in
sucrose medium

Sucrose

Sucrose
translocator

No growth

Isolation of clone

Amplification

Isolation of
plasmid DNA

Excision by
restriction
endonuclease

Gene of the sucrose
translocator from
leaf cells

Figure 22.10 Identification of a plant gene by complementation of a yeast mutant deficient in sucrose uptake.

synthesize a flower pigment. In such a case a gene probe based on the known transposon sequence is used to identify by DNA hybridization the region of the genome in which the transposon has been inserted. By using the cloning and screening procedures already mentioned, it is possible to identify a gene, in our example a gene for the enzyme of flower pigment synthesis, and to determine the amino acid sequence of the corresponding enzyme by DNA sequence analysis. Labelling a gene with an inserted transposon is called *gene tagging*.

Instead of transposons Ti plasmids can also be used to tag a gene. The next section will describe in detail how Ti plasmids insert T-DNA into genes. A gene that has lost its function by the insertion of T-DNA may be identified by a gene probe for T-DNA.

22.2 Agrobacteria have the ability to transform plant cells

Gram-negative soil bacteria of the species *Agrobacterium tumefaciens* induce a tumour growth at wounding sites in various plants, often on the stem, which can lead to the formation of *crown galls*. The tumour tissue from these galls continues to grow as a callus in cell culture. As already described in section 19.3, mature differentiated plant cells, which normally no longer divide, can be stimulated to unrestricted growth by the addition of the phytohormones *auxin* and *cytokinin*. In this way a tumour, in the form of a callus, can be formed from a differentiated plant cell. The gall is formed by agrobacteria in basically the same way. The stricken plants are forced to produce high concentrations of auxin and cytokinin, resulting in a proliferation of the plant tissue. Since the capacity for increased cytokinin and auxin synthesis is inherited after cell division by all the succeeding cells, a callus culture of crown gall tissue can be multiplied without adding phytohormones.

The crown gall tumour cells have acquired yet another ability: they can produce a variety of products, called *opines*, by condensing amino acids and α-ketoacids or amino acids and sugars. These opines are formed in such high amounts that they are secreted from the crown galls. Figure 22.11 shows octopine, lysopine, nopaline, and agropine as examples of opines. Each *Agrobacterium* strain induces the synthesis of only a single opine. The synthesis of the first three opines mentioned above proceeds via condensation to a Schiff base and a subsequent reduction by NADPH. The opines are so stable that they cannot be metabolized by most soil bacteria. *Agrobacterium tumefaciens* strains have specialized in utilizing these opines. Normally, a particular opine can only be catabolized by that strain which has induced its synthesis in the plant. In such a way the stricken plants are used to produce a special nutrient which can only be consumed by the corresponding *Agrobacterium*.

The conversion of the differentiated plant cells to opine-producing tumour cells

(a)   **(b)**

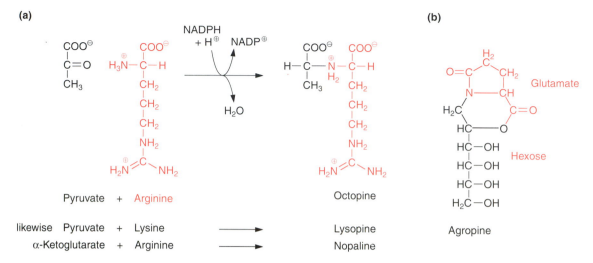

Pyruvate + Arginine →	Octopine
likewise Pyruvate + Lysine →	Lysopine
α-Ketoglutarate + Arginine →	Nopaline
	Agropine

Figure 22.11 Opines are formed from amino acids and ketoacids (a) or from an amino acid and a hexose (b).

occurs without the agrobacteria entering those cells. This reprogramming of the plant cells is caused by the transfer of functional genes from the bacterium to the genome of the stricken plant cells. The agrobacteria have acquired the ability to transform plants in order to use them as a production site for their nutrient. Jeff Schell named this novel parasitism *genetic colonization*.

The Ti plasmid contains the genetic information for tumour formation

The phytopathogenic function of *A. tumefaciens* is encoded in a tumour-inducing *Ti plasmid*, with a size of about 200 kbp (Fig. 22.12). Strains of *A. tumefaciens* containing no Ti plasmids are unable to induce the formation of crown galls.

Plants are infected by bacteria at wounds, frequently occurring at the stem base. When wounded, plants defend themselves against pathogens by secreting phenolic substances (Chapter 18), which are used by agrobacteria as a signal to initiate the attack. These phenols stimulate the expression of 7–8 virulence (*vir*) genes, located within a region of the Ti plasmid. These *vir* genes encode virulence proteins, which enable the transfer of the bacterial tumour-inducing genes to the plant genome. From a 12–14 kbp long section of the Ti plasmid, called T-DNA (T, transfer), a single strand is excised by a vir nuclease. The cleavage sites are defined by *border sequences* present at both ends of the T-DNA. The transfer of the T-DNA single strand from the bacteria to the plant cell nucleus (Fig. 22.13) proceeds by a very complicated process of which the details are not yet fully resolved. The DNA transfer probably occurs in basically the same way as in conjugation, a form of

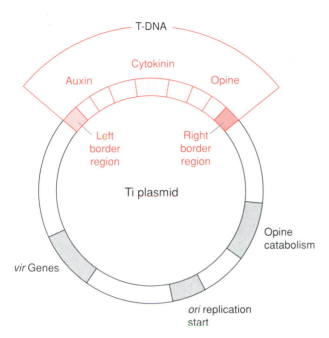

Figure 22.12 Diagram of a Ti plasmid (not to scale). The T-DNA which is transferred to the genome of the plant, representing about 7–13 per cent of the Ti plasmid, is defined by its left and right border sequence. The T-DNA contains genes encoding enzymes for the synthesis of the phytohormones cytokinin and auxin, and of a specific opine. The T-DNA is only transcribed and translated in the plant cell. The remaining part of the plasmid is transcribed and translated in the bacterium and contains several *vir* genes and also a single or several genes for opine catabolism. The *ori* region represents the replication start (After Glick and Pasternak 1994.)

bacterial sexual propagation. The missing T-DNA single strand in the plasmid is subsequently replaced by replication. The T-DNA strand which had been transferred to the plant cell moves on to the nucleus. *Vir*-encoded proteins protect the T-DNA from being attacked *en route* by DNA-degrading plant enzymes and also facilitate the transport through the nuclear pores to the nucleus. The mechanism for integration into the plant nuclear genome is still unknown, but the right border region of the T-DNA seems to have an important function in this process. The T-DNA is integrated randomly into chromosomes and, when inserted in a gene, can eliminate its function. This can be used to identify a gene by gene tagging in an analogous way to gene tagging by the transposons described at the end of section 22.1.

The T-DNA integrated in this way in the nuclear genome has the properties of a eukaryotic gene and is inherited in a Mendelian fashion. It is replicated by the plant cell as if it were its own DNA and, since it contains eukaryotic promoters, is also transcribed. The resultant mRNA corresponds to a eukaryotic mRNA and is translated as such. The T-DNA encodes cytokinin synthase, a key enzyme for the

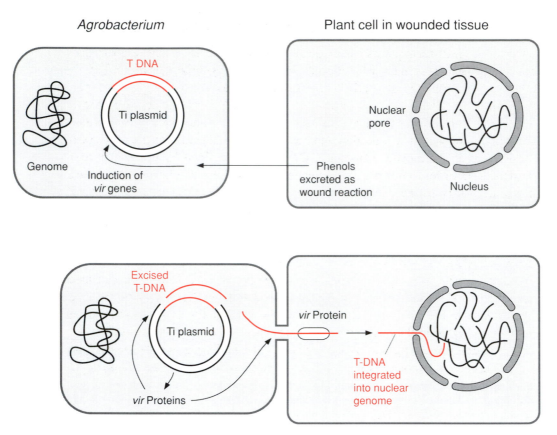

Figure 22.13 Agrobacteria transform plant cells to make them produce opines as their nutrient. Wounded plant tissues induce the expression of virulence genes on the Ti plasmid of the agrobacterium, resulting in the synthesis of virulence proteins which cause a single-stranded DNA segment, known as T-DNA, to be excised from the plasmid, transferred to the plant cell, and integrated in the nuclear genome.

synthesis of the cytokinin *zeatin* (Fig. 19.6), and also two enzymes for the synthesis of the auxin *indoleacetic acid* (IAA) from tryptophan. This bacterial IAA synthesis proceeds in a different manner than plant IAA biosynthesis (Fig. 19.2), but details of this pathway will not be considered here. Moreover, the T-DNA encodes one or two enzymes, different from strain to strain, for the synthesis of a special opine. The T-DNA integrated in the plant genome thus carries the genetic information for synthesis of cytokinin and auxin to induce tumour growth, and also the information for the plant to synthesize an opine.

The enzymes required for catabolism of the corresponding opine are encoded in that part of the Ti plasmid which remains in the bacterium. Thus, in parallel to the transformation of the plant cell, the enzymes for opine degradation are synthesized by the bacteria.

22.3 Ti plasmids are used as transformation vectors

Its ability to transform plants has made *A. tumefaciens* an excellent tool for integrating foreign genes in their functional state in a plant genome. It was necessary, however, to modify Ti plasmids before using them as vectors (Fig. 22.14). The genes for auxin and cytokinin synthesis were removed to prevent tumour growth on the transformed plants. Since synthesis of an opine is unnecessary for a transgenic plant and would be a burden on its metabolism, the genes for the opine synthesis were also removed. Thus the T-DNA is only defined by the two border sequences. In order to insert a foreign gene between these two border sequences, it was necessary to incorporate a DNA sequence containing cleavage sites for several restriction endonucleases (known as a *polylinker sequence* or a multicloning site) within the T-DNA region of the Ti plasmid. The Ti plasmid could then be cleaved in the T-DNA region by a certain restriction endonuclease. A foreign DNA sequence, excised by the same restriction endonuclease, can be inserted in this cleavage site (see Fig. 22.3). Since the polylinker sequence contains cleavage sites for several restriction endonucleases, several DNA molecules can be inserted sequentially in the T-DNA. In Fig. 22.14 a promoter, enabling the expression of the inserted DNA in the host cell, is located to the left of the cleavage site.

In modern transformation systems, vectors derived from the Ti plasmid no longer contain any *vir* genes and are therefore unable to transform a plant cell on

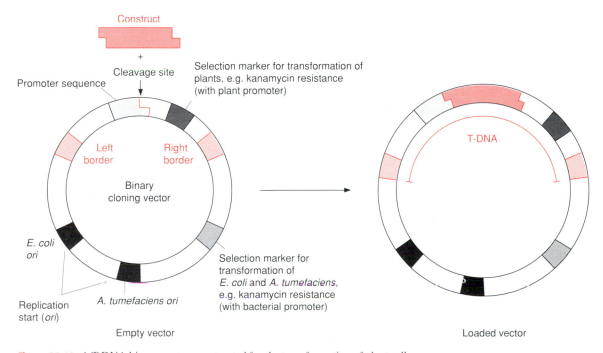

Figure 22.14 A T-DNA binary vector constructed for the transformation of plant cells.

their own. For transformation they require the assistance of a second so-called *helper plasmid*, which contains the *vir* genes but no T-DNA and is therefore also unable to transform a plant on its own. Since both these vectors are required simultaneously for a transformation, they are called *binary vectors*. A large variety of such vectors have been designed for special applications and are available commercially.

In order to transform a plant, sufficient amounts of vectors containing the gene to be transferred must be present. It proved to be advantageous to first amplify the vectors by cloning them in *E. coli* (Fig. 22.15). Since *E. coli* does not recognize the replication start site of the natural Ti plasmid (*A. tumefaciens ori*), a second replication start (*E. coli ori*) is introduced into the plasmid.

The plasmid is provided with a selection marker in order to select those *E. coli* cells which have been transformed by the Ti plasmid. For this a gene encoding *neomycin phosphotransferase* is frequently used. This enzyme degrades the antibiotic *kanamycin*, thus rendering the cell resistant towards this antibiotic. Kanamycin is only very seldomly used in medicine, which is an important aspect in regard to transgenic plants containing this antibiotic resistance gene. The kanamycin resistance gene is provided with a bacterial promoter and, after transformation of the vector, is therefore expressed in *E. coli* cells. When kanamycin is added to the culture medium of the bacteria, only the transformed bacteria survive, which are protected from the antibiotic by the resistance gene on the vector. Thus the Ti plasmid can be propagated efficiently by cloning in *E. coli*.

As hosts for the binary vectors, *A. tumefaciens* strains are used that do not contain complete Ti plasmids, but only the helper plasmids mentioned above. These encode the vir proteins required for the transfer of the T-DNA from the binary vector into the plant genome (Fig. 22.15).

A new plant is regenerated following transformation of a leaf cell

After transformation, the few transformed plant cells are selected from the large number of non-transformed cells by another selection marker contained in the T-DNA vector. Mostly the kanamycin resistance gene described above is also used for this purpose, but in this case provided with a plant promoter.

It was mentioned in the introduction to this chapter that *A. tumefaciens* attacks plants at wounds. Leaf discs therefore, with their cut edges, are a good target for performing transformation (Fig. 22.16). The leaf discs are immersed in a suspension of *A. tumefaciens* cells transformed by the vector. After a short time the discs are transferred to an agarose-containing culture medium which, besides nutrients, contains the phytohormones cytokinin and auxin, to induce the cells of the leaf disc to grow a callus. The addition of the antibiotic kanamycin kills all plant cells except the transformed ones, which are protected from the antibiotic by the resistance gene. The remaining agrobacteria are killed by another antibiotic specific for bacteria. Calli of

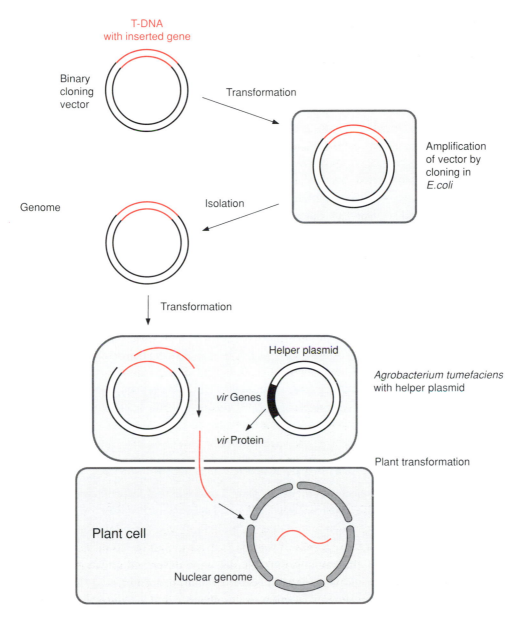

Figure 22.15 Transformation of a plant cell by means of a binary cloning system. After bacterial transformation, the T-DNA vector containing the DNA insert is propagated in *E. coli*, isolated and transferred into *Agrobacterium*. A helper plasmid, already present in *Agrobacterium*, encodes the virulence proteins required for the transformation of plant cells.

transformed cells develop at the cut edges of the leaf discs. When the concentrations of cytokinin and auxin are appropriate, these calli can be propagated almost without limit in tissue culture on agarose culture media. In this way transformed plant cells can be kept and propagated in tissue culture for very long periods of time. If required, new plants can be regenerated from these tissue cultures.

To regenerate new plants, cells of the callus culture are transferred to a culture medium containing more cytokinin than auxin, and this induces the callus to form shoots. Root growth is then stimulated by transferring the shoots to a culture medium containing more auxin than cytokinin. After plantlets with roots have developed and grown somewhat, they can be transplanted to soil and then, in many cases, grow to be completely normal plants capable of being multiplied by flowering and seed production.

The pioneering work of Jeff Schell, Marc van Montagu, Patricia Zambryski, Robert Horsch, and several others has developed the *A. tumefaciens* transformation system to a very easy method for transferring foreign genes to cells of higher plants. Nowadays it is often possible even for a student to produce several hundred different transgenic tobacco plants with no great difficulty.

Using this method, more than a hundred different plant species have been transformed successfully. Initially it was very difficult, or even impossible, to transform monocot plants with the *Agrobacterium* system. Recently, this transformation method has been improved to such an extent that it can now also be applied successfully to transform several monocots, such as rice. An alternative way to transform plant cells is a physical gene transfer, the most successful being the bombardment of plant cells by microprojectiles.

Plants can be transformed by a modified shotgun

Transformation by bombardment of plant cells with microprojectiles was developed in 1985. The microprojectiles are small spheres of tungsten or gold with a diameter of 1–4 μm, which are coated with DNA. A *gene gun* (similar to a shotgun) is used to shoot the pellets into plant cells (Fig. 22.17). Initially gunpowder was used as propellant, but nowadays the microprojectiles are often accelerated by compressed air, helium, or other gases. The target materials include calli, embryonic tissues, and leaves. In order to penetrate the cell wall of the epidermis and mesophyll cells, the velocity of the projectiles must be very high and can reach about 1500 km h^{-1} in bombardments with a gene gun in a vacuum chamber. The cells in the centre of the line of fire may be destroyed during such a bombardment but, because the projectiles are so small, whereas the cells nearer the periphery survive. The DNA transferred to the cells by these projectiles can be integrated in the nuclear genome. Transformed cells are again selected on an agarose medium containing the antibiotic and, by the addition of phytohormones, induced to form a callus. A number of crop plants have been transformed successfully using the

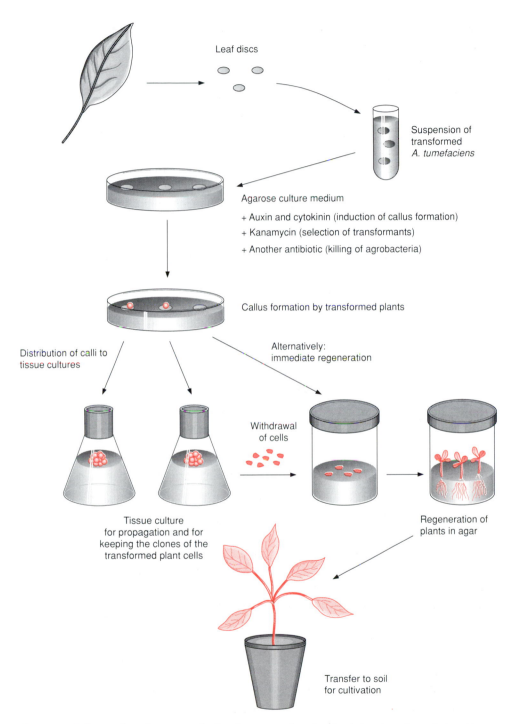

Leaf discs

Suspension of
transformed
A. tumefaciens

Agarose culture medium

+ Auxin and cytokinin (induction of callus formation)

+ Kanamycin (selection of transformants)

+ Another antibiotic (killing of agrobacteria)

Callus formation by transformed plants

Distribution of calli to
tissue cultures

Alternatively:
immediate regeneration

Withdrawal
of cells

Tissue culture
for propagation and for
keeping the clones of the
transformed plant cells

Regeneration of
plants in agar

Transfer to soil
for cultivation

Figure 22.16 Generation of a transgenic plant by means of an *Agrobacterium* transformation system.
See text.

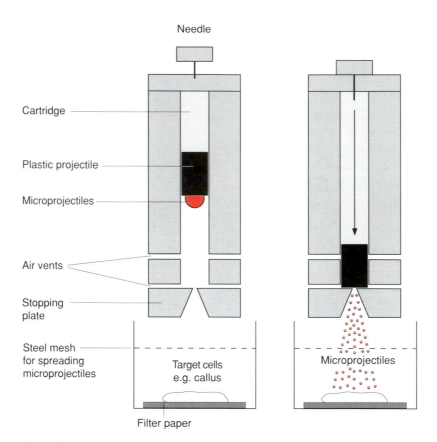

Needle

Cartridge

Plastic projectile

Microprojectiles

Air vents

Stopping plate

Steel mesh for spreading microprojectiles

Target cells e.g. callus

Microprojectiles

Filter paper

Figure 22.17 Transformation of a plant by a gene gun. Gold or tungsten spheres are coated with a thin DNA layer by a deposit of $CaCl_2$. The spheres are inserted in front of a plastic projectile into the barrel of the gun. The gun and the target are in an evacuated chamber. When the gun is fired, the plastic projectile is driven to the stopping plate, by which the microprojectiles are driven through holes and shot at a high velocity into the cells of the plant tissue. The DNA carried in this way into the plant cells contains an antibiotic resistance gene (e.g. for kanamycin resistance) as well as the transferred gene. Transformed cells can be selected by this marker in the same way as for cells transformed with *A. tumefaciens*. The transformed cells are propagated by callus tissue culture and, from these, plants are regenerated. (After Hess. Biotechnologie der Pflanze, Verlag Ulmer, Stuttgart (1992)).

gene gun. Thus by bombardment of embryogenic callus cells of sugar cane 10–20 different transformed plant lines are obtained routinely by one shot. Using this method, it is now also possible to transfer genes to mitochondria and chloroplasts, opening the way for transforming their genomes.

Protoplasts can be transformed by the uptake of DNA

The transformation of protoplasts is another way to transfer foreign genetic information to a plant cell. Protoplasts can be obtained from plant tissues by digestion of the cell walls (section 1.1), and are able to take up foreign DNA in the presence of CaCl$_2$ or polyethylene glycol, to be integrated in the genome. This transformation resembles that of bacteria by plasmids. In protoplast transformation, the gene to be transferred is linked with a selection marker encoding resistance to an antibiotic. After the antibiotic has been added, only the transformed protoplasts survive. In principle, the protoplasts of all plants can be transformed in this way since there is no host specificity involved as in the case of transformation by *A. tumefaciens*. However, the use of this method is restricted to a few plant species where it has been possible to regenerate intact, fertile plants from protoplasts. Transgenic maize plants have been generated by protoplast transformation.

22.4 Selection of appropriate promoters enables the defined expression of an inserted gene

Any foreign gene transferred to a plant can only be expressed when it has been provided with a suitable promoter sequence (section 20.2). The selection of the promoter determines where, when, how much, and under what conditions gene expression takes place. Usually these promoters are already included in the commercially available vectors (see Fig. 22.14).

To have a high expression of an inserted foreign gene in all parts of the plants, the *CaMV 35S promoter* from the cauliflower mosaic virus (section 20.5) is often used. This promoter contains transcription enhancers which enable a particularly high transcription rate in different tissues of very many plants. The *nos promoter* from the Ti plasmid of *A. tumefaciens*, normally regulating the gene for nopaline synthesis, is also often used as a non-specific promoter in transgenic plants.

The DNA sequences of plant promoters can be determined by the analysis of plant genes obtained from a genomic gene library (section 22.1). A promoter for specific gene expression in the potato tuber, for instance, has been identified by analysis of the DNA sequence of the gene for patatin, the storage protein of the potato (Chapter 14).

A *reporter gene* can be used to determine whether an isolated promoter sequence is tissue specific. Such reporter genes encode proteins which are easily detected in a plant, for instance the *green fluorescent protein* (GFP) from the jellyfish *Aequorea victoria*, which can be seen directly by confocal microscopy in intact leaves. Frequently also the gene for the *β-glucuronidase* (GUS) enzyme from *E. coli* is used as reporter gene (Fig. 22.18). This enzyme resembles β-galactosidase mentioned in section 22.1 and does not occur in plants. It hydrolyses a synthetic X-glucuronide to give a blue-coloured hydrolysis product, which can be identified readily under the microscope. When a potato plant is transformed with the GUS reporter gene fused to the patatin promoter mentioned above, after addition of X-glucuronide a deep blue colour develops only in cuts of the tubers, but not in other tissues of the plant. This indicates that the patatin promoter acts as a specific promoter for gene expression in the potato tubers. Many promoters have been

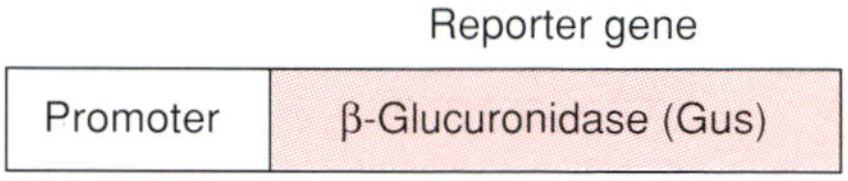

Reporter gene

Promoter	β-Glucuronidase (Gus)

Figure 22.18 In order to check the function of a promoter, it is linked to a reporter gene. In the example shown, the reporter gene encodes the enzyme β-glucuronidase from *E. coli*. X-Glucuronide is hydrolysed by this enzyme and a blue product is formed, analogously to the β-galactosidase reaction (Fig. 22.6).

isolated that are only active in certain organs or tissues, such as leaves, roots, phloem, flowers, or seed. Moreover, promoters have been isolated that control the expression of gene products only under certain environmental conditions, such as light, high temperatures, water stress, or pathogenic infection. Frequently, promoters are also active in heterologous plants, although the extent of expression by a certain promoter can vary between different plant species.

Gene products are directed into certain subcellular compartments by targeting sequences

In transgenic plants the expression of a foreign gene can be restricted to certain tissue or cell types by the use of defined promoters. Furthermore, a protein encoded by a foreign gene can be directed to a particular subcellular destination, e.g. the chloroplast stroma or the vacuolar compartment, by the presence of additional amino acid presequences which serve as *targeting signals* for transfer via the various subcellular membrane transport systems (sections 14.5, 21.3). A large number of targeting sequences is now available, making it possible to direct a protein encoded by a foreign gene to a defined subcellular compartment, such as the vacuolar compartment or the chloroplast stroma.

22.5 Genes can be turned off by transformation

As well as producing transgenic plants with new properties by the transfer and expression of foreign genes, it is of great interest to decrease or even eliminate certain properties by inhibiting the expression of the corresponding gene, for instance the activity of a translocator or of an enzyme. A decrease in gene expression can be achieved by inactivating the encoding mRNA by the synthesis of a complementary RNA, called *antisense RNA* (Fig. 22.19). Normally, mRNA occurs as a single strand, but in the presence of a complementary mRNA strand, can form a double-stranded RNA, which is unstable and degraded rapidly by ribonucleases. This may be the reason why mRNA loses the ability to encode protein synthesis when the corresponding antisense RNA is present.

To synthesize the antisense RNA, the corresponding gene is first isolated according to the methods already described, and inserted as cDNA in *reverse orientation* in a vector with which the plants are transformed. As the orientation is

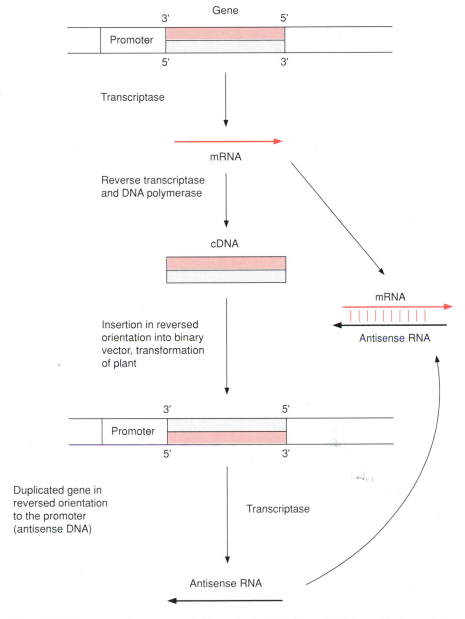

Figure 22.19 Decrease of gene expression by antisense RNA. The mRNA formed by transcription of a gene is reverse transcribed to give a double-stranded cDNA (see Fig. 22.2), and then inserted in opposite orientation into a vector (Fig. 22.14) for plant transformation. The transformed plant now contains in its genome, besides the normal gene, a duplicate, the antisense DNA, in which the promoter causes transcription of the originally non-coding DNA strand. Transcription of both genes results in an mRNA double strand due to base pairing and this duplex is degraded.

reversed in this inserted gene, transcription results in the formation of the anti-sense RNA. Since its introduction in 1985 this antisense technique has become a very important tool in plant genetic engineering for specifically decreasing gene expression, but it does not always work. An alternative method to reduce the activity of a gene is *co-suppression*, where an additional copy of an endogenous gene is introduced by transformation. The overexpression of the same or similar genes frequently results in a decrease in activity not only of the inserted gene but also of the gene already present. The mechanism of this effect is not yet fully understood. Co-suppression can be more efficient than antisense inhibition, but functions only in certain cases.

22.6 Plant genetic engineering can be used for many different purposes

In the 12 years since the first publication of plant transformations via the *Agrobacterium* system, this method of genetic engineering has produced revolutionary results in basic as well as in applied plant science. In basic science it has led in a very short time to the identification and characterization of many new proteins, such as enzymes and translocators. When investigating the function of a protein in a plant, it is now common to increase or decrease the expression of the protein by molecular genetic transformation. From the effects of these changes conclusions can be drawn about the role of the corresponding protein in metabolism.

In agriculture, plant genetic engineering has been utilized in many ways to augment protection against pests and to increase the qualitative and quantitative yield of crop plants. Up until 1995 approximately 1800 registered controlled field tests with genetically engineered plants have been carried out world-wide. A number of examples of practical applications of plant molecular genetics have already been described in the preceding chapters. In 1995 approval was given to grow commercially as many as 14 different crop species transformed by genetic engineering. Most of them had been developed to increase plant protection, e.g. resistance against insects, viruses, and herbicides. Transformed cultivars of tomato were produced to improve the quality and storage properties of tomato fruits, and transformed rapeseed cultivars to produce short chain fatty acids. Genetic engineering is also used on a commercial scale to generate male sterile plants for producing hybrids (section 20.7), e.g. for rapeseed breeding.

Plants are selectively protected against some insects by the BT protein

Field crops are in great danger of being attacked by insects. Some examples may illustrate this:

1. The Colorado beetle, originating in North America, can cause complete defoliation of potato fields.

2. The larvae of the corn borer, penetrating maize shoots, cause much crop damage by feeding inside the shoots.

3. In a similar way, the cotton borer prevents the formation of cotton flowers by feeding inside shoots.

It has been estimated that about one-sixth of the global plant food production is lost by insect pests. In order to avoid serious crop losses, the farmer often has no option but to use chemical pest protection. In former times chlorinated hydro-carbons such as DDT or aldrin were in use as a very potent means of protection against insects. Since these substances are degraded only very slowly and there-fore accumulate in the food chain, they cause damage to the environment and are now restricted in their use, or even forbidden by law in many countries. Nowadays, organophosphorous compounds are used as insecticides, which, as cholinesterase inhibitors, impair nerve function at the site of the synapses. These compounds are readily degraded, but unfortunately destroy not only pests but also useful insects such as bees, and are also poisonous for humans. The threat to humans lies not so much in the pesticide residues in consumed plant material, but primarily to the people applying the insecticide.

For more than 30 years preparations from *Bacillus thuringiensis* have been in use as alternative biological insecticides. These bacteria form toxic peptides (*BT proteins*) which bind to receptors in the intestine of certain insects, thus impairing the absorption of food. This inactivation of the intestinal function causes the insect to starve to death. More than 100 bacterial strains are known that form different BT proteins with a relatively specific toxicity towards certain insects. Toxicological investigations have shown that BT proteins are not harmful to humans. For many years now bacterial suspensions containing the BT protein have been sprayed to protect crops from insects. Unfortunately, these prepara-tions are relatively expensive and are easily washed off the leaves by rain. Spraying also has the disadvantage that it does not reach larvae inside plant shoots, e.g. the corn borer and cotton borer.

The genes for various BT proteins have been cloned and used to transform a number of plants. Although transgenic plants produce only very low amounts of the toxic BT protein (0.1 per cent of total protein), it is more than enough to deter insects from eating the plant. The BT protein is decomposed in the soil and degraded in the human digestive tract just like all other proteins. On this basis, insect-resistant, transformed lines of potato, maize, and cotton have been now licensed for agricultural use in the USA. The generation of insect-resistant tobacco transformants will be completed soon. Although these transgenic plants do not fully replace the use of customary insecticides, they are expected to allow a substantial reduction in the use of insecticides, easing threats to the entire fauna spectrum. The use of such transgenic plants thus may contribute to preserving the environment.

The insertion by genetic engineering of foreign genes encoding *proteinase inhibitors*, is an alternative way to protect plants from insect pests (section 14.4). After wounding, e.g. by insect attack or also by fungal infection, the formation of proteinase inhibitors, which inhibit specifically proteinases of animals and micro-organisms, is induced in many plants. Insects feeding on these plants consume the inhibitor, whereby their digestive processes are disrupted with the result the insect pests starve to death. The synthesis of the inhibitors is not restricted to the wound site, but often occurs in large parts of the plant and thereby protects them from further attacks. The introduction of suitable foreign genes into transgenic potato, lucerne, and tobacco plants enabled a high expression of proteinase inhibitors in these plants, efficiently protecting them from being eaten by insects. This strategy has the advantage that the proteinase inhibitors are not specific to certain insect groups. These proteinase inhibitors are contained naturally in many of our foods, sometimes in relatively high concentrations, but are destroyed by cooking.

The expression of an *amylase inhibitor* in pea seeds, which prevents storage losses caused by the larvae of the pea beetle (section 14.4), is another example of how genetic engineering of plants offers protection from insect damage.

Plants can be protected against viruses by gene technology

Virus diseases can result in catastrophic harvest losses. Many crop plants are threatened by viruses. Infection with the cucumber mosaic virus can lead to the total destruction of pumpkin, cucumber, melon, and zucchini (courgette) crops. In contrast to fungal or animal pests, viruses can not be combated directly by the use of chemicals. Traditional procedures, such as decreasing the propagation of the viruses by crop rotation, are not always successful. Another way to control virus infections has been to attack the virus-transferring insects, especially aphids, with pesticides.

It has long been known that after infection with a weakly pathogenic strain of a certain virus, a plant may be protected against infection by a more aggressive strain. This phenomenon has been applied successfully in the biological plant protection of squash plants. It was presumed that a single molecular constituent of the viruses caused this protective function and this has been verified by molecular biology: the introduction of the *coat protein* gene of the tobacco mosaic virus into the genome of tobacco plants makes them resistant to this virus. This has been confirmed for many other viruses: if a gene for a coat protein of a particular plant virus is expressed sufficiently in a plant, the plant usually becomes resistant to infection by this pathogen. Although the mechanism of this protection is not yet fully understood, this principle has already been used several times with success to generate virus-resistant plants by genetic engineering. In the USA a virus-resistant zucchini variety, generated in this way, has been licensed for cultivation.

The generation of fungus-resistant plants is still at an early stage

The use of gene technology to generate resistance to fungal infections in plants is still at an early stage. An attempt is being made to utilize the natural protective mechanisms of plants. Some plants protect themselves against fungi by attacking the cell wall of the fungi. The cell walls of most fungi contain chitin, an *N*-acetyl-D-glucosamine polymer, which does not occur in plants. Some plants or their seeds contain *chitinases* which lyse the cell wall of fungi. This protective function has been transferred to other plants. Tobacco plants have been generated by transformation which contain a chitinase gene from beans, thereby gaining an increased resistance against certain fungi. Another strategy lies in the expression of enzymes for the synthesis of fungicide phytoalexins, e.g. stilbenes (section 18.4). However, it may be some time before fungus-resistant plants are ready for cultivation.

Non-selective herbicides can be used as selective herbicides following the generation of herbicide-resistant plants

The importance of herbicides for plant protection has already been discussed in section 3.6, and examples of the effects of various herbicides have been dealt with in sections 3.6, 10.4, and 15.3. The most economically successful herbicide is glyphosate (trade name Roundup, Monsanto) (Fig. 10.18) which specifically inhibits the synthesis of aromatic amino acids via the shikimate pathway at the EPSP synthase step (Fig. 10.19). Since the shikimate pathway is not present in animals, animal metabolism is not markedly affected by glyphosate. Because of its simple structure, it is very rapidly degraded by soil bacteria and thus leaves no residues in the soil. As a non-selective herbicide glyphosate even destroys many of the very persistent weeds. For example, it is widely used to clear vegetation from railway tracks and to control the weeds on the ground of fruit and wine plantations, and to kill weeds before the crops are planted. In order to apply this powerful herbicide as a selective *post-emergence* herbicide, glyphosate-resistant transformed plants have been generated for a number of crop plants by means of genetic engineering. To generate glyphosate resistance in plants, genes were isolated from bacteria encoding EPSP synthases, which are less sensitive towards glyphosate than the plant enzymes. Transformant plants which express bacterial EPSP synthase activity acquired protection against the herbicide. Glyphosate-resistant cotton, rapeseed, and soybean are already available to the farmer. In a similar way crop plants have been made resistant to the herbicide glufosinate (trade name Basta, AgrEvo, Fig. 10.7) by the expression of bacterial detoxifying enzymes. Basta-resistant maize and rapeseed have already been licensed for cultivation.

Plant genetic engineering is used for the improvement of the yield and quality of crop products

In the application of genetic engineering for generating resistance against pests or herbicides usually only one additional gene is transferred into the plant. For altering the quality or the yield of harvested products, however, a transfer of several genes is often required and therefore is more difficult to achieve.

A promising way to increase crop yields is the generation of *hybrids* from genetically engineered male sterile plants as described in section 20.7. The hybrid of a transformed rapeseed variety is already licensed for cultivation in the European Union. Another strategy for the improvement of crop yield is to alter the partitioning of biomass between the harvestable and non-harvestable organs of the plants. In transgenic potato plants an improvement of the tuber yield has been observed (section 13.3), but these results still have to be confirmed by field trials.

Genetic engineering is being utilized in many ways to modify the quality of harvested products to maximize their use as food or industrial raw material. Some examples of potential applications of gene technology are given below.

By expressing an ADP-glucose pyrophosphorylase (section 9.1) from *E. coli* it was possible to increase the starch content and thus decrease the water content in potato tubers. This is of interest for the use of potatoes in the food industry. On the other hand, by antisense inhibition (section 22.5) of ADP-glucose pyrophospho-rylase, potato tubers were obtained which contained only 3–5 per cent of the normal starch content and which mainly stored sucrose. The purpose of this trans-formation was not to convert potato into sugar beet, but to convert this sucrose into polyglucans needed for industrial raw material by the introduction of further genes. It was also possible to reduce the synthesis of amylose in potato tubers to such an extent that the resulting starch consisted uniformly of only the branched amylopectin, which may be advantageous for using this starch as an industrial raw material.

Generation of transformed tomatoes, in which the ripening process was altered in an attempt to improve the quality and storage stability, has been described in section 19.5. Transgenic plants producing certain fats with either short chain fatty acids as a raw material for the detergent industry, or those with a high content of erucic acid, as a precursor for the synthesis of plastic materials, or a high content of saturated fatty acids for use in deep frying, have been discussed in section 15.5. Work is also in progress to alter the amino acid composition of storage proteins to increase their nutritional value. An attempt is being made to increase the methionine content in the soybean in order to improve its quality as fodder (section 14.3)

Transgenic plants are also suitable for producing peptides and proteins used as pharmaceuticals, such as enkephalins, human serum albumins, or interferons. The production of the cancer therapeutic taxol, which at the moment requires 9000 kg of yew tree bark for the isolation of 1 g, is being attempted in transgenic plants in order to make this pharmaeutical available to a larger number of patients. The production in transgenic plants of vaccines against various illnesses, in order to reduce the production costs, is at present under investigation. It is even envisaged that oral vaccination by the consumption of transgenic bananas could be possible. Many other examples of applications of plant genetic engineering could be mentioned here, and it appears certain that not all the attempted applications will be successful or economically viable; however, one has to take into account that, at the time when this textbook is being written, the technique of plant genetic engineering is just 12 years old.

Genetic engineering provides a chance for increasing the protection of crop plants against environmental stress

In the preceding chapters various mechanisms have been described by which a plant protects itself against environmental stresses, such as heat (section 21.2), cold (sections 3.10, 15.1), drought and soil salinity (Chapter 8, section 10.4), xenobiotic and heavy metal pollution (section 12.2), and oxygen radical production (sections 3.9, 3.10). Genetic engineering opens the prospect of increasing the resistance of cultivated plants to these stresses by overexpression of enzymes participating in the stress responses. Thus, an increase of the number of double bonds in the fatty acids of membrane lipids through genetic engineering, has improved the cold tolerance of tobacco (section 15.1).

Strategies have been proposed for making plants salt tolerant by increasing the synthesis of osmotically compatible substances, such as mannitol, betaine, or proline (section 10.4), although it may still take considerable time until this is put into practice. With the growth of the world's population, the availability of sufficient arable land becomes an increasing problem. Because of their high salt content, often caused by inadequate irrigation management, large areas of the globe can no longer be utilized for agriculture. If plant genetic engineering were to succeed in generating salt-resistant crop plants, this would be a very important contribution to the world food supply.

At the moment plant genetic engineering raises, on the one hand, exaggerated expectations, and on the other hand, induces in parts of the population fear of its negative consequences. Responsible application of plant genetic engineering requires that for each plant licensed for cultivation, a risk analysis is made according to strict scientific criteria as to whether the corresponding plant represents a hazard to the environment. Amongst other criteria, it has to be examined whether crossing between the released transformants and wild plants is possible, and what the potential consequences of this are for the environment. For example, crossing can occur between transgenic rapeseed and other Brassicaceae, such as wild mustard. Moreover, transgenic plants themselves could grow in the wild. In this way herbicide-resistant weeds may develop from herbicide-resistant cultivated plants. However, it should be noted that in the conventional application of herbicides, herbicide-resistant weeds have also evolved by natural selection (section 10.4). Experiments in the laboratory, as well as controlled field tests, are required for such risk analyses. It is beyond the scope of this book to deal with the criteria set by legislation to evaluate the risk by the release of a transgenic cultivar. It is expected that, with responsible use, plant genetic engineering may contribute to the improvement of crop production and the provision of sustainable raw materials for industry. If plant protection based on plant genetic engineering were simply to have the result that 'less chemicals are put on the field' then this would be an improvement for the environment.

Further reading

Chrispeels, M. J. and Sadava, D. E. (1994). *Plant, genes and agriculture*. Jones and Bartlett, Boston.

De Block, M., Herrera-Estrella, L., Van Montagu, M., and Schell, J. (1984). Expression of foreign genes in regenerated plants and their progeny. *EMBO Journal*, **3**, 1681–9.

Gibson, S. and Somerville, C. (1993). Isolating plant genes. *Trends in Biotechnology*, **11**, 306–13.

Glick, B. R. and Pasternak, J. J. (1994). *Molecular biotechnology. Principles of recombinant DNA*. ASM Press, Washington, DC.

Hehl, R. (1994). Transposon tagging in heterologous host plants. *Trends in Genetics*, **10**, 385–6.

Hiatt, A. (1993). *Transgenic plants*. Marcel Dekker, New York.

Horsch, R. B., Fraley, R. T., Rogers, S. G., Sanders, P. R., Lloyd, A., and Hoffman, N. (1984). Inheritance of functional foreign genes in plants. *Science*, **223**, 496–8.

Matzke, M. A. and Matzke, A. J. M. (1995). How and why do plants inactivate homologous (trans) genes? *Plant Physiology*, **107**, 679–85.

Moffat, A. S. (1995). Exploring transgenic plants as a new vaccine source. *Science*, **268**, 658–60.

Nicholl, D. S. T. (1994). *An introduction into genetic engineering*. Cambridge University Press, Cambridge.

Plant Cell and Environment (1994). Special issue on transgenic plants. *Plant Cell and Environment*, **17**, 465–680.

Portrykus, I. (1991). Gene transfer to plants: assessment of published approaches and results. *Annual Review of Plant Physiology and Plant Molecular Biology*, **42**, 205–25.

Riesmeier, J. W., Willmitzer, L., and Frommer, W. B. (1992). Isolation and characterization of a sucrose carrier cDNA from spinach by functional expression in yeast. *EMBO Journal*, **11**, 4705–13.

Tinland, B. (1996). The integration of T-DNA into plant genomes. *Trends in Plant Science*, **1**, 178–84.

Watson, J. D., Gilman, M., Witkowski, J., and Zoller, M. (1992). *Recombinant DNA*. W. H. Freeman, New York.

Willmitzer, L. (1993). Transgenic plants. In Biotechnology, a multi-volume comprehensive treatise, (ed. H. J. Rehm, G. Reed, A. Pühler, and P. Stadler), Vol. 2, pp. 627–59. Verlag Chemie, Weinheim.

Zambryski, P. C. (1992). Chronicles from the *Agrobacterium*–plant cell DNA transfer story. *Annual Review of Plant Physiology and Plant Molecular Biology*, **43**, 465–90.

index